NTA NEET (UG)

BIOLOGY

(Line by line Chapterwise and Topicwise based on NCERT Textbook)

Author

Rakshita Singh

Published by
Invincible Publishers

201A, SAS Tower, Sector 38, Gurugram – 122003
Phone: +91-124-4034247, +91 9599066061
www.invinciblepublishers.com
Sales: Office No. 4760-61/23 Basement, Pratap Street,
Ansari Road Daryaganj, New Delhi. 110002
Phone: +91-11-40198405
Email: *invinciblepublishers@gmail.com*

First edition – 2023

ISBN: 978-81-962775-9-8

CONTENT

SECTION A: BOTANY

SECTION B: ZOOLOGY

Chapter 1

The Living World

MULTIPLE CHOICE QUESTIONS

Topic 1 — Characteristics of Living Organisms

1. All living organisms are linked to each other because they
(a) show a common cellular organization.
(b) possess common genetic material of the same type.
(c) share common genetic material but to varying degrees.
(d) all of these.

2. Assertion: Non-living objects also grow.
Reason: They grow by accumulation of material on the surface.
(a) Both assertion and reason are true and reason is the correct explanation of assertion.
(b) Both assertion and reason are true but reason is not the correct explanation of assertion.
(c) Assertion is true but reason is false.
(d) Both assertion and reason are false.

3. An isolated metabolic reaction occurring in-vitro are not living things, but surely living reactions. It proves that 'A' of the body is the defining feature of life forms. Here 'A' is
(a) reproduction
(b) cellular organization
(c) metabolic reactions
(d) growth

4. Which one is the defining characteristics of living organisms?
(a) Growth
(b) Ability to make sound
(c) Reproduction
(d) Response to external stimuli

5. Assertion: Reproduction is not a defining property of living organisms.
Reason: Mules do not reproduce.
(a) Both assertion and reason are true and reason is the correct explanation of assertion.
(b) Both assertion and reason are true but reason is not the correct explanation of assertion.
(c) Assertion is true but reason is false.
(d) Both assertion and reason are false.

6. Which one of the following is affected by photoperiod?
(a) reproduction
(b) metabolism
(c) growth
(d) cellular organization

7. Assertion: Non-living objects also exhibit metabolism.
Reason: Isolated metabolic reactions occurring, invitro are living things.
(a) Both assertion and reason are true and reason is the correct explanation of assertion.
(b) Both assertion and reason are true but reason is not the correct explanation of assertion.
(c) Assertion is true but reason is false.
(d) Both assertion and reason are false.

8. Reproduction occurs in yeast and hydra, by
(a) conjugation (b) sporulation
(c) budding (d) none of these

9. The organism(s) that can be multipled by fragmentation is/are

(a) fungi (b) filamentous algae
(c) protonema of moss (d) all of these

10. In Amoeba, reproduction is synonymous with
(a) digestion (b) growth
(c) locomotion (d) none of these

11. Which one is considered as the twin characteristics of growth?
(a) increase in mass
(b) increase in number
(c) increase in length
(d) both A. and B.

12. Unicellular organisms are grown by
(a) cell elongation
(b) cell division
(c) accumulation of material on the surface
(d) none of these

13. Mountains, boulders and sand mounds also grow, but they are not considered as living organisms because they grow by
(a) Accumulation of material on their inner surface.
(b) the division of their particles.
(c) accumulation of material on their outer surface.
(d) both A. and B.

14. Which one is the correct statement about growth in plants?
(a) Growth occurs continuously throughout their lifespan by cell division.
(b) The growth is seen only upto a certain age.
(c) Growth occurs by the accumulation of material on the upper surface of their cells.
(d) Growth occurs only in certain cells.

15. Living organisms that do not reproduce are known as
(a) mule
(b) worker bees
(c) infertile human couples
(d) all of these

16. Match Column-A against Column-B and choose the correct option from the codes given below.

	Column -A Organisms		Column -B Method of organisms
a	Planaria	1	Fragmentation
b	Hydra	2	Regeneration
c	Fungi	3	Binary fission
d	Amoeba	4	Budding

Codes

	a	b	c	d
A	2	4	1	3
B	4	2	3	1
C	3	1	4	2
D	1	3	2	4

17. The sum total of all the chemical reactions occurring in our body is known as
(a) anabolism (b) catabolism
(c) metabolism (d) none of these

18. An isolated metabolic reaction outside the body of an organism, performed in a test-tube is
(a) living
(b) non-living
(c) neither living nor non-living
(d) sometimes living and sometimes non-living depending upon environmental factors

19. Which one is the incorrect statement from the following?
(a) All living organisms exhibit metabolism.
(b) Metabolism is the sum total of all chemical reactions occurring in our body.
(c) Metabolism is a defining feature of all living organisms.
(d) Metabolic reactions cannot be demonstrated outside the body in cell-free systems.

20. Living organisms respond to environmental stimuli which could be
(a) physical (b) chemical
(c) biological (d) all of these

21. Match Column-A with Column-B and choose the correct option from the codes given below:

	Column -A		Column -B
a	Growth	1	Production of progeny
b	Reproduction	2	Sum of all chemical reactions occurring in body
c	Metabolism	3	Sense and respond to environmental stimuli
d	Con-sciousness	4	Increase in mass and number

Codes

	a	b	c	d
A	3	2	4	1
B	4	1	2	3
C	1	3	2	4
D	2	4	3	1

Topic 2 Diversity in the Living World

22. Which one does not match according to rules of nomenclature.

(a) Biological names are generally given in Latin.

(b) The first word in biological name represents the genus while the second component denotes the specific epithet.

(c) When hand written, both words of biological names are separately underlined.

(d) Both the words of biological name start with capital letter.

23. Which of the following is against the rules of ICBN?

(a) Handwritten scientific names should be underlined.

(b) Every species should have a generic name and a specific epithet.

(c) Scientific names are in latin and should be italized.

(d) Generic and specific names should be written starting with small letters.

24. Biological names are generally written in

(a) Greek language (b) Latin language

(c) English language (d) Hindi language

25. Assertion: Binomial nomenclature given by Linnaeus is being practiced by biologists all over the world.

Reason: Each name of this system has two components, the generic name and the specific epithet. Pg-8, Medium

(a) Both Assertion and Reason are true and Reason is correct explanation of Assertion.

(b) Both Assertion and Reason are true, but Reason is not the correct explanation of Assertion.

(c) Assertion is true, but Reason is false.

(d) Assertion is false, but Reason is true.

26. Assertion: The process of classification of organisms is taxonomy.

Reason: It is merely based on the external features of organisms.

(a Both Assertion and Reason are true and Reason is correct explanation of Assertion.

(b) Both Assertion and Reason are true, but Reason is not the correct explanation of Assertion.

(c) Assertion is true, but Reason is false.

(d) Assertion is false, but Reason is true.

27. Assertion: Systematics is the study of organisms, their diversities but not the relationships among them.

Reason: Systematics is derived from a English word 'systema'.

A) Both Assertion and Reason are true and Reason is correct explanation of Assertion.

B) Both Assertion and Reason are true, but Reason is not the correct explanation of Assertion.

C) Assertion is true, but Reason is false.
D) Assertion is false, but Reason is true.

28. The process by which organisms are grouped into convenient categories based on some easily observable characters, is called
(a) taxonomy (b) identification
(c) classification (d) nomenclature

29. The scientific term used for the categories of organisms to study them is
(a) taxa (b) biological name
(c) systematics (d) none of these

30. The process of classification is called
(a) systematics (b) taxonomy
(c) nomenclature (d) identification

31. The modern taxonomic studies are based on
(a) cell structures
(b) external and internal structure
(c) development process
(d) all of these

32. Which one is not the basic to taxonomy among the following process?
(a) Identification
(b) Classification
(c) Collection of specimen
(d) Nomenclature

33. Systematics refers to
(a) diversities of different kinds of organisms and their relationship.
(b) identification and study of organ systems of organisms.
(c) identification and preservation of organisms.
(d) study of habitat of organisms and their classification.

34. Match Column-A with Column-B and choose the correct option from the codes given below.

	Column -A		Column -B
a	Mammalia	1	Specific epithet
b	*Mangifera*	2	Branch of study
c	*indica*	3	Taxa
d	Systematics	4	Generic name

Codes

	a	b	c	d
A	3	4	1	2
B	2	1	3	4
C	1	2	4	3
D	4	3	2	1

35. The process of naming of living organisms is called
(a) systematics (b) taxonomy
(c) nomenclature (d) none of these

36. Identification of an organism is its
(a) correct description
(b) correct classification
(c) correct nomenclature
(d) all of these

37. ICBN stands for
(a) International Code for Biological Naming
(b) International Centre for Botanical Nomenclature
(c) Indian Code for Biological Nomenclature
(d) International Code for Botanical Nomenclature

38. Choose the correct statement about scientific names.
(a) They ensure that each organism has only one name.
(b) Description of any organism shouldenable the people to arrive at the same name.
(c) They ensure that such a name has not been used for any other organism.
(d) All of these.

39. The system of providing a name with two components is called
(a) trinomial nomenclature
(b) binomial nomenclature
(c) uninominal nomenclature
(d) none of these

40. The naming system, which is practised by

biologists all over the world, was given by
(a) Carolus Linnaeus (b) Whittaker
(c) Haeckel (d) Woese

41. Match the Column-A with Column-B and choose the correct option from the codes given below.

	Column -A		Column -B
a	Biodiversity	1	Correct description of an organism
b	*Nomenclature*	2	The variety of living organisms
c	*Identification*	3	A system of providing a name with two components
d	Binomial nomenclature	4	Naming of living organisms

Codes

	a	b	c	d
A	2	4	1	3
B	4	3	2	1
C	1	2	4	3
D	3	1	4	2

42. Select correctly written scientific name of mango which was first described by Carolus Linnaeus.
(a) Mangifera indica Car Linn.
(b) Mangifera indica Linn.
(c) Mangifera indica
(d) Mangifera Indica

43. In *Mangifera indica* Linn., the specific epithet is
(a) Mangifera (b) indica
(c) Linn. (d) Both A. and B.

Topic 3 Taxonomic Categories

44. Choose the correct statements from the following:
I. In case of plants, classes with a few similar characters are aligned to a higher category called phylum.
II. Sub-categories have also beendeveloped in the taxonomic hierarchy to facilitate more sound and scientific placement of various taxa.
III. Class includes related orders.
IV. Convolvulaceae family is included in polymoniales order on the basis of its floral characters.

Choose the correct option.
(a) I and IV (b) II and III
(c) II, III and IV (d) All of these

45. Match Column-A with Column-B for housefly classification and select the correct option using the codes given below.

	Column -A		Column -B
a	Family	(1)	Diptera
b	Order	(2)	Arthropoda
c	Class	(3)	Muscidae
d	Phylum	(4)	Insecta

Codes

	a	b	c	d
(A)	4	3	2	1
(B)	4	2	1	3
(C)	3	1	4	2
(D)	3	2	4	1

46. Assertion: Genus may have one or more
Reason: Genus comprises a group of
(a) Both Assertion and Reason are true Assertion.
(b) Both Assertion and Reason are true, but Reason is not the correct explanation of Assertion.
(c) Assertion is true, but Reason is false.
(d) Assertion is false, but Reason is true.

47. Assertion: Order is the assemblage of families which exhibit a few similar characters.

Reason: Plant family polymoniales is included in the order solanaceae based on the floral characters.

(a) Both Assertion and Reason are true and Reason is correct explanation of Assertion.
(b) Both Assertion and Reason are true, but Reason is not the correct explanation of Assertion.
(c) Assertion is true, but Reason is false.
(d) Assertion is false, but Reason is true.

48. Assertion: Wheat belongs to the family poaceae.
Reason: Wheat is a member of order poales.
(a) Both Assertion and Reason are true and Reason is correct explanation of Assertion.
(b) Both Assertion and Reason are true, but Reason is not the correct explanation of Assertion.
(c) Assertion is true, but Reason is false.
(d) Assertion is false, but Reason is true.

49. Which one of the following options represents the correct classification for the animal shown below?

	phy-lum	cla-ss	ord-er	fam-ily	gen-us	spe-cies
A	cho-rda-ta	ver-teb-rata	chi-rop-tera	feli-dae	can-is	tig-ris
B	cho-rda-ta	ma-mma-lia	car-ni-vora	feli-dae	pan-thera	tig-ris
C	ver-teb-rata	mam ma-lia	car-ni-vora	feli-dae	pan-thera	tig-ris
D	ma-mm-lia	feli-dae	car-ni-vora	feli-ac-eae	pan-thera	leo

50. All the taxonomic categories together constitute the
(a) taxon (b) family
(c) kingdom (d) hierarchy

51. The lowest taxonomic category is BOTANY
(a) agenus (b) species
(c) class (c) family

52. The basic requirement for placing anorganism in various categories is the knowledge of
(a) characters of an individual
(b) characters of group of organisms
(c) binomial nomenclature
(d) both A and B

53. ***Solanum*** **includes species**
(a) nigrum (b) melongena
(c) tuberosum (d) all of these

54. The taxonomic category, genus represents
(a) an individual organism
(b) a collection of organisms
C) a group of closely related species of organisms
D) none of these

55. Match Column-I and Column-II and choose the correct option from the codes given below.

	Column -1		Column -II
a	Lion	1	Panthera tigris
b	Leopard	2	Panthera leo
c	Tiger	3	Solanum nigrum
d	Potato	4	Panthera pardus

Codes -

	a	b	c	d
A	2	4	1	3
B	4	2	1	3

C	1	2	4	3
D	3	4	2	1

56. 'Suffix' used for a unit of classification in plants to indicate 'family' taxonomic category is
(a) -ales (b) -onae
(c) -aceae (d) ae

57. The axonomic category 'order' lies between
(a) Genus and species
(b) Genus and family
(c) Family and class
(d) Class and phylum

58. Choose the incorrect statement from the following.
(a) Each genus may have one or more than one specific epithets representing different organisms.
(b) Potato and brinjal are two different species but both belong to the genus Solanum.
(c) 'Families' are characterized on the basis of merely reproductive features of plant species.
(d) 'Order' is a higher taxonomic category and is identified on the basis of aggregates of characteristics.

59. Fishes, amphibians, reptiles and birds represent which taxonomic category?
(a) Class (b) Phylum
(c) Order (d) Kingdom

60. The 'phylum' taxon of animal classification is equivalent to which taxon of plant classification?
(a) Class (b) Division
(c) Order (d) Family

61. Which one shows hierarchical arrangement of taxonomic categories of plants in descending order?
A. Kingdom←Division←Class←Order←Family←Species←Genus
B. Kingdom← Division←Order←Class ←Family← Genus←Species
C. Kingdom→ Division→ Order→ Class→ Family→Genus→Species
D. Kingdom → Division→ Class→Order →Family→ Genus→ Species

62. The kingdom that comprises all plants from various divisions is
(a) Plantae (b) Animalia
(c) Chordata (d) Mammal

63. Match the Column-A with Column-B and choose the correct option from the codes given below:

	Column -A		Column -B
a	Order	(1)	Solanum
b	Kingdom	(2)	Solanaceae
c	Family	(3)	Plantae
d	Genus	(4)	tuberosum
e	Species	(5)	Polymoniales

Codes

	a	b	c	d	e
(A)	5	3	2	1	4
(B)	4	2	5	3	1
(C)	1	5	3	4	2
(D)	3	4	1	5	2

64. In a taxonomic hierarchy, on moving from species to kingdom, the number of common characteristics
(a) will increase
(b) remain same
(c) will decrease
(d) may increase or decrease

Topic 4 Taxonomical Aids

65. The contrasting characteristics generally in a pair used for identification of animals in Taxonomic key are referred to as
(a) Lead (b) Couplet
(c) Doublet (d) Alternate

66. Taxonomic key is used in the preparation of the
(a) monographs (b)) flora
(c) both A and B (d) none of these

67. Each statement in a key is known as a
(a) couplet (b) lead
(c) monograph (d) none of these

68. Match items given in Column-I with those in Column-II and select the correct option given below.

	Column -A		Column -B
a	Herbarium	(1)	It is place having a collection of preserved plants and animals.
b	key	(2)	A list that enumerates methodically all the species found in an area with brief.
c	Museum	(3)	It is a place where dried and pressed plant specimens mounted on sheets are kept.
d	Catalogue	(4)	A booklet containing a list of characters and their alternates which are helpful in identification of various taxa.

Codes

	a	b	c	d
(A)	1	4	3	2
(B)	3	2	1	4
(C)	2	4	3	1
(D)	3	4	1	2

69. Choose the correct statements about taxonomic keys.
I. It is used for the identification of only plants.
II. Each statement in the key is called a lead.
III. No separate taxonomic keys are required for different taxonomic categories.
(a) II and III (b) I and II
(c) III and I (d) II only

70. Which one is not a taxonomic aid among the following?
(a) Flora (b) Manual
(c) Taxon (d) Key

71. Which one contains information on any one taxon among the following?
(a) Flora (b) Monograph
(c) Manual (d) Catalogue

72. Assertion: Herbarium is a storehouse of collected plant specimens that are dried, pressed and preserved on sheets.

Reason: Herbaria serve as quick referral systems in taxonomical studies.
(a) Both Assertion and Reason are true and Reason is correct explanation of Assertion.
(b) Both Assertion and Reason are true, but Reason is not the correct explanation of Assertion.
(c) Assertion is true, but Reason is false.
(d) Assertion is false, but Reason is true.

73. Assertion: Museums have collection of live plants and animals.

Reason: In museums, as far as possible, conditions similar to their natural habitat are provided.
(a) Both Assertion and Reason are true and Reason is correct explanation of Assertion.
(b) Both Assertion and Reason are true, but Reason is not the correct explanation of Assertion.
(c) Assertion is true, but Reason is false.
(d) Assertion is false, but Reason is true.

74. Assertion: The taxonomical keys are based on the contrasting characters.

Reason: Each statement in the key is called a couplet.

(a) Both Assertion and Reason are true and Reason is correct explanation of Assertion.
(b) Both Assertion and Reason are true, but Reason is not the correct explanation of Assertion.
(c) Assertion is true, but Reason is false.
(d) Assertion is false, but Reason is true.

75. Taxonomic studies of various organisms are useful in
(a) agriculture (b) forestry
(c) industry (d) all of these

76. Which one of the following taxonomic studies require?
(a) correct classification
(b) correct identification of organisms
(c) intensive laboratory and field studies
(d) all of these

77. A store house of collected plant specimens that are dried, pressed and preserved on sheets is a
(a) herbarium (b) botanical garden
(c) zoological park (d) catalogue

78. Which information is not provided by the herbarium sheet about a plant among the following ?
(a) Collector's name
(b) Place of collection
(c) Economic importance of plant species
(d) Botanical name of the plant

79. Which one of the specialised gardens having collections of living plants for reference?
(a) Herbarium
(b) Botanical gardens
(c) Zoological parks
(d) Museum

80. Match the Column-A and Column-B and choose the correct option from the codes given below:

	Column -A		Column -B
a	Kew Botanical Garden	(1)	Dehradun
b	Indian Botanical Garden	(2)	Lucknow
c	National Botanical Research Institute	(3)	England
d	Forest Research Institute	(4)	Howrah

Codes

	a	b	c	d
(A)	3	4	2	1
(B)	4	2	3	1
(C)	1	3	4	2
(D)	2	1	3	4

81. At which one of the following place National Botanical Research Institute is situated?
(a) Howrah (b) Lucknow
(c) Dehradun (d) Delhi

82. Which one is the incorrect statement?
(a) Herbarium is a storehouse of collected plant specimens that are dried, pressed and preserved on sheets.
(b) Herbaria serve as quick referral systems in taxonomical studies.
(c) Botanical gardens have collection of preserved plant materials also.
(d) Indian Botanical Garden is situated at Howrah, India.

83. Biological museums have collection of
(a) preserved plant specimens
(b) preserved animal specimens
(c) live plants and animals
(d) both A and B

84. In biological museums insects are preserved
(a) in preservative solutions
(b) after stuffing
(c) after collecting, killing and pinning
(d) as skeletons

85. Match Column-A with Column-B and choose the correct option from the codes given below:

	Column -A (Organism)		Column -B (Method of Preservation Skeleton)
a	Birds	(1)	Skeleton
b	Human	(2)	Preservative solution
c	Insects	(3)	Stuffing
d	Small animals	(4)	In boxes afterkilling and pinning

Codes

	a	b	c	d
(A)	3	1	4	2
(B)	4	3	2	1
(C)	1	2	3	4
(D)	2	4	1	3

86. The place where wild animals are kept in protected environment is
(a) botanical garden (b) zoological park
(c) museum (d) herbarium

87. Zoological parks are also called
(a) museum (b) garden
(c) zoo (d) none of these

88. Zoological parks and botanical gardens have a collection of
(a) exotic living species only
(b) endemic living species only
(c) both A and B
(d) only local plants and animals

89. Which one of the following keys are used for identifying plants and animals?
(a) morphology
(b) anatomy
(c) similarities and dissimilarities
(d) all of these

90. The contrasting characters in a key which are generally in a pair is called
A) couplet B) doublet
C) triplet D) all of these

ANSWER KEY

1. (c)	2. (a)	3. (c)	4. (d)	5. (a)	6. (a)	7. (d)	8. (c)	9. (d)	10. (d)
11. (d)	12. (b)	13. (c)	14. (a)	15. (b)	16. (d)	17. (c)	18. (b)	19. (d)	20. (d)
21. (b)	22. (d)	23. (d)	24. (b)	25. (a)	26. (c)	27. (d)	28. (c)	29. (a)	30. (b)
31. (d)	32. (b)	33. (a)	34. (a)	35. (c)	36. (d)	37. (d)	38. (d)	39. (b)	40. (a)
41. (a)	42. (b)	43. (b)	44. (b)	45. (a)	46. (a)	47. (c)	48. (b)	49. (b)	50. (d)
51. (b)	52. (d)	53. (d)	54. (c)	55. (a)	56. (c)	57. (c)	58. (c)	59. (b)	60. (b)
61. (b)	62. (a)	63. (a)	64. (c)	65. (b)	66. (c)	67. (b)	68. (b)	69. (d)	70. (c)
71. (b)	72. (d)	73. (c)	74. (c)	75. (d)	76. (d)	77. (a)	78. (c)	79. (b)	80. (a)
81. (a)	82. (c)	83. (d)	84. (c)	85. (c)	86. (b)	87. (c)	88. (c)	89. (c)	90. (a)

2 Chapter

Biological Classification

MULTIPLE CHOICE QUESTION

1. Fungi has cell wall composed of–
(a) Cellulose
(b) Non – cellulosic + amino acid
(c) Chitin
(d) Absence of cell wall

2. How many kingdom from R.H. Whittaker system does have exclusive autotrophic mode of nutrition
(a) Zero (b) One
(c) Two (d) Three

3. Unicellular eukaryotic are categorised in-
(a) Monera (b) Protista
(c) Plantae (d) Animalia

4. How many of the following does belong to Protista Amoeba, Spirogyra, Chlamydomonas, Chlorella, Paramecium
(a) 5 (b) 4
(c) 3 (d) 2

5. In five kingdom classification multicellularity began from –
(a) Animalia (b) Plantae
(c) Protista (d) Fungi

6. Choose the correct with respect to earliest for scientific basis of classification
(a) It was proposed by Aristotle
(b) Plants were divided as trees, shrubs & herbs on the basis of their morphological characters
(c) Animals were classified into two groups that are those which have red blood and those that did not
(d) All of these

7. Linnaeus system of classification did not deal with –
(a) Eukaryotes and prokaryotes
(b) Unicellular & multicellular
(c) Photosynthetic & non – photosynthetic
(d) All of these

8. How many kingdom according to five kingdom classification and Linnaeus system of classification is/are dedicated for prokaryotes exclusively
(a) 1, 0 (b) 1, 1
(c) 2, 0 (d) 3, 1

9. Moneran cell wall is composed by-
(a) Polysaccharide (Non cellulose) only
(b) Polysaccharide (cellulose)
(c) Polysaccharide (chitin)
(d) Amino acid and Non cellulosic polysaccharide

10. Chemosynthetic mode of nutrition is found is –
(a) Monera (b) Protist
(c) Plantae (d) Fungi

11. R.H Whittaker classification is/are based upon –
(a) Cell structure & body organisation
(b) Mode of nutrition & reproduction
(c) Phylogentic relationship
(d) All of these

12. Five kingdom classification was proposed in –
(a) 1969 (b) 1996
(c) 1699 (d) None of these

13. Choose the correct about 3 – domain system
(a) Two domain are dedicated for prokaryotic while one domain is dedicated for eukaryotic
(b) One domain is dedicated for prokaryotic while two domains are for eukaryotic
(c) It has seven kingdom which are categorised

in 3 – domain
(d) It has six kingdom of which one kingdom is in first and third domain while 5 – kingdom is second domain.

14. Earlier classification system included bacteria, BGA (blue green algae) fungi, mosses, ferns under 'Plants' on basis of-
(a) Mode of nutrition
(b) Body organisation & nuclear structure
(c) Presence of cell wall
(d) Nature of cell wall.

15. How many of following are prokaryotes: Bacteria, Mosses, ferns, fungi, pteridophyta, blue green algae, gymnosperms angiosperm
(a) 1 (b) 2
(c) 3 (d) More than 4

Topic 1 Kingdom Monera

16. On the basis of shape; bacteria are grouped under_____ categories
(a) Four (b) Five
(c) Three (d) None of these

17. Choose the correctly stated statement
(a) Bacterial structure and behaviour are complex.
(b) Bacterial structure and behaviour are simple
(c) Bacterial structure is complex while behaviour is simple
(d) Bacterial structure is simple while behaviour is complex

18. Synthesis of own food from inorganic substrate is occur in –
(a) Autotrophic nutrition
(b) Chemosynthetic autotroph
(c) Photosynthetic autotroph
(d) All of these

19. Identify shape of bacteria

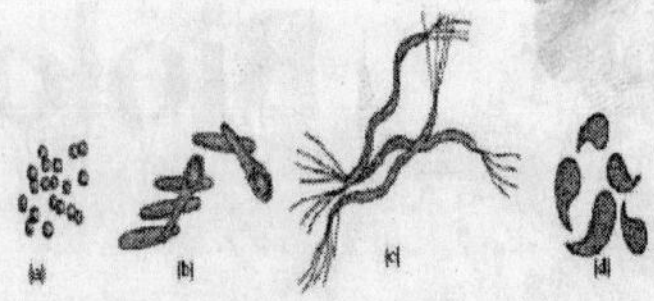

(a) a = cocci, b = rod – shaped, c = bacilli, d = comma – shaped
(b) a = spherical coccus, B = Bacilli, c = spirilla, d = vibrio
(c) a = cocci, b = spirilla, c = vibrio, d = Bacilli
(d) a = vibrio, b = spirilla, c = bacilli, d = coccus

20. Choose the correct statement:
(a) Bacteria are sole members of kingdom monera.
(b) Bacteria are abundant macro – organism
(c) Bacteria occurrence is limited to some area.
(d) Bacteria can't live in extreme habitat like desert

Topic 2 Archaebacteria

21. Archaebacteria differ from other bacteria in having –
(a) Definite nuclear structure
(b) Cell wall structure
(c) Adaptability cytoplasmic oncentration
(d) Some membranous cell organelles

22. Survival of archaebacteria in extreme condition is achieved by –
(a) Cell wall structure
(b) Some membranous cell organelles
(c) Adaptability & cytoplasm
(d) All of these

23. Which of following statement is/are false
(a) Methanogens are present in alimentary canal of several ruminant animals like cow & buffaloes
(b) Methanogens are responsible for production of biogas from dung of ruminant animals

(c) Methanogens are present in gut of several non – ruminant like cow & buffaloes
(d) A & B

24. Match Column-A against Column-B and choose the correct option from the codes given below.

	Column -A Organisms		Column -B Method of organisms
a	Halophiles	1	Marshy area
b	Thermo-acidophiles	2	Salty area
c	Methanogens	3	Hot springs

(a) i) – c, ii) – b, iii – a
(b) i) – c, ii) – a, iii – b
(c) i) – b, ii) – c, iii – a
(d) i) – b, ii) – a, iii – c

Topic 3 Eubacteria

25. Choose the wrong statement for chemosynthetic autotroph bacteria
(a) They oxidise various inorganic substrate- such as nitrates, nitrites & ammonia and use the released energy for their ATP production
(b) They play great role in recycling nutrient like nitrogen phosphorous, iron & sulphur
(c) For their energy production they utilize solar energy
(d) They can prepare their food from inorganic substrate.

26. Citrus canker is –
(a) Plant disease cause by bacteria
(b) Human disease cause by bacteria
(c) Pet disease cause by bacteria
(d) None of these

27. Which of following is not economic importance of heterotrophic bacteria
(a) Making curd from milk
(b) Antibiotic production
(c) N2 fixing in legumes root
(d) N2 fixing in Anabaena

28. Choose the incorrect option about bacterial reproduction –
(a) Bacteria reproduce mainly by fission
(b) Under unfavourable condition they produce spores
(c) They also reproduce by sexual reproduction
(d) They show a sort of sexual reproduction

29. Here are few statement given below, Identify organism on basis of statement
i. Lack cell wall
ii. Smallest living cell known
iii. Can survive without oxygen
iv. Pathogenic in animal & plants.
(a) Nostoc (b) Anabaena
(c) Mycoplasma (d) Chlorella

30. Label A , B and identify organism (c)

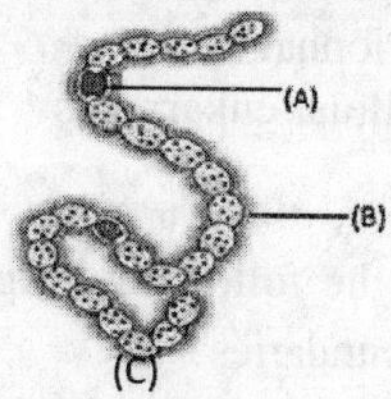

(a) A = Heterocyst B = Mucilagenous sheath C = Nostoc, an archaebacteria
(b) A = Heterocyst B = Mucilagenous sheath C = Nostoc
(c) A = Mucilagenous, B = Heterocyst, C = Nostoc
(d) A = heterocyst, B = Mucilagenoussheath, C = Nostoc, a filamentous algae

31. Choose the correct about blue green algae
i. Also known as cyanobacteria
ii. Presence of chlorophyll a, b similar to green plants
iii. Photosynthetic autotroph
iv) May be unicellular, colonial or filamentous
v. Occur in aquatic as well as terrestrial
(a) i), iii), iv), v) (b) i), ii), iii), iv), v)
(c) i), ii), iv), v) (d) None of these

32. Nitrogen fixation is done by –
(a) Specialised vegetative cell i.e. Heterocyst of Nostoc & Anabaena

(b) Specialised reproductive cell i.e. Heterocyst of Nostac & Anabaena
(c) Specialised vegetative as well as reproductive cell i.e. Heterocyst of Nostoc & Anabaena
(d) None

Topic 4 Kingdom Protista

33. Protista form a link with –
(a) Plants only
(b) Animals only
(c) Fungi only
(d) Plants, animals and fungi

34. Protista includes -
(a) Unicellular prokaryotes
(b) Bacteriophages
(c) Unicellular eukaryotes
(d) B.G.A

35. Which of the following kingdoms has no well defined boundaries?
(a) Monera
(b) Protista
(c) Fungi
(d) Metaphyta and Metazoa

36. Members of Protista are primarily
(a) Parasites (b) Terrestrial
(c) Aquatic (d) Photosynthetic

37. Nearly all protists are –
(a) Aerobic
(b) Anaerobic
(c) Aerobic or anaerobic
(d) Photosynthetic

38. Nutritionally, protists are-
(a) Photoautotrophs
(b) Heterotrophs
(c) Saprotrophs
(d) Photoautotrophs, heterotrophs or autotrophs

39. Based upon the modes of nutrition, protists are grouped into –
(a) Plant-like protists (algae) and ingestive, animal-like protists (protozoa); and absorptive, fungus like protists
(b) Chrysophytes, Dinoflagellates and Euglenoids only
(c) Slime moulds and fungi only
(d) Flagellated protozoans and sporozoans only

40. Which of the following are placed under Protista-?
(a) Chryosophytes and Dinoflagellates
(b) Euglenoids
(c) Slime moulds and protozoans
(d) All

41. Locomotory structures in protists are –
(a) Flagella (b) Cilia
(c) Pseudopodia (d) All

Topic 5 Chrysophytes

42. The diatoms do not easily decay like most of the other algae because
(a) They have highly siliceous wall
(b) They have water proof cells
(c) Their cell wall are mucilaginous
(d) Cell wall is virus-resistant

43. Diatomaceous earth is used for all except
(a) Polishing
(b) Filtration of oils and syrups
(c) Sound and fire proof room
(d) Biogas

44. Chrysophytes are -
(a) Planktons (b) Nektons
(c) Benthonic (d) Active swimmers

45. Chief producers in ocean are -
(a) Dinoflagellates (b) Diatoms
(c) Euglenoids (d) Green algae

46. Photosynthetic protists are -
(a) Euglenoids, Diatomsand Dinoflagellates
(b) Euglenoids and slime moulds
(c) Diatoms and Zooflagellates
(d) Desmids +Ciliates

47. Chrysophytes include –
(a) Diatoms and desmids (golden algae)
(b) Euglenoids
(c) Dinoflagellates
(d) Slime moulds

48. Which of the following modes of reproduction can be found in at least some protists?
(a) Binary fission
(b) Sexual reproduction
(c) Spore formation
(d) All

49. Select the following statement that does not apply to diatoms –
(a) Diatom cell wall may be impregnated with silicon
(b) Cell wall is made up of 2 half-shells fit tightly together
(c) Diatom is a chrysophyte
(d) Diatom is multiflagellate

50. Silica gel (Keieselghur)/Diatomite/Diatomaceous earth is obtained by –
(a) Diatoms (b) Dinoflagellates
(c) Euglenoids (d) Brown algae

Topic 6 Dinoglagellates

51. Dinoflagellates have –
(a) A single flagellum in the transverse groove between the cell plates
(b) A single flagellum in the longitudinal groove between the cell plates
(c) Two flagella one lies longitudinally and the other transversely in a furrow between the wall plates
(d) No flagella

52. In which of the following the cell wall has stiff cellulose plate on the outer surface –
(a) Dinoflagellates (b) Desmids
(c) Diatoms (d) Euglenoids

53. Which of the following releases toxins that may even kill other marine animals like fishes –
(a) Gonyaulax (b) Paramecium
(c) Euglenoids (d) Sporozoans

54. Dinoflagellates are mostly-
(a) Marine (b) Fresh water
(c) terrestrial (d) Saprophytes

55. Red tides in warm coastal water develop due to super abundance of-
(a) Dinoflagellates
(b) Euglenoid forms
(c) Diatoms and desmids
(d) Chlamydomonas nivalis

56. Red tide is caused by –)
(a) Ceretium (b) Noctiluca
(c) Gonyaulax (d) All of these

Topic 7 Euglena

57. Euglenoids e.g. Euglena are found –
(a) In fresh running water
(b) In fresh stagnant water
(c) In marine environment
(d) In both fresh and marine water

58. Which of the following statements about Euglena is true? (Pg. 21, E)
(a) Euglenoids are flagellates
(b) Euglena placed in continuous darkness loses their photosynthetic activity and die
(c) The pigments of Euglena are quite different from those of green plants
(d) Euglena is a marine protist

59. Which of the following statement is true about Euglena?
(a) They show flagellar locomotion
(b) They have a rigid cell wall
(c) They have no chloroplast
(d) They are obligate autotroph

60. i. Instead of a cell wall they have a protein rich pellicle making their body flexible.
ii. They have 2 flagella, a short and a long one.
iii. They have mixotrophic nutrition
iv. In light they are photosynthetic, but act as heterotroph (predating other smaller organism) when they are in dark.
v. They are connecting link between plants and animals.

The above statements are assigned to –
(a) Dinoflagellates
(b) Slime mould
(c) Desmids and Diatoms
(d) Euglena

Topic 8 Slime Moulds

61. Slime moulds –
(a) Are parasite
(b) Do not produce fruiting bodies
(c) Do not produce spores
(d) Saprophytic protists

62. The slimy mass of protoplasm with nuclei forms the body of slime moulds is called –
(a) Plasmodium
(b) Myxamoeba
(c) Sporocytes
(d) Periplasmodium

63. Which of the following is correct about the slime mould?
I. Its thalloid body, plasmodium, has pseudopodia for locomotion and engulfing organic matter
II. During unfavourable conditions plasmodium differentiates and produces fruiting bodies, sporangium
III. Spores possess no true cell wall.
IV. They are dispersed by air current.
V. Being extremely resistant, spores survive for many years
VI. Plasmodium can grow upto several feet.
(a) I, II, IV, V, VI
(b) I, II , III
(c) I, II , III, VI
(d) II, III , VI

Topic 9 Protozoans

64. Which one is correct about Trypanosoma?
(a) They are flagellated protozoan
(b) They are parasite
(c) They cause sleeping sickness
(d) All

65. Paramecium-
(a) Is a ciliated protozoan
(b) Shows water current movement by cilia which helps the food to be steered into gullet
(c) Has a cavity (gullet) that opens to the outside of the cell surface
(d) All

66. Plasmodium (malarial parasite)
(a) Is a ciliated protozoan
(b) Shows water current movement by cilia which helps the food to be steered into gullet
(c) Causes malaria
(d) All

67. Which of the following always produce an infectious spore like stage in their life cycles?
(a) Ciliated protozoans
(b) Flagellated protozoans
(c) Sporozoans
(d) None

68. Protozoans are not included in kingdom Animalia because –
(a) Mostly asymmetrical
(b) Unicellular eukaryotes
(c) Heterotrophic nature
(d) Multicellular prokaryotes

69. All protozoans are –
(a) Saprophytes only
(b) Parasites only
(c) Predators only
(d) Heterotrophs (parasites or predator) only

70. Which of the following is considered to be primitive relatives of animals -?
(a) Dinoflagellates (b) Slime moulds
(c) Protozoa (d) Protochordata

71. How many major groups protozoan have?
(a) 3 (b) 4
(c) 2 (d) 8

72. Which of the following are protozoans?
(a) Diatoms, flagellates, ciliates
(b) Desmids, flagellates, ciliates
(c) Amoeboid, flagellates, ciliates, sporozoans
(d) Amoeba, ·Paramecium, dinoflagellates, Plasmodium

73. Which of the following statements is wrong about the amoeboid protozoans?
(a) They live in freshwater, sea water or moist soil
(b) Amoeba has pseudopodia for locomotion and capture prey
(c) Entamoeba show holozoic nutrition
(d) Marine forms are shelled with silica

74. Flagellated protozoans are –
(a) Free living
(b) Parasites
(c) Either free living or parasites
(d) Pseudopodia

Topic 10 Kingdom Fungi

75. Which one is unicellular fungus?
(a) Puccinia (b) Toad stool
(c) Penicillium (d) Yeast

76. Coenocytic hypha is –
(a) Uninucleate hypha
(b) Multicellular hypha
(c) Multinucleate hypha without septae
(d) Hypha in coelom

77. Many fungi are in _____ association with photosynthetic organisms to form mycorrhizae or lichens –
(a) Parasitic (b) Symbiotic
(c) Photosynthetic (d) Saprobic

78. Fungi can be parasites on –
(a) Animals (b) Human being
(c) Plants (d) All

79. Fungi prefer to grow in –
(a) Cold and dry places
(b) Hot and dry places
(c) Sea water
(d) Warm and humid places

80. Fungi occur-
(a) In air and soil
(b) In water
(c) On plants and animals
(d) All

81. Fungi show a great diversity in –
(a) Morphology
(b) Habitat
(c) Both a and b
(d) Nutrition

82. Mode of nutrition in fungi is not –
(a) Parasitic (b) Saprophytic
(c) Autotrophic (d) Heterotrophic

83. All of the following are fungi except –
(a) Yeast (b) Penicillium
(c) Plasmodium (d) Puccinia

84. Which of the following is odd?
(a) Toad stool (b) Puccinia
(c) Alternaria (d) Mushroom

85. Cell walls of all fungi consist of the polysaccharide –
(a) Chitin (b) Cellulose
(c) Silica (d) Pectin

86. The body of multicellular fungus is called a –
(a) Monokaryon (b) Hyphae
(c) Rhizoids (d) Dikaryon

87. The cells of the body of a multicellular fungus arc organised into rapidly growing individual filaments called –
(a) Mycelium (b) Rhizoids
(c) Hyphae (d) Dikaryon

88. Reproduction in fungi can take place by all of the following vegetative methods except-
(a) Gemmae (b) Fragmentation
(c) Fission (d) Budding

89. Fungi show asexual reproduction by all of the following spores except-
(a) Conidia (b) Oospore
(c) Sporangiospore (d) Zoospores

90. Sexual reproduction in fungi is by all of the following except-
(a) Oospores (b) Ascopores
(c) Zoospores (d) Basidiospores

91. Select the correct statements below that correctly apply to the Kingdom Fungi-
(a) Some fungi form beneficial interrelationships with plants
(b) Certain fungi are natural sources of antibiotics
(c) The fungal life cycle typically includes a spore stage
(d) All

92. The below diagram shows a generalized life cycle of a fungus. The appropriate terms for A to E are-

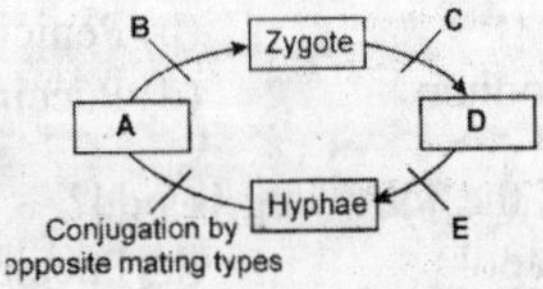

(a) Spores are absent in air
(b) Spores are present in the bread
(c) Spores are in the air
(d) The bread gets decomposed·

93. Which of the following is the correct sequence of 3 steps in the sexual cycle of fungi-
(a) Mitosis→ Meiosis → Fertilization
(b) Plasmogamy→ Karyogamy→ Meiosis
(c) Meiosis→ Plasmogamy→ Karyogamy
(d) Karyogamy→Plasmogamy→ Meiosis

94. Fungi are classified on the basis of –
(a) Morphology of mycelium
(b) Development of fruiting bodies
(c) Mode of spore formation
(d) All

95. Dikaryophase I Dikaryon formation is a specific characteristic of-
(a) All fungi
(b) Phycomycetes and ascomycetes
(c) Only basidiomycetes
(d) Ascomycetes and basidiomycetes

96. Coenocytic, multinucleate and branched mycelial habit is found in-
(a) Basidiomycetes
(b) Phycomycetes
(c) Ascomycetes
(d) Deuteromycetes

97. Match the following Columns

Column A	Column B
A. Phycomycetes	I. Sac fungi
B. Ascomycetes	II. Algal fungi
C. Basidiomycetes	III. Fungi imperfecti
D. Deuteromycetes	IV. Club fungi

The correct matching is -
(a) A-II, B-I, C- IV, D-III
(b) A- II, B - IV, C - I, D – III
(c) A- IV, B - I, C - II, D – III
(d) A- IV, B - III, C - II, D I

Topic 11 Phycomycetes

98. Members of phycomycetes are found-
I. In aquatic habitat
II. On decaying wood
III. On moist and damp places
IV. As obligate parasite on plants
(a) None of the above
(b) I and IV
(c) II and III
(d) All of the above

99. In phycomycetes asexual reproduction occurs by-
(a) Zoospores (motile)
(b) Aplanospores (non-motile)
(c) Both
(d) Aplanogamete

100. Which of the following spores are produced endogenously?
(a) Zoospores and Conidia
(b) Conidia and aplanospores
(c) Aplanospores and zoospores
(d) Aplanospore, zoospores and conidia

101. In Phycomycetes sexual reproduction occurs by

(a) Isogamy and anisogamy
(b) Isogamy, oogamy
(c) Isogamy, anisogamy and oogamy
(d) Oogamy and anisogamy

102. All the following belong to phycomycetes except –
(a) Penicillium
(b) Rhizopus (bread mould)
(c) Mucor
(d) Albugo

103. Which of the following is parasite on mustard?
(a) Albugo (b) Puccinia
(c) Yeast (d) Ustilago

Topic 12 Ascomycetes

104. Which of the following is not true about ascomycetes?
(a) Mode of nutrition saprophytic, decomposer,coprophilous (growing on dung) and parasitic
(b) Includes unicellular (e.g. yeast) and multicellular forms
(c) Mycelium is coenocytic
(d) Aspergillus, Claviceps, Neurospora are important members of Ascomycetes

105.
I. It includes unicellular as well as multicellular fungi
II. In multicellular forms hyphae are branched and septate
III. Conidiophore producces conidia (spores) exogenously in chain
IV. Sexual spores are ascopores produced endogenously in Ascus
V. Fruiting body is called ascocarp

Which of the above characters are show by-?
(a) Phycomycetes (b) Sac fungi
(c) Club fungi (d) Fungi imperfecti

106. Which of the following are edible ascomycete's delicacies?
(a) Morels + Mushroom
(b) Truffles + Toadstool
(c) Morels + Truffles
(d) Puffball + Mushroom

107. Which one is used extensively in biochemical and genetical work?
(a) Agaricus (b) Alternaria
(c) Neurospora (d) Mucor

108. Which of the following ascomycetes is thesource of antibiotic?
(a) Neurospora (b) Penicillium
(c) Claviceps (d) None

Topic 13 Basidiomycetes

109. Basidiomycetes include -
(a) Mushroom, Toadstool, Puffball and bracket fungi
(b) Smut fungi and rust fungi
(c) Both a and b
(d) Bread mould, sac fungi and algal fungi

110. Which of the following are common parasite basidiomycetes
(a) Puccinia (rust) and Ustilago (smut)
(b) Sac fungi
(c) Puffballs
(d) Agaricus (mushroom)

111. Where does meiosis occur in mushroom?
(a) Basidiospore
(b) Basidium
(c) Basidiocarp
(d) Ascus mother cell

112.
I. Mycelium is branched and septate
II. No asexual spores are generally formed
III. Vegetative reproduction by fragmentation is common
IV. Sex organs are absent but sexual reproduction takes place by nsomatogamy
V. Karyogamy and meiosis occur in basidium to form haploid exogenous 4 basidiospores

VI. Basidia are arranged in basidiocarp.

The above characters are assigned to –
(a) Sac fungi
(b) Club fungi
(c) Algal fungi
(d) Fungi imperfect

113. Plasmogamy in fungi is the fusion of-
(a) Two haploid gamete cells and their nuclei at once
(b) Two haploid nuclei
(c) Two haploid gamete cells
(d) Two diploid vegetative cells with nuclei

114. Karyogamy is -
(a) Fusion of two protoplasts
(b) Fusion of two nuclei
(c) Fusion of two plasma membranes
(d) All of these

Topic 14 Deuteromycetes

115. Which of the following is false about deuteromycetes?
(a) They reproduce only by asexual spores (conidia)
(b) Mycelium is branched and septate
(c) They have only parasitic forms
(d) They have no sexual stage (perfect stage)

116. Which of the following is correct aboutclass Deuteromycetes?
(a) Some members are saprophytes or parasites
(b) A large number of members are decomposers of litter and help in mineral cycling
(c) Alternaria, Colletotrichum and Trichoderma are deuteromycetes
(d) All

117. Sexual reproduction is found in all except –
(a) Deuteromycetes
(b) Ascomycetes
(c) Phycomycetes
(d) Basidiomycetes

118. If sexual stage is discovered in a member of deuteromycetes, it is moved to-
(a) Phycomycetes
(b) Basidiomycetes
(c) Ascomycetes
(d) Both band c

Topic 15 Diagram Based Questions

119. Identify the diagram.

(a) (i) Mucor (ii) Aspergillus (iii) Agaricus
(b) (i) Aspergillus (ii) Mucor (iii) Agaricus
(c) (i) Agaricus (ii)Aspergillus (iii) Mucor
(d) (i) Agaricus (ii) Mucor (iii) Aspergillus

120. Identify the diagram.

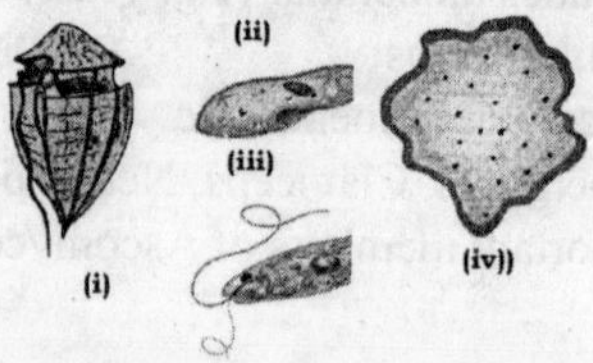

(a) (i) Dinoflagellates (ii) Euglena
(b) (i) Dinoflagellates (ii) Paramoceium
(c) (i) Euglena (ii) Dinoflagellates
(d) (i) Slime mould (ii) Paramecium

121. Kingdom plantae includes-
i. All eukaryotic chlorophyllous organisms
ii. Some prokaryotic chlorophyllous organisms
iii. Few eukaryotic partial heterotrophic plant
iv. Few prokaryotic partial heterotrophic plant

(a) i, iii (b) ii, iv
(c) i, ii, iii (d) i, iii, iv

122. Plantae does not includes how many of following:

Algae, Fungi, Bryophyte, Bladderwort, Pteridophyta, Gymnosperm, Angiosperm
(a) Zero (b) One
(c) Two (d) Three

123. Life cycle of angiosperms plant have
(a) Diploid sprophyte and diploid gametophyte
(b) Diploid gametophyte & haploid sporophyte
(c) Diploid sporophyte & haploid gametophyte
(d) Haploid sporophyte & haploid gametophyte

124. How many of following enlisted are correct about plantae
I. Cells have eukaryotic structure
II. Prominent chloroplast
III. Cellulosic cell wall
IV. Life cycle has three distinct phase
V. Show alteration of generation
(a) One (b) Two
(c) Three (d) Four

Topic 16 Kingdom Animalia

125. Kingdom Animalia are characterized by-
(a) Heterotrophic eukaryotic unicellular & multicellular organism that lack cell wall
(b) Holozoic ,digest food in an internal cavity and store food as complex carbohydrates or fat
(c) Higher as well as lower forms show elaborate sensory mechanisms
(d) All of the above

126. How many of following term is correct about Animalia- Heterotroph, eukaryotic, prokaryotic, unicellular, multicellular, store food as glycogen, presence of elaborated neuromotor mechanism without any exception, embryological development
(a) 6 (b) More than 6
(c) 5 (d) Less than 3

Topic 17 Viruses, viroids, prions, & lichens

127. Genetic material of bacteriophage is –
(a) ds DNA (b) ss RNA
(c) ds RNA (d) ss DNA

128. bacteriophage is –
(a) bacteria that infect virus
(b) virus that infect bacteria
(c) bacteria that infect cellular organism
(d) virus that infect other than bacteria

129. The protein coat called ___(A)___ made of small subunit called ___(B)___ that protect ___(C)___ of virus
(a) A = capsomere, B = capsid, C= genetic material
(b) A = capsid, B = capsomere, C = genetic material
(c) A = capsid, B = capsomere, C = enzyme and mineral
(d) A = capsomere, B = capsid, C = enzyme and mineral

130. Head of bacteriophage is –
(a) Helical (b) Polyhedral
(c) Icosahedral (d) A & B

131.

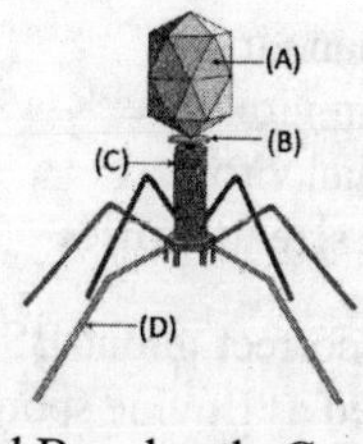

(a) A = head B = sheath, C = tail fibers, D = Collar
(b) A = head B = collar C = sheath, D = tail fibers
(c) A = collar B = head C = tail fibers D = sheath
(d) A = tail fibers B = sheath C = head D = collar

132. Viroid was discovered by –
(a) T.O. Diener (1971)
(b) W.M. Stanley (1935)
(c) T.O diener (1935)
(d) W.M. Stanley (1971)

133. Choose the correct on basis of size :
(a) Bacteria < virus < viroid
(b) Viroid < virus < bacteria
(c) Viroid > bacteria < virus
(d) Bacteria > viroid > virus

134. Choose the correct statements given below. (i-vi)
Correct set (i-vi) is
i. Viroid = virus-capsid
ii. Potato spindle disease cause by prions
iii. Viroid havc free DNA
iv. Viroid have free RNA
v. DNA of viroid was of low molecular weight
iv. RNA of viroid was of light molecular weight
(a) i,iv only (b) i, vi, iii
(c) i, iv, vi (d) i, iii, v

135. Prion cause-
(a) BSE in cattle and CJD in human
(b) BSE in human and CJD in cattle
(c) BSE and CJD cause in cattle
(d) BSE and CJD cause in human

136. Prions are-
(a) Smaller than virus
(b) Larger than virus
(c) Smaller than viroid
(d) Similar in size to viruses

137. Choose the incorrect about BSE
(a) It expanded as Bovine spongiform encephalophathy
(b) Caused by prion
(c) Its analogous variant is CJD
(d) Its homologous variant is CJD

138. Lichen are –
(a) Saprotroph only
(b) Symbiotic
(c) Parasitic only
(d) A & C

139. Lichen are mutual association of-
(a) Mycobiont (fungal) and phycobiont (algae)
(b) Gymnosperm root & fungi
(c) Algae & gymnosperm root
(d) All of these

140. Mycobiont and phycobiout belongs to ___ and___respectively
(a) Autotrophic & heterotrophic
(b) Autotrophic & autotrophic
(c) Heterotrophic & autotrophic
(d) Heterotrophic & heterotrophic

141. The function of fungal part of a lichen is/are
(a) Absorption of water
(b) Absorption of mineral
(c) Provide shelter
(d) All of these

142. Lichen cannot grow in –
(a) Polluted area
(b) Area where there is no pollution
(c) Association between fungi and algae is unpolluted region
(d) All of these

143. In R.H Whittaker system, viroids, prions & lichens are grouped into-
(a) Monera (b) Protista
(c) Protista and fungi (d) None of these

144. Viruses did not place in classification due to-
(a) Lack in study of viruses
(b) They are not considered truly 'living'
(c) Lack of genetic material
(d) All of these

145. Viruses are not-
(a) Non-cellular organism
(b) Inert crystalline structure outside the living cell
(c) Active crystalline structure outside the living cell
(d) Once they infect a cell they take over the machinery of host cell to replicate themselves, killing the host

146. The name viruses-
(a) which means venom was given by Dmitri Ivanowsky
(b) which means venom was given by M.W. Beijerinek
(c) which means venom was given by Stanley
(d) which means venom was given by Pasteur

147.

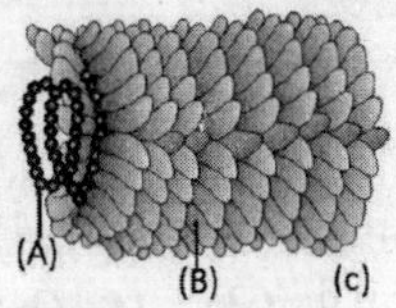

Identify a, b & organism (c)
(a) a=DNA, b=capsid, c=TMV
(b) a=RNA, b=capsid, c=TMV
(c) a=capsid, b=DNA, c=bacteriophage
(d) a=capsid, b=RNA, c=bacteriophage

148. Choose the correct statement –
(a) Genetic material of mosaic disease of tobacco causing organism is DNA
(b) Viruses were found to be smaller than bacteria but they can passed through bacteria proof filters
(c) M.W Beijerinek (1898) demonstrated that he extract of infected plant of tobacco could cause infection in healthy plants
(d) Viruses were found to be smaller than bacteria and they can passed through bacteria proof filters.

149. Contagium vivum fluidum was stated by –
(a) Dmitri lavanowsky (1898)
(b) M.W. Beijerinek (1892)
(c) W.M. Stanley (1935)
(d) None of these

150. Who showed that viruses could be crystallized and crystals outside host-
(a) W.M. Stanley(1935)
(b) M.W.Beijerinek (1898)
(c) Dmitri lvanowsky (1892)
(d) M.W. Stanley (1898)

151. Which one of following is a major constituent in crystallined virus structure –
(a) Carbohydrate (b) Protein
(c) Fat (d) Nucleic acid

152. Viruses belong to
(a) Autotroph
(b) Obligate parasite
(c) Saprotroph
(d) Holozoic

153. Genetic material of viruses are/is –
(a) DNA
(b) RNA
(c) DNA and RNA both in an individual virus
(d) DNA or RNA in an individual virus

154. The infection material of viruses is/are
(a) Protein coat
(b) Genetic material
(c) Nucleoprotein
(d) All of these

155. In general viruses that infect plants have-
(a) ds RNA (b) ss RNA
(c) ds DNA (d) ss DNA

156. Animal infection viruses are not generally
(a) ss RNA (b) ds RNA
(c) ds DNA (d) ss DNA

ANSWER KEY

1. (c)	2. (d)	3. (b)	4. (a)	5. (d)	6. (d)	7. (d)	8. (a)	9. (d)	10. (a)
11. (d)	12. (a)	13. (a)	14. (d)	15. (a)	16. (a)	17. (d)	18. (d)	19. (b)	20. (a)
21. (c)	22. (a)	23. (d)	24. (c)	25. (c)	26. (b)	27. (d)	28. (c)	29. (c)	30. (d)
31. (a)	32. (a)	33. (d)	34. (c)	35. (b)	36. (c)	37. (a)	38. (d)	39. (a)	40. (d)
41. (d)	42. (a)	43. (d)	44. (a)	45. (b)	46. (a)	47. (a)	48. (d)	49. (d)	50. (a)
51. (c)	52. (a)	53. (a)	54. (a)	55. (a)	56. (c)	57. (b)	58. (a)	59. (a)	60. (d)
61. (d)	62. (a)	63. (a)	64. (d)	65. (d)	66. (c)	67. (c)	68. (b)	69. (d)	70. (c)
71. (b)	72. (c)	73. (d)	74. (a)	75. (d)	76. (c)	77. (b)	78. (d)	79. (d)	80. (d)
81. (c)	82. (c)	83. (c)	84. (c)	85. (a)	86. (d)	87. (c)	88. (a)	89. (b)	90. (c)
91. (d)	92. (c)	93. (d)	94. (d)	95. (b)	96. (a)	97. (a)	98. (d)	99. (c)	100. (c)
101. (c)	102. (a)	103. (a)	104. (c)	105. (b)	106. (c)	107. (c)	108. (b)	109. (c)	110. (a)
111. (b)	112. (b)	113. (c)	114. (c)	115. (d)	116. (a)	117. (d)	118. (c)	119. (b)	120. (a)
121. (c)	122. (c)	123. (c)	124. (c)	125. (d)	126. (c)	127. (a)	128. (b)	129. (b)	130. (d)
131. (b)	132. (a)	133. (b)	134. (c)	135. (a)	136. (c)	137. (d)	138. (d)	139.(a)	140. (c)
141. (d)	142. (a)	143.(a)	144. (c)	145. (c)	146. (a)	147. (b)	148. (d)	149. (b)	150. (a)
151. (b)	152. (b)	153. (d)	154. (b)	155. (b)	156. (d)				

Chapter 3

Plant Kingdom

MULTIPLE CHOICE QUESTIONS

1. Phylogenetic classification based on –
   (a) evolutionary relationship
   (b) organism belonging to some taxa haven't a common ancestor
   (c) giving equal weightage to vegetative & sexual character but not on evolutionary relationship
   (d) A and B both
2. Artificial classification system is –
   (a) based on mainly on vegetative character and on the androecium structure
   (b) based on ultrastructure, anatomical, embryological characters
   (c) based on external and internal features
   (d) based on chromosome number.
3. Which one of the system is given by George Bentham and Joseph Dalton Hooker?
   (a) Artificial classification system
   (b) Phylogenetic classification
   (c) Natural classification system
   (d) A and B respectively
4. Which one is incorrectly placed?
   (a) Numerical taxonomy – Number and code are assigned to all the character and the data are then processed
   (b) Cytotaxonomy – Based on cytological information
   (c) Chemotaxonomy – Based on phytochemistry
   (d) Natural classification – Linnaeus

Topic 1 | Algae

5. Eudorina show –
   (a) Fusion of flagellate similar size gamete
   (b) Fusion of non – flagellate similar size gamete
   (c) Oogamous
   (d) Anisogamous
6. How many of following is an example of isogamous Ulothrix, Spirogyra, Volvox, fucus, Polysiphonia
   (a) 1 (b) 2
   (c) 3 (d) 4
7. Identify organism and label A, B, C, D

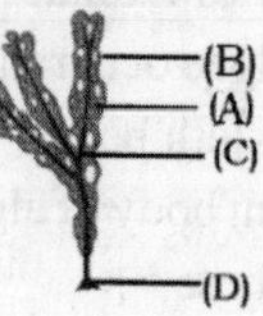

   (a) Laminaria A = leaf B = air bladder C = stripe D = holdfast
   (b) Fucus A = frond B = air bladder C = stripe D = Hold fast
   (c) Fucus A = air bladder B = frond, C = midrib D = holdfast
   (d) Laminaria A = leaf C = midrib D = petiole
8. Algae are useful to man in –
   (a) Fixation of almost half of total CO2 on earth
   (b) Primary producer
   (c) Increase level of oxygen
   (d) All of these

9. Hydrocolloids are produced by –
   (a) Brown algae (algin), carrageen(red algae), Agar(brown algae)
   (b) Brown algae (algin), Red algae (carrageen)
   (c) Brown algae (algin, agar), Red algae (carrageen)
   (d) None of these

10. Choose correct statement –
   (a) Chlorella, a multicellular alga rich in protein
   (b) Chlorella & Spirulina are astronaut food because of their high carbohydrate, vitamin mineral but less protein
   (c) The product obtained by Gracilaria are used to grow microbes
   (d) Laminaria, Sargassum a member of Rhodophyceae are among 70 species of marine algae used as food

11. Blue – green algae placed in which kingdom according to R.H. Whittaker
   (a) Monera (b) Protista
   (c) Fungi (d) Plantae

12. Choose the correct statement algae:
   (a) Algae are chlorophyllous, autotrophic member of Plantae
   (b) Some algae occur in association with fungi and on sloth bear
   (c) The plant body of algae lack root, stem, leaf
   (d) All of these

13. Colonial form alga is –
   (a) Ulothrix (b) Volvox
   (c) Kelp (d) Spirogyra

14. Zoospore is –
   (a) Sexual spore in algae
   (b) Asexual spore in algae
   (c) Develop in zoosporangium in number of four
   (d) Non flagellated spore

15. Fusion between one large static female gametes and smaller motile male gamete is termed as ___ as seen is ___
   (a) Isogamous, Spirogyra
   (b) Oogamous, Volvox
   (c) Anisogamous, Fucus
   (d) Oogamous, Ulothrix

Topic 2 Chlorophycease

16. Chara is
   (a) Common stonewort
   (b) Marine green algae
   (c) Unisexual algae
   (d) None of these

17. Flagellation in green algae is –
   (a) 2 – 8, equal, apical
   (b) 2, unequal, lateral
   (c) 2 – 8, unequal, lateral
   (d) Absent

18. Chlorophyceae are commonly called as-
   (a) Green algae
   (b) Blue – green algae
   (c) Brown algae
   (d) Red algae

19. Major pigment of Chlamydomonas –
   (a) Chlorophyll a, b
   (b) Chlorophyll a, c
   (c) Chlorophyll a, d
   (d) Fucoxanthin, phycoerythrin

20. Choose incorrect statement about green alga-
   (a) The chlorophyll localised in definite chloroplast
   (b) Spirogyra have spiral chloroplast
   (c) Most member have one or more storage bodies i.e. pyrenoid localised in chloroplast
   (d) The cell wall is made of outer layer that is of cellulose and inner layer of pectose

21. Reproduction in green algae is/are –
   (a) Isogamous (b) Anisogamous
   (c) Oogamous (d) All of these

Topic 3 Phaeophyceae

22. The color of brown algae depend upon
(a) Amount of xanthophyll
(b) Fucoxanthin present in them
(c) Phycoerythrin and fucoxanthin ratio
(d) (a) & (b) both

23. Choose the correct about cell of brown algae –
(a) Cellulosic cell wall cover outside by align
(b) Cellulosic cell wall with pectin and polysulphate esters
(c) They have two flagella, equal sized and laterally inserted
(d) A and C both

24. Dictyota is member of –
(a) Same member of Ectocarpus, Gelidium, fucus
(b) Same member those having Caminaria or mannitol as stored food
(c) Same member of Laminaria, Porphyra, fucus
(d) Same member those having phycoerythrin as accessory pigment

25. Gametes of Sargassum are- (Pg 33, E)
(a) Pyriform (b) Cup – shaped
(c) Ribbon – shaped (d) Discoid

26. Phaeophyceae is commonly named as –
(a) Green alga (b) Brown alga
(c) Red algae (d) None

27. Choose the correct statement from following –
(a) Ectocarpus is filamentous forms while kelps is profusely branched from
(b) Kelps may reach a height of average 100cm
(c) The plant body of brown algae is attached to substratum by stripe
(d) Leaf – like photosynthetic organ of brown algae is stripe

28. Major pigment found in Fucus is/are
(a) Chlorophyll a, c
(b) Chlorophyll a, d
(c) Chlorophyll a, b
(d) Fucoxanthin and phycoerythrin

Topic 4 Rhodophyceae

29. Porphyra show –
(a) Isogamous (b) Anisogamous
(c) Oogamous (d) All of these

30. Rhodopyceae is called red algae because of –
(a) Predominance of red pigment
(b) Abundance if d – phycoerythrin
(c) (a) & (b) both
(d) None of these

31. The stored food in Polysiphonia is ___A___ which is very similar to ___B___ and ___C___ in structure
(a) A = floridean starch B = amylopectin C = glycogen
(b) A = floridean starch B = chitin C = glycogen
(c) A = mannitol B = floridean starch C = amylopectin
(d) None of these

32. Member of Rhodophyceae reproduce by–
(a) Non – motile asexual spore and motile sexual gametes
(b) motile asexual spore and motile sexual gametes
(c) Non – motile asexual spore and non – motile sexual gametes
(d) motile asexual spore and non – motile sexual gametes

Topic 5 Bryophyta

33. First organism to colonize rock are –
(a) Mosses (b) Lichen
(c) Liverwort (d) (a) & (b) both

34. For trans – shipment of living material which of

following is more suitable to use
(a) Marchantia (b) Funaria
(c) Sphagnum (d) Riccia

35. Which of the following is obtained from Sphagnum as coal:
(a) Bituminous (b) Peat
(c) Lignite (d) Anthracite

36. Bryophyta include –
(a) Hornwort (b) Liverwort
(c) Mosses (d) All of these

37. Identify given plant diagram and label its parts:
(a) Funaria, A = gametophyte B = sporophyte
(b) Sphagnum, A = gametophyte B = sporophyte
(c) Funaria, A = sporophyte B = gametophyte
(d) Sphagnum, A = sporophyte B = gametophyte

38. Bryophytes are –
(a) Amphibians of plant kingdom
(b) Reptilians of plant kingdom
(c) First vascular bundles containing plant
(d) (a) & (c) both

39. The body organization of bryophytes have –
(a) Unicellular or multicellular rhizoid
(b) Less differentiation than algae
(c) They have true root stem and leaves
(d) (a) & (c) both

40. The main plant body of bryophyte is __A__ that produce __B__
(a) A = diploid B = gametes
(b) A = haploid B = gametes
(c) A = haploid B = spores
(d) A = diploid B = spores

41. Choose the correct statement
(a) Sex organs in bryophytes are unicellular and jacketed
(b) Male sex organ is antheridium that produce flagellate (four flagella) antherozoids
(c) Female sex organ is archegonium i.e. flask – shaped and produce single egg
(d) Water is required for travelling of egg from archegonium to antheridium

42. In bryophyta, meiosis occur –
(a) During development of gametes
(b) Immediately after zygote formation
(c) After sometime of zygote formation
(d) In gameophytic stage

43. Identify the given diagram and label
(a) Sphagnum, a liverwort A = archegonia branch B = antheridial branch
(b) Sphagnum, a moss A = archegonia branch B = antheridial branch
(c) Funaria, a moss A = antheridial branch B = archegonia branch
(d) Sphagnum, a liverwort A = antheridial branch B = archegonia branch

44. Choose the correct statement with regard to bryophyta
(a) Sporophyte is free – living but attached to photosynthetic gametophyte derives nourishment from it
(b) Sporophyte is not free – living but attached to photosynthetic gametophyte and derives nourishment from it
(c) Gametophyte is not free – living but attached to photosynthetic sporophyte and derives nourishment from it
(d) Gametophyte is free living but attached to photosynthetic sporophyte and derives nourishment from it

Topic 6 Liverwort

45. In Marchantia
(a) Male and Female sex organs are produced on same thalli
(b) Male and female sex organs are produced on different thalli
(c) Gametophytes is differentiated into foot seta and capsule
(d) Spores geminates to form free – living sporophyte

46. Choose the correct statement :
(a) The thalloid plant body of liverwort is dorsiventrally appressed closely to substrate
(b) The leafy members have tiny true leaf in two rows on the stem like structure
(c) The leafy membrane have tiny leaf like appendage in four rows on the stem like structure
(d) The thalloid plant body of liverwort is isobilaterally appressed closely to substrate

47. Asexual reproduction in bryophytes is not take place by –
(a) Fragmentation
(b) Gemmae
(c) Budding in secondary protonema
(d) Oogamous

48. Gemmae are –
(a) Green, unicellular, asexual bud, develop in small receptacles i.e. gemma cup
(b) Green, multicellular, asexual bud develop in small receptacles i.e. gemma cup
(c) Non – green unicellular, asexual bud, develop in small receptacles i.e. gemma cup
(d) Green, multicellular, sexual bud develop in small receptacles i.e. gemma cup n

Topic 7 Mosses

49. The predominant stage of life cycle of a moss is-
(a) Gametophytes (b) Sporophytes
(c) Protonema stage (d) Frothallus stage

50. The gametophyte of moss is divided into-
(a) Two stage, first protonema stage which develops directly from gamete.
(b) Two stage, second leafy stage which develop from secondary protonema as a lateral bud.
(c) Two stage, first leafy stage and second protonema stage
(d) Two stage, first protenema stage which develops directly from spore and second leafy stage which develop from spore germination as terminal bud.

51. Protonema stage is –
(a) Creeping, green unbranched and frequently filamentous stage
(b) Prostate, green, branched and frequently filamentous stage
(c) Creeping, green, branched and frequently filamentous stage
(d) Prostate, non – green, unbranched and frequently stage

52. Choose the correct statement about leafy stage of mosses
(a) They consist, upright, slender axes bearing spirally arranged leaves.
(b) They are attached to soil through multicellular and branched rhizoid
(c) This stage bear sex organ
(d) All of these

53. In sexual reproduction which of following is not seen in mosses
(a) Sex organ are produced at apex of leafy stage
(b) After fertilization zygote develop into sporophyte
(c) Development of embryo
(d) All of these

54. The sporophyte of mosses –
I) Is more elaborate than that is liverwort
II) Consisting of foot, seta and capsule
III) Spores present in capsule
IV) Spore produce after meiosis
V) Elaborate mechanism of spore dispersal
VI) Presence of peristomic teeth.
(a) All are correct
(b) I), II), III) only
(c) IV), V), VI) only
(d) I), III), V) only

55. Choose incorrect matched (Pg36, M)

Column – A	Column – B
(a) Hornwort	i) Marchantia
(b) Bryopsida	ii) Polytrichum
(c) Liverwort	iii) Marchantia
(d) Mosses	iv) Sphagnum

Topic 8 Pteridophytes

56. Heterosporous pteridophytes is/are-
(a) Selaginella (b) Salvinia
(c) Psilotum (d) (a) & (b) both

57. Pteridophytes with all similar kind of spores is in
(a) Terror of Kashmir
(b) Psilotum
(c) Selaginella
(d) (a) & (b) both

58. Seed habit reported for first time is
(a) Blue – green algae
(b) Pteridophyte
(c) Angiosperm
(d) Bryophyta

59. Pteridophyte classification into –
(a) 4 classes (b) 4 orders
(c) 4 families (d) All of these

60. Adiantum is member with –
(a) Pteris (b) Equisetum
(c) Lycopodium (d) Selaginella

61. Match the following: (Pg36, M)

Column – I	Column – II
i) Sphenopsida	(a) Dryopteris
ii) Lycopsida	(b) Selaginella
iii) Psilopsida	(c) Psilotum
iv) Pteropsida	(d) Equisetum

(A) a – iv, b – ii, c – iii, d – i
(B) a – iii, b – ii, c – iv, d – i
(C) a – ii, b – iii, c – i, d – iv
(D) a – i, b – iv, c – ii, d – iii

62. Identify following pteridophytes –
(a) A = Salvinia, B = horsetail, C = fern, D = Selaginella
(b) A = Selaginella, B = Salvinia, C = fern, D = horsetail
(c) A = Equisteum, B = fern, C = Selaginella, D = horsetail
(d) A = Selaginella, B = Salvia, C = Dryopteris, D = Equisteum

63. Label A, B, C, D, E in following diagram:
(a) A = strobilus, B = rhizome,
(b) A = cone, C = Node, D = internode
(c) A = strobilus, B = rhizome, C = node, D = internode, E = branch
(d) None of these

64. Pteridophytes includes –
(a) Horsetail (b) Ferns
(c) Polytrichum (d) (a) & (b) both

65. First terrestrial vascular plant is –
(a) Algae
(b) Bryophyta (liverwort & hornwort)
(c) Pteridophyta
(d) Bryophyta (Mosses)

66. Choose the correct statement from following
(a) The plant body is differentiated into true root, only true prostrate stem as in Selaginella and true leaf
(b) The leaves of pteridophytes are small as in Selaginella or macrophyll in ferns.
(c) Pteridophytes possess xylem, phloem
(d) All of these

67. In pteridophyta –
(a) The main plant body is a sporophyte
(b) The main plant body is a gametophyte
(c) The main plant body is a gametophyte on which sporophytic phase is partially dependent
(d) (a) & (b)

68. Choose the correct with regard to reproduction in pteridophyte
(a) Sporophyte bear sporangia that are subtended by sporophyll
(b) Gametophyte bear sporangia that are subtended by sporophyll
(c) Sporophyll compact to form strobili as in fern
(d) The sporangia produce spores by mitosis in spore mother cell

69. Gametophyte of pteridophyte is –
(a) Small but multicellular, free living, mostly photosynthetic, differentiated into root, stemand leaf
(b) Small inconspicuous but multicellular dependent mostly photosynthetic thalloid body
(c) Small but multicellular, free living mostly photòsynthetic thalloid structure
(d) Small inconspicuous but multicellular free – living mostly non – photosynthetic thalloid body

70. Water needed for fertilization in –
(a) Eucalyptus (b) Bryophytes
(c) Pteridophyptes (d) (b) & (c) both

71. Sex organ bear on –
(a) Sporophytes
(b) Gametophyte
(c) On both gametophytes & sporophyte
(d) None

Topic 9 Gymnosperm

72. The endosperm of gymnosperm represent
(a) Female gametophyte
(b) Triploid structure
(c) Diploid structure
(d) (a) & (c)

73. Read the following statements and choose the incorrect response with respect to gymnospermous reproduction
(a) Pollen grains are carried by air currents
(b) Pollen tube carries the male gametes to archegonia
(c) Following fertilization, zygote develop but embryo stage is lacking
(d) Ovule develops into seed

74. All the given structure of Pinus and Cycas are haploid, except
(a) Pollen grain (b) Egg
(c) Nucellus (d) Endosperm

75. Gymnosperm is example of –
(a) Vascular, embryophyte with ovule enclosed is ovary
(b) Vascular, non-embryophyte
(c) Non-vascular, non-embryophyte
(d) Vascular, embryophyte

76. Vascular archegoniates with diplontic lifecycle are –
(a) Bryophytes (b) Gymnosperm
(c) Pteridophytes (d) (b) & (c)

77. Gymnosperms are plants in which –
(a) Ovules are enclosed by any ovary wall both before and after fertilization
(b) Ovules are not enclosed by any ovary wall both before and after fertilization
(c) Ovules are enclosed by any ovary wall before fertilization but not after fertilization
(d) Ovules are not enclosed by any ovary wall before fertilization but after fertilization

78. Tallest tree species belongs to –
(a) Angiosperm (b) Gymnosperm
(c) Pteridophyte (d) Algae

79. Fungi show symbiotic association with gymnosperm in form of –
(a) Mycorrhiza in Pinus
(b) Mycorrhiza in cycas
(c) Coralloid rest in Pinus
(d) Coralloid rest in cycas

80. The stem of –
(a) Cycas is unbranched

(b) Pinus is branched
(c) Cedrus is branched
(d) All of these

81. Needle-like leaves, thick cuticle, sucken stomata are character of –
(a) Cycas (b) Pinus
(c) Gnetum (d) Ginkgo

82. Gymnosperms are –
(a) Heterosporous, haploid microspores and haploid megaspore
(b) Homosporous, both spores are haploid
(c) Heterosporous, both spores(microspores & megaspores) are diploid
(d) None of these

83. Choose the correct statement –
(a) The male and female cones borne on same plant as in Cycas
(b) The male and female cones borne on different plant as in Cycas
(c) The male and female cones borne on same plant as in Pinus
(d) Both (a) & (c)

84. Choose the correct about female cone of gymnosperm:
(a) The nucleus is protected by bitegmic structure
(b) The megaspore mother call divides mitotically to form four megaspores
(c) One of four megaspores, enclosed within the megasporangium which develop into a multicellular female gametophyte that bear one archegonia
(d) Ovule is unitegmic

85. Statement-I : The cones bearing megasporophyll with ovules are female cone

Statement-II : The strobili bearing microsporangia are called male cone
(a) Both stated statement are correct
(b) Both stated statement are incorrect
(c) Statement-I is correct while statement- II is incorrect
(d) Statement-I is incorrect while statement-II is correct

86. Identify given plant diagram and choose correct response
(a) Ginkgo, a living fossil
(b) Cycas, a living fossil
(c) Taxus
(d) Gnetum

87. What is difference between bryophytic and gymnospermous & gametophytes
(a) Bryophytic gametophytes is independent free-living structure while gametophytes of gymnosperm is dependent
(b) Gametophyte of gymnosperm remain within the sporangia retained on sporophytes
(c) Both (a) & (b)
(d) None of these

88. Choose the correct set about given figure:
i) Pinnate leaves
ii) Palmate leaf
iii) Branched stem
iv) Branching is same as in Cedrus
v) Unbranched
vi) Bear male cone and female cone on same plant
vii) Bear male cone & female cone on different plant
viii) It is living fossil along with Ginkgo
(a) i, iii, vi, viii (b) i, v, vii, viii
(c) ii, v, vi (d) i, iv, vii, viii

89. Anthoceros thallus and coralloid root of Cycas are
(a) Similar in morphological structure
(b) Performing N2-fixing
(c) Presence of vascular bundle
(d) (b) & (c)

90. Gametophytes is parasitic over sporophytes is
(a) Cycadales (b) Coniferales
(c) Monocot (d) All of these

Topic 10 Angiosperm

91. Zygote is result of-
(a) Syngamy (b) Double fertilization
(c) Triple fusion (d) Both (a) & (c)

92. Fusion of 2nd male gamete with diploid secondary nucleus result in formation of-
(a) PEN (b) Embryo
(c) Both (a) & (b) (d) Sporophyte

93. Double fertilization is-
(a) Fusion of two nuclei of polar nuclei
(b) Fusion of male gamete with egg
(c) Fusion of male gamete with secondary nuclei
(d) Both (b) & (c)

94. PEN provide-
(a) Protection of embryo
(b) Nourishment to embryo
(c) Anchorage to embryo
(d) None of these

95. Which of following structure degenerate after fertilization-
(a) Synergid (b) Antipodal cell
(c) (a) & (b) (d) Embryo

96. Angiosperm differ with gymnosperm-
(a) In presence of true root, stem & leaf
(b) Seed enclosed in fruit
(c) Ovary enclosed in ovule
(d) Both (b) & (c)

97. Ovule develop into _____ and ovaries develop into ______ of angiosperm
(a) Seed, fruit (b) Fruit, seed
(c) Fruit, fruit (d) Seed, seed

98. Pistil is-
(a) Female sex organ of flower
(b) Male sex organ of flower
(c) Non-reproductive organ of flower
(d) Divided into two part that are anther and filament.

99. Kelp, Polysiphonia, Ectocarpus, Fucus, Wolffian, Volvox How many of following are show haplontic, haplodiplontic and diplontic life cycle respectively
(a) 1, 3, 2 (b) 3, 1, 2
(c) 1, 2, 3 (d) 2, 3, 1

100. Mitosis is observed in-
(a) Haploid plant cell
(b) Diploid plant cell
(c) Both (a) & (b)
(d) Only vegetative cell

101. Choose correct statement about haplontic life cycle-
i) Sporophytic generation is represented by single cell zygote
ii) Free-living sporophyte
iii) Sporophyte is parasite on gametophyte
iv) Gametophyte arise from gametes after mitotical division
v) Example are Spirogyra and some species of Chlamydomonas
vi) Gametophyte arise from meiosis occur in spore-
(a) i, ii, v, vi (b) i, iii, v, vi
(c) iii, iv, v (d) i, iii, iv

102. Eucalyptus show-
(a) Diploid dominant sporophyte that is photosynthetic and independent phase
(b) Gametophyte is represent by few diploid cell
(c) Dominant phase is gametophyte
(d) All of these

103. Gymnosperms are-
(a) Haplontic (b) Diplontic
(c) Haplo-diplontic (d) Diplo-haplontic

104. Bryophytes and Pteridophyte exhibit-
(a) Multicellualr sporophyte
(b) Multicellular gametophyte
(c) Unicellular sporophyte
(d) (a) & (b) both

105. Bryophytes and pteridophytes differ in their –
(a) Stage of meiosis
(b) Dominant phases
(c) Stage of syngamy
(d) Stage of gametogenesis

106. In bryophytes –
(a) Sporophyte totally or partially dependent on the gametophyte for its anchorage and nutrition
(b) Gametophyte totally or partially dependent on the sporophyte for its anchorage and nutrition
(c) A dominant, independent, photosynthetic, thalloid haploid Sporophyte alternate with gametophyte
(d) (a) & (c) both

107. Choose the correct response with respectto pteridophyte lifecycle
(a) Diploid gametophyte alternate with sporophyte
(b) Sporophyte and gametophyte are independent
(c) Sporophyte show saprophytic
(d) Meiosis occur in gametophyte

108. The sporophyll of gymnosperms arranged ____ on axis to from cones
(a) Spirally (b) Alternately
(c) Decussate (d) Superposed

109. Identify life cycle pattern
(a) A = haplontic, B =haplo – diplontic, C = diplontic
(b) A = haplontic, B = diplontic, C = haplo – diplontic
(c) A = haplo – diplontic, B = haplontic, C = diplontic
(d) A = as in Volvox and angiosperm, B = as in Ectocarpus, C = as in gymnosperm

110. Bryophyte attached to substratum by –
(a) Holdfast (b) Rhizoid
(c) Root (d) (a) & (c)

111. Brown algae focus attached to substratum by –
(a) Holdfast (b) Stipe
(c) Frond (d) Rhizoid

112. The plant body of liverwort is __A__ whereas mosses have ____B___ bearing ____C___ arranged leaves
(a) A = dorsiventral, B = upright, slender axes, C = alternally
(b) A = isobilateral, B = upright, slender axes, C = spirally
(c) A = dorsiventral, B = isobilateral axes, C = alternately
(d) A = dorsiventral, B = upright, slender axes, C = spirally

113. Embryophytes doesn't includes
(a) Algae, Bryophytes
(b) Bryophyte, Pteridophytes
(c) Gymnosperm, angiosperm
(d) Algae only

114. Double fertilization does not occur in –
(a) Pteridophyte, some gymnosperm,
(b) Monocot, dicot
(c) Dicot, some gymnosperm
(d) Bryophytes, pteridophyte, some gymnosperm & monocot

115. Identify following life cycle pattern and that pattern shown in
(a) Haplontic life cycle eg: Volvox
(b) Haplodiplontic lifecycle eg: Ectocarpus, Psilotum
(c) Haplodiplontic lifecycle eg: Fucus, Marchantia
(d) Diplontic lifecycle eg: Bryophytes, Pteridophytes

116. Tallest and smallest plant species belonging to angiosperm is –
(a) Sequoia and Wolffia
(b) Eucalyptus and Wolffia
(c) Sequoia and duck-weed

(d) None of these

117. Dicotyledons and monocotyledons are two _____ of angiosperm

(a) Family (b) Class
(c) Order (d) Division

118. How many of following is correct about dicotyledons and monocotyledons respectively Seed with two cotyledons, trimerous, pentamerous, parallel veination Seed with one cotyledons, tetramerous, reticulate veination

(a) 4, 3 (b) 3, 4
(c) 2, 5 (d) 5, 2

119. A group of plant flower with having three members in each whorl is placed is-

(a) Monocot (b) Dicot
(c) Tetramerous (d) Both (b) & (c)

120. Choose the correct statement

(a) Embryo sac develop from one functional megaspore(diploid) which result from mitosis and degeneration of megaspore mother cell
(b) Embryo sac of consist of one egg apparatus, three antipodal cell and two polar nuclei
(c) Polar nuclei, antipodal cells, egg are diploid structure of embryo sac of angiosperm
(d) Secondary nuclei is haploid

121. Secondary nuclei result from fusion is

(a) Polar nuclei and 1st male gamete
(b) Polar nuclei and 2nd male gamete
(c) Both nuclei of polar nuclei
(d) Egg apparatus and polar nuclei

122. Choose the correct sequence

(a) Gamete formation → pollination → fertilization → embryo → new plant
(b) Gamete formation → transfer of gamete → fertilization → pollination → embryo → new plant
(c) Pollination → gametogenesis → fertilization → embryo → new plant
(d) None of these

123. Microspore of angiosperm represent-

(a) Sporophytic phase
(b) Gametophytic phase
(c) Both (a) & (b)
(d) Female gamete

124. Pollen tube in angiosperm discharge-

(a) One male gamete is embryo sac
(b) Two male gamete is embryo sac
(c) Three male gamete is embryo sac
(d) More than one option is correct

125. Syngamy is-

(a) Fusion of egg and 1st male gamete
(b) Fusion of egg and 2nd male gamete
(c) Fusion of polar nuclei & 1st male gamete
(d) Both (b) & (c)

ANSWER KEY

1. (d)	2. (a)	3. (c)	4. (d)	5. (d)	6. (b)	7. (b)	8. (d)	9. (b)	10. (d)
11.(a)	12. (d)	13.(b)	14. (b)	15. (b)	16.(a)	17.(a)	18. (a)	19.(a)	20. (d)
21. (d)	22. (d)	23.(d)	24.(b)	25.(a)	26.(b)	27. (a)	28. (a)	29. (c)	30. (a)
31.(a)	32.(c)	33.(d)	34.(c)	35. (b)	36.(d)	37.(c)	38.(a)	39. (a)	40.(b)
41.(c)	42.(c)	43.(b)	44.(b)	45. (b)	46.(a)	47.(d)	48.(b)	49. (a)	50. (b)
51. (c)	52.(d)	53.(c)	54. (a)	55.(a)	56. (d)	57.(d)	58. (b)	59.(a)	60. (a)
61. (a)	62.(b)	63.(d)	64.(d)	65. (c)	66. (d)	67. (a)	68.(a)	69.(c)	70. (d)
71.(b)	72.(d)	73.(c)	74. (c)	75. (d)	76. (b)	77. (b)	78. (b)	79.(a)	80. (d)
81. (b)	82.(a)	83.(b)	84.(d)	85. (a)	86. (a)	87. (c)	88.(b)	89.(b)	90.(d)
91.(a)	92. (a)	93.(d)	94.(b)	95. (c)	96.(b)	97. (a)	98.(a)	99.(a)	100.(c)
101.(b)	102.(a)	103.(b)	104.(d)	105.(b)	106.(a)	107.(b)	108.(a)	109.(c)	110.(b)
111.(a)	112(d)	113.(a)	114.(a)	115.(b)	116.(b)	117.(b)	118.(a)	119.(a)	120.(b)
121.(c)	122.(a)	123.(b)	124.(b)	125.(a)					

4 Chapter Morphology of Flowering Plants

MULTIPLE CHOICE QUESTIONS

1. Statement – I: Esau got National Academy of Science in 1957
   Statement – II: In 1989, Esau received National Medal of Science in 1989.
   (a) Statement – I & statement – II are both correct
   (b) Statement – I & statement – II are both incorrect
   (c) Statement – I is correct and statement – II is incorrect
   (d) Statement – I is incorrect and statement – I is correct

2. Morphology is study of
   (a) External structure of an organism
   (b) Internal structure of an organism
   (c) Systematics
   (d) (a) & (b) booth

3. Curly top virus spreads a plant via-
   (a) Xylem (b) Phloem
   (c) Vascular bundle (d) None of these

4. The book 'Plant Anatomy' was published by Esau in –
   (a) Same year as she did her doctorate
   (b) 1960
   (c) 1954
   (d) 1957

5. Which of referred as 'Webster's of plant biology' – an encyclopedia
   (a) Plant anatomy
   (b) Anatomy of angiospermic plant
   (c) Anatomy of seed plants
   (d) (a) & (b) both

6. Esau was ______ woman to receive 'National Academy of science '
   (a) 7th (b) 6th
   (c) 5th (d) 1th

Topic 1 The Root

7. Root is characterized by mainly
   (a) presence of node & internode
   (b) (-ve) phototropism
   (c) (-ve) geotropism
   (d) (-ve) hydrotropism

8. Which of the following is not the main function of root system is/are
   (a) Absorption of sap from soil
   (b) Providing proper anchorage to plant parts.
   (c) Synthesis of plant growth regulators
   (d) None of these

9. Identify given diagram

	(a)	(b)	(c)
(a)	Tap root	Fibrous root	Adventitious root
(b)	Tap root	Adventitious root	Fibrous root
(c)	Adventitious root	Fibrous root	Tap root
(d)	Fibrousroot	Tap root	Adventitious root

10. Radical form-
    (a) Root system of plant
    (b) Floral part of plant

(c) Shoot system of plant
(d) (a) & (b) both

11. The lateral roots arise from primary root is-
(a) Primary root (b) Secondary root
(c) Tertiary root (d) (a) & (b) both

12. Choose the given statement which is suitable for following figure
(a) It comprises of primary & secondary root
(b) Such roots are observed in mustard
(c) These roots are replace by large number root
(d) (a) & (b) both

13. From given set of example choose, how many of following are example of fibrous root and adventitious root respectively. Sweet potato, carrot, turnip, wheat, grass, Monstera, banyan tree
(a) 1, 4 (b) 1, 3
(c) 2, 3 (d) 3, 2

14. Adventitious roots arise from-
(a) Radicle
(b) Base of stem in tuft as in wheat
(c) Part of plant other than radicle as in mustard
(d) Secondary root

Topic 2 Regions of the Root

15. In aquatic plant the apex of root is covered by
(a) Thimble parenchymatous root cap
(b) Root pocket
(c) Coleorhiza
(d) Coleoptile

16. Identify region of root tip (Pg. 67, M)
(a) A = Region of maturation, B = Region of elongation, C = Region of meristematic activity, D = Root cap
(b) A = Region of elongation, B = Region of meristematic activity, C = Root cap, D = Protective covering
(c) A = Region of meristem, B = Region of maturation, C = Region of elongation, D = Root cap
(d) A = Region of growing cell, B = Region of mature cell, C = Region of dividing cell, (D = Protective covering

17. Root hair arise from –
(a) Cortical cell of region of maturation
(b) Epidermal cell of region of maturation
(c) Cortical cell of region of elongation
(d) Epidermal cell of region of elongation

18. Choose mismatch pair (Pg. 67, H)

Column – I	Column – II
(a) Region of meristematic	Small thin wall dense
(b) Region of elongation	Responsible for growth of root in length
(c) Region of maturation	Proximal to region of elongation
(d) Root hair	Differentiated and mature cell proximal to region of maturation

Topic 3 Modification of Root

19. Pneumatophores are helpful in-
(a) Transpiration
(b) Getting oxygen for respiration
(c) Absorption of water
(d) Assimilation of food

20. Silt roots and pneumatophores are observed in-
(a) Maize, Rhizophora
(b) Maize, Rhizopus
(c) Sugarcane Rhizopus
(d) (a) & (b) both

21. Mechanical root observed in –
(a) Sugarcane (b) Maize
(c) Banyan tree (d) All of these

22. For food storage root get modified in –
(a) Potato (b) Sweet potato
(c) Ginger (d) (a) & (b) both

23. Match the following – (Pg. 67, H)

Column – I	Column – II
(a) Conical root	(I) Raddish
(b) Napiform root	(II) Turnip
(c) Tuberous root	(III) Sweet potato
(d) Fusiform root	(IV) carrot

	a	b	c	d
(a)	IV	II	III	I
(b)	IV	III	II	I
(c)	III	IV	I	II
(d)	III	IV	I	II

24. Modification of root Asparagus is meant for –
(a) Storage of food
(b) Mechanical support
(c) Respiration
(d) Climbing support

25. Slit root arise from –
(a) Lower nodes of Zea mays
(b) Lower internode of sugarcane
(c) Lower internode of Zea mays
(d) Upper node of sugarcane

26. Pneumatophores are
i) Positive geotropism
ii) Negative geotropism
iii) Grown in marshy area
iv) Found in mangroves
v) Positive phototropism
vi) Negative phototropism
(a) i, iii, iv, vi (b) ii, iii, iv, v
(c) i, iii, v (d) ii, iv, vi

Topic 4 Stem

27. Stem distinguish from root in –
(a) Presence of node & internode
(b) Absence of node & internode
(c) Presence of hairs for water absorption
(d) Absence of bud

28. Stem are develop from –
(a) Radicle of germinating seed
(b) Plumule of germinating seed
(c) Cotyledons of germinating seed
(d) Coleoptile

29. The region of stem where leaves are born are______
(a) Nodes
(b) Internode
(c) Both node & internode
(d) Floral bud

30. Stems are generally –
(a) (+ve) geotropism, (-ve) hydrotropism, (+ve) phototropism
(b) (-ve) geotropism, (-ve) hydrotropism, (+ve) phototropism
(c) (+ve) geotropism, (+ve) hydrotropism, (+ve) phototropism
(d) (+ve) geotropism, (-ve) hydrotropism, (–ve) phototropism

Topic 5 Modification of Stem

31. In pineapple –
(a) The lateral branches originate from basal and underground portion of main stem, grow horizontally beneath the soil and then come out obliquely upward giving rise to leafy shoot.
(b) The lateral branch arises time arch downward to touch the ground growing aerially for some time arch downward to touch the ground
(c) A lateral branch with short internode and each node bearing a rosette of leaves and a tuft of roots.
(d) None of these

32. In Oxalis stem is modified for –
(a) Storage
(b) Support

(c) Protection
(d) Vegetative propagation

33. Lateral branch with short internode & each node bearing a rosette of leaves and a tuft of root found in –
(a) Pistia (b) Eichhornia
(c) Grasses (d) (a) & (b) both

34. Underground modified stem of potato is known as-
(a) Tuber (b) Rhizome
(c) Corm (d) Bulb

35. Stem store food for—
(a) Favourable condition growth
(b) Unfavourable condition growth
(c) Flowering condition
(d) (a) & (c) both

36. Choose odd on with respect to stem modification ______
(a) Zaminkand (b) Colocasia
(c) Bougainvillea (d) Turmeric

37. How many of following stem modification does develop from axillary buds Colocasia, grapevines, cucumber, pumpkin, Opuntia, Citrus, Watermelon, Bougainvillea
(a) 7 (b) 6
(c) 5 (d) 4

38. Ginger and turmeric are example of –
(a) Rhizome (b) Rhizoid
(c) Corm (d) Roots

39. Photosynthetic green flattened modified stem xerophyte is in –
(a) Acacia (b) Euphorbia
(c) Opuntia (d) Hydrilla

40. Stem is modified for protection in –
(a) Citrus thorn
(b) Bougainvillea spine
(c) Opuntia thorn
(d) A and C

41. Statement – I: Some plants of arid region modify their stems into fleshy cylindrical structure as in Euphorbia

Statement – II: In grapevines, stem tendril are for help plant to climb
(a) Statement – I and Statement – II are correct.
(b) Statement – I is correct while statement – II is not correct
(c) Statement – I is incorrect while statement – II is correct
(d) Statement – I and statement – II are incorrect

42. Stem tendril of pumpkin develop from-
(a) Accessory bud
(b) Axillary bud
(c) Extra – axillary bud
(d) Floral bud

43. Choose the correct statement about stem modification of mint
(a) A slender lateral branch arises from base of main axis and after growing underground for some time arch upward to touch the ground.
(b) A slender lateral branch arises from base of main axis and after growing aerially for some time arch downwards to touch the ground.
(c) Stem modification is same as in strawberries
(d) Stem modification mint is known as sucker

44. Match the following:

Column – I	Column – II
I) Strawberry	A. Sucker
II) Jasmine	B. Offset
III) Pistia	C. Runner
IV) Pineapple	D. Stolon

(a) I – C, II – D, III – B, IV – A
(b) I – B, II – C, III – A, IV – D
(c) I – C, II – A, III – B, IV – D
(d) I – A, II – B, III – C, IV – D

45. Choose odd one with respect to stem modification-
(a) Chrysanthemum (b) Banana
(c) Pineapple (d) Strawberry

Topic 6 The Leaf

46. Choose the correct response:
    (a) Leaf develop at the node and bears a bud in its axile
    (b) Leaves originate from SAM are arranged in acropetal orders.
    (c) Leaf is lateral gernerally flattened vegetative structure for photosynthesis
    (d) All of these

47. Stipules are –
    (a) Two lateral small leaf like structure
    (b) Four lateral small leaf like structure
    (c) One lateral small leaf like structure
    (d) Many lateral small leaf like

48. The leaf base expanded into a sheath crossing the stem partially or wholly in–
    (a) Monocot
    (b) Dicot
    (c) All angiosperms plant
    (d) Gymnosperms

49. Pulvinus is –
    (a) Swollen leaf base of legume
    (b) Swollen petiole of legume and china Rose
    (c) Swollen lamina
    (d) Swollen stipule

50. Label – A, B, C, D, E (Pg. 70, M)

	A	B	C	D	E
(a)	Lamina	Stipule	Petiole	Axillary bud	Leaf base
(b)	Lamina	Stipule	Petiole	Axillary bud	Leaf base
(c)	Lamina	Pulvinus	Pedicel	Axillary bond	Leaf base
(d)	Lamina	Stipule	Pedicel	Extra-axillary bond	Leaf base

Topic 7 Venation

51. Arrangement of vein & veinlet in lamina of leaf
    (a) Venation (b) Phyllotaxy
    (c) Aestivation (d) None of these

52. Leaves of dicotyledonous plants generally characterized by -
    (a) Presence of parallel venation
    (b) Veins which are parallel to each other within a lamina.
    (c) Presence of reticulate venation
    (d) (a) & (b) both

53. Identify the leaf venation and type of leaf.
    (a) Parallel venation; monocot mainly
    (b) Parallel venation; dicot mainly
    (c) Reticulate venation; dicot mainly
    (d) Reticulate venation; monocot mainly

54. A leaf is simple
    (a) When its lamina is entire
    (b) When its lamina is incised, the incision do not touch the midrib
    (c) (a) & (b) both
    (d) None of these

55. When the incisions of lamina reach to midrib breaking leaf into a number of leaflet is not-
    (a) Compound leaf
    (b) Simple leaf
    (c) Pinnate leaf
    (d) Palmate leaf

56. Identify A and B (Pg. 70, M)
    (a) A = pinnately compound leaf; Neem B = palmately compound leaf; Silk cotton
    (b) A = palmately compound leaf; Silk cotton B = pinnately compound leaf; Neem
    (c) A = pinnately compound leaf; Silk cotton B = palmately compound leaf; Neem
    (d) A = palmately compound leaf; Neem B = pinnately compound leaf; Silk cotton

57. Midrib of pinnately compound leaf is –

(a) Mid-vein (b) Rachis
(c) Petiole (d) None of these

58. Leaflet of pinnately compound leaf arise on-
(a) Common point i.e. at tip of petiole
(b) Common axis
(c) Common point i.e. at tip of rachis
(d) (a) & (c) both

59. Leaflet of ________ arise on common point i.e. at tip of petiole
(a) Pinnately compound leaf
(b) Palmately compound leaf
(c) Simple leaf
(d) All of these

Topic 8 Phyllotaxy

60. Phyllotaxy is pattern of arrangement of____ on the____
(a) Leaf, stem (b) Phloem, stem
(c) Vein, leaf (d) None of these

61. Identify types of phyllotaxy shown by given diagram

	A	B	C
(a)	Opposite	Alternate	Whorled
(b)	Alternate	Opposite	Whorled
(c)	Alternate	Whorled	Opposite

(d) None of these

62. Choose correct statement
(a) In alternate type; a single leaf arises at each node.
(b) In opposite type; a pair leaves arises at each node.
(c) In whorled type; more than two leaves arises at each node.
(d) All of these

63. Sunflower show-
(a) Alternate phyllotaxy
(b) Opposite phyllotaxy
(c) Whorled phyllotaxy
(d) None of these

Topic 9 Modification of Leaves

64. In Australian acacia
(a) Lamina modification
(b) Petiole modified
(c) Stipule modified
(d) All of these

65. Select the correct option:
(a) Both (a) & (b) are modified by leaves
(b) A is tendrils for climbing
(c) B is spines for defence
(d) All of these

66. Pitcher of pitcher plant is modified –
(a) Leaf (b) Stem
(c) Root (d) Fruit

67. Flower is modified –
(a) Node (b) Internode
(c) Leaf (d) Shoot

68. Choose the correct statement
(a) In flower, SAM changes to floral meristem
(b) In flower, internode do not elongate
(c) The axis get condensed in flower.
(d) All of these

69. The arrangement of flowers on the floral axis is ______
(a) Phyllotaxy (b) Inflorescence
(c) Aestivation (d) Placentation

70. On the basis whether floral apex gets develop into flower or continues to grow, inflorescence are mainly of-
(a) 3 types (b) 4 types
(c) 2 types (d) None of these

71. In racemose-
(a) Main axis continues to grow
(b) Flower are in basipetal order
(c) Main axis terminate into flower
(d) B & C both

72. Choose the correct statement about given figure
(a) It is of racemose type inflorescence
(b) Flowers are in basipetal order
(c) Flowers are in acropetal order
(d) Example of Cassia

73. Given diagram is of –
(a) Racemose inflorescence
(b) Cymose inflorescence
(c) Cymose inflorescence of Cassia
(d) B & C both

Topic 10 The Flower

74. A complete flower consist of –
(a) One whorl (b) Two whorls
(c) Three whorls (d) Four whorls

75. Flower stalk is known as –
(a) Pedicel (b) Thalamus
(c) Petiole (d) Stipules

76. Thalamus is not –
(a) Swollen end of pedicel
(b) Different whorl arranged on it
(c) Accessory whorl
(d) Receptacle for different whorl

77. Choose the correct statement-
(a) Calyx, corolla, are accessory organ
(b) Androecium, gynoecium are reproductive organ
(c) Perianth present in lily
(d) All of these

78. Perianth is
(a) Indistinct calyx & corolla
(b) Fused corolla & androecium
(c) Reproductive organ
(d) None of these

79. Bisexual flowers is –
(a) When a flower has both androecium & gynoecium
(b) Present in Solanaceae, Liliaceae
(c) Present in mustard and Pea
(d) All of these

80. How many of following show Actinomorphic, Zygomorphic respectively.

Mustard, datura, chilli, Pea, Canna, bean, gulmohur, Cassia
(a) 3, 4 (b) 4, 3
(c) 4, 4 (d) None of these

81. Statement – I: when a flower can be divided into two equal radial halves in any radial plane passing through the centre it is actinomorphic flower

Statement – II: when a flower can be divided into two similar halves only in one particular vertical plane, it is zygomorphic
(a) Statement – I & II are correct
(b) Statement – I is correct
(c) Statement – II is correct only
(d) Statement – I & II are incorrect

82. Cassia show -
(a) Racemose inflorescence, zygomorphic
(b) Racemose inflorescence, actinomorphic
(c) Cymose inflorescence, actinomorphic
(d) Cymose inflorescence, zygomorphic

83. Flower with leaf that found the base of pedicel are –
(a) Bracteate (b) Ebracteate
(c) Petiolate (d) Sessile

84. Flower with floral appendages 3 or multiple of 3 are said –
(a) Tetramerous (b) Trimerous
(c) Triploid (d) Pentamerous

85. In hypogynous flower which of following floral part takes highest position
(a) Calyx (b) Corolla
(c) Androceium (d) Pistil

86. Which of following is mismatched

Column-I Column-II

A)  1. Mustard

B)  2. Brinjal

C) 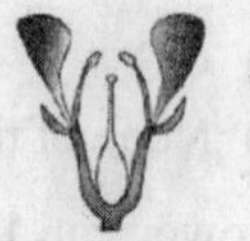 3. Peach

D)  4. Cucumber

87. Superior ovary found in –
   (a) Hypogynous flower
   (b) Perigynous flower
   (c) Epigynous flower
   (d) Cucumber

88. Choose the correct about perigynous flower –
   (a) Gynoecium is situated in centre
   (b) Apart from gynoecium, rest parts are located on rim of thalamus almost at same level
   (c) Ovary is half inferior
   (d) All of these

89. How many of following are example of perigynous, hypogynous and epigynous respectively.

   Mustard, china Rose. Brinjal, plum, peach, rose, guava, cucumber, ray floret sunflower, Pea, Asparagus

   (a) 3, 3, 5 (b) 3, 3, 3
   (c) 3, 5, 3 (d) 5, 3, 3

90. 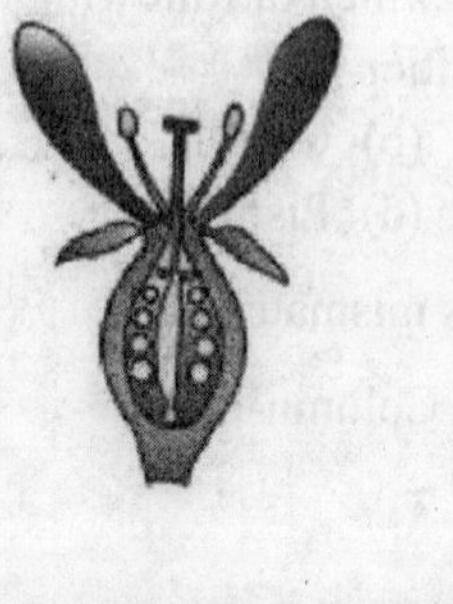

   (a) Hypogynous flower
   (b) Epigynous
   (c) Perigynous
   (d) China rose

Topic 11 Parts of Flower, Calyx, Corolla

91. Flower consist of –
   (a) Four reproductive whorl
   (b) Four whorl
   (c) Four accessory whorl
   (d) All of these

92. The outermost whorl of flower is –
   (a) Calyx (b) Corolla
   (c) Bract (d) Thalamus

93. Choose the correct statement-
   (a) Sepals are members of calyx
   (b) Petals are members of calyx
   (c) Sepal are plural of corolla
   (d) None of these

94. Sepals united in ________ and sepals are free in _____ condition
   (a) Gamosepalous, Polysepalous
   (b) Polysepalous, Gamosepalous
   (c) Polysepalous, Polysepalous
   (d) Gamosepalous, Gamosepalous

95. Corolla are –
   (a) Composed of petal
   (b) United by sepals
   (c) Composed of tepals
   (d) Usually for bud protection

96. Polypetalous is condition with _____ while gamopetalous is for _____
   (a) Free petal; fused petal
   (b) Fused petal; free petal
   (c) Free petal; free petal
   (d) Fused petal; fuced petal

97. Label (i), (ii), (iii), (iv), (v)

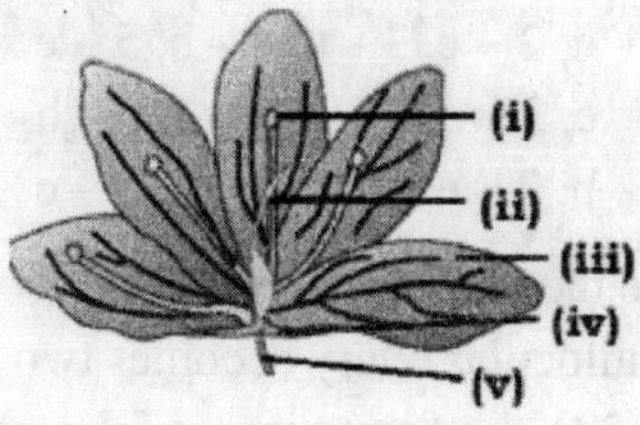

	(i)	(ii)	(iii)	(iv)	(v)
(a)	Gynoecium	Androecium	Pedicel	Corolla	Calyx
(b)	Gynoecium	Androecium	Corolla	Calyx	Pedicel
(c)	Androecium	Gynoecium	Calyx	Corolla	Pedicel
(d)	Androecium	Gynoecium	Corolla	Calyx	Pedicel

98. The mode of arrangement of sepals or petals in floral bud with respect to the other members of same whorl is termed as———
(a) Placentation (b) Aestivation
(c) Phyllotaxy (d) Inflorescence

99. Given diagram represent –
(a) Twisted aestivation
(b) Imbricate aestivation
(c) Vexillary aestivation
(d) Valvate aestivation

100. In Calotropis-
(a) Sepals or petals in a whorl just touch one another at the margin, without overlapping
(b) One margin of the appendage overlaps that of the next one
(c) Margin of sepals or petals overlap one another but not in particular direction
(d) None of these

101. "Keel" present in –
(a) Valvate (b) Imbricate
(c) Papilionaceous (d) Twisted

102. In Pea find odd one out –
(a) 'Standard' is largest petals
(b) 'Standard' overlaps the two lateral Keel.
(c) 'Keel' are smallest anterior petals.
(d) Keel are fused

103. The aestivation in gulmohur is ———
(a) Valvate (b) Twisted
(c) Imbricate (d) Vexillary

104. Find odd one with respect to aestivation
(a) China rose (b) Cassia
(c) Lady's finger (d) Cotton

Topic 12 Androecium

105. Androecium composed of –
(a) Sepals (b) Petal
(c) Stamen (d) Carpel

106. Each anther is usually _____ and each lobe has _____ chambers, pollen sacs
(a) Bilobed; two (b) Bilobed; four
(c) Tetralobed; four (d) None

107. Staminode is –
(a) Fertile stamen (b) Sterile stamen
(c) Both (a) & (b) (d) None of these

108. How many of following statements are true.
(i) Stamens united into one bundle i.e. monoadelphous
(ii) Monoadelphous is in china Rose, diadelphous is in Pea and polydephous is in Citrus
(iii) Variation in the length of filaments within a flower as in Salvia & mustard
(iv) Two bundle of stamens are diadelphous and when stamen are united into two or more bundle i.e. polyadelphous
(a) 1 (b) 2
(c) 3 (d) 4

Topic 13 Gynoecium

109. Female reproductive part of flower is –
(a) Androecium (b) Gynoecium
(c) Petal (d) Sepal

110. Pollen grains receptive surface is –
(a) Stigma (b) Style
(c) Ovary (d) Ovule

111. Placenta attach-
(a) Ovule to ovary
(b) Ovary to thalamus

(c) Ovary and other floral part
(d) None of these

112. Apocarpous is ——
i) Free carpel
ii) Fused carpel
iii) Present in rose
iv) Present in lotus
v) Present in tomato
(a) i, iii, iv (b) i, iii, v
(c) ii, iii, iv (d) ii, iv, v

113. After fertilization, the ovary develop into _____ and ovule matures into a _____.
(a) Fruit; fruit (b) Seed; fruit
(c) Fruit; seed (d) Seed; seed

114. Placentation is arrangement of _____ within the _____.
(a) Ovary; ovule
(b) Placenta; embryosac
(c) Ovule; ovary
(d) None of these

115. 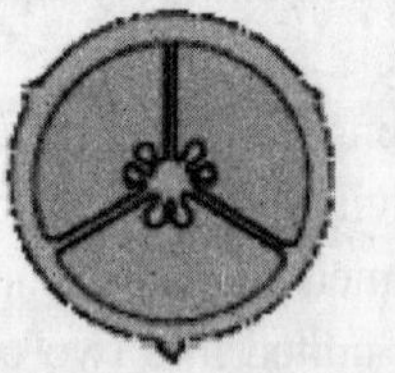

(a) Such placentation seen in Argemone
(b) The placenta is axial and the ovules are attached to it in an unilocular ovary
(c) Such placentation seen in china rose
(d) The placenta is axial and the ovules are attached to it in multilocular ovary as in Dianthus

116. Match the column I and column II

Column I		Column II	
1	Parietal	a	Pea
2	Axile	b	Lemon
3	Marginal	c	Argemone
4	Basal	d	Primrose
5	Free-central	e	Sunflower

(a) 1 – c, 2 – b, 3 – a, 4 – e, 5 – d
(b) 1 – d, 2 – c, 3 – a, 4 – b, 5 – e
(c) 1 – e, 2 – d, 3 – a, 4 – c, 5 – b
(d) 1 – b, 2 – e, 3 – a, 4 – d, 5 – c

117. Choose the correct statement –
(a) Unilocular ovary becomes two chambered due to the formation of false septum as in mustard
(b) In Argemone ovary is two chambered due to the formation of true septum
(c) Axile placentation found in multilocular ovary as in tomato
(d) (a) & (c) both

118. Dianthus have –
(a) 
(b) 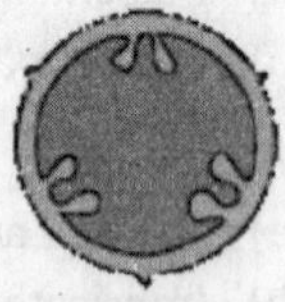
(c) 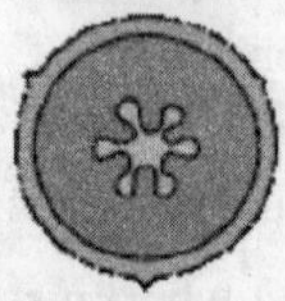
(d) 

119. In Marigold –
(a) Same placentation found in sunflower
(b) Placenta develop at base of ovary
(c) Single ovule is attached to ovary
(d) All of those

Topic 14 — The Fruit

120. Parthenocarpic fruit is –
(a) Develop after fertilization from ovary
(b) Develop without fertilization
(c) Develop after fertilization from thalamus
(d) (a) & (c) both

121. Pericarp differentiated into –
(a) Outer thin epicarp, middle fleshy edible mesocarp and an inner stony hard endocarp in Mango
(b) Outer fleshy epicarp, middle stony hard

endocarp in mango
(c) Outer thin epicarp, middle stony hard mesocarp and an inner seed in mango
(d) None of these

Topic 15 The Seed, Structure of a Dicotyledonous, Monocotyledonous Seed

122. Seed of wheat is made up of –
(a) A radicle, an embryonal axis & one cotyledon
(b) A radicle, an embryonal axis & two cotyledon
(c) Embryo only
(d) Only one cotyledon

123. Find odd one with respect to endosperm.
(a) Pea (b) Gram
(c) Castor (d) Bean

124. How many are correct statement about dicot seed?
i) Testa, an inner layer is one of two layers of seed coat
ii) Seed were attached to fruit by hilum
iii) Micropyle is small pore below hilum
iv) Castor is endospermic seed
(a) 1 (b) 2
(c) 3 (d) 4

125. How many of following is wrong stated statement?
i) Generally monocot seeds are nonendospermic seed
ii) Orchid is example of dicot seed
iii) In maize, seed coat fused with fruit wall
iv) Orchid is endospermic seed
(a) 1 (b) 2
(c) 3 (d) 4

126. Label a, b, c, d, e
(a) Scutellum Coleorhiza Plumule Radicle Coleoptile
(b) Scutellum Coleorhiza Radicle Plumule Coleoptile
(c) Scutellum Coleoptile Radicle Plumule Coleorhiza
(d) Scutellum Coleoptile Plumule Radicle Coleorhiza

127. Aleurone layer is –
(a) Carbohydrate enrich layer
(b) Proteinous layer
(c) Lipid enrich layer
(d) A and B

128. Scutellum present in
(a) Orchid (b) Castor
(c) Pea (d) Gram

Topic 16 Semi-technical Description of a Typical Flowering Plant

129. Number of androecium in mustard is –
(a) 2 (b) 4
(c) 6 (d) 5

130. How many of following is incorrect about Brassicaceae (mustard) actinomorphic, zygomorphic, bisexual, K4, superior ovary, C2+2, C(4)
(a) 1 (b) 2
(c) 3 (d) 4

Topic 17 Fabaceae

131. Fabaceae was earlier called as –
(a) Leguminosae (b) Papilionoideae
(c) Both (a) & (b) (d) Fabaceae

132. Given diagram is-

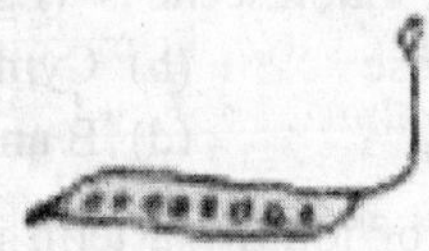

(a) L.S of carpel of pea
(b) Fruit of pea
(c) T.S. of carpel of pea
(d) Androecium of Pea

133. Calyx of fabaceae show-
(a) Polypetalous
(b) Polysepalous

(c) Valvate aestivation
(d) Both (b) & (c)

134. Androecium of Fabaceae is –
(a) Ten in number (b) 9 are united
(c) 1 is free (d) All of these

135. How many of following is endospermic seed- Arhar, groundnut, Indigofera, muliathi, Sesbania, Trifolium
(a) O (b) 1
(c) 2 (d) 3

136. The correct floral formula of sunhemp is-
(A) $\oplus$ ⚥ $k_{(5)}\ C_{1+2+2}\ A_{(9)+1}\ G_1$
(B) % ⚥ $k_{(5)}\ C_{1+2+2}\ A_{(9)+1}\ G_1$
(C) % ⚥ $k_{(5)}\ C_5\ A_{10}\ G_2$
(D) $\oplus$ ⚥ $k_{3+3}\ A_{3+3}\ G_3$

Topic 18 Solanaceae

137. Which of the following is potato family?
(a) Fabaceae (b) Solanaceae
(c) Liliaceae (d) Brassicaceae

138. Find out one with respect to Solanaceae
(a) Alternate phyllotaxy
(b) Exstipulate
(c) Reticulate venation
(d) Pulvinate

139. In Solanum, inflorescene is- (Pg. 80, E)
(a) Racemose (b) Cymose
(c) Solitary (d) B and C

140. How many of following term is not correctly stated about tobacco's family. Bicarpellary, obligately placed, apocarpous, superior ovary, bilocular, placenta swollen with many ovules, free – central placentation, drupe fruit
(a) 0 (b) 1
(c) 2 (d) 3

141. Persistant calyx found in-
(a) Brinjal (b) Pea
(c) Onion (d) Colchicine

142. $\oplus$ ⚥ $K_{(5)}\ C_{(5)}\ A_5\ G_{(2)}$

is floral formula of how many of following- Aloe, belladonna, ashwagandha, muliathi, sunhemp, Indigofera, Gloriosa (Pg. 80, E)
(a) 1 (b) 2
(c) 3 (d) 4

143. Makoi plant –
(a) Solanum nigrum
(b) Solanum tuberosum
(c) Allium
(d) Petunia

Topic 19 Liliaceae

144. Given diagram is –

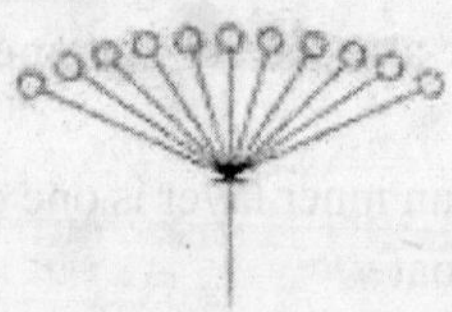

(a) Flower of Allium
(b) Inflorescence of Allium
(c) Inflorescence of dicot family
(d) Racemose

145. How many of following are endospermous seed. Aloe, Asparagus, Tulip, Potato, Tomato, Pea, Petunia, Chilli, Sesbania, Trifolium, Lupin, Muliathi, Ashwagandha, Colchicine, Gloriosa.
(a) 10 (b) 8
(c) 15 (d) 5

146. Onion show-
(a) Axile placentation
(b) Parietal placentation
(c) Free central placentation
(d) Basal placentation

147. Gynoceium of Aloe is not-
(a) Tricarpellary (b) Apocarpous
(c) Syncarpous (d) Superior ovary

148. Floral formula of Colchicum autumnale does not show-
(a) $Br \oplus$ ⚥ $k_{(5)}$
(b) $\widehat{P_{(3+3)} A_{(3+3)}}$
(c) G(3)
(d) All of these

149. Choose mismatched –

Column-I	Column-II
(a) 	sparagus (vegetables)
(b) 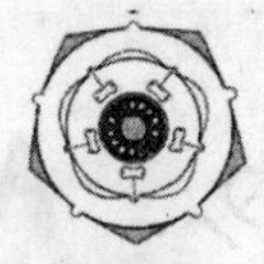	Mustard
(c) 	Pisum sativum
(d) 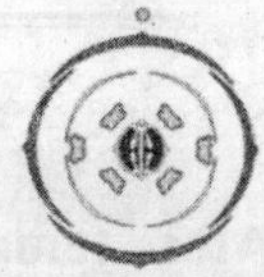	Brassicaceae

150. The floral feature of angiosperm represented in summarized form as-
(a) Floral diagram (b) Floral formula
(c) A and B (d) None of these

ANSWER KEY

1. (a)	2. (a)	3. (b)	4. (c)	5. (c)	6. (b)	7. (d)	8. (b)	9. (b)	10. (a)
11. (b)	12. (d)	13. (b)	14. (c)	15. (b)	16. (a)	17. (b)	18. (d)	19. (b)	20. (a)
21. (d)	22. (b)	23. (a)	24. (a)	25. (a)	26. (b)	27. (a)	28. (b)	29. (a)	30. (b)
31. (a)	32. (d)	33. (d)	34. (a)	35. (b)	36. (c)	37. (b)	38. (a)	39. (c)	40. (a)
41. (a)	42. (b)	43. (b)	44. (a)	45. (d)	46. (d)	47. (a)	48. (a)	49. (a)	50. (a)
51. (a)	52. (c)	53. (a)	54. (c)	55. (b)	56. (a)	57. (b)	58. (b)	59. (b)	60. (a)
61. (b)	62. (d)	63. (a)	64. (b)	65. (d)	66. (a)	67. (d)	68. (d)	69. (b)	70. (c)
71. (a)	72. (b)	73. (a)	74. (d)	75. (b)	76. (c)	77. (d)	78. (a)	79. (d)	80. (a)
81. (a)	82. (a)	83. (a)	84. (b)	85. (d)	86. (b)	87. (a)	88. (d)	89. (b)	90. (b)
91. (b)	92. (a)	93. (a)	94. (a)	95. (a)	96. (a)	97. (d)	98. (b)	99. (d)	100. (a)
101. (c)	102. (b)	103. (c)	104. (b)	105. (c)	106. (a)	107. (b)	108. (d)	109. (b)	110. (a)
111. (a)	112. (a)	113. (c)	114. (c)	115. (c)	116. (a)	117. (d)	118. (c)	119. (d)	120. (b)
121. (a)	122. (a)	123. (c)	124. (b)	125. (c)	126. (a)	127. (b)	128. (a)	129. (c)	130. (d)
131. (b)	132. (a)	133. (c)	134. (d)	135. (d)	136. (b)	137. (b)	138. (d)	139. (b)	140. (d)
141. (a)	142. (b)	143. (a)	144. (b)	145. (a)	146. (a)	147. (b)	148. (b)	149. (b)	150. (c)

5 Chapter

Anatomy of Flowering Plants

MULTIPLE CHOICE QUESTIONS

Topic 1 The Tissue: Meristematic Tissues

1. Primary body of plant is formed by-
   (a) Meristem
   (b) Vascular cambium
   (c) Both (a) & (b)
   (d) None

2. Lateral meristem are-
   (a) Type of primary meristem
   (b) Appearing early in life of plant
   (c) Responsible for producing secondary tissues
   (d) Both (a) & (b)

3. Secondary meristem include-
   (a) Fascicular vascular cambium
   (b) Cork cambium
   (c) Secondary phloem
   (d) Both (a) & (b)

4. Meristem that occur in mature region of root and shoot of plant-
   (a) Apical meristem
   (b) Intercalary meristem
   (c) Lateral meristem
   (d) None of these

5. Apical meristems
   (a) Occur at root tip
   (b) Produce primary tissues
   (c) Regenerate parts of plant
   (d) Both (a) & (b)

6. During leaf formation and stem elongation, some cells of apical meristem left behind form-
   (a) Primary cell.
   (b) Intercalary meristem
   (c) Axillary bud
   (d) Interfascicular cambium

7. Intercalary meristem
   (a) Occur in grasses
   (b) Occur between mature tissue
   (c) Both (a) & (b)
   (d) None

8. Primary meristem
   (a) Appear later in life of plant
   (b) Appear early in life of plant
   (c) Regenerates parts of plant
   (d) Both (b) & (c)

9. Identify the correct labels-

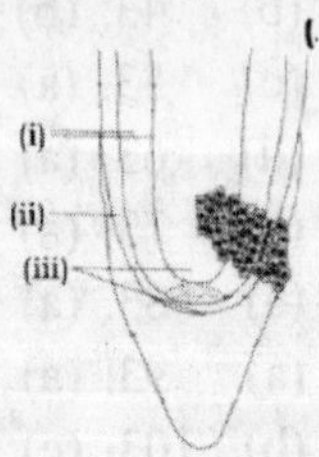

   (a) (i) – cortex, (ii) – Protoderm, (iii) – initial of central cylinder & cortex
   (b) (i)– Protoderm, (ii) – cortex, (iii) – central cylinder
   (c) (i) – central cylinder, (ii) – cortex, (iii) – Protoderm
   (d) (i) – central cylinder, (ii) – Protodrem, (iii) – cortex

10. Identify the axillary bud in given figure –

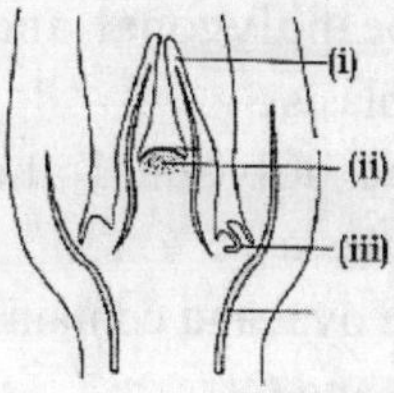

(a) (i) (b) (ii)
(c) (iii) (d) Both (ii) and (iii)

Topic 2 Permanent Tissue

11. Cell of permanent tissue
   (a) Divide regularly to repair damage
   (b) Divide occasionally
   (c) Do not divide generally
   (d) Both (b) and (c)

12. Simple tissue are –
   (a) Meristematic tissues having all cells similar in structure and function
   (b) Meristematic tissues having different types of cells
   (c) Permanent tissues having all cells similar in structure and function
   (d) Permanent tissues having many different type of cells

13. Complex tissues are –
   (a) Meristematic tissues having all cells similar in structure and function
   (b) Permanent tissues having all cells similar in structure and function
   (c) Meristematic tissues having different types of cells
   (d) Permanent tissues having different types of cells.

Topic 3 Simple Tissue

14. Choose the best option
   (a) All collenchymatous cells Assimilate food
   (b) No collenchymatous cells assimilate food
   (c) Some collenchymatous cells do not assimilate food
   (d) All collenchymatous cells do not assimilate food

15. Collenchyma provide mechanical support to –
   (a) Young stem
   (b) Petiole of leaf organs only
   (c) Organs only
   (d) All of these

16. Identify the given figure

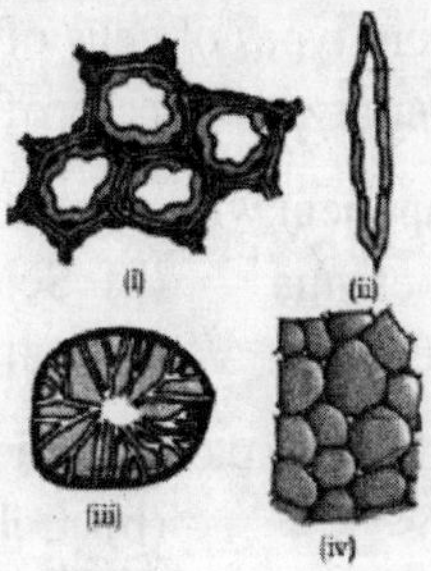

   (a) (i) – parenchyma, (ii) – fibre, (iii) – sclereid, (iv) – collenchyma
   (b) (i) – sclereids, (ii) – fibre, (iii) – parenchyma, (iv) collenchyma
   (c) (i) – collenchyma, (ii) – sclerids, (iii) – fobres, (iv) – parenchyma
   (d) (i) – collenchyma, (ii) fibre, (iii) – sclereids, (iv) – parenchyma

17. Sclenenchyma cells are –
   (a) Usually dead with protoplast
   (b) Usually dead without protoplast
   (c) Usually living with protoplast
   (d) Usually living without protoplast

18. Read the given statements –
   (i) Sclereids are found in leaves of tea.
   (ii) Fibres generally occur single in various plant parts.
   (iii) Sclerenchyma provides mechanical support to young stems.
   (iv) Parenchyma cells have thick walls.
   (v) Collenchyma cells are thickened at corners.

How many are correct
(a) 2 (b) 3
(c) 4 (d) 1

19. Pulp of pear has which type of sclerenchyma cells-
(a) Sclereids (b) Fibres
(c) Tracheids (d) Trichomes

20. Simple tissues are made of
(a) Some types of cells of similar origin
(b) Only one type cells
(c) Different types of cells of same origin
(d) Different types of cells of different origin

21. Major component within organs is formed by –
(a) Collenchyma (b) Sclerenchyma
(c) Parenchyma (d) All of these

22. Walls of parenchyma are made of–
(a) Proteose (b) Cellulose
(c) Keratin (d) Pectin

23. Parenchyma performs functions –
(a) Photosynthesis (b) Storage
(c) Secretion (d) All of the above

24. Parenchyma cells are generally –
(a) Of varying diameters, with no intercellular space
(b) Of similar diameters, with no intercellular space
(c) Of similar diameters, with small intercellular space
(d) Both B and C

25. Where does collenchyma occur?
(a) Below endodermis in most monocots
(b) Below epidermis in most monocots
(c) Below epidermis in most dicots
(d) Below endodermis in most dicots

26. Cell of collenchyma are thickened at corners due to deposition of
(a) Cellulose (b) Hemicellulose
(c) Pectin (d) All of these

27. Collenchyma cells –
(a) May be polygonal and never contain chloroplasts.
(b) May be polygonal and often contain chloroplasts
(c) May be oval and contain chloroplasts
(d) Both B and C

28. Collenchyma cells
(a) Have no intercellular spaces
(b) Have large intercellular spaces
(c) May or may not have intercellular spaces
(d) None of these

29. Mechanical support in plants is provides by
(a) Parenchyma (b) Collenchyma
(c) Sclerenchyma (d) Both B and C

Topic 4 Complex Tissues

30. In roots –
(a) Protoxylem lies towards centre and metaxylem towards periphery, called endarch
(b) Protoxylem lies towards centre and metaxylem towards periphery, called exarch
(c) Metaxylem lies towards centre and protoxylem towards periphery called endarch
(d) Metaxylem lies towards centre and protoxylem towards periphery called exarch

31. A maturãe sieve elements –
(a) Have peripheral nucleus
(b) Have peripheral cytoplasm and no nucleus
(c) Have no vacuole and no nucleus
(d) Have large vacuole and peripheral nucleus

32. Phloem fibres –
(a) Are made of parenchyma
(b) Are made of collenchyma
(c) Present in primary phloem
(d) Present in secondary phloem

33. Which of the statements about Phloem is correct?
   (a) Protoploem consists of narrow sieve tube
   (b) Metaphloem consists of narrow sieve tubes
   (c) Protopholem consists of bigger sieve tubes
   (d) Both protophloem and metaphloem have bigger sieve tubes.

34. Complex tissues are –
   (a) Made of one of cells, working as unit
   (b) Made of many types of cells, working as a unit
   (c) Made of one type of cells, working separately
   (d) Made of many types of cells, working separately

35. Xylem has following functions except–
   (a) Conducting water from roots to upper plant part
   (b) Conducting minerals from leaves to roots
   (c) Providing mechanical strength to plant parts
   (d) Conducting sap from roots to leaves

36. Xylem tissue consists of-
   (a) Sieve tube, companion cells, fibres, parenchyma
   (b) Sieve cells, vessels, fibres, parenchyma
   (c) Vessels, tracheids, sieve tube, fibres
   (d) Vessels, tracheid, fibres, parenchyma

37. Gymnosperms lack-
   (a) Xylem vessels
   (b) Companion cells
   (c) Sieve tubes and companion cells
   (d) All of the above

38. Phloem of gymnosperms possess-
   (a) Albuminous cells
   (b) Companion cells
   (c) Sieve tube
   (d) Both ((b) and (c)

39. Xylem has all dead cells except-
   (a) Xylem parenchyma
   (b) Xylem fibres
   (c) Xylem vessels
   (d) Xylem tracheids

40. Ray parenchymatous cells help in –
   (a) Radial conduction of food
   (b) Axial conduction of water
   (c) Axial conduction of food
   (d) Radial conduction of water

41. Food materials can be stored in xylem parenchyma in all of these forms except –
   (a) Starch (b) Fat
   (c) Tannin (d) None

42. In stems,
   (a) Protoxylem lies towards centre and metaxylem towards periphery, called endarch
   (b) Protoxylem lies towards centre and metaxylem towards periphery, called exarch
   (c) Metaxylem lies towards centre and protoxylem towards periphery called endarch
   (d) Metaxylem lies towards centre and protoxylem towards periphery called exarch

Topic 5 The Tissue System

43. The three types of tissue systems – epidermal ground and vascular systems are classified based on their-
   (a) Function (b) Location
   (c) Structure (d) Both (b) and (c)

Topic 6 Epidermal Tissue System

44. Outer layer of primary plant body is –
   (a) Epiblema (b) Epidermis
   (c) Epicarp (d) Ectodermis

45. Waxy layer on epidermis-

(a) is called trichome
(b) is called epiblema
(c) is absent in roots
(d) help in exchange of gases

46. Consider the following statements –
(i) Epidermal cells are parenchymatous.
(ii) Epidermis is usually two – layered.
(iii) Stomata are usually present in epidermis of stem.
(iv) Outer walls of guard cells are thick and inner walls are thin.
(v) Subsidiary cells are epidermal cells.

How many of these statements are incorrect?
(a) 2 (b) 1
(c) 3 (d) 4

47. Stomatal apparatus consists of –
(a) Stomatal aperture only
(b) Stomatal aperture and guard cells
(c) Subsidiary cells
(d) Both (b) and (c)

48. Epidermal cells modify to form
(a) Trichomes only
(b) Trichomes, Root hairs, Stomata
(c) Trichomes, Root hair, Subsidiary cells
(d) Root hairs only

49. Trichomes –
(a) Present on stem and are multicellular
(b) Present on root and are multicellular
(c) Present on stem and are unicellular
(d) Present on root and are unicellular

Topic 7 The Ground Tissue System

50. Ground tissue includes all tissues except –
(a) Cortex (b) Pith
(c) Pericycle (d) Epidermis

51. In leaves, mesophyll is present in –
(a) Epidermal tissue system
(b) Ground tissue system
(c) Vascular tissue system
(d) Both (a) & (B)

Topic 8 The Vascular Tissue System

52. Which condition is present in dicots stem?
(a) Cambium present between xylem & phloem, known as closed type vascular bundle
(b) Cambium absent between xylem & phloem, known as closed type vascular bundle.
(c) Cambium present outside xylem & phloem, known as open type vascular bundle
(d) Cambium present between xylem & phloem, known as open type vascular bundle.

53. A : Monocot have closed type of vascular bundles
R : monocots do not show secondary growth
(a) Both A and R are correct and R is correct explanation of A
(b) Both A & R are correct and R is not the explanation of A
(c) A is correct and R is incorrect
(d) Both A & R are incorrect

54. Read given statements in context of given figure

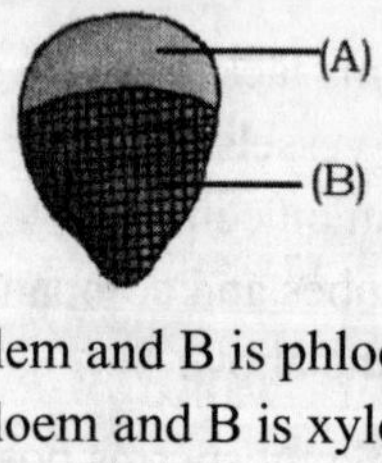

(i) A is xylem and B is phloem
(ii) A is phloem and B is xylem
(iii) Primary xylem in figure is endarch type.
(iv) Primary in figure is exarch type.

Choose the correct statements
(a) (i) and (iii) (b) (i) and (iv)
(c) (ii) and (iii) (d) (ii) and (iv)

55. The given figure can be vascular bundle of

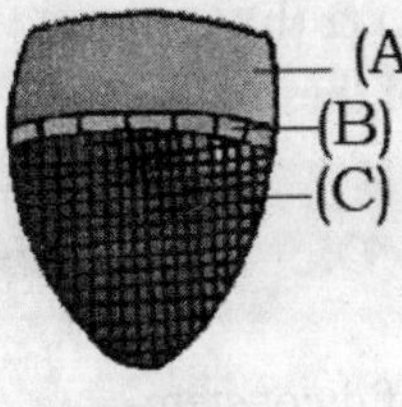

(a) Shoot of sunflower
(b) Shoot of grass
(c) Root of sunflower
(d) Root of grass

Topic 9 Anatomy of Dicot & Monocot Plants

56. For understand the tissue organization of roots, stems and leaves better, it is convenient to study–
(a) Longitudinal section of young and growing zones of organs
(b) Transverse section of young & growing zones of organs
(c) Longitudinal section of mature zones of organs
(d) Transverse section of mature zones of organs

Topic 10 Dicotyledonous Root

57. Innermost layer of cortex in dicot root is –
(a) Pericycle (b) Hypodermis
(c) Endodermis (d) Pith

58. Parenchyma cells are generally thin walled. An example of thick-walled parenchyma in dicot root is
(a) Pith (b) Pericycle
(c) Endodermis (d) Hypodermis

59. Stele includes
(a) Endodermis, pericycle, pith
(b) Endodermis, pericycle, vascular bundles
(c) Pericycle, vascular bundle, pith
(d) Endodermis, vascular bundle, pith

60. Identify the figure (i) & (ii)

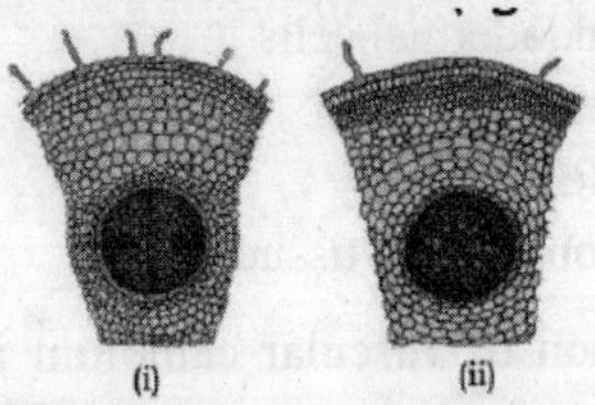

(a) (i)– T.S of dicot root (ii) – T.S of monocot root
(b) (i) – T.S of dicot stem (ii) – T.S of monocot stem
(c) (i) – T.S of monocot root (ii) – T.S of dicot root
(d) (i) – L.S of monocot stem (ii) – L.S of dicot root

61. Identify the correct labels of monocot root T.S
(a) (i)- cortex, (ii)- endodermis, (iii)- pericycle
(b) (ii)- cortex, (i)- endodermis, (iii)- pericycle
(c) (iii)- cortex, (ii)- endodermis, (i)- pericycle
(d) (i)- cortex, (iii)- endodermis, (ii)- pericycle

62. Choose correct order of cells from outside to inside in a sunflower root-
(a) Epidermis – endodermis – cortex – pericycle
(b) Epiblema – cortex – endodermis – pericycle
(c) Epiblema – cortex – pericycle – endodermis
(d) Epidermis – endodermis – pericycle – cortex

63. Suberin is deposited on –
(a) Tangential walls of epidermal cells
(b) Radial walls of cortical cells
(c) Tangential walls of endodermal cells
(d) radial walls of epidermal cells

65. The substance that casparian strips is made up of is –
(a) waxy
(b) water – impermeable
(c) suberin
(d) all of these

65. Initiation of lateral roots in dicot during

secondary growth occurs in –
(a) Endodermal cells
(b) Pericycle
(c) Medullary ray
(d) Conjunctive tissue

66. Initiation of vascular cambium in dicot root during secondary growth occurs from–
(a) Thin walled parenchymatous cells
(b) Thick walled collenchyma cells
(c) Thinn walled endodermal cells
(d) Thick walled parenchyma cells

67. Which of the following is true about
(a) Parenchymatous and lie outside phloem
(b) Parenchymatous and lie outside endodermis
(c) Collenchymatous and lie between xylem and phloem
(d) Parenchymatous and lie between xylem & phloem

68. Endodermis is present in dicot root in
(a) Two layer with little intercellular spaces
(b) Two layer without any intercellular spaces
(c) Single layer with little intercellular spaces
(d) Single layer without any intercellular spaces

69. Cortex of dicot root consists of –
(a) Multi layers of thick walled parenchyma
(b) Multi layers of thin walled parenchyma
(c) Single layer of thick walled parenchyma
(d) Single layer of thin walled parenchyma

Topic 11 Monocotyledonous Root

70. Xylem bundles in monocot root-
(a) Are fewer than dicot root
(b) Are less than six
(c) Are polyarchy
(d) All of the above

71. Secondary growth in monocot roots occur-
(a) By vascular cambium
(b) By interfascicular cambium
(c) Both (a) & (b)
(d) None of these

Topic 12 Dicotyledonous Stem

72. Epidermis of dicot stem-
(a) Is called epiblema
(b) Lacks stomata
(c) Has a thin layer of cuticle
(d) Lacks trichomes

73. Cortex in dicot stem is found between-
(a) Epidermis and endodermis
(b) Endodermis and pericycle
(c) Pericycle and pith
(d) Endodermis and pith

74. Hypodermis of dicot stem is made of-
(a) Parenchyma (b) Collenchyma
(c) Sclerenchyma (d) All of these

75. Starch sheath is found in dicot stem in- -
(a) Endodermis (b) Cortex
(c) Pericycle (d) Pith

76. Cortical cells dicot stem has-
(a) No intercellular spaces
(b) Inconspicuous intercellular spaces
(c) Conspicuous intercellular spaces
(d) Very large intercellular spaces

77. Pericycle of dicot stem is present in the form of-
(a) Semi – square patches of collenchyma
(b) Semi – lunar patches of sclerenchyma
(c) Semi – lunar patches of collenchyma
(d) Semi – lunar patches of parenchyma

78. Medullary rays are –
(a) Axially placed, parenchymatous
(b) Axially placed, collenchymatous
(c) Radially placed, parenchymatous
(d) Radially placed, collenchymatous

79. Location of medullary rays –
(a) Above endodermis
(b) Between endodermis and pericycle
(c) Between pericycle and cortex

(d) Between vascular bundles

80. In sunflower stem, vascular bundle is-
    (a) Conjoint, closed, exarch protoxylem
    (b) radial, open, endarch protoxylem
    (c) conjoint, open, exarch protoxylem
    (d) conjoint, open, endarch protoxylem

81. which of these is incorrect about pith of dicot stem?
    (a) Parenchymatous cells
    (b) No intercellular space
    (c) Central portion of stem
    (d) Large intercellular space

82. Identify the correct labels

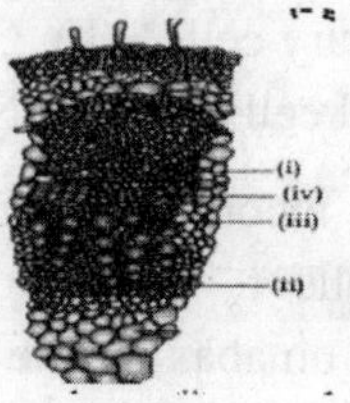

(a) (i)- protoxylem, (ii)- cambium, (iii)- phloem, (iv)- metaxylem
(b) (ii)- protoxylem, (iv)- cambium, (i)- phloem, (iii)- metaxylem
(c) (iv)- protoxylem, (i)- cambium, (ii)- phloem, (iii)- metaxylem
(d) (iii)- protoxylem, (iv)- cambium, (i)- phloem, (ii)- metaxylem

83. Identify endodermis in the given figure-

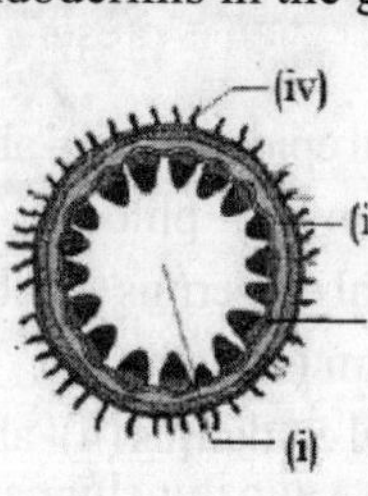

(a) (i) (b) (ii)
(c) (iv) (d) (iii)

Topic 13 Monocot Stem

84. Select the correct match of columns (a) & (b)

	Column A		Column B
i	Hypodermis of grasses stem	1	parenchyma
ii	Hypodermis of sunflower stem	2	Collenchyma
iii	Bundle sheath of grasses stem	3	Sclerenchyma
iv	Ground tissue of grasses stem		

(a) (i)- 2, (ii)- 3 (b) (iv)- 1, (iii)- 1
(c) (iii)- 3, (i)- 3 (d) (ii)- 1, (iv)- 3

85. In monocot stem,
    (a) Peripheral vascular bundles are generally smaller than central ones
    (b) Central vascular bundles are generally smaller than peripheral ones
    (c) Both peripheral and central are almost same sized
    (d) None of these

86. Phloem parenchyma is absent in-
    (a) Gymnosperms (b) Monocots
    (c) Both (d) None

87. The given figure is

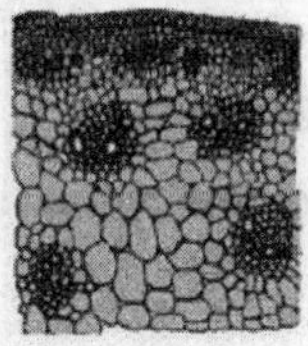

(a) Monocot root (b) Dicot root
(c) Monocot stem (d) Dicot stem

Topic 14 Dorsiventral Leaf (Dicot)

88. Read the given statements and choose the number of correct statements

(i) Leaf of dicot lack cuticle
(ii) Stomata on adaxial side of epidermis is more in number than abaxial side
(iii) Mesophyll is the ground tissue in dicot leaf
(iv) The adaxial epidermis may lack stomata
(a) 1 (b) 2
(c) 3 (d) 4

89. In the leaf of sunflower, mesophyll lies-
(a) Between epidermis and cortex
(b) Between adaxial epidermis and abaxial epidermis
(c) Between endodermis and pericycle
(d) Between pericycle and vascular bundles

90. Which of the given statements about dicot leaf is incorrect?
(a) The abaxial palisade parenchyma is made of elongated cells
(b) Spongy parenchyma is oval or round
(c) The spongy parenchyma has large spaces between cells
(d) The parenchyma on adaxial side of leaf are arranged vertically & parallel to each other

91. Consider the statements given below-
(a) Size of vascular bundle in leaf depend upon size veins
(b) Vascular bundles in leaf are surrounded by bundle sheath cells
(a) (a) is correct & (b) is incorrect
(b) (a) is incorrect & (b) is correct
(c) Both are correct
(d) Both ate incorrect

92. Identify the correct option in context of given figures

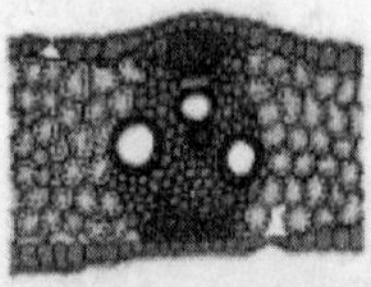
(i)

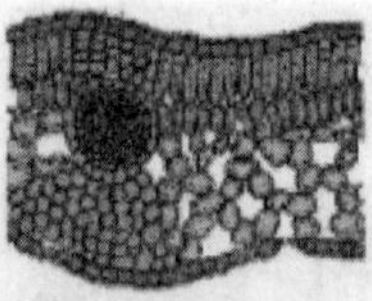
(ii)

(a) (i)- dicot stem, (ii)- monocot stem
(b) (i)- dicot leaf, (ii)- monocot leaf
(c) (i)- monocot stem, (ii)- dicot stem
(d) (i)- monocot leaf, (ii)- dicot leaf

Topic 15 Isobilateral Leaf

93. Which of the following is correct for isobilateral leaves?
(a) Present in all angiosperms
(b) Two different types of mesophyll are found
(c) Stomata on both surfaces of mesophyll
(d) Has similar sizes of vascular bundles

94. In grasses, large, empty, colourless cells are called-
(a) Subsidiary cells (b) Complementary cells
(c) Cortical cells (d) None of these

95. Identify the incorrect statement in regards to bulliform cells-
(a) Present on abaxial side
(b) Empty cells
(c) Makes leaf curl inward when flaccid
(d) Helps to minimize water loss

96. Identify correct labels for given figure.

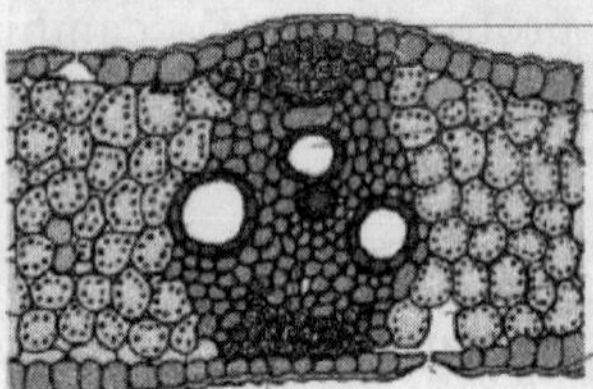

(a) (i)- adaxial epidermis (ii)- abaxial epidermis, (iii)- xylem (iv)- phloem
(b) (ii)- adaxial epidermis (i)- abaxial epidermis, (iii)- xylem (iv)- phloem
(c) (i)- adaxial epidermis (ii)- abaxial epidermis, (iv)- xylem (iii)- phloem
(d) (ii)- adaxial epidermis (i)- abaxial epidermis, (iv)- xylem (iii)- phloem

Topic 16 Secondary Growth

97. Increase in girth of plant-
   (a) Involves lateral meristem
   (b) Involves intercalary meristem
   (c) Involves apical meristem
   (d) All of these

Topic 17 Vascular Cambium

98. Vascular cambium-
   (i) Is meristematic
   (ii) Present in patches between xylem and phloem in young stem
   (iii) Present as a single layer between xylem and phloem in young stem
   (iv) Forms complete ring later

   How many of the above statements are correct-
   (a) 1 (b) 2
   (c) 3 (d) 4

Topic 18 Formation of Cambial Ring

99. In dicot stem, cambium cells present between xylem & phloem is-
   (a) Intrafasicular cambium
   (b) Interfascicular cambium
   (c) Cork cambium
   (d) Cortical cambium

100. Interfasicular cambium is formed by-
   (a) Pericycle cells (b) Endodermal cells
   (c) Medullary cells (d) Complementary cells

Topic 19 Activity of Cambial Ring

101. Cambial ring cuts off new cells-
   (a) Towards inner side only
   (b) Towards outer side only
   (c) Towards inner and outer side both
   (d) Along its own axis

102. Cambial ring cuts off new cells –
   (a) Towards pith, called secondary phloem
   (b) Towards pith, called secondary cambium
   (c) Towards pith, called secondary medullary rays
   (d) Towards pith, called secondary xylem

103. Cambial ring cut off –
   (a) More cells on outer side
   (b) More cells on inner side
   (c) Equal cells on both sides
   (d) Cells randomly

104. Assertion: secondary xylem form a compact mass.

   Reason: cambium is lesser active on outer side comparatively.

   Choose the best option-
   (a) Assertion & Reason both are correct and Reason is correct explanation for Assertion.
   (b) Assertion & Reason both are correct and Reason is not the correct explanation for Assertion
   (c) Assertion is correct but Reason is incorrect
   (d) Assertion is incorrect but Reason is correct

105. Secondary medullary rays are-
   (a) Narrow bands of parenchyma
   (b) Narrow bands of meristem
   (c) Wide bands of parenchyma
   (d) Wide bands of meristem

106. Which of these is correct about activity of cambial ring?
   (a) Secondary xylem crushes primary xylem
   (b) Secondary xylem crushes primary phloem
   (c) Secondary xylem crushes secondary phloem
   (d) Both (b) & (c)

107. Identify the secondary xylem in the figure-

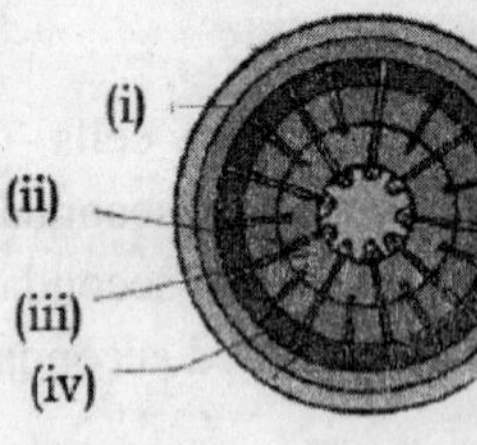

(a) (i) (b) (ii)
(c) (iii) (d) (iv)

Topic 20 Spring Wood and Autumn Wood

108. Activity of cambium is under control of-
(a) Physiological factors
(b) Environmental factors
(c) Both (a) & (b)
(d) Depend on season only

109. In spring, cambium produceBOTANY
(a) Less xylary elements, having vessels with wider cavities
(b) More xylary elements, having vessels with wider cavities
(c) Less xylary elements, having vessels with narrow cavities
(d) More xylary elements, having vessels with narrow cavities

110. Select the characters of autumn wood from the list-
(i) Light in colour (ii) Dark in colour
(iii) Low density (iv) High density
(v) Wider vessels (vi) Narrow vessels
(a) i, iii, v (b) i, iv, vi
(c) ii, iv, vi (d) ii, iii, v

111. Annual rings are constituted by-
(a) Alternate concentric rings of 3 types of woods
(b) continuous concentric rings of 3 types of woods
(c) Alternate concentric rings of 2 types of woods
(d) continuous concentric rings of 2 types of woods.

Topic 21 Heartwood and Sapwood

112. Heartwood is-
(a) Light in colour
(b) Dark in colour
(c) Alternately light & dark in colour
(d) None of these

113. Consider the following statements about heartwood-
(i) Lighter in colour
(ii) Comprises dead elements
(iii) Suberized walls
(iv) Resistant to attack of micro-organisms
(v) Conducts water and provide mechanical support to plant

How many of the statements are correct?
(a) 2 (b) 3
(c) 4 (d) 5

Topic 22 Cork Cambium

114. Cork is impervious to water due to
(a) Lignin (b) Suberin
(c) Keratin (d) Cellulose

115. Bark includes –
(a) Secondary xylem and periderm
(b) Secondary phloem and periderm
(c) Pericycle and vascular cambium
(d) Pith and stele

116. Phlloderm is –
    (a) Parenchymatous (b) Collenchymatous
    (c) Sclerenchymatous(d) Meristematic

117. Lenticles are
    (a) Circle – shaped (b) Rectangular
    (c) Lens – shaped (d) Polygonal shaped

118. Select the correct labels –

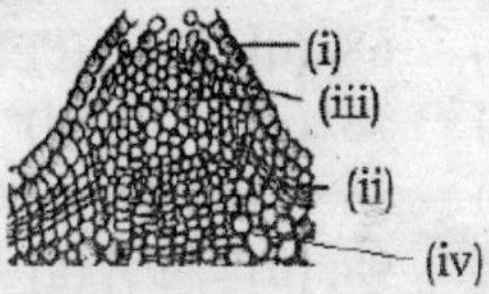

    (a) (i) – complimentary cells
    (b) (ii) – cork cambium
    (c) (iii) – secondary cortex
    (d) (iv) – epidermis

119. Assertion: Cork cambium is needed due to activity of vascular cambium.

    Reason: Phellogen is present below endodermis.

    Select the appropriate answer-
    (a) Both Assertion & Reason are correct
    (b) Assertion is correct and Reason is wrong
    (c) Assertion is wrong and Reason is correct
    (d) Both Assertion and Reason are wrong

120. Phellogen is made of-
    (a) Narrow, thick-walled, meristematic cells
    (b) Narrow, thin-walled, parenchyma
    (c) Narrow, thick-walled, parenchyma
    (d) Narrow, thin-walled, meristem

121. Phellogens cuts-
    (a) Cork on inner side and phelloderm on outside
    (b) Phellem on inner and secondary cortex on outside
    (c) Bark on outside and secondary cortex on inside
    (d) Phellem on outside and phelloderm on inside

Topic 23 Secondary Growth in Roots

122. In sunflower root. Vascular cambium is originated from tissues –
    (a) Below phloem bundle
    (b) Of pericycle
    (c) Of interfascicular cambium
    (d) Both (a) & (b)

123. Secondary growth does not occur in –
    (a) Gymnosperm stem
    (b) Gymnosperm root
    (c) Monocot
    (d) All of these

124. Identify the cambial ring –

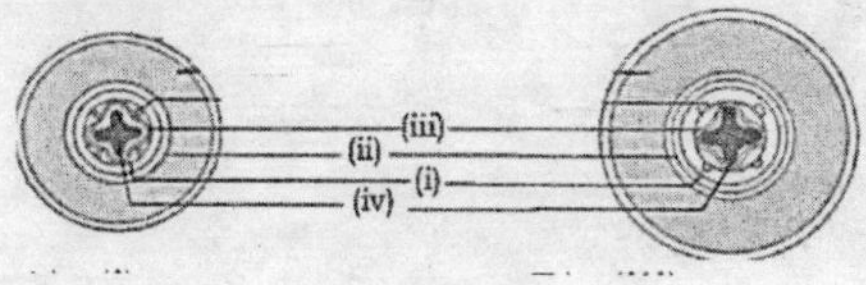

    (a) (i) (b) (iii)
    (c) (iv) (d) (ii)

ANSWER KEY

1. (a)	2. (c)	3. (d)	4. (c)	5. (d)	6. (c)	7. (c)	8. (d)	9.(a)	10. (c)
11. (d)	12. (c)	13. (d)	14. (c)	15. (a)	16. (d)	17. (c)	18. (a)	19.(a)	20. (c)
21. (c)	22. (c)	23. (d)	24. (d)	25. (c)	26. (d)	27. (d)	28. (a)	29.(d)	30. (d)
31. (c)	32. (d)	33. (a)	34. (c)	35. (c)	36. (d)	37. (a)	38. (a)	39.(a)	40. (d)
41. (d)	42. (a)	43. (d)	44. (c)	45. (c)	46. (c)	47. (d)	48. (c)	49.(a)	50. (d)
51. (c)	52. (d)	53. (c)	54. (c)	55. (a)	56. (d)	57. (c)	58. (c)	59. (c)	60. (c)
61. (c)	62. (c)	63. (c)	64. (d)	65. (c)	66. (d)	67. (d)	68. (d)	69. (c)	70. (c)
71. (d)	72. (c)	73. (a)	74. (c)	75. (a)	76. (c)	77. (c)	78. (c)	79. (d)	80. (d)
81. (c)	82. (c)	83. (d)	84. (c)	85. (a)	86. (c)	87. (c)	88. (c)	89. (c)	90. (a)
91. (c)	92. (d)	93. (d)	94. (d)	95. (a)	96. (c)	97. (a)	98. (d)	99. (a)	100. (c)
101. (c)	102. (d)	103. (c)	104. (a)	105. (a)	106. (d)	107. (c)	108. (c)	109. (c)	110. (c)
111. (c)	112. (c)	113. (a)	114. (c)	115. (c)	116. (a)	117. (c)	118. (c)	119. (c)	120. (d)
121. (d)	122. (d)	123. (c)	124. (c)						

6 Chapter Cell: The Unit of Life

MULTIPLE CHOICE QUESTIONS

Topic 1 Cell Structure and Function

1. Which one of the following is not correct statement about about physico-chemical approach:
   (a) Established by analysis of living tissue for element and compounds.
   (b) Explains what type of organic compounds is present in living organism.
   (c) Explains the abnormal process that occur during any diseased condition.
   (d) This approach is known as forward biology.

2. The detailed description in living organisms that brings out their knowledge of diversity is about
   (a) Their form
   (b) Their appearance
   (c) Both
   (d) None

3. Which one of the following brings out the unit of diversity in the cellular organisation of all life form:
   (a) Theory of evolution
   (b) Species theory
   (c) Cell theory
   (d) Darwinian theory

Topic 2 Cell: The Unit of Life, Cell Theory

4. The hypothesis that the bodies of animals and plant are composed of cells and their products was proposed by:-
   (a) Schleiden and Schwann
   (b) Rudolf Virchow
   (c) Schwann only
   (d) Virchow and Schleiden

5. Scientist who gave the final shape to cell theory?
   (a) Schleiden
   (b) Schwann
   (c) Virchow
   (d) Schleiden & Schwann

6. Which of the following is related to cell theory:-
   (i) All living organisms are composed of cells and product of cells.
   (ii) Proposed by Schleiden and Schwann.
   (iii) Modified by Rudolf Virchow
   (iv) All cells arise from pre – existing cell.
   (v) "Omnis cellula – e – cellula"
   (a) Only one of the above
   (b) Only two of the above
   (c) Only four of the above
   (d) All five

7. Unicellular organism are capable of
   (a) Independent existence
   (b) Performing the essential functions of life.
   (c) Both
   (d) Does not ensure independent living

8. Living cell was firstly seen and described by:-
   (a) Robert Hooke
   (b) Anton von Leeuwenhoek
   (c) Robert Koch
   (d) Robert Brown

9. Cell theory was proposed by:-
   (a) Matthias Schleiden and Theodore Schwann
   (b) Schleiden; Schwann and Virchow.
   (c) Rudolf Virchow
   (d) Sutton and Boveri

10. All the plants are composed of different kinds of cells which forms the tissue of the plant, this statement was given by:-
   (a) A German botanist ; Rudolf Virchow .
   (b) A British zoologist ; Matthias Schleiden
   (c) A British zoologist ; Theodore Schwann
   (d) A German botanist; Matthias Schleiden

11. Who studied the different types of animal cells to propose cell theory:-
   (a) A British zoologist; Matthias Schleiden
   (b) A German botanist; Theodore Schwann.
   (c) A physicist; Rudolf Virchow.
   (d) A British zoologist; Theodore Schwann.

12. A thin outer layer studied by Theodore Schwann nowadays known as:-
   (a) Plasma membrane
   (b) Cell wall
   (c) Glycocalyx
   (d) Middle lamella

13. Based on studies of Matthias Schleiden; what is the unique character of plant cell?
   (a) Cell wall (b) Middle lamella
   (c) Glycocalyx (d) None of these

Topic 3 An Overview of Cell

14. Which of the following is not incorrect?
   (a) Mycoplasma is the smallest cell > 0.3 μm in width.
   (b) Bacteria could be 3 μm to 5 μm in length
   (c) Human RBCs are about 7.0 mm in diameter.
   (d) Cell's shape is independent of their work they perform.

15. What is the delimiting boundary around a human cheek cell?
   (a) Cell membrane (b) Protoplasm
   (c) Protoplast (d) Cell wall

16. What is the Semi – fluid matrix inside the cell?
   (a) Cell membrane (b) Protoplast
   (c) Cytoplasm (d) Nucleus

17. How many of the following statements are not true:-
   (i) All cells have membrane bound nuclei and nucleolus.
   (ii) Nucleus contains the chromosome
   (iii) DNA is the Genetic material.
   (iv) Cytoplasm is the main arena of cellular activities in plant and animal cells.
   (a) Only (ii), (iii), & (iv)
   (b) Only (ii) & (iv)
   (c) Only (i) & (iii)
   (d) Only (i)

18. Besides the nucleus; the ____ cell have other membrane bound distinct structures.
   (a) Eukaryotic (b) Prokaryotic
   (c) Both (a) and (b) (d) None of these

19. What is the non – membranous organelle present in both Eukaryotic as well as Prokaryotic cell
   (a) Endoplasmic reticulum
   (b) Protein
   (c) Mitochondria
   (d) Ribosomes of 80s' type

20.. Animal cells have another non – membrane bound cellular organelle known as:-
   (a) Microbodies (b) Nucleus
   (c) Lysosome (d) Centrosome

Topic 4 Prokaryotic Cell

21. A special form of cell membrane ; which is the characteristic of prokaryotes is:-
   (a) Plasmid (b) Cell wall
   (c) Cell membrane (d) Mesosomes.

22. Which of the following is membrane less bodies other than Ribosomes.

(a) Cell wall (b) Inclusion
(c) Mesosomes (d) Chromatophores

23. Which of the following is the essential infolding's of cell membrane
(a) Inclusion (b) Mesosome
(c) Chromatophores (d) Plasmid

24. The prokaryotic cells are represented by:-
(a) Bacteria
(b) BGA
(c) Mycoplasma & PPLO
(d) All of these

25. All prokaryotic cell have this cellular boundary surrounding the cell – membrane except in mycoplasma
(a) Glycocalyx (b) Protoplast
(c) Cell wall (d) Cytoplasm

26. Which of the following is related to prokaryotic cell:-
(a) Have no well defined nucleus
(b) Have basically naked genomic material.
(c) An addition to genomic DNA; the extra – genomic DNA is also present known as plasmid.
(d) All of the above

27. Which of the following confirms certain unique phenotypic characters to some bacteria
(a) Chromosomal material
(b) Extra chromosomal material
(c) Mitochondrial DNA
(d) Genetic material present in chloroplast

Topic 5 Cell Envelope and it's modification

28. Longest portion of flagellum is:-
(a) Basal body (b) Hook
(c) Filament (d) None of the above

29. Which of the following structure helps in motility in bacterial cell:-
(a) Cell membrane (b) Pili
(c) Fimbriae (d) Flagella

30. Which of the following is not a surface structure:-
(a) Fimbriae (b) Pili
(c) Flagella (d) Inclusion

31. Which of the following is small bristle like fibres sprouting out of the cell:-
(a) Pili (b) Cilia
(c) Flagella (d) Fimbriae

32. Which of the following is elongated tubular proteinaecious structure:-
(a) Pili (b) Inclusion
(c) Mesosome (d) Fimbriae

33. Which of the following help the bacteria attach to rocks in streams:-
(a) Inclusion (b) Mesosome
(c) Fimbriae (d) Pili

34. What is the sequence of cell envelope in most of the prokaryotic cell (Outer to Inner)
(a) Glycocalyx > cell membrane > cell wall.
(b) Cell membrane > cell wall > Glycocalyx
(c) Cell wall > Glycocalyx > cell membrane
(d) Glycocalyx > cell wall > cell membrane.

35. The prokaryotic cell have a single protective unit made up of
(a) Glycocalyx + cellulosic cell wall + cell membrane
(b) Peptidoglycan cell wall + cell membrane + Glycocalyx
(c) Chitinous cell wall + cell membrane + Glycocalyx
(d) Silicous cell wall + Glycocalyx + cell membrane

36. How many of the following statements are correct:-
(i) Glycocalyx is outermost layer.
(ii) All three layer have same function.
(iii) Bacteria can be classified on the basis of differences in the cell envelope.

(iv) Bacteria can be classified on the basis of response to the staining procedure
(a) Only one (b) Only two
(c) Only three (d) All four

37. The bacteria that take up gram stain are
(a) Gram positive type.
(b) Gram negative type.
(c) Both type
(d) Neither gram positive nor gram negative.

38. The bacteria that do not take up gram stain are
(a) Gram positive type.
(b) Gram negative type.
(c) Either gram positive or gram negative
(d) Neither gram positive nor gram negative

39. Which of the following in a bacterial envelope is a loose sheath of slimy layer
(a) Glycocalyx (b) Cell wall
(c) Cell membrane (d) None of the above

40. Glycocalyx could be a thick and tough layer and known as:-
(a) Slimy layer
(b) Cyst
(c) Capsule
(d) None of the above

41. Which of the following determines the shape of a bacteria cell:-
(a) Glycocalyx (b) Capsule
(c) Cell membrane (d) Cell Wall

42. How many of the following in not incorrect regarding a cell membrane in prokaryotes
(i) Selectively permeable in nature
(ii) Structurally similar to eukaryotic cell membrane
(iii) Interacts with outer world.
(iv) Innermost layer of cell envelope
(v) Living layer.
(a) Only (ii), (iii) & (iv)
(b) Only (i), (ii), (iii), (iv) & (v)
(c) Only (i), (iii), (iv) & (v)
(d) Only (i), (iv) & (v)

43. How many of the following is the membranous extensions into the cell of bacteria:-

Mesosomes, Tubules, Vesicles, Lamellae, Chromatophores,Inclusions
(a) 6 (b) 3
(c) 5 (d) 4

44. How many functions from the following, the mesosomes can perform
(i) DNA replication
(ii) Respiration
(iii) DNA distribution to daughter cells
(iv) Secretion
(v) Increases surface area
(vi) Contains enzymatic content.
(a) Only four (b) Only Three
(c) All six (d) Only five

45. In cyanobacteria, there are some another membranous extensions except mesosomes are:
(a) Inclusion (b) Fat globules
(c) Chromatophores (d) All of the above

46. What are structures related to Bacterial flagellum:-
(a) Basal body & filament
(b) Basal body, Hook & filament.
(c) Hook & filament
(d) Filament only.

Topic 6 Ribosomes and Inclusion Bodies

47. Ribosomes are associated with the structures in a bacterial cell:-
(a) t – RNA strand (b) Golgi body
(c) Cell membrane (d) E.R

48. Ribosomes in the bacterial cell are
(a) 20nm to 30 nm in size.
(b) Made up of two subunits(Larger 60s' &smaller 40s')
(c) Made up of two subunits(Larger 50s' & smaller 30s')

(d) Associated with E.R and cell membrane

49. A polysome is:-
(a) Several mRNA bound to a single Ribosome.
(b) Several subunits of ribosomes attached to each other.
(c) Several ribosomes attached to a single strand of mRNA
(d) Several mRNA attached to each other.

50. Which of the following structure translate the mRNA into proteins: in a bacterial cell:-
(a) Inclusions of cytoplasm
(b) Ribosomes of E.R
(c) Ribosomes of Polysome.
(d) Polysomes of Ribosome.

51. Inclusion bodies in a prokaryotic cell are:-
(a) Reserve material containing structure
(b) Cell membrane infoldings
(c) Membrane bound structure
(d) All of the above

52. What are example of inclusion bodies:-
(i) Mesosome
(ii) Chromatophores
(iii) Gas vacuole
(iv) Phosphate granules
(v) Cyanophycean granules
(vi) Glycogen granules
(a) Only (ii), (iii) & (iv)
(b) Only (iii), (iv) & (v)
(c) Only (iii), (iv), (v) & (vi)
(d) Only (iv), (v) & (vi)

53. Inclusion bodies can be found in
(a) All type of cells
(b) All eukaryotic cell
(c) BGA & green photosynthetic bacteria
(d) Prokaryotic cell.

Topic 7 Eukaryotic Cell

54. All of the above except are eukaryotic except:-
(a) Protista (b) Plants
(c) Monera (d) Animals

55. How many of the following statements are true regarding Eukaryotic cell.
(i) Cytoplasm has extensive compartmentalization
(ii) Presence of membrane bound organelle
(iii) Organised nucleus
(iv) A variety of complex locomotory and cytoskeletal structures.
(v) Genetic material is organised into chromosomes
(a) 2 (b) 3
(c) 4 (d) 5

56. Statement – I:- Plant cells differs from animals cells.

Statement – ii:- The former one posses cell walls, plastids & a large vacuole which is absent in latter one.
(a) Both statements are correct.
(b) Both statement are incorrect.
(c) Statement – I is correct but statement– II is incorrect.
(d) Statement – I is incorrect but statement – II is correct.

57. Centrioles are present in
(a) Animal cells
(b) Plant cells
(c) Both animal and plant cells.
(d) All other than plant cells.

58. Which of the following correctly explain the diagram.

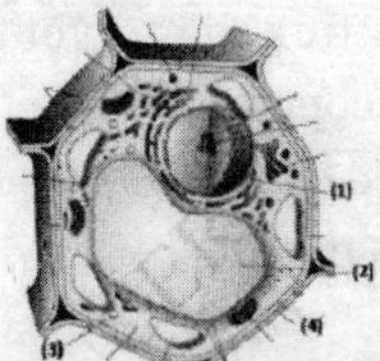

(a) 1 > Nuclear membrane 2 > vacuole 3 > peroxisome 4 > cell wall
(b) 1 > Microvilli 2 > cell wall 3 > cytoplasm 4 > Plasma membrane

(c) 1 > Vacuole 2 > cytoplasm 3 > Microvilli 4 > cell wall
(d) None of the above

Topic 8 Cell Membrane

59. What ability explains the fluidity of cell membrane:-
(a) Quasi – fluid nature of cell membrane.
(b) Lateral movement of proteins.
(c) Cell growth, formation of intercellular junctions; secretion; endocytosis; cell division
(d) All of the above.

60. The plasma membrane is:-
(a) Semi – permeable in nature
(b) Impervious in nature
(c) Impermeable in nature
(d) Selectively permeable in nature.

61. How many of the following functions the cell membrane can perform:- Active transport; Osmosis; Passive transport.
(a) Only one (b) Only two
(c) All (d) None

62. Na+ - K+ pump transports molecules
(a) By passive transport
(b) By active transport
(c) By utilisation of ATP
(d) Both B & C

63. The detailed structure of the membrane was studied:-
(a) Only after the advent of the electron microscope in 1950s.
(b) Enabled to deduce the possible structure of plasma membrane
(c) Both
(d) None

64. Which of the following cell's study enabled the scientists to deduce the possible structure of Plasma membrane?
(a) RBC (b) Cork cell
(c) WBCs (d) Bacterial Cell

65. Cell membrane is mainly composed of
(a) Lipids and Proteins
(b) Proteins & Cholesterols
(c) Lipids & Carbohydrates
(d) Carbohydrates & Proteins

66. What is the correct arrangement of Lipid molecules in the cell membrane
(a) Polar head > Outside, non – polar tails > Inner side.
(b) Non – polar head > Outside Polar tail > Inner side
(c) Polar tail > Outside non – polar head > Inner side
(d) Polar tail > inner side non – polar head > outer side

67. What ensures that the non – polar tail is protected from aqueous environment?
(a) Polar head > Outside non – polar tails > Inner side.
(b) Non – polar head -> Outside Polar tail > Inner side
(c) Polar tail -> Outside non – polar head > Inner side
(d) Polar tail > inner side non – polar head > outer side

68. The constituents of cell membrane are:-
(i) Phospholipid (ii) Carbohydrate
(iii) Proteins (iv) Cholesterol
(v) Phosphoproteins
(a) Only (i), (ii) & (iv)
(b) Only (ii), (iv) & (v)
(c) Only (ii), (iii), (iv) & (v)
(d) Only (i), (ii), (iii) & (iv)

69. Which of the following study revealed that cell membrane also contains proteins & carbohydrate:-
(a) Electron microscopic study.

(b) Phase – Contrast microscopic study.
(c) Biochemical investigation study
(d) Cobalt – chloride paper test study.

70. Which of the following statement is incorrect:-
(a) The tail is hydrophobic of saturated hydrocarbons.
(b) The tail is hydrophilic of saturated hydrocarbons.
(c) The tail is hydrophobic of unsaturated hydrocarbons.
(d) The tail is hydrophilic of unsaturated hydrocarbons

71. Which of the following statement is incorrect:-
(a) The ratio of proteins and lipids varies considerably in different cells.
(b) In erythrocytes; it has approximately 52% proteins and 40% lipids.
(c) On the basis of ease of extraction membrane proteins are of extrinsic and intrinsic type.
(d) None of the above

72. The improved model of the structure of cell membrane was proposed by:-
(a) Messelson & Stahl
(b) Schleiden & Schwann
(c) Anton von Leeuwenhoek
(d) Singer and Nichloson

73. The quasi–fluid nature of lipid enables:-
(a) Flip–flop movement of proteins within the lipid bilayer.
(b) Lateral movement of proteins within the lipid bilayer.
(c) Flip–flop movement of lipid crossing the protein bilayer.
(d) lateral movement of lipid crossing the protein bilayer.

74. One of the most important function of the plasma membrane is:-
(a) Transport of molecules across it.
(b) Flip – flop movement.
(c) Secretion
(d) Cell enlargement.

Topic 9 Cell Wall

75. The outer covering of fungi and plants is:-
(a) Glycocalyx (b) Cell wall
(c) Cell membrane (d) All

76. What is the function of cell wall:-
(a) Gives shape to the cell
(b) Protects the cell
(c) Cell – to – cell interaction
(d) All of the above

77. What are chemical composition of algal cell wall Cellulose, Galactans, Mannans, Calcium carbonate, Chitin
(a) Only two of them
(b) Only three of them
(c) Only four of them
(d) All five of them

78. Cell wall of plants consists of:-
(a) Cellulose & Pectin's only
(b) Cellulose, hemicellulose & Pectin's only
(c) Cellulose, hemicellulose, Pectin & Proteins.
(d) Hemicellulose & Proteins only.

79. Which of the following is capable of growth
(a) Primary cell wall
(b) Secondary cell wall
(c) Tertiary cell wall
(d) All of them

80. Secondary cell wall is formed
(a) Outside the primary cell wall.
(b) Inside the cell membrane
(c) Inside the plasmodesmata.
(d) Inside the primary cell wall.

81. Which of the following in plant acts as glue between neighbouring plant cells:-
(a) Ca – Pectate
(b) Mg – Pectate
(c) Ca & Mg – Pectate
(d) None of the above

82. Which of the following is traversed by plasmodesmata:-
(a) Cell wall & cell membrane
(b) Cell membrane & Glycocalyx
(c) Cell membrane, cell wall, Glycocalyx & Middle lamella.
(d) Cell wall & middle lamella.

Topic 10 Endomembrane System

83. What are the constituent of Endomembrane system:-
(a) Endoplasmic reticulum
(b) Golgi body & E.R.
(c) E.R; Golgi body; Lysosome & Vacuole.
(d) E.R, Golgi body & Lysosome.

84. Why Mitochondria, Chloroplast & Peroxisome are not the part of Endo – system:-
(a) They are autonomous organelles.
(b) They are semi – autonomous organelles.
(c) They are not coordinated with Endomembrane system.
(d) They have their own genetic material.

85. Which of the following is the network of tiny-tubular structure scattered in cytoplasm:-
(a) E.R (b) Golgi body
(c) Lysosome (d) Vacuole

86. Which of the following structure divides the intercellular space into two compartments:-
(a) E.R
(b) Golgi body
(c) Lysosome
(d) None of the above

87. The extra luminal & luminal compartment represents:-
(a) Cytoplasm & inside ER
(b) Inside ER & cytoplasm
(c) Outside ER & cytoplasm
(d) Cytoplasm & outside ER

88. The ER having Ribosomes attached to its outer surface is known as
(a) RER (b) SER
(c) Both (d) None

89. RER is frequently observed in cells, actively involved in:-
(a) Protein Synthesis
(b) Lipid synthesis
(c) DNA synthesis
(d) Glucose synthesis

90. Which of the following is continuous with the outer membrane of nucleus:-
(a) R.E.R (b) S.E.R
(c) Golgi body (d) Lysosome

91. Steroidal hormones are synthesised by:-
(a) R.E.R (b) Lysosome
(c) S.E.R (d) Ribosome

92. Golgi body was firstly observed by
(a) Camillo Golgi in 1898
(b) Camillo Golgi in 1897
(c) Camillo Golgi in 1895
(d) Camillo Golgi in 1993.

93. Golgi body is
(i) Reticular structure.
(ii) Densely stained structure
(iii) Made up of cisternae, Tubule & Vesicle
(iv) Concentric cisternae
(a) Only (i) & (iii)
(b) Only (ii), (iii) & (iv)
(c) All of the above
(d) Only (iii) & (iv)

94. What is the diameter of cisternae of Golgi body:-
(a) 0.5μm to 1. 0μm
(b) 0. 1 μm to 2. 0 μm
(c) 0. 2 μm to 2. 5 μm
(d) 0. 3 μm to 2. 0 μm

95. The convex – face of cisternae of Golgi body is also known as:-
(i) Cis – face (ii) Forming face
(iii) Trans – face (iv) Maturing face
(a) (i) & (ii) (b) (ii) & (iii)

(c) (iv) & (iii) (d) (i) & (iv)

96. Which of the following statement is correct:-
(a) Cis & Trans faces are same but inter connected.
(b) Cis & Trans faces different & not inter connected
(c) Cisternae is 0.1 to 2.0 μm in diameter.
(d) None of the above

97. Golgi body principally performs the functions of:-
(a) Secretion
(b) Packaging of materials.
(c) Both
(d) None

98. Materials to be packed in the ______ Fuses with the _____ face:-
(a) Cis – face and Trans – face
(b) Trans – face and cis – face
(c) E.R and cis – face
(d) E.R and trans – face

99. A number of proteins synthesized by ribosomes on the _____(i)_____ are modified in the ____ (ii)____ of the ____(iii)_____
(a) (i) ER (ii) Golgi body (iii) cisternae
(b) (i)Golgi body (ii) cisternae (iii)ER
(c) (i) cisternae (ii) RE (iii) Golgi body
(d) (i) ER (ii) cisternae (iii) Golgi body

100. The vesicular structure formed by the process of packing in Golgi apparatus is:-
(a) `Vacuole (b) ER
(c) Lysosome (d) All

101. The isolated lysosomal vesicle have been found to be very rich in
(i) Lipases
(ii) Proteases
(iii) Carbohydrases
(a) Only i) & ii) (b) Only ii) & iii)
(c) Only i) & iii) (d) All

102. Enzymes present in lysosomes are accumulatively known as:-
(a) Acid proteases (b) Lipases
(c) Acid hydrolases (d) Carbohydrases

103. The membrane bound space in cytoplasm is known as:-
(a) ER (b) Golgi body
(c) Lysosome (d) Vacuole

104. Vacuole contains hydrolases; lipases; proteases; water; sap; excretory products & material not useful for the cell
(a) Only four of the above
(b) Only three of the above
(c) Only five of the above
(d) All of them.

105. The membrane of vacuole is
(a) Single membrane
(b) Tonoplast
(c) Both
(d) none

106. In a plant cell vacuole can occupy up to _____% space of cell
(a) 70 (b) 80
(c) 90 (d) 50

107. In plant tonoplast facilitates the transport of a number of ___(i)___; ____(ii)_____ the concentration gradient.
(a) (i) Solutes (ii) Along
(b) (i) ions (ii) Along
(c) (i) ions (ii) against
(d) (i) solutes (ii) against

108. How many of the following statements are not wrong:-
(i) Concentration of same ions inside the vacuole is significantly higher.
(ii) In amoeba contractile vacuole helps in osmoregulation & excretion.
(iii) In Protistans, food vacuoles are formed by engulfing the food particle.
(a) Only two (b) Only one
(c) All three (d) None

Topic 11 Mitochondria

109. Which of the following statement is correct about mitochondria:-
(a) Easily visible under the microscope; without stain.
(b) Number of mitochondria per cell is invariable
(c) Number of mitochondria depends on the physiological activity of cell.
(d) All of the above.

110. How many of the following statement is correct regarding mitochondria :-
i) A sausage – shaped str.
ii) Diameter is 0.2 – 1.0 μm
iii) Avg. Diameter is 0.5 μm
iv) Length is 1.0 – 4.1 μm
(a) One (b) Two
(c) Three (d) Four

111. Each mitochondria is ___X____ membrane bound structure; dividing its lumen into ____Y___ distinct compartment
(a) X → single Y → one
(b) X → double Y → one
(c) X → single Y → two
(d) X → double Y → two

112. Matrix of mitochondria is:-
(a) Filled with a dense homogenous substance.
(b) Outer aqueous compartment
(c) Space present between Inner and Outer membrane of Mitochondria
(d) Present within the outer membrane of mitochondria

113. The outer membrane of mitochondria forms the ____ limiting boundary of the organelle, while the inner membrane forms a number of_____
(a) Discontinuous ; infoldings
(b) Infoldings; Cristae
(c) Continuous ; Cristae
(d) Cistae ; Infoldings

114. Which of the following increase the surface area:-
(a) Matrix (b) Inner membrane
(c) Outer membrane (d) Cristae

115. How many of the following statements are correct:-
i) Only outer membrane has enzyme for ETS
ii) Only inner membrane has enzymes.
iii) Outer membrane is devoid of enzymes.
iv) Mitochondria matrix has enzyme of kerb's cycle.
v) Mitochondria is the site of aerobic respiration
vi) Matrix also possess SS – DNA molecule & few RNA molecules.
(a) Only two (b) Only four
(c) Only five (d) Only three

116. The matrix of mitochondria possess:- Single circular DNA molecules; A few RNA molecules; 70s' ribosomes; Components required for the synthesis of proteins.
(a) Only two of them
(b) Only three of them
(c) All of them
(d) None of them

117. Mitochondria divides by:-
(a) Endomitosis (b) Meiosis
(c) Budding (d) Fission

Topic 12 Plastids

118. Plastids are found in:-
(a) Only plants cells
(b) Only Euglenoids
(c) Both Plants and Euglenoids
(d) Plants; Euglenoids & Cyanobacteria.

119. Classification of plastids are based on-
(a) Chromatophores (b) Mesosomes
(c) Inclusions (d) Pigments

120. Which of the following is responsible for trapping of light energy
(a) Chlorophyll like a, b, c etc.
(b) Carotenoids
(c) Chlorophylls & carotenoids
(d) Chromosomes

121. Carotenoids is group of
(a) Chlorophyll pigments
(b) Chlorophylls & carotene
(c) Carotenes and xanthophyll's
(d) Carotenes ; xanthophyll's & other pigments.

122. Leucoplast is :-
(a) Unmodified plastids
(b) Contains stored nutrients
(c) Imparts colour to the plant cell
(d) Imparts colour to the cyanobacteria

123. What are types of chloroplast:-
i) Chromoplast ii) Leucoplast
iii) Amyloplast iv) Aleuroplast
v) Elaioplast
(a) Three of the above
(b) Four of the above
(c) Five of the above
(d) None of the above

124. Elaioplast contains
(a) Proteins and fats
(b) Fats and starch
(c) Fats and oils
(d) Fats ; Protein and oils.

125. Aleuroplast contains
(a) Proteins and fats (b) Fats and oils
(c) Proteins & starch (d) Protein only

126. Majority of chloroplast of the green plants are found in :-
(a) Mesophyll cells of roots
(b) Mesophyll cells of stems
(c) Mesophyll cells of leaves
(d) Mesophyll cells of flowers.

127. Mesophyll cells are:-
(a) Lens – shaped; Oval; Spherical only
(b) Oval &spherical only
(c) Discoidal & ribbon – shaped
(d) None of them

128. What is dimension of chloroplast :-
(a) Length 2 – 4 μm & width 5 – 10 μm
(b) Length 1 – 2 μm & width 2 – 4 μm
(c) Length 5 – 10 μm & width 2 – 4 μm
(d) Length 2 – 4 μm & width 1 – 2 μm

129. Number of chloroplast per cell may vary from_______ per cell of chlamydomonas to ________ per cell in mesophylls.
(a) 20 – 40; 1 – 5
(b) 1 ; 20 – 40
(c) 10 – 20; 20 – 40
(d) 5; 10 – 20

130. Common features of mitochondria & chloroplasts are :-
(a) Number of membrane & type of DNA molecules only
(b) Number of membrane; Ribosomes type and DNA molecule type
(c) Types of thylakoid & genetic material.
(d) Types of thylakoid, genetic material and permeability of membrane.

131. What are types of thylakoid inside the chloroplast:-
(a) Intergranal thylakoid and stroma lamellae
(b) Granum thylakoid only
(c) Stroma thylakoid only
(d) None of the above

132. Flat membranous tubules connecting the thylakoids in chloroplast is known as:-
(a) Granal thylakoid
(b) Grama
(c) Stroma thylakoid / lamellae
(d) All of the above

133. The membrane of chloroplast encloses a space known as:-
(a) Matrix (b) Cytoplasm
(c) Lumen (d) All of them

134. The stroma of chloroplast contains:-
   (i) Enzyme for carbohydrate & proteins synthesis.
   (ii) Small single stranded DNA molecule.
   (iii) Ribosomes of 70's type.
   (a) Only one the above
   (b) Only two of the above
   (c) Only three of the above
   (d) None of the above

135. Chlorophyll pigments are present in the:-
   (a) Matrix (b) Stroma
   (c) Membrane (d) Thylakoid

136. The ribosomes of chloroplast are:-
   (a) Same as Eukaryotic cell
   (b) 70's type with single subunit
   (c) 70's type with two subunits
   (d) All of the above

Topic 13 Ribosomes

137. Which of the following statements are true regarding ribosomes :-
   (i) Granular structure
   (ii) First observed as dense particles by George Palade in 1953
   (iii) Composed of m – RNA & proteins.
   (iv) Surrounded by a single unit membrane
   (a) Two of them (b) Three of them
   (c) All of them (d) Only one of them

138. What are the types of Ribosomes in a Prokaryotic and Eukaryotic cell.
   (a) 70s' and 80s' (b) 80s' and 70s'
   (c) 70s' and 70s' (d) 80s' and 80s'

139. How many subunits are presents in a ribosome
   (a) Two; one large and one smaller subunits
   (b) Three; two large and one smaller subunits
   (c) Only one subunits
   (d) Three; one large and two smaller subunits.

140. Subunits 50s' and 30s' are found in
   (a) 60s' type (b) 70s' type
   (c) 80s' type (d) 90s' type

141. What is sedimentation co-efficient
   (a) Svedberg unit
   (b) Measurement of density
   (c) Measurement of size
   (d) All of these

142. What type of ribosome are found in Eukaryotic cell
   (a) 70s' type only
   (b) 80s' type only
   (c) Both 70s' and 80s' type
   (d) 70s' ; 80s' & 60s' type

Topic 14 Cytoskeleton

143. Cytoskeleton refers to the :-
   (a) Cilia and flagella only
   (b) Network of filamentous proteinaecious structure
   (c) Microtubules only
   (d) Both (a) & (C)

144. Microtubules; microfilaments & intermediate filaments are constituents of:-
   (a) Ribosomes (b) Central sheath
   (c) Cytoskeleton (d) Cytolamellae

145. Cytoskeleton in a cell is involved in functions like
   (a) Mechanical supports
   (b) Motility
   (c) Maintenance of the shape of cell
   (d) All of the above

Topic 15 Cilia and Flagella

146. Which of the following statements in untrue:-
   (a) Cilia and flagella are hair like outgrowth
   (b) Cilia are small and works like oars.
   (c) Flagella are longer and responsible for cell movement.

(d) None of them

147. Statement – (I): both eukaryotic and prokaryotic cells contains flagella.

Statement – (II): eukaryotic flagella are structurally different from prokaryotic flagella.

(a) Both statements are correct
(b) Both statements are not correct
(c) Statement – (I) is correct but statement – (II) is wrong
(d) Statement – (I) is wrong but statement – (II) is correct

148. The core of cilia and flagella is known as
(a) Central sheath
(b) Central microtubule
(c) Axoneme
(d) Bridge

149. The microtubules in the cilia and flagella:-
(a) Runs parallel to each other.
(b) Forms the axoneme and outer membrane
(c) Both (a) & (B)
(d) Arranged centrally only

150. What is arrangement of microtubules in the cilium and flagellum
(a) 9 – peripheral & 3 – central
(b) Two – peripheral & 9 – central
(c) 9 – peripheral & two central
(d) All peripheral

151. The central sheath is:-
(a) Connected to inter doublet bridges
(b) Encloses peripheral doublets
(c) Connected to peripheral microtubules
(d) All of the above

152. Which of the following statement regarding cilia and flagella are not correct:-
(a) Peripheral doublets are inter connected by linker
(b) Linker are also known as inter doublet bridge
(c) Both emerges out from a centriole like structure
(d) Linker are also known as basal body

Topic 16 Centrosome and Centrioles

153. Centrosome and centrioles can be found in:-
(a) Animal cells only
(b) Plant cells only
(c) Both animal & plant cells
(d) In plant & Bacterial cells

154. Centrioles in the centrosome are:-
(a) Parallely arranged to each other
(b) Perpendicularly arranged to each other
(c) Arranged like a cart wheel
(d) Made up of triplets of centrally arranged microtubules

155. The basal body of centriole has micro tubular arrangement of:-
(a) 9 + 0 (b) 9 + 2
(c) 9 + 3 (d) 3 + 9

156. The central part of the proximal region of the centriole is:-
(a) Known as radial spoke
(b) Known as a central hub
(c) Connected to the peripheral doublets
(d) All of the above

157. The structure that give rise to the spindle fibers during cell division in animal cell is:-
(a) Cilia (b) Flagella
(c) Both (d) Centriole

Topic 17 Nucleus

158. A human cell has approximately _____ meters long thread of DNA, distributed among its _______ chromosomes:-
(a) 4; 46 (b) 2; 46
(c) 4; 23 (d) 2; 23

159. Each chromosome

(a) Has primary constriction
(b) Is visible only in dividing cells.
(c) Has disc shaped structure known as kinetochore
(d) All of the above

160. Function of centriole is:
(a) Provides site of attachment to the spindle fibers on chromosome
(b) Holds two chromatids of a chromosome
(c) Both (a) & (b)
(d) None

161. What is type of chromosome having a middle centromere:-
(a) Metacentric (b) Sub – metacentric
(c) Acrocentric (d) Telocentric

162. What is the type of chromosome having its centromere near the telomere
(a) Metacentric (b) Sub – metacentric
(c) Telocentric (d) Acrocentric

163. Chromosomes having centromere slightly away from the middle is:-
(a) Metacentric (b) Sub – metacentric
(c) Telocentric (d) Acrocentric

164. Chromosome having one long and one short arm are:-
(a) Metacentric & sub – metacentric
(b) Sub – metacentric & acrocentric
(c) Acrocentric & telocentric
(d) Telocentric & metacentric

165. A non – staining is present on a few chromosome
(a) Secondary constriction or centromere
(b) Satellite or centromere
(c) Secondary constriction or satellite
(d) None of the above

166. i) Nucleus as an organelle was first described by Robert brown
ii) Stained by the basic dyes, the material is known as chromatin by Robert brown
iii) Double membrane bound structure

How many of the above statement are not true about the nucleus & its material:-
(a) Only one (b) Only two
(c) Only three (d) Only four

167. The nucleus has highly extended and elaborate nucleoprotein fibers known as:-
(a) Nucleoli (b) Chromosome
(c) Chromatin (d) Nuclear matrix

168. The contents of an inter phase nucleus are:-

Nucleoli ; chromatin ; nuclear matrix; two membranes
(a) Only two of the above
(b) Only three of the above
(c) Only four of the above
(d) Only of the above

169. What forms the barrier between the cytoplasmic content and nuclear matrix:-
(a) The outer membrane only
(b) The inner membrane only
(c) The perinuclear space
(d) All of the above

170. i) The outer membrane of nucleus is continuous with rest of the cellular organelles
ii) The inner membrane is continuous with E.R
iii) Their are interruption known as pores present in outer membrane of nucleus
iv) Pores are formed by the fusion of both of the membranes.

How many of the above statements are incorrect:-
(a) 2 (b) 1
(c) 3 (d) 4

171. The nuclear pores facilitates :-
(a) Movement of RNA & protein molecules in only one direction
(b) Only proteins in both direction

(c) Proteins in one direction & RNA in both directions
(d) None of the these

172. Few of the mature cells have no any nucleus:-
(a) Their function are not specific
(b) Are dead cells with cytoplasm
(c) Their function are controlled by some another cells.
(d) All of the above

173. Statement – (I): The nucleus per cell varies per cell.
Statement – (II): Normally there is only one nucleus per cell.
(a) Both (I) & (II) are true & (II) is correct explanation of (I)
(b) Both (I) & (II) are true but (II) is not the correct explanation of (I)
(c) (II) is wrong but (I) is true.
(d) (I) is wrong but (II) is true.

174. The nucleus matrix contains:-
(a) Nucleoplasm and chromatin
(b) Nucleoplasm, Chromatin and Mitochondria
(c) Nucleoplasm, chromatin &E.R
(d) None of the above

175. What is not true about the nucleolus:-
(a) Spherical structure present in the nucleoplasm '
(b) Membrane less structure.
(c) Also known as Ribosomal factory of the cell.
(d) None of the above

176. At which phase of cell cycle the nucleolus has a loose and indistinct network of nucleoprotein fibers known as chromatin:-
(a) Prophase (b) Anaphase
(c) Interphase (d) Metaphase

177. Cell show structured chromosome during:-
(a) All phases except anaphase
(b) All phases except metaphase
(c) All phases except Inter phase
(d) All phases except m – phase

178. Chromatin contains
(a) Histones; Non – histones & RNA
(b) Histones & non – histone proteins only
(c) DNA & some basic proteins
(d) Both (a) & (C)

Topic 18 Microbodies

179. Membrane bound minute vesicles that contain various enzymes are present in both plant and animal cells called:
(a) Chloroplast (b) Mitochondria
(c) Ribosomes (d) Microbodies

ANSWER KEY

1. (d)	2. (c)	3. (c)	4. (a)	5. (c)	6. (d)	7. (c)	8. (b)	9.(a)	10. (d)
11. (d)	12. (a)	13. (a)	14. (b)	15. (a)	16. (c)	17. (a)	18. (a)	19. (c)	20. (d)
21. (c)	22. (d)	23. (b)	24. (d)	25. (c)	26. (d)	27. (b)	28. (c)	29. (d)	30. (d)
31. (d)	32. (d)	33. (c)	34. (d)	35. (b)	36. (c)	37. (a)	38. (b)	39. (a)	40. (c)
41. (d)	42. (d)	43. (d)	44. (c)	45. (c)	46. (b)	47. (c)	48. (c)	49. (c)	50. (c)
51. (a)	52. (c)	53. (d)	54. (c)	55. (d)	56. (a)	57. (c)	58. (a)	59. (d)	60. (d)
61. (c)	62. (b)	63. (c)	64. (a)	65. (a)	66. (a)	67. (a)	68. (d)	69. (c)	70. (a)
71. (a)	72. (d)	73. (b)	74. (a)	75. (b)	76. (d)	77. (c)	78. (c)	79. (a)	80. (d)
81. (a)	82. (d)	83. (c)	84. (c)	85. (a)	86. (a)	87. (a)	88. (a)	89. (a)	90. (a)
91. (c)	92. (a)	93. (c)	94. (a)	95. (a)	96. (b)	97. (b)	98. (c)	99. (d)	100. (c)
101. (d)	102. (c)	103. (d)	104. (a)	105. (c)	106. (c)	107. (c)	108. (c)	109. (c)	110. (d)
111. (d)	112. (b)	113. (c)	114. (d)	115. (a)	116. (c)	117. (d)	118. (c)	119. (d)	120. (c)
121. (d)	122. (b)	123. (d)	124. (c)	125. (d)	126. (c)	127. (d)	128. (c)	129. (b)	130. (b)
131. (a)	132. (c)	133. (c)	134. (a)	135. (d)	136. (c)	137. (b)	138. (a)	139. (a)	140. (b)
141. (d)	142. (b)	143. (b)	144. (c)	145. (d)	146. (d)	147. (a)	148. (c)	149. (a)	150. (c)
151. (c)	152. (d)	153. (c)	154. (b)	155. (a)	156. (b)	157. (d)	158. (b)	159. (d)	160. (c)
161. (a)	162. (c)	163. (b)	164. (c)	165. (c)	166. (a)	167. (c)	168. (c)	169. (d)	170. (b)
171. (d)	172. (c)	173. (b)	174. (a)	175. (d)	176. (c)	177. (c)	178. (d)	179. (d)	

7 Chapter Cell Cycle and Cell Division

MULTIPLE CHOICE QUESTIONS

Topic 1 Cell Cycle

1. All cells reproduce by dividing into______, with each parental cells giving rise to______ cells each time they divide.
   (a) One ; four daughter
   (b) Two ; two daughter
   (c) One ; two daughter
   (d) Two ; four daughter

2. A cell cycle made up of all the listed events, except:
   (a) Cell growth
   (b) DNA replication
   (c) Transcription
   (d) Cell division

3. Consider the following statements-

   Statement-I: cell growth (in terms of cytoplasmic increases) is a contingent process which occur during cell cycle.

   Statement-II: DNA synthesis occur only during one specific stage in the cell cycle.

   Statement-III: The event of cell cycle are under genetic control.
   (a) Statement-I is false & statement-II and III are true
   (b) Statement-I and II are false & statement-III are true
   (c) All statement are true
   (d) None of the above stated statement are true.

Topic 2 Phases of Cell Cycle

4. The G2 of cell cycle is pronounced by-
   (a) Cell growth and division
   (b) Cell duplication
   (c) Protein synthesis & centriole duplication
   (d) Cell growth & protein synthesis

5. Find mismatch column.

	Column-I	Column-II
(a)	Karyokinesis	Separation of daughter chromosome
(b)	cytokinesis	Division of cytoplasm
(c)	Interphase	Smallest phase of cell cycle
(d)	M-phase	Mitosis phase

6. The inactive stage of cell cycle is-
   (a) Quiescent stage (b) G1
   (c) S – Phase (d) A & B

7. Choose the correct statement with respect to G0 phase:
   (a) Also known as quiescent stage and start after G2 phase
   (b) Cell of this stage remain unactive and no longer proliferation
   (c) Cell of this stage remain in active but no longer proliferation unless called to do so depending on the requirement of organism
   (d) Cell of this stage remain active and proliferation till death without any condition

8. How many chromosome does onion somatic cell have –
(a) 12 (b) 14
(c) 16 (d) 20

9. What number of chromosome does onion somatic cell have in G, S, G2 & M – phase respectively
(a) 32, 16, 16, 32 (b) 16, 32, 16, 16
(c) 16, 16, 16, 16 (d) None of these

10. Mitotic division occur in –
(a) Diploid somatic cell
(b) Haploid male honey bee
(c) A & B
(d) Gametes

11. Match the following column:

Column – I	Column – II
(a) G_1 Phase	(i) Metabolically active cell, do not proliferate
(b) S Phase	ii) Content of DNA doubled
(c) G_0 phase	iii) Protein synthesised
(d) G_2 Phase	iv) Metabolically active cell grows continuously

(a) a – iv), b – ii), c – i), d – iii)
(b) a – i), b – ii), c – iv), d – iii)
(c) a – iv), b – iii), c – i), d - ii
(d) None of these

12. Human cell divides once in approximately every
(a) 60 minutes (b) 90 minutes
(c) 24 hours (d) None of these

13. Read the following statements and choose the correct option.
Statement A: The M-phase represents the phase when actual cell division occurs
Statement B: Interphase represents the phase between two successive M-phases
(a) Only statement A is correct
(b) Only statement B is correct
(c) Both the statements are incorrect
(d) Both the statements are correct

14. Match the columns and choose the correct option

Column I	Column II
(a) G1 phase	(i) Quiescent stage of the cell cycle.
(b) G2 phase	(ii) DNA denoted as 2C, increases to 4C
(c) Synthesis phase	(iii) Proteins are synthesized in preparation for mitosis
(d) G0 phase	(iv) Cell contain initial amount of DNA i.e., 2C

(a) a-iv, b-iii, c-i, d-ii
(b) a-iv, b-iii, c-ii, d-i
(c) a-iii, b-i, c-iv, d-ii
(d) a-ii, b-iv, c-ii, d-i

15. An average duration of yeast cell cycle is-
(a) 60 minutes (b) 90 minutes
(c) 20 minutes (d) One day

16. Cell cycle is divided into how many basic phases
(a) One (b) Two
(c) Four (d) Six

17. Which of following is/are enlisted as basic phases of cell cycle?
(a) G0 phase (b) S phase
(c) Interphase (d) Metaphase

18. The phase of cell cycle during which mitosis occur is-
(a) Interphase (b) M-phase
(c) G-phase (d) S-phase

19. The phase between two successive Mphase is-
(a) Interphase (b) G-phase
(c) S-phase (d) M-phase

20. The time span of interphase and M-phase is an average human cell cycle is-
(a) 12 hours each

(b) 95% M-phase & one hour interphase
(c) 8 hour M-phase & 16 hour interphase
(d) One hour M-phase & 23 hour interphase

21. The correct sequence of cell is-
(a) M → G2 → S → G1
(b) S → G2 → G1 → M
(c) M → G1 → G2 → S
(d) G1 → S → G2 → M

22. The process which mark as start & usually end of M-phase are-
(a) Division of cytoplasm & Karyokinesis respectively
(b) Cytokinesis and division of cytoplasm respectively
(c) Separation of daughter chromosome & cytokinesis respectively
(d) Karyokinesis & karyokinesis respectively

23. Resting phase of cell-cycle is-
(a) M-phase (b) Interkinesis
(c) G1 & G2 phase (d) Interphase

24. Interphase is divided into ___ phases further.
(a) 4 (b) 3
(c) 2 (d) 5

25. Which of following stage corresponds to the interval between mitosis & initiation of DNA replication?
(a) S-phase (b) G2-phase
(c) M-phase (d) G1-phase

26. Select the correct statement about G1 phase-
(a) Cell is metabotically inactive
(b) DNA does not replicate
(c) DNA replicate
(d) Chromosome number is doubled

27. Correct sequence of phase of M-phase is-
(a) Cytokinesis > Prophase > Metaphase > Anaphase > Telophase
(b) Prophase > Anaphase > Metaphase > Telophase > Cytokinesis
(c) G0 > G1 > S > G2
(d) None of these

28. What would be amount of DNA (C) and number of chromosome (N) in animal cell just after completion of S phase if the initial amount is 2C and 2N?
(a) 2C and 2N respectively
(b) 4C and 4N respectively
(c) 4C and 2N respectively
(d) 2C and 4N respectively

29. Duplication of centriole occur in-
(a) M-phase (b) G2-phase
(c) S-phase (d) G0-phase

30. The S-phase of animal cell marked by-
(a) DNA replication
(b) Centriole duplication
(c) Cell growth and protein synthesis
(d) A & B

Topic 3 M-Phase

31. M – phase belongs to –
(a) Metaphase (b) Meiosis
(c) Karyokinesis (d) A & B both

32. Which one of the following is the most dramatic period of cell cycle?
(a) Gap 1 only (b) M-phase
(c) S-phase only (d) Interphase

33. Which cell equational division refer to –
(a) Meiosis
(b) Mitosis
(c) Number of cell chromosome in parent & progeny cell is same
(d) B & C

34. Karyokinesis of mitosis is divided into ______ stages
(a) 2 (b) 3
(c) 4 (d) 8

35. Which one is the correct order of mitotic division?
(a) Metaphase → Anaphase → Prophase → Telophase

(b) Prophase → Metaphase → Anaphase → Telophase

(c) Anaphase → Telophase → Metaphase → Prophase

(d) Telophase → Prophase → Anaphase → Metaphase

Topic 4 Prophase

36. How do the chromosome appear during prophase of animal cell during mitosis
    (a) Consisting of four chromatid which remain attached to centromere
    (b) Consisting of two chromatid which remain attached to centromere
    (c) Consisting of four chromatid without centromere
    (d) As chromatin material without any defined structure

37. Asters formed during prophase are –
    (a) Composed of microtubules originate from centromere
    (b) Composed of protein which secreted by golgi body
    (c) Highly condensed area of chromosome
    (d) None of these

38. What difference would indicate earlyprophase & late prophase of animal cell.

	Early Prophase	**Late Prophase**
(a)	Nucleolus & nuclear membrane present	Nucleolus & nuclear membrane are absent
(b)	Chromosomes are highly condensed	There is no condensation of chromosome
(c)	Nucleolus & centrosome are present	Nucleolus & centromere are absent
(d)	Other organelles like ER, golgi body complexes are not observed	Other organelles like ER, golgi complex are observed.

39. Identify correct stage of given diagram.

(i) (ii)

	(i)	(ii)
(a)	Late prophase	Early prophase
(b)	Early prophase	Late prophase
(c)	Metaphase	Prophase
(d)	Prophase	Metaphase

40. Select the correct option:
    (I) Prophase is first stage of Karyokinesis.
    (II) It occur after completion of protein synthesis during cell cycle
    (a) Both (I) & (II) are true
    (b) Both (I) & (II) are false
    (c) (I) is true but (II) is false
    (d) (I) is false but (II) is true

41. During prophase, which of the following occurs?
    (a) Condensation of chromosomal material
    (b) Chromosomal material become tangled
    (c) Centrosome duplication
    (d) Movement of both centriole at one pole of cell

42. Choose the incorrect match
    (a) Beginning of movement of chromosome to opposite poles – Prophase
    (b) Two asters with spindle – Mitotic apparatus
    (c) Attachment of spindle – Metaphase
    (d) Chromosome move to opposite poles – Metaphase

43. Mitotic apparatus consist of –
    (a) Four asters with spindle fibres
    (b) One asters with spindle fibres
    (c) Two asters with spindle fibres
    (d) Centrosome with their microtubules without

spindle fibres.

44. How many of following structures are observed when cells are viewed under the microscope at end of prophase Golgi body, ER, Nucleolus, Nuclear envelop, centrosome
(a) Zero (b) One
(c) Three (d) All of these

Topic 5 Metaphase

45. Kinetochores are
(a) Precursors of microtubules
(b) Sites of attachment of spindle fibres
(c) Site for origination of spindle fibres
(d) Small disc – shaped structure at telomere of chromosome

46. Metaphase is characterised by –
(a) Some chromosomes coming to lie at the pole
(b) One chromatid of each chromosome connected by its centromere to spindle fibres from one pole
(c) Sister chromatid connected by its kinetochore to spindle fibres from opposite poles
(d) All of these

47. Identify stage

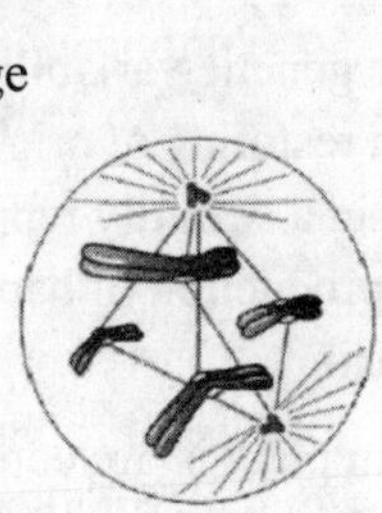

(a) Transition to metaphase
(b) Anaphase
(c) Metaphase
(d) Telophase

48. The complete disintegration of nuclear envelop marks start of:
(a) Late prophase (b) Metaphase
(c) Anaphase (d) None of these

49. Identify stage of given diagram

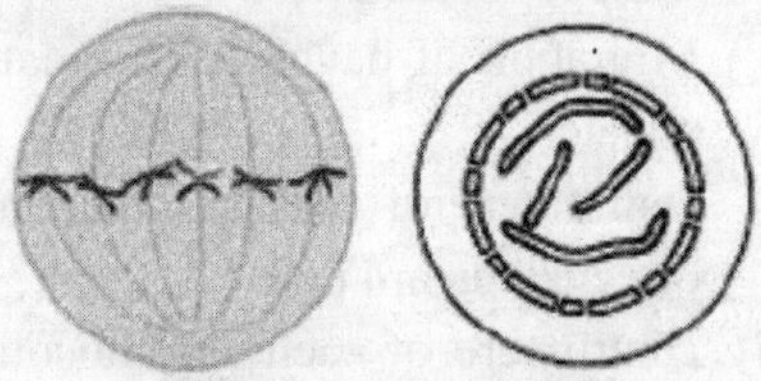

(ii) (i)

(a) Early prophase, metaphase
(b) Late prophase, transition to metaphase
(c) Early prophase, transition to metaphase
(d) Late prophase, metaphase

50. Which stage of cell cycle is best to study chromosome morphology
(a) Late prophase (b) Early prophase
(c) Anaphase (d) Metaphase

51. Condensation of chromosome is completed in –
(a) Stage where centrosome is duplicated
(b) Stage where DNA content doubled
(c) Stage where complete integration of nuclear envelope occurs
(d) Stage where complete disintegration of nucleus envelope occurs

52. Metaphase chromosome is made up of –
(a) Two non – sister chromatid which are held together by centromere
(b) Four sister chromatid which are held together by centromere
(c) Two sister chromatid which are held together by centromere
(d) Four non – sister chromatid which are held together by centromere

Topic 6 Anaphase

53. At the onset of anaphase, each chromosome split into –
(a) One chromatid
(b) Four daughter chromatids

(c) Two daughter chromosomes
(d) Eight chromatids

54. Anaphase is characterised by –
i) Migration of daughter chromatid toward equator.
ii) centromere of each chromosome remain directed toward pole
iii) centromere of each chromosome remain directed toward equator
iv) Chromatid split and centromere separate
v) Chromatid separate after centromere split
(a) i, ii, v (b) ii, v
(c) iii, v (d) ii, iv

55. Identify stage –

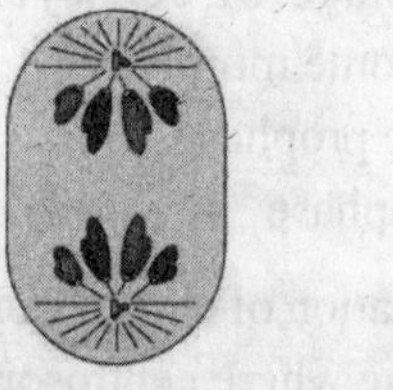

(a) Anaphase (b) Telophase
(c) Interphase (d) Metaphase

Topic 7 Telophase

56. During telophase:
(i) Chromosome cluster at opposite spindle poles
(ii) Two daughter nuclei formed
(iii) Chromosomes lose their individuality
(iv) It is reversal of prophase
(v) Nucleolus is not reformed
Choose the incorrect statement:-
(a) i), (ii) (b) (iii), (iv)
(c) (v) only (d) none of these

57. Match the following column –

	Column I		Column II
a	Syncytium	i	Divide the cytoplasm of animal cell
b	Cell-plate	ii	Occur in liquid endosperm of coconut
c	Cell furrow	iii	Method of cytokinesis in plant cell

(a) a-iii, b-ii, c-i
(b) a-ii, b-iii, c-i
(c) a-i, b-ii, c-iii
(d) a-ii, b-i, c-iii

58. Cell plate represent –
(a) Primary lamella
(b) Middle lamella
(c) Both
(d) formation of plate by lysosome

Topic 8 Cytokinesis

59. Mitosis usually results in the production of
(a) haploid daughter cells with identical genetical complement
(b) growth of multicellular organism
(c) diploid daughter cells without identical genetical complement
(d) haploid daughter cells without identical genetical complement

60. Which one is not a significance of meiosis division?
(a) Increase genetic variability in organisms
(b) Helps in restoring of original chromosome number in a sexually reproducing species.
(c) Ensure production of haploid phase
(d) Cell repair

61. The growth in plant is/are contributed by
(a) Mitotic division in apical meristem
(b) Meiotic division in lateral meristem
(c) Meiotic division in apical meristem
(d) A & B both

62. (a) The nucleo-cytoplasmic ratio in organism is restore by mitosis
(b) The cells of the upper layer of the epidermis, cells of lining of gut, and blood cells are being constantly replaced by Mitotic

division.

Choose the correct option from following :-

(a) Statement (a) is true but (b) is false
(b) Statement (b) is true but (a) is false
(c) Statement (a) & (b) are true
(d) Statement (a) & (b) are false

Topic 9 Meiosis

63. Meiosis result in
(a) production of gametes
(b) reduction in number of chromosomes
(c) introduction of variation
(d) all of these

64. Meiosis ensure the production of .... phase in life cycle of sexually reproduction organisms whereas fertilization restore .... phase.
(a) haploid & haploid respectively
(b) haploid & diploid respectively
(c) diploid & diploid respectively
(d) diploid & haploid respectively

65. Which of the following statement is correct?
(a) Meiosis involves single cycle of nuclear and cell division
(b) Doubling of chromosomes occur once during s-phase
(c) Recombination between sister chromatid of non-homologous chromosome
(d) Pairing of homologous chromosome

66. At the end of meiosis-II, how many haploidcells are formed?
(a) One (b) Two
(c) Zero (d) Four

67. Recombination occurs between –
(a) sister chromatid of non-homologous chromosome
(b) non-sister chromatid of non homologous chromosome
(c) sister chromatid of homologous chromosome
(d) non-sister chromatid of homologous chromosome

Topic 10 Meiosis-I

68. Diakinesis marked by –
(a) Terminalisation of chiasmata
(b) Chromosomes are fully condensed
(c) Meiotic spindle assembled
(d) All of these

69. Meiotic spindle assembled to prepare –
(a) Non homologous chromosomeseparation.
(b) Formation of aster ray.
(c) Homologous chromosome separation.
(d) Both B & C

70. Match the following

	Column I		Column II
I	Leptotene	a	Compaction of chromosome
II	Zygotene	b	Separation of chromosome except at crossover
III	Pachytene	c	Terminalisation of chiasmata
IV	Diakinesis	d	Appearance of recombination of nodules
V	Diplotene	e	Synapsis

(a) I-a, II-e, III-d, IV-c, V-b
(b) I-a, II-b, III-d, IV-c, V-e
(c) I-c, II-d, III-a, IV-e, V-b
(d) None of these

71. Identify stage

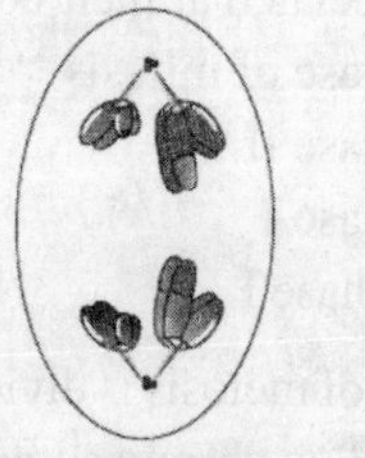

(a) Homologous chromosome separate, while sister chromatid remain associated at centromere.

(b) Homologous chromosome along with sister chromatid separate.

(c) Spindle attached to Kinetochore in this stage.

(d) This stage followed by diakinesis.

72. Spindle fibre attach to kinetochores of homologous chromosome in –

(a) Metaphase – I of meiosis

(b) Metaphase – II of meiosis

(c) Both A & B

(d) Anaphase of mitosis

73. Identify stage

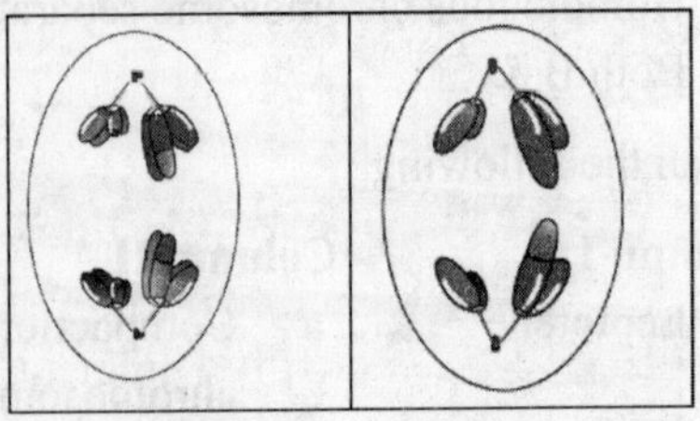

a b

(a) a = Anaphase I, b = Anaphase II

(b) a = Anaphase II, b = Metaphase II

(c) a = Anaphase II, b = Anaphase I

(d) a = Anaphase I, b = Anaphase II

74. Dyads of cells are formed in –

(a) Telophase – I (b) Telophase – II

(c) Diakinesis (d) Both A & B

75. Longest phase of meiosis is :

(a) Prophase-I (b) Prophase-II

(c) Metaphase-I (d) Telophase-II

76. During which of the given phases, homologous chromosomes separate, while sister chromatids remain associated at their centromere?

(a) Anaphase of mitosis

(b) Anaphase II

(c) Anaphase I

(d) Metaphase I

77. Prophase-I of meiosis is divided into .... phase based on chromosomal behaviour.

(a) 2 (b) 3

(c) 4 (d) 5

78. Identify correct sequence of prophase-I.

(a) leptotene, Diplotene, Zygotene

(b) Zygotene, pachytene, leptotene

(c) Diplotene, Zygotene, Pachytene

(d) None of these

79. A bivalent is

(a) Pair of non-homologous chromosomes

(b) The complex formed by a pair of synapsed homologous chromosomes.

(c) Formed during pachytene statge

(d) More clearly visible at zygotene Stage

80. Synaptonemal complex dissolves during-

(a) Leptotene (b) Diakinesis

(c) Zygotene (d) Diplotene

81. During which phase of meiosis centromere splits?

(a) Anaphase I (b) Anaphase II

(c) Telophase II (d) Telophase I

82. Choose the correct option with respect to leptotene:

(i) It is the foremost and the short-lived stage of prophase

(ii) It begins when the process of compaction of chromosome is accomplished

(iii) chromosome become visible under light microscope

(iv) It followed by zygotene

(a) One statement that is (ii) is incorrect

(b) i, ii and iii are correct while (iv) is incorrect

(c) iii & iv are correct while i, ii are incorrect

(d) All statement are correct

83. Zygotene is characterized by –

(i) chromosome start pairing

(ii) non-homologous chromosome paired

(iii) synapsis occurs between nonhomologous chromosomes

(iv) formation of synaptomeal complex in homologous chromosomes

(v) formation of synaptonemal complex in non homologous chromosomes

(a) i, ii, v (b) i, iii, iv

(c) i, iv (d) i, ii, iii, v

84. Bivalent stage is –
(a) complex formed by a pair of synapsed homologous chromosomes
(b) complex formed by a pair of synapsed non-homologous chromosomes
(c) complex formed by four pair of synapsed homologous chromosomes
(d) complex formed by four pair of synapsed non-homologous chromosomes

85. Crossing over occurs in –
(a) leptotene (b) zygotene
(c) Pachytene (d) diplotene

86. Pachytene is stage that is/are :-
(a) long lived than zygotene
(b) two chromatid of each bivalent chromosomes becomes distinct
(c) short lived than leptotene
(d) long lived than leptotene & short lived than zygotene

87. Choose the correct statement from following:
(a) Pachytene is characterised by appearance of recombination nodule
(b) Recombination nodule is site of crossing over
(c) Both A & B
(d) Recombination nodule formed in diplotene

88. Given below are statements (I - VI). Choose correct set with respect to crossing over.
(I) It occurred between sister chromatid of homologous chromosomes.
(II) It is enzyme mediated process.
(III) Recombinase enzyme involved in it.
(IV) It occurs at recombination nodules.
(V) It occurs between non sister chromatid of non-homologous chromosomes.
(VI) It occurs between sister chromatid of non-homologous chromosomes.
(a) I, II, III & IV (b) V, II, III & IV
(c) II, III, IV & VI (d) II, III & IV

89. Diplotene is not characterized by
(a) Dissolution of synaptonemal complex.
(b) Tendency of recombined homologous chromosomes of tetrad to separate from each other, except at sites of crossover crossovers.
(c) Formation of chiasmata
(d) Tendency of recombined non homologous chromosome of bivalent to separate from each other, except at sites of crossover.

90. Chiasmata is –
(a) X – shaped structures
(b) Formed by recombined chromosome yet to be separated
(c) Site of cross over
(d) All of these

91. Which stage of Meiosis – I last for months or year in some vertebrate oocytes?
(a) Diakinesis (b) Diplotene
(c) Pachytene (d) Zygotene

Topic 11 Meiosis-II

92. Meiosis – II initiated immediately after
(a) Telophase – I
(b) Prophase
(c) Cytokinesis – I
(d) Chromosome have fully elongated

93. Which one of the following resembles mitosis when sister chromatids split'
(a) Meiosis – I (b) Meiosis – II
(c) Both (d) None of these

94. In the beginning of Meiosis – II, a cell contain four chromatid. What number of chromatid is expected to be in each daughter cell at end of telophase – II
(a) 4 (b) 2
(c) 8 (d) 16

95. Find mismatched column

	Column I	Column II
(a)	Metaphase – II	Chromosomes align at equator and microtubule from opposite poles of spindle get attached to kinetochores of non-sister chromatid
(b)	Prophase – II	Nuclear membrane disappear
(c)	Telophase – II	Formation of tetrad of cells
(d)	Anaphase – II	Splitting centromere which hold sister chromatid together, allow them to move toward opposite pole of cells

96. Movement of chromatid toward opposite pole is achieved by:
(a) Shortening of microtubules attached to centromere
(b) Shortening of microtubules attached to kinetochores
(c) Elongating of microtubules attached to kinetochores
(d) Elongating of microtubules attached to centromere

97. Which one of the following mechanism by which the conservation of specific chromosome number of each species is achieved across generations in sexually reproducing organism is done by –
(a) Mitosis
(b) Meiosis only
(c) Meiosis & Mitosis
(d) None of these

98. Choose the correct statement about meiosis
(a) Increase genetic variability of an individual of an organism
(b) decrease genetic variability of an organism from one generation to other
(c) Reduction of chromosome by onefourth
(d) Play an important role in evolution

ANSWER KEY

1. (b)	2. (c)	3. (a)	4. (d)	5. (c)	6. (a)	7. (c)	8. (c)	9. (c)	10. (a)
11. (a)	12. (c)	13. (d)	14. (d)	15. (b)	16. (b)	17. (c)	18. (a)	19. (a)	20. (d)
21. (d)	22. (c)	23. (d)	24. (b)	25. (d)	26. (b)	27. (a)	28. (c)	29. (c)	30. (d)
31. (c)	32. (b)	33. (d)	34. (c)	35. (b)	36, (d)	37. (a)	38. (b)	39. (b)	40. (a)
41. (a)	42. (a)	43. (c)	44. (a)	45. (b)	46. (a)	47. (c)	48. (b)	49. (d)	50. (d)
51. (d)	52. (c)	53. (c)	54. (b)	55. (a)	56. (b)	57. (b)	58. (b)	59. (a)	60. (d)
61. (a)	62. (c)	63. (d)	64. (d)	65. (d)	66. (d)	67. (c)	68. (d)	69. (c)	70. (a)
71. (a)	72. (a)	73. (a)	74. (a)	75. (a)	76. (c)	77. (d)	78. (c)	79. (b)	80. (d)
81. (b)	82. (c)	83. (a)	84. (a)	85. (c)	86. (a)	87. (c)	88. (d)	89. (d)	90. (b)
91. (d)	92. (c)	93. (b)	94. (c)	95. (c)	96. (b)	97. (b)	98. (c)		

8 Chapter Transport in Plants

MULTIPLE CHOICE QUESTIONS

1. Transport of minerals in plants is
   (a) unidirectional only
   (b) multidirectional
   (c) Both unidirectional & multidirectional
   (d) Dependent on the hormones released

2. Nutrients are re-exported from
   (a) nascent leaves to senescent leaves
   (b) senescent leaves to nascent leaves
   (c) all parts to senescent leaves
   (d) not re-exported

3. Melvin Calvin earned Nobel prize in 1961 for
   (a) mapping pathway of carbon assimilation in respiration
   (b) mapping pathway of carbon absorption in photosynthesis
   (c) mapping pathway of carbon assimilation in photosynthesis
   (d) mapping pathway of carbon absorption in respiration.

4. Cytoplasmic streaming helps to move substances over
   (a) long distances (b) small distances
   (c) Both (a) and (b) (d) None

5. Translocation is transport over
   (a) long distance (b) short distance
   (c) Both long and short distance
   (d) None of these

Topic 1 Diffusion

6. Diffusion rates are affected by
   (a) Concentration gradient
   (b) pressure
   (c) temperature
   (d) All of these

7. Statement (a) Diffusion can't occur in dead cell.
   Statement (b) Diffusion is the only means for gaseous movement in plant body.
   Choose the best option
   (a) Statement A is correct and Statement B is wrong
   (b) Statement A is wrong and Statement B is correct
   (c) Both are correct
   (d) Both are wrong

8. Movement by diffusion is
   (a) active with energy expenditure
   (b) passive with energy expenditure
   (c) both active and passive
   (d) without energy expenditure

9. In diffusion, molecules
   (a) move in a fixed fashion, from high to low concentration
   (b) move in random fashion, from high to low concentration
   (c) move in fixed fashion, from low to high concentration
   (d) move in random fashion, from low to high concentration.

10. Diffusion
    (a) in solid is more likely rather than of solid
    (b) of solid is more likely than in solid
    (c) Both (a) and (b)
    (d) Does not occur in solids at all.

Topic 2 Facilitated Diffusion

11. Porins are found in
    (a) outer membrane of plastids & mitochondria
    (b) Inner membrane of plastids & mitochondria
    (c) both outer & inner membrane of mitochondria
    (d) membrane of lysosoems

12. Water channels are made up of ____ different types of aquaporins
    (a) 6 (b) 8
    (c) 10 (d) 12

13. 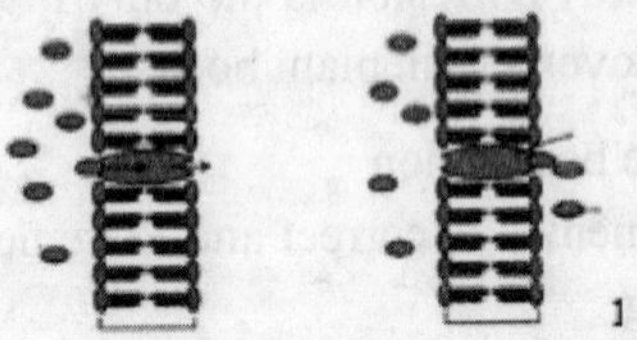

    Identify the correct label for the figure given
    (a) It shows facilitated diffusion
    (b) shows simple diffusion
    (c) shows sample of water channels
    (d) Both (a) and (c)

14. Diffusion rate
    (a) depend on size and larger substance diffuse faster
    (b) depend on size and smaller substance diffuse faster
    (c) doesn't depend on size
    (d) depend on size and do not substance diffuse larger.

15. Hydrophilic moiety substances
    (a) diffuse through a membrane easily
    (b) do not diffuse through a membrane easily
    (c) need no membrane facilitation to diffuse through
    (d) Both (a) and (c)

16. Read the following statements
    (i) Membrane proteins aid in transport of hydrophobic substance.
    (ii) Membrane proteins do not set up a concentration gradient.
    (iii) Diffusion through membrane proteins is called facilitated diffusion.
    (iv) Membrane proteins always need ATP to transport substances across membrane.

    How many of the above statements are wrong?
    (a) 1 (b) 2
    (c) 3 (d) 4

17. In facilitated diffusion
    (a) special proteins & ATP are involved
    (b) only special proteins are involved
    (c) only ATP are involved
    (d) neither special proteins nor ATP are involved.

18. Facilitated diffusion cannot
    (a) transport molecules from high concentration to low concentration
    (b) transport molecules from low concentration to high concentration
    (c) Both (a) and (b)
    (d) Concentration does not matter

19. Facilitated diffusion transport is
    (i) Saturable (ii) unsaturable
    (iii) inhibitable (iv) non-inhibitable
    (v) selective (vi) non-selective
    (vii) uphill (viii) downhill

    Which is the correct combination
    (a) i, iii, v, vii (b) ii, iv, vi, viii
    (c) i, iii, v, viii (d) i, iv, vi, viii

Topic 3 Passive Symports and Antiports

20. 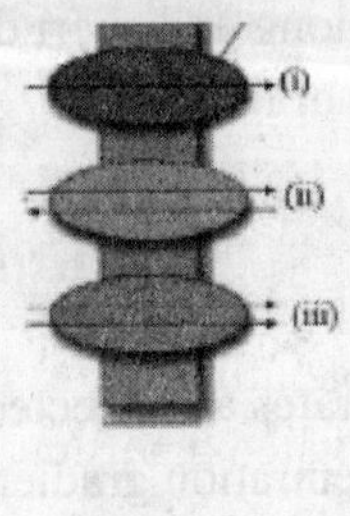

(i) Symport will not work if single type of molecule is there.
(ii) Uniport allows two molecules to move in a unidirection together.
(iii) Antiport allows two types of molecules to move in opposite direction.

How many of the above statements is correct
(a) 0 (b) 1
(c) 2 (d) 3

21. Identify correct statements about the figure given
(a) (i) – It shows simple diffusion via uniport
(b) (ii) – It shows simple diffusion via antiport
(c) Both (a) & (b) are correct
(d) None of these

Topic 4 Active Transport

22. Active transport is
(a) Uphill (b) downhill
(c) unsaturable (d) both (a) and (b)

23. Pumps are
(a) energy using transport proteins
(b) non-energy using transport protein
(c) energy using transport lipids
(d) non-energy using transport lipids

24. Active transport rate reaches maximum when
(a) All molecules are loaded on carrier proteins
(b) 80% of the molecules are loaded on carrier proteins
(c) 50% of the molecules are loaded on carrier protcins
(d) Can be any of these

Topic 5 Comparison of Different Transport Processes

25. Match the columns A & B

A	B
(i) Simple diffusion	(I) Uphill
(ii) Facilitated diffusion	(II) Saturable
(iii) Active transport	(III) Selective
	(IV) Passive
	(V) Carrier protein

(a) (i)–III, (ii)–IV, (iii)–i
(b) (i)–IV, (ii)–II, (iii)–III
(c) (i)–V, (ii)–IV, (iii)–II
(d) (i)–IV, (ii)–I, (iii)–II

Topic 6 Plant-water Relations

26. Most herbaceous plants have about
(a) 10-15% of its dry weight as fresh matter.
(b) 85-90% of its fresh weight as dry matter.
(c) 10-15% of its fresh weight as dry matter.
(d) 85-90% of its fresh weight as water.

27. Why is water often limiting factor for plant growth & productivity?
(a) Due to high respiration
(b) Due to high photosynthesis
(c) Due to low availability of water
(d) Due to transpiration of water

Topic 7 Water Potential

28. Kinetic energy possessed by water molecules is represented directly by its
(a) pressure potential
(b) water potential
(c) soluble potential
(d) osmotic potential

29. Pure water has
(a) Lowest water potential at all pressures
(b) Zero water potential at all pressures.
(c) Water potential at all pressures highest
(d) Both (b) and (c)

30. Water moves from system containing water at (i) to one with (ii)
(a) (i) – low ψ_w (ii) – high ψ_w
(b) (i) – zero ψ_w (ii) – zero ψ_w
(c) (i) – positive ψ_w (ii) – zero ψ_w
(d) (i) – high ψ_w (ii) – low ψ_w

31. Which of the given equations is correct?
(a) $\psi_w = \psi_s + \psi_p$
(b) $\psi_s = \psi_w + \psi_p$
(c) $\psi_p = \psi_s + \psi_w$
(d) $\psi_p = \psi_s + \psi_w$

32. ψ_s is
(a) always positive
(b) always negative
(c) sometimes negative
(d) mostly zero

33. ψ_p is
(a) always positive
(b) always negative
(c) usually positive, sometimes negative
(d) usually negative, sometimes positive

34. The more the solute
(a) the lower (less negative) the ψ_s
(b) the higher (less negative) the ψ_s
(c) the lower (more negative) the ψ_s
(d) the higher (more negative) the ψ_s

35. If pure water (pH = 7) is kept in open vessel at room temperature, its water potential

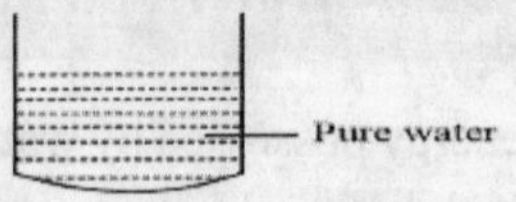

(a) zero (b) positive
(c) negative (d) can't say

36. Water potential represents kinetic energy of water molecules.

When $\psi_s = 0$.
(a) kinetic energy of molecules of water is zero.
(b) kinetic energy of molecules of water is not zero.
(c) kinetic energy of molecules of water is negative
(d) kinetic energy of molecules of water is positive

Topic 8 Osmosis

37. If one chamber has a of –2000 kPa and the other –1200 kPa, which is chamber with higher
(a) B (b) A
(c) C (d) Can't say

38. If one of the solutions has and other has what will be direction of water movement?
(a) a to b
(b) b to a
(c) No net movement
(d) Random movement

39.

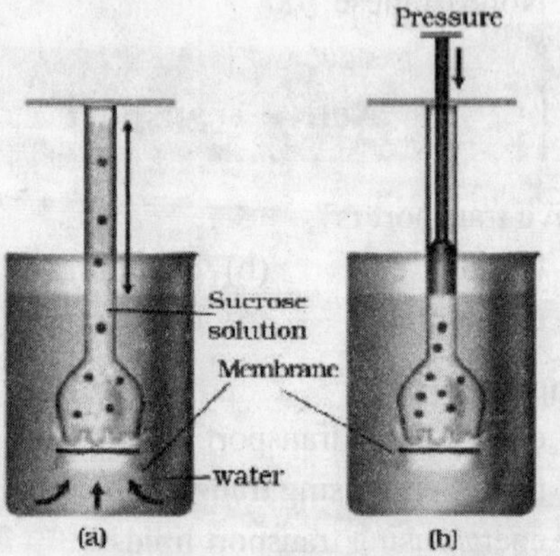

To prevent water from diffusing in pressure is applied. Which of the following statements is correct?
(a) The more the solute, the greater will be pressure required.
(b) The more the solute, the lesser will be pressure required
(c) The pressure is equal to osmotic potential exactly.
(d) Both (a) & (c)

40. Osmotic pressure is (ii) & osmotic potential is (ii)
(a) (i) – positive, (ii) negative
(b) (i) –negative, (ii) – positive
(c) (i), – positive, (ii) – positive
(d) (i) – negative, (ii) – negative

41. Cell wall is
   (a) impermeable to water and substances in solution
   (b) permeable to water and substances in solution
   (c) permeable to water but not to substance in solution.
   (d) impermeable to water but not to substances in solution.

42. Read the following statements
   (i) vacuolar sap contribute to solute potential of cell
   (ii) cell membrane & tonoplast together are importance determinants of movement of molecules in or out of cells.
   (iii) Osmosis occurs spontaneously in response to a driving force

   How many of the above statements are Incorrect
   (a) 1 (b) 2
   (c) 3 (d) None of these

43. Osmosis refers to
   (a) diffusion of solute across permeable membrane
   (b) diffusion of solute across ifferentially permeable membrane
   (c) diffusion of water across differentially permeable membrane
   (d) diffusion of water across permeable membrane

44. Water moves under osmosis from
   (a) higher chemical potential to lower chemical potential
   (b) lower chemical potential to higher chemical potential.
   (c) lower potential concentration to higher concentrations.
   (d) more than one option is correct.

45. In potato osmometer, if potato tuber is placed in water
   (a) water exits the cavity of potato tuber via simple diffusion.
   (b) water enters the cavity of potato tuber via simple diffusion.
   (c) water exits the cavity of potato tuber via osmosis
   (d) water enters the cavity of potato tuber via osmosis

Questions 46 to 52 are to be solved in reference to the given figure.

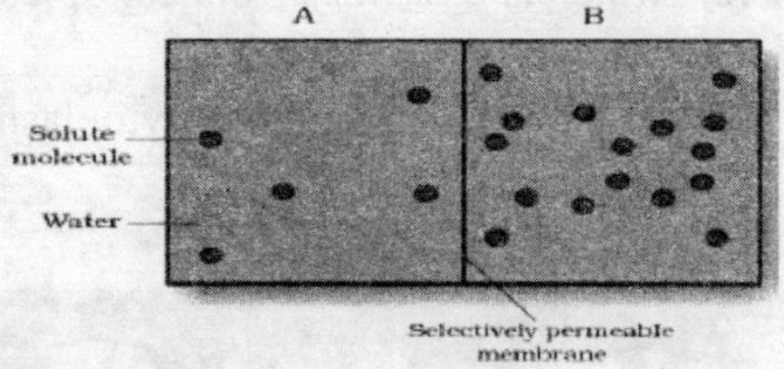

46. Which chamber has a lower water potential?
   (a) A (b) B
   (c) Both are equal (d) Can't say

47. Solution of which chamber has morenegative solute potential?
   (a) A (b) B
   (c) Both are equal (d) can't say

48. In which direction will osmosis occur?
   (a) from A to B
   (b) from B to A
   (c) No net movement
   (d) can't say (insufficient data)

49. Which solution has a higher solute potential?
   (a) B (b) A
   (c) Both are equal (d) can't say

50. At equilibrium which chamber will have lower water potential?
   (a) B (b) A
   (c) Both are equal (d) Insufficient data

Topic 9 Plasmolysis

51. Plasmolysis occurs
   (a) in hypotonic solution
   (b) when water moves into the cell
   (c) when solution has more solute than protoplasm

(d) when cell membrane becomes turgid

52. When water moves out of a cell placed in hypertonic solution.
(a) water is first lost from vacuole, then cytoplasm
(b) water is first lost from cytoplasm, then vacuole
(c) water is first lost from tonoplast, then cytoplasm
(d) water is first lost from cytoplasm, then tonoplast

53. 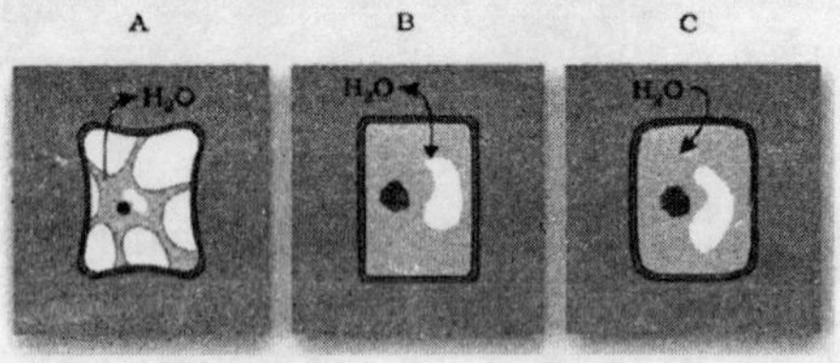

Identify A, B & C in given figure.
(a) A – plasmolysed, B – isotonic, C – Turgid
(b) B – flaccid, C – hypertonic, A – Turgid
(c) A – hypotonic, C – hypotonic, B – Turgid
(d) A – turgid, B – flaccid, – hyper tonic

54. The process of plasmolysis is
(a) always irreversible
(b) always reversible
(c) usually reversible
(d) always temporary

55. Plant cells do not rupture in hypotonic solution due to
(a) turgor pressure
(b) pressure potential
(c) cell membrane
(d) cell wall

56. What will be of flaccid cell
(a) positive (b) negative
(c) zero (d) any of the above

57. Match the columns

A	B
(i) Isotonic	I. external solution is more dilute
(ii) Hypotonic	II. external solution is more concentrated
(iii) Hypertonic	III. external solution balances the osmotic pressure of cytoplasm

(a) (i)-I, (ii)-III, (iii)-II
(b) (i)-III, (ii)-I, (iii)-II
(c) (i)-III, (ii)-II, (iii)-I
(d) (i)-II, (ii)-I, (iii)-III

58. Cells
(a) swell in hypotonic, shrink in isotonic
(b) swell in isotonic, shrink in hypertonic
(c) swell in hypertonic, shrink in hypotonic
(d) swell in hypotonic, shrink in hypertonic

Topic 10 Imbibition

59. Imbibition
(a) Causes reduction in volume
(b) is a type of active transport
(c) is along the concentration gradient
(d) occurs in gases

Topic 11 Long Distance Transport of Water

60. Bulk movement of substances through vascular tissues of plants is called
(a) Active transport
(b) Facilitated diffusion
(c) Transportation
(d) Translocation

61. Substances in mass flow
(a) are swept at some pace in solution
(b) are swept at speed depending upon size.
(c) are swept at some pace in suspensions
(d) Both (a) and (c)

62. Bulk flow is achieved by
(a) positive water pressure gradient

(b) negative water pressure gradient
(c) Both (a) and (b)
(d) It rarely depends on pressure gradient

63. Xylem is associated with translocation of mainly
(i) Water (ii) organic solutes and sucrose
(iii) mineral salts (iv) organic nitrogen
(v) inorganic solute (vi) hormones

Choose correct combination
(a) i, v, iii (b) ii, vi, iv
(c) iii, vi, iv (d) i, ii, iii

Topic 12 How do Plants Absorb Water?

64. Water movement through root tissues
(a) can be initially symplastic and finally apoplastic
(b) has to be ultimately apoplastic
(c) can be completely apoplastic
(d) None of the above is correct.

65. Read the given statements
(i) Mycorrhiza is symbiotic bacterial association of root system.
(ii) Mycorrhiza helps in water absorption.
(iii) Mycorrhiza may penetrate root cells
(iv) Mycorrhiza may from a network around young root.

Which of these statements is correct?
(a) 1 (b) 2
(c) 3 (d) 4

66. Which of the following statements about mycorrhiza is correct?
(a) Fungus provides N-containing compounds to plant roots.
(b) Roots provide minerals & water to mycorrhizae.
(c) Pinus seeds cannot germinate without mycorrhiza
(d) Both (a) & (c)

67. Absorption of water along with mineral solutes by root hairs is
(a) purely by facilitated diffusion
(b) purely by diffusion
(c) purely by active transport
(d) by a combination of diffusion andactive transport

68. Apoplast is continuous throughout the plant, except at
(a) endodermis (b) casparian strips
(c) plasmodesmata (d) tracheids

69. Apoplastic movement involves
(a) crossing the cell membrane
(b) crossing the tonoplast
(c) crossing the cell wall
(d) Both (a) and (c)

70. Symplastic system of cells is connected through
(a) nucleoplasmic strands extending through phragmosomes
(b) nucleoplasmic strands extending through plasmodesmata
(c) cytoplasmic strands extending through phragmosomes
(d) cytoplasmic strands extending through plasmodesmata

71. Cytoplasmic streaming
(a) helps in long distance transport
(b) can be seen in Hydrilla leaf
(c) may be part of symplastic movement
(d) Both (b) & (c)

72. Most of the water flow in roots occurs via
(a) symplast through living cells
(b) symplast through cortical cells
(c) apoplast through intercellular spaces
(d) apoplast through endodermal cells

Topic 13 Water Movement a Plant

73. 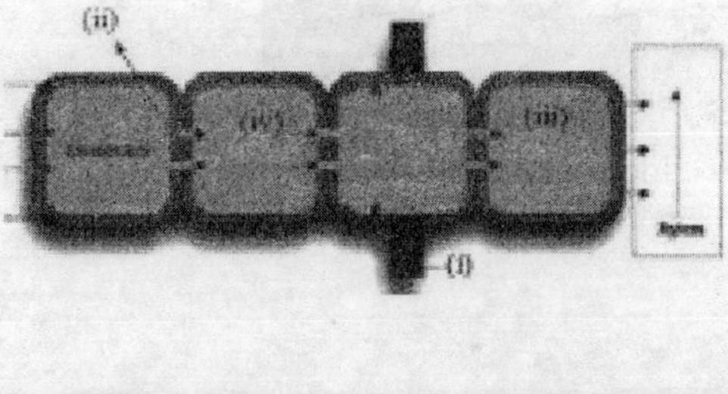

Identify the correct labels in given figures

(a) (i)-Cortex, (ii)-Plasmodesmata, (iii)-Endodermis, (iv)-pericycle

(b) (i)-Casparian strips, (ii)- Plasmodesmata, (iii)-Pericycle, (iv)- Cortex

(c) (i)-Cytoplasmic extension, (ii)-Cortex, (iii)-Endodermis, (iv)-Pericycle

(d) (i)-Endodermis (ii)-Cytoplasmic extension, (iii)-Cortex, (iv)-Pericycle

Topic 14 Root Pressure

74. Root pressure is caused by
   (a) active transport of ions into roots
   (b) passive transport of water into roots
   (c) active transport of water into roots
   (d) both (a) & (b)

75. Root pressure is
   (a) always positive (b) always negative
   (c) mostly positive (d) mostly negative

76. Guttation occurs
   (a) due to negative root pressure and low evaporation
   (b) due to positive root pressure and low evaporation
   (c) due to negative root pressure and high evaporation
   (d) due to positive root pressure and high evaporation

77. Root pressure contributes in
   (a) majority of plant water transport
   (b) transpirational pull
   (c) re-establishing the continuous chain of water in xylem
   (d) Both (a) & (c)

78. 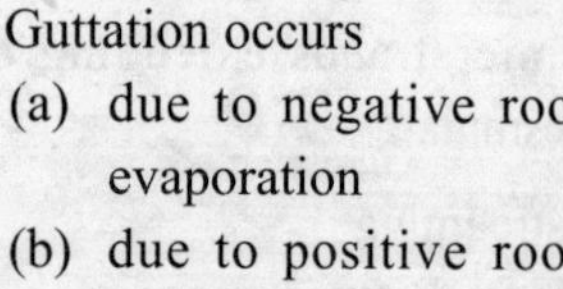

Identify the correct labeles for given figure.

(i)	(ii)	(iii)	(iv)
(a) Apoplastic Pathway	symplastic pathway	xylem	phloen
(b) symplastic Pathway	apoplastic pathway	xylem	phloen
(c) Apoplastic Pathway	symplatic pathway	phloem	xylen
(d) symplastic Pathway	Apoplastic pathway	Phloem	xylen

Topic 15 Transpiration Pull

79. Which of the statements is correct?
   (a) About 30% water reaching leaves is transpired
   (b) About 80% water reaching leaves is transpired
   (c) About 90% water reaching leaves is transpired
   (d) About 99% water reaching leaves is transpired

Topic 16 Transpiration

80. Why does water diffuse into the surrounding?
   (a) Due to lower concentration of water vapou in atmosphere.
   (b) Due to lower concentration of water vapou in the sub-stomatal cavity.
   (c) due to lower concentration of water vapou in the intercellular spaces.
   (d) Due to lower concentration of water vapou in the xylem element.

81. 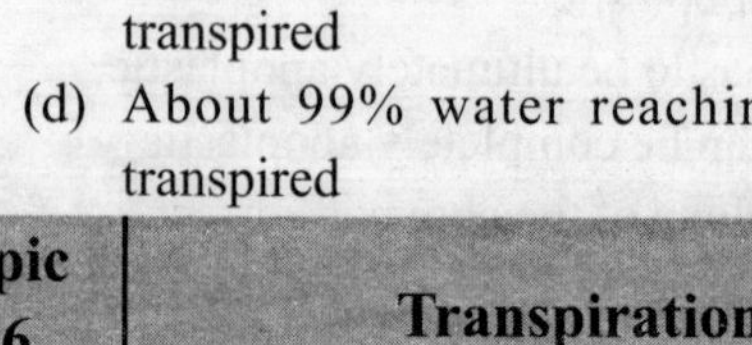

Choose correct option
   (a) The figure shows water movement in a monocot leaf.
   (b) The water moves due to a negative pressure

created by xylem.

(c) The water moves due to a push force created by the water diffusing into surrounding

(d) The figure shows water movement in a dicot leaf.

82. The immediate cause of opening and closing of stomata is

(a) change in CO2 concentration

(b) change in turgidity of guard cell

(c) change in concentration in complementary cell

(d) change in water (moisture) content of air.

83. Stomatal aperture is

(a) cell between guard cells

(b) Pore between guard cell

(c) Inner wall of guard cell

(d) Inner wall of subsidiary cell

84. Inner wall of each guard cell is

(a) thin and inelastic

(b) thin and elastic

(c) thick and inelastic

(d) thick and elastic

85. Microfibrils of cell wall of guard cell are

(a) proteinaceous and radial

(b) proteinaceous and longitudinal

(c) cellulosic and radial

(d) cellulosic and longitudinal

86. Which of these statements about stomata opening or closing is correct?

(a) When guard cells are turgid, stoma closes

(b) When guard cells are flaccid, stoma opens

(c) When guard cells lose turgor, elastic outer walls regain their original shape and stoma closes

(d) None of these

87. How many of the following are plant factors affecting transpiration? Water status of plant, number of stomata, wind speed, light, canopy structure, humidity, temperature

(a) 7 (b) 3

(c) 4 (d) 5

88.

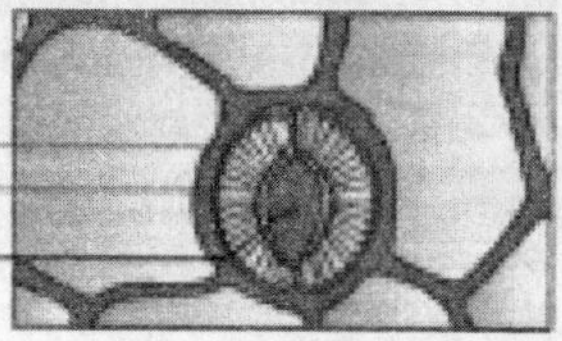

Choose the correct labels for the given figure

	(i)	(ii)	(iii)
(a)	Guard cell	Microfibril	Stomatal aperture
(b)	Microfibril	Guard cell	Stomatal aperture
(c)	Stomatal aperture	Microfibril	Guard cell
(d)	Guard cell	Stomatal aperture	Microfibril

89. Which of the following statements is incorrect?

(a) Cohesion is the property of xylem sap.

(b) Adhesion is the attraction of water molecules to surface of tracheary elements

(c) Adhesion is a property of xylem wall.

(d) Capillarity is the ability of water to ascend in thin tubes.

90. Water is moved from xylem to leaf cells by

(a) a push force of continuously upcoming xylem sap

(b) a push force of water incoming from phloem

(c) a pull force of cohesion between water molecules

(d) a pull force of water diffusing into air.

Topic 17 Transpiration and Photosynthesis --A Compromise

91. Transpiration

(a) supplies water for photosynthesis

(b) decreases water available for photosynthesis

(c) has no relation with photosynthesis

(d) Both (a) & (b)

92. Read the given statements

(i) Transpiration enables mineral transport

(ii) Transpiration maintains shape of plants.
(iii) Transpiration enables light absorption by plants.
(iv) Transpiration causes cooling effect.

Choose the appropriate answer
(a) Statement (i) & (ii) are correct, (iii) & (iv) are wrong
(b) Statement (iii) & (iv) are correct, (i) & (ii) are wrong
(c) Statement (i), (ii), (iii) are correct (iv) is wrong
(d) Statement (i), (ii), (iv), are correct (iii) is wrong

93. Transpiration cools leaf surface by upto
(a) 2–3°C (b) 5–6°C
(c) 10–15°C (d) 18–20°C

94. A limiting factor for photosynthesis is
(a) Water due to evaporation by high light intensity
(b) Water due to transpiration
(c) Sunlight due to cooling effect of transpiration
(d) CO_2 due to excess transpiration

95. Humidity of rainforests is largely due to
(a) Cycling of nutrients from leaf to root
(b) Cycling of water from leaf to root
(c) Cycling of nutrients from root to leaf
(d) Cycling of water from root to leaf

96. Read the given statements
(i) C4 plants loses double the water lost by C3 plant for same amount of CO2 fixed
(ii) C4 plants are more efficient in making sugar than C3 plants.

Choose the best option
(a) I and II are correct
(b) I is correct and II is incorrect
(c) I is incorrect and II is correct
(d) Both I & II are incorrect

Topic 18 Uptake of Mineral Ions

97. Quantity & types of solute reaching xylem are decided at
(a) control points of epidermal cells as the transport starts there
(b) control point of pericycle cells as they surround the xylem
(c) control point of cortical cells as they have large intercellular spaces.
(d) control point of endodermal cells as they have specific transport proteins and suberin.

98. Layer of suberin present in root
(a) in cortical possess ability to passively transport ions in one direction only.
(b) can actively transport selected ions in both the directions.
(c) can transport ions actively in one direction only.
(d) can transport ions in multi directions actively

99. All minerals cannot be passively absorbed by roots because
(a) They are present as ions which cannot more a cross cell membrane.
(b) Concentration of minerals in soil is higher than the concentration of minerals in roots
(c) Both (a) & (b)
(d) all minerals are passively absorbed

100. Which of the given statements is correct about uptake of mineral ions?
(a) Active uptake of ions is helps in uptake of water actively
(b) Passive uptake of ions helps in uptake of water actively
(c) Active uptake of ions help in uptake of water passively.
(d) Passive uptake of ions help in uptake of water passively.

101. Ions are absorbed from soil by
(a) active transport only
(b) passive transport only
(c) mostly passive transport
(d) both active & passive transport

102. The transport proteins embedded in the plasma membrane of endodermal cells
(a) allow all types of solute to pass to xylem.
(b) allow some solutes to cross the membrane
(c) shows no selectively
(d) Both (a) & (c)

Topic 19 Translocation of Mineral Ions

103. Which of the following is correct
(a) most of the P and S are carried as organic compounds
(c) most of P and S are carried as inorganic compounds.
(c) Little of P and S are carried as organic compounds.
(d) Both (b) & (c)

104. Read the given statements
(i) Some exchange of materials occur between xylem and phloem
(ii) We cannot say that xylem transports only inorganic nutrients.

Choose the correct option.
(a) I and II both are correct and II is correct explanation for I
(b) I and II both are correct but II does not explain I
(c) I is correct and II is incorrect
(d) I and II both are incorrect

105. After the ions reach xylem, their further transport to all parts of plant is through
(a) diffusion (b) active transport
(c) translocation (d) All of the above

106. Chief sinks for mineral elements are the following except
(a) root hairs (b) fruits
(c) seeds (d) shoot tip

107. Unloading of mineral ions occurs at the fine vein ending through (i) and uptake by cells is (ii)

	(i)	(ii)
(a)	active transport	diffusion
(b)	diffusion	actively
(c)	active transport	passive
(d)	diffusion	passive

108. Mineral ions are
(a) rarely remobilized, from mature parts to young parts.
(b) rarely remobilized from young leaves to old parts
(c) frequently remobilized from older parts to young parts
(d) frequently remobilized from young parts to older parts

109. Elements most readily mobilized
(a) phosphorous (b) potassium
(c) calcium (d) both (a) & (b)

110. Most of the nitrogen travels through xylem as
(a) inorganic ions
(b) inorganic complex
(c) organic compounds
(d) both (a) & (b)

Topic 20 Phloem Transport--Flow from Source to Sink

111. Which of the following statements about source and sink is incorrect
(a) Source is the part which produces food, like leaf
(b) Sink is the part which needs food
(c) Leaf can never be a sink
(d) Roots may act as a source

112. Choose the correct option
(a) Movement in phloem is bidirectionalleft & right, while movement in xylem is unidirectional upward

(b) Movement in phloem is unidirectional upward while movement in xylem is bidirectional left & right
(c) Movement in phloem is bidirectional up & down while movement in xylem is unidirectional up
(d) movement in phloem is unidirectional upward while movement in xylem is bidirectional up & down.

113. Phloem sap includes
(a) water, sucrose, hormones
(b) sucrose only
(c) water and sucrose but no hormones
(d) sucrose and hormones but no water

114. Amino acids are transported by
(a) xylem only
(b) phloem only
(c) both xylem & phloem
(d) neither xylem nor phloem

Topic 21 The Pressure Flow or Mass Flow Hypothesis

115. Choose the correct option for phloem transport
(a) loading is active process and unloading is passive
(b) loading is passive process and unloading is active
(c) Both loading and unloading are active
(d) Both loading and unloading are passive

116. Choose the incorrect option for phloem transport
(a) transport occurs due to high osmotic pressure in phloem
(b) transport occurs due to low osmotic pressure in xylem.
(c) transport occurs due to water potential gradient
(d) transport occurs due to pressure potential gradient

117.

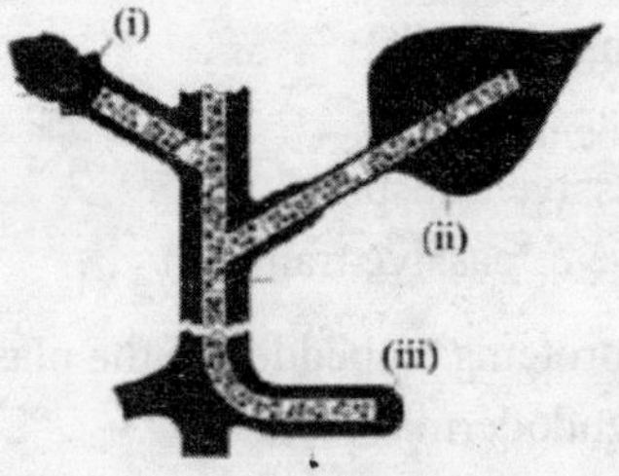

Identify the correct match
(a) Sugar enter sieve tube, water follows by osmosis
(b) Sugar leaving sieve tube, water follow by osmosis

	(i)	(ii)	(iii)
(a)	A	B	B
(b)	B	A	A
(c)	A	B	A
(d)	B	A	B

118. Girdling experiment was used to identify
(a) tissue for water transport
(b) tissue for food transport
(c) tissue for mineral transport
(d) Both (a) and (c)

119. In girdling experiment, a ring of bark is cut upto the depth of
(a) the xylem layer
(b) the phloem layer
(c) the cortex layer
(d) the endodermal layer

120. In girdling experiment, the stem swells
(a) above the ring, and transport is unidirectional towards roots
(b) above the ring, and transport is unidirectional towards shoot
(c) below the ring, and transport is unidirectional towards shoot
(d) Both above and below the ring, and transport is bidirectional towards root and shoot.

121. The sugar mainly transported is
   (a) Glucose (monosaccharide)
   (b) Glucose (disaccharide)
   (c) Sucrose (disaccharide)
   (d) Sucrose (monosaccharide)

122. Glucose is converted to sucrose at
   (a) the source
   (b) the sink
   (c) midway in xylem during transport
   (d) Sometimes at source and other times at sink

123. Select the correct pathway for sugar transport by phloem as per the mass flow hypothesis
   (a) Sucrose Sieve tube Companion cells via diffusion
   (b) Sucrose sieve tube companion cells via active transport
   (c) Sucrose Companion cells Sieve tube via diffusion
   (d) Sucrose Companion cells Sieve tube via active transport

124. Loading at source during phloem translocation
   (a) makes phloem hypertonic and attract water from xylem
   (b) makes phloem hypotonic and attract water from xylem
   (c) makes phloem hypertonic and attract water from shoot cells
   (d) makes phloem hypotonic and attract water from shoot cells

ANSWER KEY

1. (a)	2.(b)	3. (c)	4. (b)	5. (a)	6. (d)	7. (c)	8. (b)	9. (b)	10. (a)
11. (a)	12. (b)	13. (a)	14. (b)	15. (b)	16. (c)	17. (b)	18. (b)	19. (a)	20. (c)
21. (d)	22. (a)	23. (a)	24. (a)	25. (a)	26. (c)	27. (d)	28. (b)	29. (c)	30. (d)
31. (a)	32. (a)	33. (c)	34. (c)	35. (a)	36. (b)	37. (b)	38. (a)	39. (a)	40. (a)
41. (b)	42. (d)	43. (c)	44. (a)	45. (d)	46. (b)	47. (b)	48. (a)	49. (b)	50. (a)
51. (c)	52. (b)	53. (a)	54. (c)	55. (d)	56. (c)	57. (b)	58. (d)	59. (c)	60. (d)
61. (d)	62. (c)	63. (c)	64. (d)	65. (c)	66. (c)	67. (b)	68. (b)	69. (c)	70. (d)
71. (d)	72. (c)	73. (b)	74. (a)	75. (a)	76. (b)	77. (b)	78. (b)	79. (d)	80. (a)
81. (d)	82. (b)	83. (c)	84. (d)	85. (c)	86. (d)	87. (b)	88. (b)	89. (c)	90. (d)
91. (a)	92. (d)	93. (c)	94. (b)	95. (d)	96. (c)	97. (d)	98. (c)	99. (a)	100. (c)
101. (d)	102. (b)	103. (d)	104. (a)	105. (c)	106. (a)	107. (b)	108. (c)	109. (d)	110. (c)
111. (c)	112. (c)	113. (a)	114. (c)	115.(c)	116. (b)	117. (d)	118. (b)	119. (b)	120. (a)
121. (c)	122. (a)	123. (d)	124. (a)						

9 Chapter Mineral Nutrition

MULTIPLE CHOICE QUESTIONS

Topic 1 — Methods to Study the Mineral Requirements of Plants

1. Hydroponics helps in
   (a) identification of essential elements
   (b) discovery of deficiency symptoms of essential elements
   (c) growing some commercial crops like tomatoes
   (d) all of these

2. Hydroponics has been successfully employed as a technique for commercial production of
   (a) Vegetables such as tomato
   (b) Seedless cucumber
   (c) lettuce
   (d) all of these

3. Hydroponics was first time demonstrated by
   (a) Julius von Sachs, a prominent German botanist in 1860.
   (b) Julius von Sachs, a prominent French botanist in 1860.
   (c) Melvin Calvin, a prominent French botanist in 1960.
   (d) None of these

Topic 2 — Essential Mineral Elements

4. Plants growing near nuclear test sites take up
   (a) Selenium (b) Strontium
   (c) gold (d) none of these

5. Hydroponics are techniques that are able to detect the minerals even at a very low concentration upto
   (a) 10–8g/ml (b) 10–9g/ml
   (c) 108g/ml (d) 10–6g/ml

Topic 3 — Criteria for Essentiality

6. Which of following does play important roles in opening and closing of stomata
   (a) Potassium (b) Phosphorous
   (c) Calcium (d) magnesium

7. Which of following is part of chlorophyll and ATP respectively.
   (a) Manganese, phosphorous
   (b) Magnesium, phosphorous
   (c) Manganese, Potassium
   (d) Magnesium, Potassium

8. Which of following is used to categorize essential elements.
   (i) components of biomolecules
   (ii) Activate or inhibit enzymes.
   (iii) Osmotic potential of a cell role.
   (iv) Components of energy-related chemical compounds.
   (a) i, ii only (b) iii & iv only
   (c) i, iii & iv (d) all of these

9. The criteria for essentiality of an element are:
   (a) The element must be absolutely necessary for supporting normal growth and reproduction
   (b) The requirement of the element must be specific and not replaceable by another element.

(c) The element must be directly involved in the metabolism of the plant.
(d) all of these

10. Macronutrients are not
(a) Generally present is plant tissues in large amounts in excess of 10 m mole/kg of dry matter.
(b) Carbon, Hydrogen & oxygen are mainly obtained from CO_2 and H_2O.
(c) Phosphorous, sulphur, potassium, calcium are macronutrient.
(d) Manganese, which is absorbed from soil as mineral nutrition.

11. How many essential elements are classified into macronutrient and micronutrient.
(a) 9 (b) 17
(c) 8 (d) 10

12. How many essential elements are called non-mineral elements
(a) 3 (b) 5
(c) 7 (d) 9

13. Match the column-I & II

Column I	Column II
(a) Mo	(i) RubisCo
(b) Mg^{2+}	(ii) Alcohol dehydrogenase
(c) Zn^{2+}	(iii) Nitrogenase

(a) A-i, B-ii, C-iii (b) A-iii, B-i, C-ii
(c) A-ii, B-iii, C-i (d) A-iii, B-ii, C-i

14. Mg^{2+} act as
(a) activator for enzymes phosphoenol pyruvate carboxylase during photosynthetic carbon fixation.
(b) inhibitor for Ribulose bisphosphate carboxylase-oxygenase during photosynthetic carbon fixation.
(c) inhibitor for enzyme phosphoenol pyruvate carboxylase during photosynthetic carbon fixation
(d) b and c both

Topic 4 Role of Macro- and Micro Nutrients

15. Redox reaction is performed by
(a) Copper (b) Iron
(c) Both a & b (d) none of these

16. Synthesis of auxin need
(a) Zinc (b) Copper
(c) Iron (d) Chlorine

17. The best defined function of manganese is
(a) Splitting of water that liberate oxygen during photosynthesis
(b) Combination of OH– and H+ that leads to formation of water.
(c) aerobic respiration
(d) fermentation

18. Choose correct set about Boron
(i) It absorbed as B72-
(ii) It absorbed as B4O33-
(iii) It required for uptake & utilization of ZN2+
(iv) Pollen germination
(v) Carbohydrate translocation.
(a) I, ii, iii, iv, v (b) iii, iv, v
(c) iv, v (d) none of these

19. Boron is required for uptake & ultilization of
(a) Ca2+ (b) Mn2+
(c) Mg2+ (d) all of these

20. Anion-cation balance in cell maintained by
(a) Potassium (b) phosphorous
(c) Nitrogen (d) Sulphur

21. Which of following involved is protein synthesis
(a) Phosphorous (b) Potassium
(c) Calcium (d) Iron

22. Choose incorrect about function of calcium in plants
(a) It accumulate in newly formed leaf.
(b) Synthesis of cell wall.
(c) formation of spindle fibres
(d) none of these

23. How many of enlisted elements are/is major constituents of nucleic acids. Potassium, Phosphorous, Calcium, Magnesium, Nitrogen
(a) 2 (b) 3
(c) 4 (d) 1

24. Plant obtain sulphur in the form of
(a) Sulphide (b) Sulphate
(c) Sulphite (d) a & b

25. Plant obtain iron in the form of
(a) ferrous (b) ferrate
(c) ferric ions (d) a & b

26. Magnesium help in maintaing structure of
(a) cell wall (b) cell membrane
(c) Ribosome (d) Lysosome

27. In how many amino acid does sulphur present.
(a) 0 (b) 1
(c) 2 (d) 5

28. Main constituent of biotin is
(a) Sulphur
(b) Same as main constitute of thiamine
(c) Retinol
(d) a & b both

29. Plant obtain calcium is form of
(a) calcium ions (b) calcium solid
(c) calcium (d) none

30. Which of following is present in ferredoxin.
(a) Sulphur (b) Iron
(c) Both a & b (d) None

31. Coenzyme A consist of
(a) Magnesium (b) Sulphur
(c) Iron (d) Potassium

32. Which of following does plant obtain as monovalent ion
(a) Manganese (b) Zinc
(c) Copper (d) Chlorine

33. Which of following does not take part is nitrogen metabolism
(a) Nitrogenase
(b) Nitrate reductase
(c) Molybdenum
(d) nitrate oxygenase

Topic 5 — Deficiency Symptoms of Essential Elements

34. Chlorosis is
(a) loss of chlorophyll leading to yellowing of leaves
(b) caused by deficiency of elements like N, K, Mg, S, Mn, Fe, Zn, MO and Cu.
(c) A & B both
(d) none of these

35. Deficiency of Ca, Mg, Cu, K lead to
(a) Chlorosis
(b) Necrosis
(c) inhibition of cell division
(d) no of these

36. Inhibition of cell division is not related to deficiency of
(a) S (b) Cu
(c) K (d) Mo

37. Given below are list of element N, K, Mg, Ca, Cu, B, Fe, Mn, Zn, Mo

How many of them are related to necrosis, chlorosis and inhibition of cell division respectively.
(a) 8, 4, 4 (b) 4, 8, 4
(c) 4, 4, 8 (d) none of these

38. Whenever the supply of an essential elements becomes limited, plant growth is
(a) accelerated (b) retarded
(c) no effect (d) optimum

39. Critical concentration is
(a) The concentration of the essential element below which plant growth is retarded.
(b) the concentration of the non-essential element below which plant growth is accelerated.
(c) The concentration of the essential element above which plant growth is accelerated.

(d) The concentration of the essential element below which plant growth is accelerated.

40. Deficiency symptoms is
(a) Morphological changes that indicate certain element deficiencies
(b) It is not varying from element to element
(c) Never disappear when the deficient mineral nutrient is provided to plant.
(d) all of these

41. For elements that are actively mobilized within the plants, and exported to young developing tissues, the deficiency symptoms tend to appear
(a) first in meristematic tissue
(b) first in older tissue
(c) first in leaf
(d) all of these

42. Which of following element deficiencies are visible in senescent leaves
(a) Nitrogen (b) Potassium
(c) Magnesium (d) all of these

Topic 6 Toxicity of Micronutrient

43. The requirement of micronutrients is always in …….concentration their moderate decrease cause the …… and a moderate increase cause ……
(a) high, deficiency symptoms, toxicity
(b) low, deficiency symptoms, toxicity
(c) low, toxicity symptoms, deficiency symptoms
(d) high toxicity, symptoms, deficiency symptoms

44. Any mineral ion concentration is tissues that reduced the dry weight of tissues by about how much present is considered toxic.
(a) 10% (b) 20%
(c) 2% (d) more than 50%

45. Choose correct statements
(i) The prominent symptom of manganese toxicity is appearance of brown spots surrounded by chlorotic venis.
(ii) Manganese compete with magnesium for binding with enzymes.
(iii) Manganese inhibit calcium translocation in Root apex
(iv) Excess of manganese may in fact induce deficiencies of iron, magnesium & calcium
(a) i, ii, iii, iv (b) i, ii, iv only
(c) i, iii only (d) iv only

Topic 7 Mechanism of Absorption of Elements

46. Choose the correct about mechanism of absorption of elements by palnts.
(a) An initial rapid uptake of ions into the 'outer space' of cells i.e. apoplast is passive.
(b) The passive movement of ions into the apoplast usually occurs through ionchannels, the trans-membrane protein.
(c) The entry or exist of ions to and from the symplast is an active process.
(d) all of these

Topic 8 Translocation of Solutes

47. Mineral salts are transported through
(a) Xylem (b) Phloem
(c) Pericycle (d) a & b

Topic 9 Soil as Reservoir or Essential Elements

48. Soil supplies
(a) Mineral salt
(b) Nitrogen-fixing bacteria
(c) hold water
(d) all of these

49. which of following may be supplied by fertilizers
(a) Macronutrients (b) Micronutrient
(c) both A & B (d) none

Topic 10 Metabolism of Nitrogen

50. Formation of amine group of amino acid in leaf is achieved by
(a) oxidation of nitrite
(b) reduction of nitrite
(c) oxidation of nitrate
(d) reduction of nitrate

51. Denitrification is process of
(a) Reduction of nitrate
(b) Reduction of nitrite only
(c) Oxidation of nitrite only
(d) oxidation of nitrate only

52. Denitrifying bacteria are
(a) Thiobacillus & Nitrosomonas
(b) Nitrococcus & Nitrosomonas
(c) Pseudomonas & Thiobacillus
(d) None of these

53. Pseudomonas is responsible for
(a) Ammonification (b) Nitrification
(c) Denitrification (d) All of these

54. Identify A, B, C, D, E

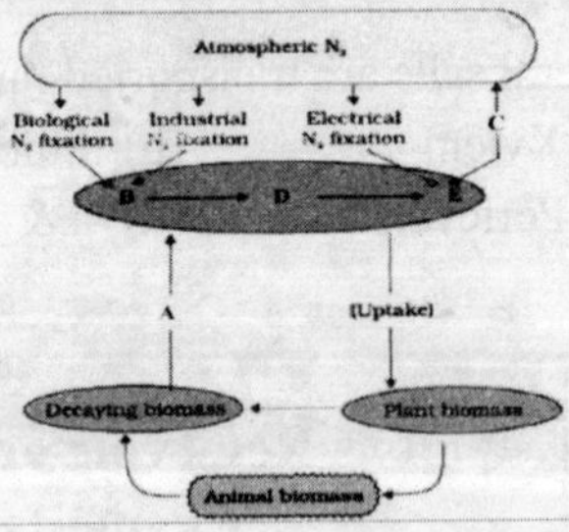

	A	B	C	D	E
A	Ammonification	NO_2-	Denitrification	NO_3-	NH_3
B	Ammonification	NH_3	Denitrification	NO_2-	NO_3-
C	Denitrification	NO_3-	Ammonification	NH_3	NO_2-
D	Denitrification	NH_3	Ammonification	NO_3-	NO_2-

55. Nitrogen is a constituent of
(a) hormones (b) Chlorophyll
(c) Proteins (d) All of these

56. How many from following statement are correct.
(i) Plants compete with microbes for limited nitrogen that is available in soil.
(ii) Nitrogen is limiting nutrient for agricultural ecosystem
(iii) Nitrogen is not limiting nutrient for natural ecosystem.
(iv) Nitrogen is absorbed mainly as NO_2-
(v) Nitrogen is example of macronutrient of plant as manganese because these present is plant tissue in large amounts (in exccss of 10 mmole kg-1 of dry matter)
(a) 4 (b) 2
(c) 3 (d) 5

57. Two nitrogen atoms joined by
(a) Three ionic bond
(b) three covalent bond
(c) three coordination
(d) a & b both

58. Nitrogen fixation is conversion of (N2) into
(a) NO_2 (b) N_2O
(c) NH_3 (d) A and C both

59. Oxidation of ammonia into nitrite is carried out by
(a) Nitrosomonas (b) Nitrobacter
(c) Pseudomonas (d) Thiobacillus

60. Nitrococcus is responsible for
(a) Oxidation of ammonia
(b) Oxidation of nitrite
(c) Reduction of nitrate
(d) formation of NO3-

61. Nitrification is related to
(a) 2NH3 + 3O22NO2- + 2H+ + $2H_2O$
(b) 2NO2-+O22NO3-

(c) Photoautotrophic bacteria
(d) A & B both

62. Plant absorb
(a) nitrate (b) nitrite
(c) Ammonia (d) All of these

Topic 11 Biological Nitrogen Fixation

63. Leg-haemoglobin is not
(a) leguminous haemoglobin
(b) oxygen producer
(c) oxygen scavengers
(d) Make red-pink color appearance of root nodule

64. Production of one molecule of ammonia require
(a) 16 ATP (b) 8 ATP
(c) 8e– (d) a & c both

65. How many electron required to convert (N2) into 2NH3
(a) 4e– (b) 8e–
(c) 16e– (d) 2e–

66. The energy require for N2 fixing comes from
(a) Nitrogenase activity of microbes
(b) Respiration of host cell
(c) Carbon assimilation of host cell
(d) All of these

67. Reductive amination require
(i) Ammonium ion
(ii) NADP
(iii) Glutamate dehydrogenasc
(iv) Water
(v) α-ketoglutaric acid
(a) i, ii, iii, v only (b) i, iii, v only
(c) ii & iv only (d) all of these

68. Transamination is
(a) Transfer of amino group from one amino acid to keto group of a keto acid.
(b) Transfer of keto group from one amino acid to amino group of a keto acid
(c) Transfer of kept group from keto acid to amino group of an amino acid.
(d) Transfer of amino group from keto acid to keto group of an amino acid.

69. Which of following is Amino donor.
(a) R_1–H | C | NH3– –COO–
(b) R_2–C–COO–
‖ O
(c) H–C–COO–
‖ O
(d) R2–H | C | N3+ – COOH

70. Asparagine & glutamine are
(a) two amide
(b) formed from two amino acid i.e. aspartic acid and glutamate respectively
(c) contain less nitrogen than amino acid.
(d) transported by phloem

71. Ureides have
(a) high carbon to nitrogen ratio
(b) high nitrogen carbon ratio
(c) N/C = 1
(d) none of them

72. Nitrogenase enzymes found exclusively is
(a) prokaryotes (b) eukaryotes
(c) both a & b (d) aquatic

73. Choose correct option.
(a) Azotobacter is free-living N2-fixing aerobic microbe.
(b) Beijerinckia is free-living N2-fixing anaerobic microbe.
(c) Rhodospirillum is aerobic and freeliving N2-fixes
(d) Azotobacter is free-living N2-fixing anerobic microbe.

74. Frankia is associated with …… to form root nodule.
(a) leguminious plants root
(b) Root of non-legumes (Alnus)
(c) A & B both
(d) none of these

75. Rhizobium is
(a) Rod-shaped (b) Coccos shaped
(c) Spirillum shaped (d) comma-shaped

76. Rhizobium is
(a) Rod-shaped
(b) symbiotic
(c) associated with alfalfa
(d) all

77. Arrange the sequence of nodule formation by Rhizobium and root of host plant.

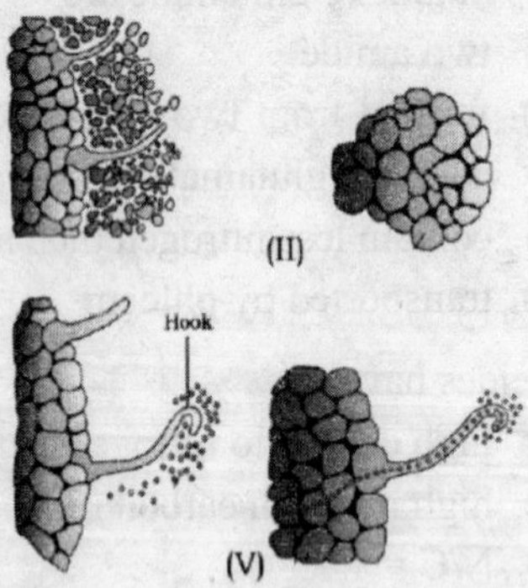

(a) I→III→V→II (b) I→III→II→V
(c) II→III→V→I (d) II→V→III→I

78. Which of following is incorrect about root nodule formation
(a) Rhizobia multiply & colonise the surrounding of roots & get attached to cortical cells.
(b) Root nodule formation initiate in cortex of root.
(c) The bacteria are released from the infection thread in to the cells which leads to the differentiation of specialized N2-fixing cells.
(d) Nodules establishes a direct vascular connection with host for exchange of nutrients.

79. The enzyme nitrogenase
(i) Mo-Fe protein
(ii) Catalyses the conversion of atmospheric nitrogen to ammonia
(iii) highly sensitive to molecular oxygen
(iv) require anaerobic condition
(a) i, ii, iii, iv are correct
(b) i, iii & iv are only correct
(c) ii, iii are incorrect
(d) I, ii are only correct.

ANSWER KEY

1. (d)	2. (d)	3. (a)	4. (a)	5. (a)	6. (a)	7. (b)	8. (d)	9. (d)	10. (d)
11. (b)	12 (a)	13. (b)	14. (d)	15. (c)	16. (a)	17. (a)	18. (c)	19. (a)	20. (a)
21. (b)	22. (a)	23. (a)	24. (b)	25. (a)	26. (c)	27. (c)	28. (b)	29. (a)	30. (a)
31. (b)	32. (d)	33. (d)	34. (c)	35. (b)	36. (b)	37. (b)	38. (b)	39. (a)	40. (a)
41. (b)	42. (d)	43. (b)	44. (a)	45. (a)	46. (d)	47. (a)	48. (d)	49. (c)	50. (d)
51. (a)	52. (c)	53. (c)	54. (b)	55. (d)	56. (c)	57. (b)	58. (c)	59. (a)	60. (a)
61. (d)	62. (a)	63. (b)	64. (b)	65. (c)	66. (b)	67. (b)	68. (a)	69. (a)	70. (a)
71. (b)	72. (a)	73. (a)	74. (b)	75. (a)	76. (d)	77. (a)	78. (a)	79. (a)	

10 Chapter Photosynthesis in Higher Plants

MULTIPLE CHOICE QUESTIONS

Topic 1 What do We Know?

1. Chlorophyll is the-
   (a) Red pigment of leaf of all plants
   (b) Blue pigment of leaf of all plants
   (c) Green pigment of root of all plants
   (d) None of these

2. In the below- mentioned experiment, when a part of the leaf is enclosed in a test tube containing KOH soaked cotton exposed to light will-
   (a) Test positive for starch
   (b) Test negative for starch due to inability to absorb light inside test tube
   (c) Test negative for starch due to inability to absorb CO_2
   (d) Test negative for starch due to absence of water

Topic 2 Early Experiments

3. ______ from a suitable oxidisable compound reduces CO_2 to Carbohydrates.
   (a) Oxygen (b) Hydrogen
   (c) Carbon (d) Both A and B

4. Cladophora is-
   (a) Purple and green bacteria
   (b) Green bacteria
   (c) Red algae
   (d) Green algae

5. A – O_2 evolved by the green plant comes from H_2O, not from carbon dioxide.
   B – This was proved by using Radio isotopic techniques.
   (a) Statement A is wrong and Statement B is right
   (b) Both Statement A and B are wrong
   (c) Statement B is wrong and Statement B is correct
   (d) Both Statement A and B are correct

6. Match the experiment objective with the scientist who performed it-

i	Priestly	I	Production glucose in photosynthesis
ii	Julius von sachs	II	Role of sunlight in photosynthesis
iii	Jan ingenhousez	III	Role of air in photosynthesis

	(i)	(ii)	(iii)
(a)	I	III	II
(b)	II	I	III
(c)	II	III	I
(d)	III	I	II

7. Who performed a series of experiments that revealed the essential role of air on the growth of green plants and when?
   (a) Cornelius van Niel (1787)
   (b) Joseph Priestly (1770)
   (c) T. Engelmann (1756)
   (d) Both A and B

8. Who discovered oxygen and when?
   (a) Joseph Priestly (1770)
   (b) T. Engelmann (1770)
   (c) Jan Ingenhousz (1787)
   (d) Joseph Priestly (1774)

9. Who showed that sunlight is essential to the plant?
   (a) T. Engelmann (b) Joseph Priestly
   (c) Jan Ingenhousz (d) Cornelius van trial
10. part of the plants that could release oxygen.
   (a) Jan Ingenhousz (b) T. Engelmann
   (c) Joseph priestly (d) None of these
11. Match the following

i	First action spectrum	I	Production glucose in photosynthesis
ii	Julius von sachs	II	Role of sunlight in photosynthesis
iii	Jan ingenhousez	III	Role of air in photosynthesis

   (a) A-1, B-4, C-1, D-3,5
   (b) A-3, B-2, C-4, D-1
   (c) A-1, B-3, C-2, D-5
   (d) A-2, B-4, C-1, D-3
12. ________ were used to detect the sites of O2 evolution.
   (a) Bacteria (b) Daze
   (c) Fungi (d) Virus
13. Cornelius van Neil studied on ___ to demonstrate that photosynthesis is essentially a light dependent reaction.
   (a) Cladophora
   (b) Purple and green bacteria
   (c) Red algae
   (d) Both A and B

Topic 3 Where does Photosynthesis Take Place?

14. Sugar is synthesized-
   (a) Non-enzymatically in grana
   (b) Non-enzymatically in stroma
   (c) Enzymatically in grana
   (d) Enzymatically in stroma
15. Which of the following is correct?
   (a) Light reaction depends on dark reaction
   (b) Dark reaction depends on light reaction
   (c) Both of the above
   (d) None of the above
16. If a plant is kept in dark for a long time-
   (a) Starch will be synthesized in chloroplast
   (b) ATP will be synthesized in chloroplast but no starch
   (c) NADPH will be synthesized in chloroplast but no starch
   (d) None of these
17. Choose the incorrect option- During daytime-
   (a) ATP will be synthesized in chloroplast by light reaction
   (b) NADPH will be synthesized in chloroplast by light reaction
   (c) Starch will not be synthesized by dark reaction
   (d) None of these
18. Where does photosynthesis take place?
   (a) Green part of leaves
   (b) Green part of stem
   (c) Brown part of stem
   (d) Both A and B
19. Assertion – Chloroplasts usually align themselves along the walls of mesophyll cells.

   Reason – They get optimum quantity of incident light by aligning along well.

   Choose the correct option.
   (a) Assertion and Reason are correct and Reason is correct explanation for Assertion
   (b) Assertion and Reason are correct but is not the explanation of Assertion
   (c) Assertion and Reason are both incorrect
   (d) Assertion is correct but Reason is incorrect
20. Chloroplast is-
   (a) Single membrane organelle

(b) Double membrane organelle
(c) Triple membrane organelle
(d) Not an organelle

21. Choose the incorrect statement from the following?
(a) There is no clear division of labour within chloroplast
(b) Chloroplast has membranous system which includes stroma as well
(c) Membrane system is responsible for trapping the light energy
(d) More than one of the above

22. Dark reaction-
(a) is not light-dependent
(b) occurs in darkness
(c) is photochemical reaction
(d) is indirectly light-dependent

23. 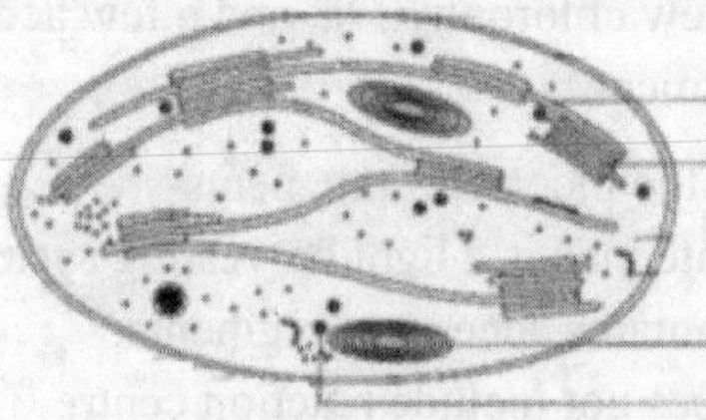

Identify correct labelling.

(a)	Starch granule	Lipid droplet	Stroma lamella	Grana
(b)	Starch granule	Lipid droplet	Grana	Stroma lamella
(c)	Lipid droplet	Starch granule	Grana	Stroma lamellae
(d)	Lipid droplet	Starch granule	Stroma lamella	Grana

Topic 4 How many types of Pigments are involved in Photosynthesis

24. Maximum absorption by chlorophyll a occors in-
(a) blue & green region
(b) ned & green region
(c) blue & red region
(d) yellow & red region

25. Assertion – Chlorophyll 'a' is the chief pigment associated with photosynthesis Reason – Chlorophyll maximum absorption coincides with maximum photosynthesis.
Choose correct option –
(a) Assertion & Reason are correct & Reason is correct explanation of Assertion
(b) Assertion & Reason are correct but Reason is not correct explanation of Assertion
(c) Assertion is correct & Reason is incorrect.
(d) Assertion & Reason are incorrect.

26. Accessory pigments include
(a) Chlorophyll a (b) Chlorophyll b
(c) both of these (d) None of these

27. Accessory pigments
(a) Pass on the energy to chl 'a'
(b) pass on the energy to primary acceptor
(c) Use energy for photolysis of water
(d) more than one option

28. Advantages of accessory pigments include –
(a) they help by photolyzing the water
(b) they protect chl 'a' from photooxidation
(c) they enable narrower range of wavelength of incoming light to be used for photosynthesis
(d) both a & b

29. The colour of leaf is due to-
(a) Chlorophyll only
(b) Chlorophyll, carotenoids only
(c) Chlorophyll, carotenoids, xanthophyll
(d) None of these

30. Leaf pigments are separated by-
(a) Crystallization
(b) Gel electrophorese
(c) Blotting
(d) Paper chromatography

31. Match the pigment with its colour –

I	II
(i) Chlorophyll a	(A) Blue green

(ii) Chlorophyll b (B) Yellow
(iii) Carotenoids (C) Yellow-green
(iv) Xanthophyll (D) Yelloworange

	(i)	(ii)	(iii)	(iv)
(a)	A	C	B	D
(b)	A	C	D	B
(c)	A	D	B	C
(d)	A	D	C	B

32. Which is the most abundant plant pigment in the world?
(a) Chlorophyll a (b) Chlorophyll b
(c) Carotenoids (d) xanthophylls

33.

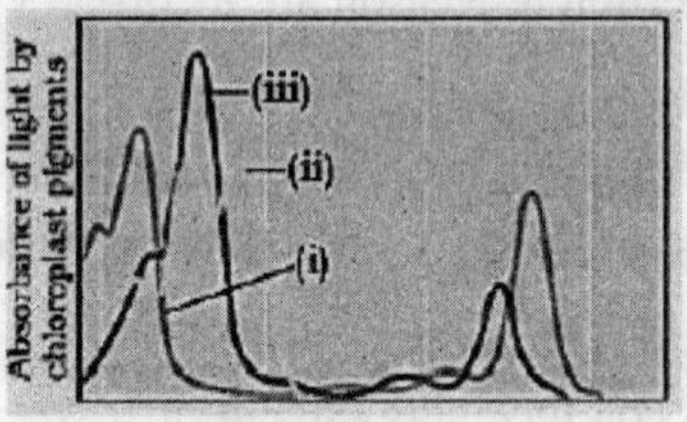

Identity correct

	i	ii	iii
A	Chl a	Chl b	Carotenoi d
B	Carotenoi ds	Chl a	Chl b
C	Chl b	Chl a	Carotenoi ds
D	Chl b	Carotenoi ds	Chl a

Topic 5 What is Light Reactions?

34. Light Reaction is also Known as-
(a) photochemical phase
(b) biosynthetic phase
(c) both of these
(d) None of these

35. Choose correct order of events in light reaction-
(i) ATP & NADPH formation
(ii) Water Splitting
(iii) Oxygen release
(iv) Light absorption
(a) III IV II I (b) IV III I III
(c) IV II III I (d) II III IV I

36. LHC stands for-
(a) Late Harvesting Complex
(b) Light Harvesting Complex
(c) Light Hanging Complex
(d) Late Hanging Complex

37. The naming of PS I & PS II was based on-
(a) their discovery order
(b) their functioning sequence
(c) the scientist who named it
(d) the components of the photosystem.

38. Which of the following is correct?
(a) PS I is called P800
(b) PS II is called P680
(c) Both a & b
(d) None of these

39. Reaction Centre is formed by- (Pg. 211, E)
(a) Only one chlorophyll 'a' molecule
(b) A few chlorophyll 'a' molecule
(c) One chlorophyll 'a' and a few accessory
(d) A few chlorophyll 'a' and a few accessory pigments.

40. Choose the incorrect statements-
(a) Antennae is a light harvesting system
(b) Contains accessory pigments
(c) Does not include reaction centre
(d) None of these

Topic 6 The Electron Transport

41. Electrons from PS-I move downhill to a molecule of energy-rich-
(a) NADP+ (b) NAD+
(c) FAD+ (d) GTP

42 The Z scheme is named so because-
(a) it was discovered by a scientist with 'Z' as initial letter of name
(b) the carriers of ETS present in thylakoid membrane are in 'Z' shape.
(c) it forms 'Z' shape when the carriers of ETS are arranged in sequence on redox potential scale.
(d) both A & C

43.

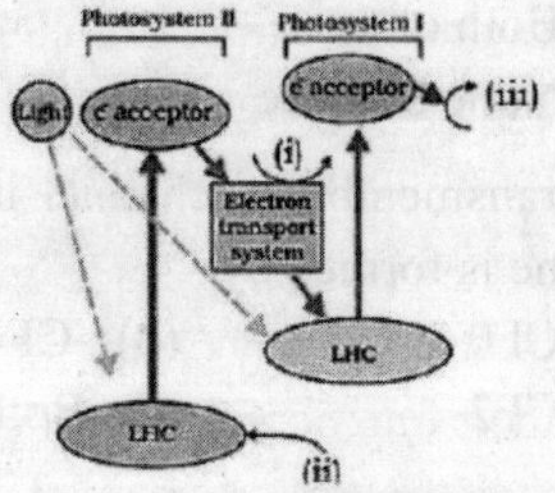

Identify the correct site for I-water splitting, II-NADP+ reduction, IIIATP synthesis

	I	II	III
(a)	I	II	III
(b)	II	I	III
(c)	III	II	I
(d)	III	I	II

44. When the light energy is absorbed by PSII, it is-
(a) Converted to mechanical energy
(b) Used to excite electrons
(c) Used to change configuration of RUBisCO
(d) Both a & c

45. The movement of excited electrons in Noncyclic Photophosphorylation:
(a) uphill in terms of reduction potential scale
(b) downhill in terms of reduction potential scale
(c) uphill and downhill in terms of oxidation-reduction potential scale
(d) both A and C

46. The electrons excited form PS II-
(a) get used up by the first electron acceptor.
(b) get used up in the middle of its ETS pathway to PS I
(c) get passed on to pigments of PS I
(d) get partially used up in ETS and the rest is passed to PS I.

47. The electrons passed on by PS I to electron acceptor are-
(a) the ones that were transferred to PS I from PS II
(b) electrons from the water splatted.
(c) electrons excited when PS I absorbs light.
(d) All of these

Topic 7 Splitting of Water

48. Splitting of water is important-
(a) for the O2 formation
(b) for the electrons released
(c) for the hydroxide ions released
(d) for the H2 released

49. Water splitting is associated with-
(a) PS I located on inner side of thylakoid membrane
(b) PS II located on inner side of thylakoid membrane
(c) PS I located on Outer stroma lamellae
(d) PS II located on outer stroma lamellae membrane

50. The O_2 is released in-
(a) Lumen of thylakoid
(b) Outer side of thylakoid
(c) Stroma
(d) Cytoplasm

51. Protons are released-
(a) Lumen of thylakoid
(b) Outer side of thylakoid
(c) Stroma
(d) Cytoplasm

Topic 8 Cyclic and Non-Cycle Photophosphorylation

52. Cyclic phosphorylation ends in formation of-
(a) ATP only
(b) Glucose only
(c) NADPH + H+ only
(d) Both A & C

53. A possible location of cyclic photophosphorylation under full light is condition.
(a) Stroma
(b) Stroma lamellae
(c) Cristate
(d) Outer membrane of chloroplast.

54. Assertion – Cyclic photophosphorylation occurs in stroma

Reason – Stroma membrane lacks PS II and NADP reductase.

Choose correct answer-
(a) Both Assertion & Reason are correct and Reason is correct explanation for A
(b) Both Assertion & Reason are correct but Reason is not correct explanation for Assertion
(c) Assertion is correct but Reason is wrong
(d) Both Assertion & Reason are wrong

55. The process of ATP synthesis in cells is/are-
(a) Photo-phosphorylation
(b) Oxidative phosphorylation
(c) Phosphosynthesis
(d) Both A & B

56. The order of working of the two photosystems is-
(a) PS I → PS II
(b) PS II → PSI
(c) Any of these depending upon location
(d) None of these

57. Non-cyclic photo-phosphorylation involves-
(a) PS I
(b) PS II
(c) Both PS I & PS II
(d) None of the these, only enzymes in stroma

58. End product of Z-scheme is-
(a) ATP (b) Glucose
(c) NADH + H+ (d) Both A & C

59. Cyclic photophosphorylation involves-
(a) PS I only (b) PS II only
(c) Both PS I & PS II (d) None of these

Topic 9 Chemiosmotic Hypothesis

60. ATP formation occurs upon-
(a) build up of potential gradient
(b) break down of potential gradient
(c) Both of these
(d) None of these

61. The transmembrane channel in ATP synthase enzyme is formed by-
(a) CF0 (b) CF1
(c) CF2 (d) Both A & B

62. The transmembrane channel allows – across membrane for ATP synthesis
(a) Osmosis of protons
(b) Simple diffusion of proton
(c) Facilitated diffusion of electron
(d) Facilitated diffusion of proton

63.

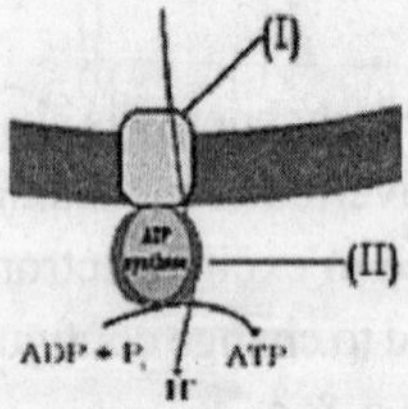

The ATP synthese has two parts I & II

	I	II
(a)	CF0	CF1
(b)	CF1	CF0
(c)	CF	CF0
(d)	None	

64. CF0 is –
(a) embedded in chloroplast membrane
(b) protruding on outer surface of chloroplast
(c) protruding on inner surface of chloroplast membrane
(d) None of these

65. CF1 is-
(a) embedded in chloroplast membrane
(b) protruding on outer surface of chloroplast membrane
(c) protruding on inner surface of chloroplast
(d) None of these

66. Conformational change in ____ makes ATP.
(a) CF0 (b) CF1
(c) Both (d) None

67. For creating proton gradient across thylakoid membrane ______ .
   (a) Energy is used
   (b) No energy is used
   (c) Energy is released
   (d) None of these

68. The end products of light reaction are-
   (a) Stored till dark reaction takes place at right
   (b) immediately used up in next round of light reaction
   (c) transferred to the stroma from lumen to be used in biosynthetic reaction occurring in stroma
   (d) None of these

69. ATP Synthesis is linked to _i_ graduate across a membrane in _ii_

	(i)	(ii)
(a)	Electron	Photosynthesis & respiration
(b)	Electron	Photosynthesis only
(c)	Proton	Photosynthesis and respiration
(d)	Proton	Photosynthesis but not respiration

70. Which of the following statements is true?
   (a) the protons accumulate towards outer side of membrane in photosynthesis
   (b) the protons accumulate to wards inner side (lumen) of thylakoid in respiration.
   (c) The protons accumulate towards outer side of membrane of respiration thylakoid
   (d) None of these

71. The proton gradient may be formed in photosynthesis due to-
   (a) Splitting of water (b) Reduction of NAD+
   (c) Both A and B (d) None of these

72. Assertion – Protons move through photosystems, protons are transported across membrane.

   Reason – Primary acceptor of electrons is an H carrier
   (a) Both Assertion & Reason are correct and Reason is explanation of Assertion
   (b) Both Assertion & Reason are correct and Reason is not the explanation for Assertion
   (c) Assertion is correct but Reason is not correct
   (d) Assertion & Reason both are incorrect

73. Which of these is an H carrier?
   (a) PS II (b) Ferredoxin
   (c) Plastocyanin (d) Plastoquinone

74. NADP reductase enzyme is located on ______ of thylakoid membrane.
   (a) Stroma side (outer side)
   (b) Lumen side (I,e. outer side)
   (c) Stroma Side (i.e. inner side)
   (d) Lumen side (i.e. inner side)

75. The proton gradient is also formed due to-
   (a) reduction of NAD+
   (b) reduction of NADP+
   (c) reduction of both NAD+ & NADP+
   (d) reduction of FAD+

76. Which of the following is true?
   (a) ATP & NADPH + H+ both are formed towards stroma.
   (b) ATP & NADPH + H+ both are formed in lumen of thylakoid
   (c) ATP is formed in lumen while NADPH + H+ is formed towards stroma.
   (d) NADPH + H+ formed in lumen while ATP is formed in stroma.

77.

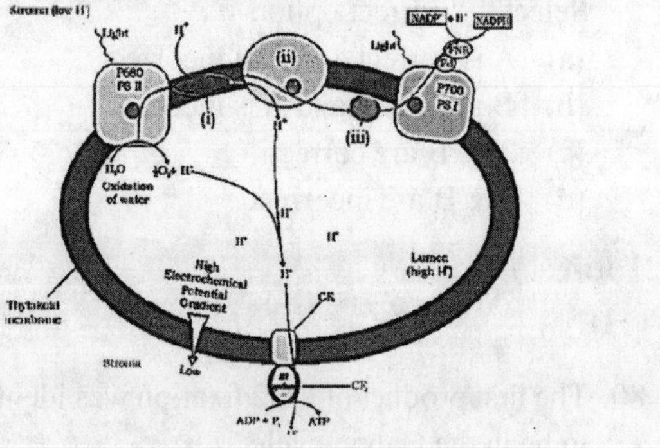

Identify correct labels-

	(i)	(ii)	(iii)
(a)	PQ	PC	Cyt B6f
(b)	Cyt B6f	PQ	PC
(c)	PC	Cyt B6f	PC
(d)	PQ	Cyt B6f	PC

78.

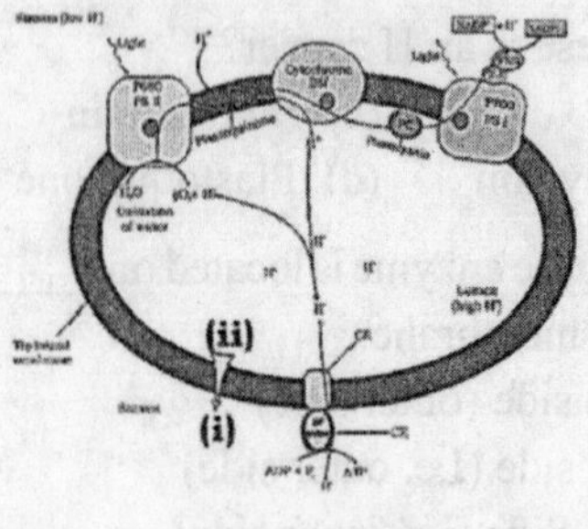

Identify correct -

(a) High electrochemical gradient — Low gradient

(b) Low electrochemical gradient — High gradient

(c) High electro-chemical gradient — High gradient

(d) Low electro-chemical gradient — Low gradient

79. Statement A – The accumulation of protons in stroma cause decrease in pH of stroma.

Statement B – The accumulation of proton in lumen cause formation of potential gradient across thylakoid membrane.

Select the correct option-

(a) A is correct and B is incorrect

(b) B is correct and A is incorrect

(c) A & B are correct

(d) A & B are incorrect

Topic 10 Where are the ATP and NADPH used?

80. The first product of CO2 fixation was identified to be in the Calvin cycle.

(a) PGA (b) RUBP

(c) Citric acid (d) OAA

81. In C4 pathway, first CO2 fixation product is same options as

(a) PGA (b) RUBP

(c) Citric acid (d) OAA

82. OAA and PGA stands for-

(a) 3-phophoglyceric acid and oxalis ascetic acid respectively

(b) 3-peptido glutaric acid and oxalis ascetic acid respectively

(c) 3-phosphas glutamic acid and oxalis ascetic acid respectively

(d) None of these

83. The products of light reaction are-

(a) ATP only

(b) ATP & NADPH

(c) ATP, NADPH, O2

(d) NADPH Only

84. O_2 is-

(a) used up in dark reaction in stroma

(b) used up in dark reaction in lumen thylakoid

(c) diffused out of chloroplast

(d) more than one option is correct

85. Statement A – Biosynthetic reaction is independent of direct presence of light

Statement B – Biosynthetic process continues for some time after the light becomes unavailable and then stops.

(a) Both A & B are correct

(b) A is correct and B is incorrect

(c) A is incorrect and B is correct

(d) A & B are in correct

86. Calvin discovered that first CO2 fixation product is-

(a) 3-carbon organic acid

(b) 4-carbon organic acid

(c) 5- carbon organic acid

(d) 6-carbon organic acid

87. To discover the first CO2 fixation product,

Calvin worked on-using-
(a) algae, radioactive C12
(b) fungi, radioactive C12
(c) algae, radioactive C14
(d) fungi, radioactive C14

Topic 11 The Primary Acceptor of CO_2

88. For a 3 carbon compound to be formed after CO2 fixation, the acceptor molecule is of-
(a) 2-carbon (b) 3-carbon
(c) 4-carbon (d) 5-carbon

89. RUBP stands for-
(a) Ribulose Bisphosphate Carboxylaseoxygenase
(b) Ribose Bisphosphate Carboxylaseoxygenase
(c) Ribulose Bisphosphate
(d) Ribose Bisphosphate Carboxylaseoxygenase

Topic 12 The Calvin Cycle

90. For formation of 1 glucose molecule, how may turns of Calvin cycle is needed?
(a) 3 (b) 1
(c) 2 (d) 6

91. Regeneration of __(i)__ takes place at expense of __(ii)__ ATP & __(iii)__ NADPH.

	(i)	(ii)	(iii)
(a)	RUBP	0	1
(b)	RUBP	1	0
(c)	RUBisCo	0	1
(d)	RUBisCo	1	0

92. Statement A – In CO2 fixation cycle, the molecules of ATP used is more than NADPH used.

Statement B – to meet the difference in number of ATP & NADPH used in dark reaction, cyclic phosphorylation take place.

Choose the correct option-
(a) A is correct but B is wrong
(b) B is correct but A is wrong
(c) A & B are correct and A explains B
(d) A & B are incorrect.

93. For formation of 1 glucose, how many molecules of ATP are required by C3 cycle?
(a) 12 (b) 16
(c) 18 (d) 10

94. For one glucose formation, how many NAD molecules are needed by C3 pathway?
(a) 10 (b) 12
(c) 16 (d) 18

95. The Calvin cycle starts with ______ ends with
(a) RUBisCo, regeneration of RUBisCo
(b) RUBP, regeneration of RUBisCo
(c) RUBP, regeneration of RUBP
(d) PGA, regeneration of PGA

96. Calvin cycle-
(a) occurs in C3 plants only
(b) occurs in C4 plants only
(c) occurs in both C3 & C4
(d) None of these

97.

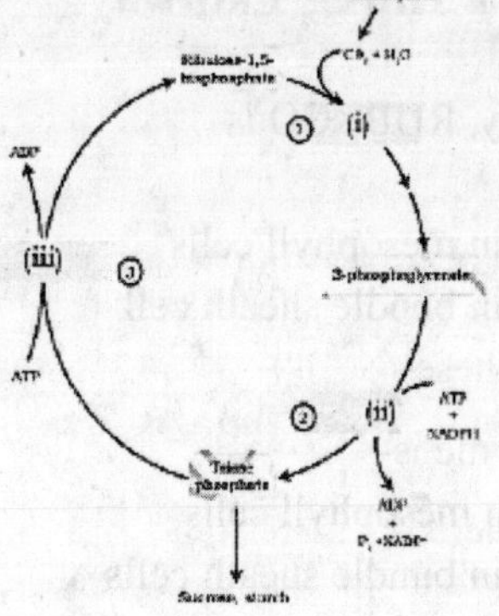

The three steps in Calvin cycle are-

	i	ii	ii
(a)	Carboxylati on	Oxidation	Regeneration
(b)	Reduction	Carboxyl ation	Regeneration

(c) Carboxylation Carboxylation Regeneration

(d) Carbonation Reduction Regeneration

98. The most crucial step of Calvin cycle is-

(a) Carbonation (b) Carboxylation
(c) Reduction (d) Regeneration

99. Carboxylation is catalyzed by the enzyme-

(a) RUBP (b) PEP
(c) NADPH (d) None of these

100. RUBisCo has the activity of-

(a) Carbonation (b) Oxygenation
(c) Oxidation (d) None of these

101. Reduction involves Use of ________ molecules of ATP for fixed CO2 reduction

(a) 1 (b) 2
(c) 3 (d) 4

102. Reduction involves use of – molecules of NADPH to reduce fixed CO2.

(a) 1 (b) 2
(c) 3 (d) 4

Topic 13 The C_4 Pathway

103. In C4 pathway, RUBisCO is-

(a) absent
(b) present in mesophyll cells
(c) present in bundle sheath cell
(d) none of these

104. PEPcase enzyme is-

(a) absent in mesophyll cells
(b) present in bundle sheath cells
(c) both A and B
(d) None of these

105. Calvin cycle takes place in __(i)__ in all C3 plants in __(ii)__ in all C4 plants

	i	ii
(a)	mesophyll cells	mesophyllcells
(b)	bundle sheath cells	mesophyll cells
(c)	mesophyll cells	bundle sheath cells
(d)	bundle sheath cells	bundle sheath cells

106. Identify correct labels-

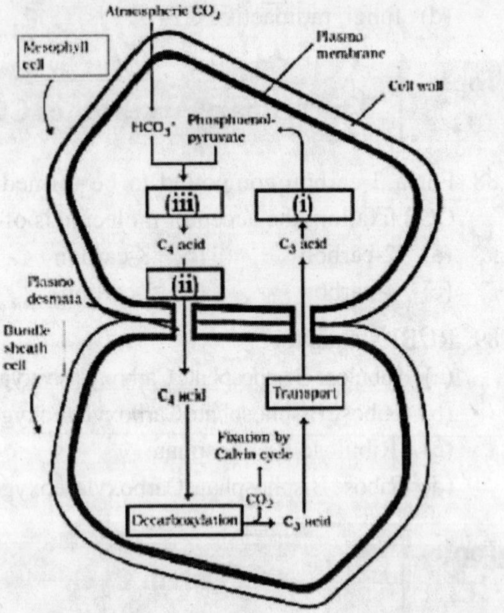

	I	II	III
(a)	Firation	Decarboxylation	Regeneration
(b)	Transport	Firation	Regeneration
(c)	Regeneration	Transport	Fixation
(d)	Regeneration	Decarboxylation	Fixation

107. Identify the A and B

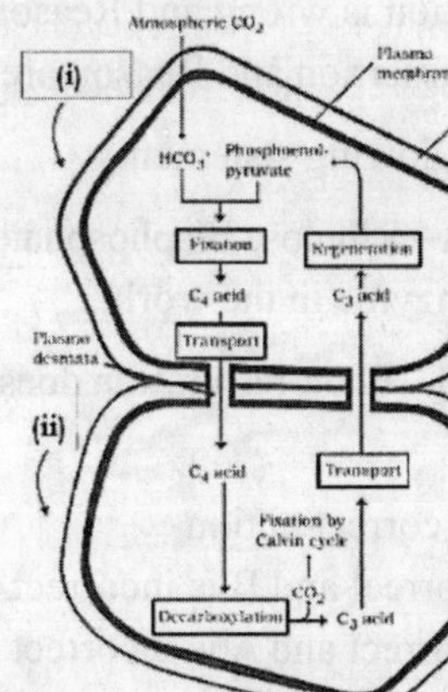

	I	II	III
(a)	Firation	Decarboxylation	Regeneration
(b)	Transport	Firation	Regeneration
(c)	Regeneration	Transport	Fixation
(d)	Regeneration	Decarboxylation	Fixation

108. C_4 plants are adaptation of plants to-
(a) wet regions (heavy rainfall)
(b) polar regions
(c) dry tropics
(d) moist rainforest

109. (i) – C4 plants lack Calvin cycle
(ii) – C4 plants lack photorespiration
(iii) – C4 plants have more productivity than C3 plants
(iv) – C4 plants cannot tolerate higher temperature
How many of the above statements are incorrect?
(a) 0 (b) 1
(c) 2 (d) 3

110. First CO2 fixation product in C4 cycle is-
(a) RBP (b) PEP
(c) OAA (d) Malate

111. Bundle sheath cells are present in ____ around ____.
(a) C4 plants, vascular bundles
(b) C3 plants, vascular bundles
(c) Both of these
(d) None of these

112. Leaves with bundle sheath cells are said to show-
(a) Krant anatomy (b) Kranz anatomy
(c) Kent anatomy (d) Krez anatomy

113. Bundle sheath cells-
(a) Allow gaseous exchange
(b) Have intercellular spaces
(c) Have large number of chloroplasts
(d) All of these

114. Example of C4 plants is-
(a) Rice (b) Maize
(c) Soyabean (d) Both A and C

115. Primary CO_2 acceptor in C4 plants is-
(a) 3-carbon molecule RUBP
(b) 3-carbon molecule PEP
(c) 4-carbon molecule PEP
(d) 4-carbon molecule OAA

116. Enzyme responsible for primary CO2 fixation in C4 plants is-
(a) RUBisCO (b) PEPCase
(c) Oxaloacetase (d) Phenolase

117. Which of the following is true?
(a) C4 plants lack RUBisCO
(b) Mesophyll cells of C4 plants lack RUBisCO
(c) Bundle sheath cells of C4 plants lack RUBisCO
(d) C3 plants lack RUBisCO

118. Primary CO2 fixation occurs in C4 plants in
(a) Bundle sheath cells
(b) Mesophyll cells
(c) Any of the above
(d) None of these

119. CO_2 fixation in C4 plants occurs in-
(a) Bundle sheath cells
(b) Mesophyll cells
(c) Both A and B
(d) None of the above

120. OAA forms other four carbon acids which are transported. They are-
(a) Malic acid and oxalic acid
(b) Malic acid and aspartic acid
(c) Succinic acid and aspartic acid
(d) Succinic acid and glutamic acid

121. The figure shows

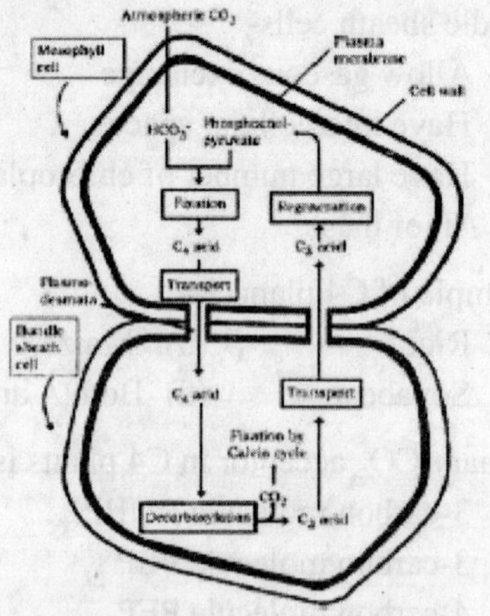

(a) Krebs cycle
(b) Calvin cycle
(c) Hatch and Slack pathway
(d) EMP pathway

Topic 14 Photorespiration

122. In photorespiration- (Pg. 220, E)
(a) Sugar and ATP are formed
(b) Sugar is formed but not ATP
(c) ATP is formed but not sugar
(d) Sugar and ATP are not formed

123. Assertion – Photorespiration doesn't occur in C4 plants.

Reason – CO_2 concentration at enzyme site is high in C4 plants.

Chose correct option-
(a) Both Assertion and Reason are correct and Reason is correct explanation for Assertion
(b) Assertion is correct and Reason is wrong
(c) Assertion is wrong and Reason is correct
(d) Both Assertion and Reason are wrong

124. Read the following statements-

Statement A – Ribulose Bisphosphate is the most abundant enzyme in the world.

Statement B – Photorespiration doesn't occur in C4 plants.

Choose the correct option-
(a) A is correct and B is incorrect
(b) B is correct and A is incorrect
(c) A and B are correct
(d) A and B are incorrect

125. RUBisCO has ________ when CO2 & O2 is equal.
(a) greater affinity for CO2 than O2
(b) greater affinity for O2 than CO2
(c) equal affinity for CO2 & O2
(d) no affinity for O2

126. In C3 plants- (Pg. 220, E)
(a) Some O2 binds to RUBisCO and CO2 fixation increases
(b) no O2 binds to RUBisCO
(c) Some O2 binds to RUBisCO and CO2 fixation decrases
(d) only O2 binds to RUBisCO

127. When O2 binds to RUBisCO, RUBP is converted to ________ and ________.
(a) Phosphoglycerate, phosphoenolpyruvate
(b) Phosphoglycerate, phosphoglycerate
(c) Phosphoenolpyruvate, phosphoglycerate
(d) Phosphoglycolate, phosphoglycerate

Topic 15 Factors Affecting Photosynthesis

128. Photosynthesis is dependent on-
(a) internal factors of plant
(b) external factors of environment
(c) both A and B
(d) None of these

129. How many of the following are factors affecting photosynthesis? age of leaves, orientation of leaf, internal CO2 concentration, amount of chlorophyll, number of leaves
(a) 4 (b) 2
(c) 3 (d) 5

130. Blackman's which law comes into effect when several factors affect any biochemical process?
(a) Law of Limited Components (1910)
(b) Law of Limiting Factors (1910)
(c) Law of Limited Components (1905)
(d) Law of Limiting Factors (1905)

131. According to Blackmann's Law, the rate of chemical process is determined by the factor which
(a) is nearest to its maximum value
(b) is nearest to its minimum value
(c) both A and B
(d) none of these

Topic 16 Light

132. The relationship between incident light and CO2 fixation rate at higher light intensity is-
(a) rate is constant with increasing intensity
(b) rate increases with increasing intensity
(c) rate decreases with increasing intensity
(d) none of these

133. At lower Light intensity, the rate of CO2 fixation-
(a) is constant with increasing light intensity
(b) increasing with increasing light intensity
(c) decreases with increasing light intensity
(d) none of these

134. Light saturation occurs at ________ of full sunlight.
(a) 5% (b) 10%
(c) 20% (d) 40%

135. Very big increase in light intensity causes-
(a) increase in photosynthesis
(b) decrease in photosynthesis
(c) no change in photosynthesis rate
(d) none of these

Topic 17 Carbon dioxide Concentration

136. C_4 plants show saturation at CO_2 concentration of-
(a) 240 μlL^{-1} (b) 360 μlL^{-1}
(c) 450 μlL^{-1} (d) 540 μlL^{-1}

137. C_3 plants show saturation for CO_2 concentration at-
(a) 450 μlL^{-1} (b) 360 μlL^{-1}
(c) 540 μlL^{-1} (d) 240 μlL^{-1}

138. Assertion – Greenhouse crops like tomatoes and bell pepper are grown in CO2 enriched atmosphere for higher yields.

Reason – C4 plants respond to higher CO2 concentration by showing increased rate of photosynthesis.

Choose the correct option.
(a) Both Assertion and Reason are correct and Reason is correct explanation for Assertion
(b) Both Assertion and Reason are correct but Reason is not the explanation for Assertion
(c) Assertion is correct but Reason is wrong
(d) Assertion and Reason are wrong

139. Which of the following is true?
(a) CO_2 is a major limiting factor for photosynthesis in nature
(b) Light is rarely a limiting factor for photosynthesis in nature
(c) Both of these
(d) None of these

140. The concentration of CO2 beyond which the CO2 becomes damaging over longer periods is-
(a) 0.03% (b) 0.04%
(c) 0.05% (d) 0.08%

141. The CO_2 fixation rates increases upto the CO2 concentration of-
(a) 300 ppm (b) 400 ppm
(c) 500 ppm (d) 800 ppm

142. At low light conditions, which of the groups respond positively to increase in CO_2-

(a) C_3 (b) C4
(c) Both (d) None

143. At high light conditions, which of the groups respond positively to increase in CO2.
(a) CO2 (b) C4
(c) Both (d) None

Topic 18 Temperature

144. Water stress causes
(a) reduced CO2 availability
(b) leaf wilting
(c) reduced surface area of leaf
(d) All of these

145.

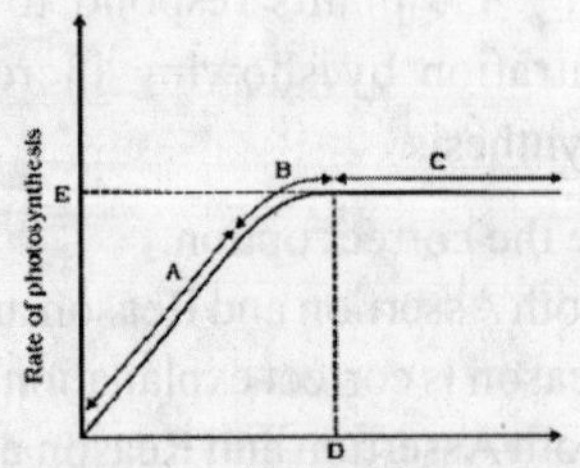

In the above graph, in which region will plant not respond to increase in CO_2 concentration?
(a) A (b) B
(c) C (d) None

146. Which of the reaction is more sensitive to temperature?
(a) Light Reaction
(b) Dark Reaction
(c) Both of then are equal
(d) None of the above

147. Which group of plants has a higher temperature optimum?
(a) C_4 (b) C_3
(c) Both (d) None

148. Which of the following is true?
(a) Tropical plants have higher temperature optimum than temperature plants
(b) Temperature plants have higher temperature optimum than tropical plants
(c) Temperature and Tropical have equal temperature optimum
(d) Polar regions plants have highest temperature optimum

ANSWER KEY

1. (d)	2. (c)	3. (b)	4. (d)	5. (d)	6. (d)	7. (b)	8. (d)	9. (c)	10. (a)
11. (d)	12. (a)	13. (b)	14. (d)	15. (b)	16. (d)	17. (c)	18. (d)	19. (a)	20. (b)
21. (a)	22. (d)	23. (c)	24. (c)	25. (a)	26. (b)	27. (a)	28. (b)	29. (c)	30. (d)
31. (b)	32. (a)	33. (d)	34. (a)	35. (c)	36. (b)	37. (a)	38. (b)	39. (a)	40. (d)
41. (a)	42. (c)	43. (d)	44. (b)	45. (c)	46. (c)	47. (c)	48. (b)	49. (b)	50. (a)
51. (c)	52. (a)	53. (b)	54. (b)	55. (d)	56. (b)	57. (c)	58. (d)	59. (a)	60. (b)
61. (a)	62. (d)	63. (a)	64. (d)	65. (d)	66. (b)	67. (a)	68. (d)	69. (c)	70. (d)
71. (a)	72. (c)	73. (d)	74. (a)	75. (b)	76. (a)	77. (d)	78. (b)	79. (b)	80. (a)
81. (d)	82. (d)	83. (c)	84. (c)	85. (a)	86. (a)	87. (c)	88. (d)	89. (c)	90. (d)
91. (b)	92. (c)	93. (c)	94. (b)	95. (c)	96. (c)	97. (c)	98. (b)	99. (d)	100. (b)
101. (b)	102. (b)	103. (c)	104. (d)	105. (c)	106. (c)	107. (c)	108. (c)	109. (c)	110. (c)
111. (a)	112. (b)	113. (c)	114. (b)	115. (b)	116. (b)	117. (b)	118. (b)	119. (c)	120. (b)
121. (c)	122. (d)	123. (a)	124. (c)	125. (a)	126. (c)	127. (d)	128. (c)	129. (d)	130. (d)
131. (b)	132. (a)	133. (b)	134. (b)	135. (b)	136. (b)	137. (a)	138. (c)	139. (c)	140. (c)
141. (c)	142. (d)	143. (c)	144. (d)	145. (a)	146. (b)	147. (a)	148. (a)		

11 Chapter Respiration in Plants

MULTIPLE CHOICE QUESTIONS

1. Compounds that are oxidised during the process of respiration is called?
   (a) Respiratory index
   (b) Reductory substrate
   (c) Respiratory quotient
   (d) Respiratory substrate

2. Statement I – Only green plants and cyanabacteria can prepare their own food by photosynthesis.

   Statement II – Only green plants and cyanobacteria can prepare their own food by converting chemical energy to light energy

   Which of the statements is/are true?
   (a) Only I (b) Only II
   (c) Both of these (d) None of these

3. "Ultimately all the food that is respired for life processes comes from photosynthesis." The above statement is –
   (a) correct
   (b) incorrect
   (c) partially correct
   (d) can't be said as it is incomplete

4. Which of the following cannot be used as respiratory substances in plants under any conditions?
   (a) fat (b) protein
   (c) carbohydrate (d) none of these

5. Respiration is defined as –
   (a) Formation of C – C bonds of complex compound
   (b) Breaking of C – C bonds of complex compound
   (c) Breaking of C – N bonds of complex compound
   (d) All of the above

6. Respiration results finally to a formation and release of which among the following?
   (a) NADPH (b) Glucose
   (c) ATP (d) Both A & C

7. The C – C bond of complex compound in broken by which process in respiration?
   (a) Oxidation
   (b) Reduction
   (c) Hydrogenation
   (d) None of the above

8. Assertion – ATP act as energy currency of cell.

   Reason – Energy released through respiration is trapped as bio-chemical energy in the form of ATP.
   (a) Only Assertion is correct
   (b) Only Reason is correct
   (c) Both Assertion and Reason is correct
   (d) Both Assertion and Reason is wrong

9. Which among the following is wrong?
   (i) Only carbohydrates are oxidised to release energy in the process of respiration.
   (ii) Energy produced in respiration is not released in a single step.
   (iii) ATP can be broken down, as and when energy needs to be utilised.
   (a) Only ii (b) Only iii
   (c) Only i (d) None of the above

10. ATP stands for?
   (a) Adenosine 3′ – triphosphate
   (b) Adenosine – 3′ – trio phosphite
   (c) Adenosine 5′ – triphosphate
   (d) Adenosine 5′ – triophosphite

Topic 1 Do Plants Breathe?

11. Respiration organs for plants are-
   (a) Lenticels
   (b) Stomata
   (c) Woody Bark
   (d) Both of the above A and B
12. Which among the following is wrong?
   (a) Roots, Leaves and Stem respire a for lower than animal do
   (b) For plants to respire, availability of O_2 is a problem as O_2 is not released within the cell during photosynthesis
   (c) There is very little transport of gases from one plant part to another
   (d) None of the above
13. What are the byproducts of Respiration process?
   (a) Oxygen (b) Water
   (c) Carbon dioxide (d) Both B and C
14. Respiration is a ________ process.
   (a) Anabolic
   (b) Catabolic
   (c) Both Anabolic as well as catabolic
   (d) None of the above
15. Choose the correct equation-
   (a) $C_6H_{12}O6 + 12O_2 \rightarrow 6H_2O + 6H_2O +$ Energy
   (b) $C_6H_{12}O_6 + 3O_2 \rightarrow 2CO_2 + 3H_2O +$ Energy
   (c) $C_6H_{12}O_6 + 6CO_2 + 6H_2O +$ Energy
   (d) $C_6H_{12}O_6 + 6O_2 \rightarrow 6CO_2 + 6H_2O +$ Energy

Topic 2 Glycolysis

16. In glycolysis, fructose 1, 6 – bisphosphate get split into which of the following?
   (a) PGAL and BPGA
   (b) Glyceraldehyde – 3 – phosphate and 3 – phosphoglyceric acid
   (c) Glyceraldehyde – 3 – phosphate and Dihydroxy acetone phosphate
   (d) None of the above
17. What is the net gain of ATP from one molecule of glucose in one complete glycolysis?
   (a) 4 (b) 3
   (c) 5 (d) 2
18. How many ATP are utilized in complete process of glycolysis of one glucose molecule?
   (a) 2 (b) 1
   (c) 3 (d) 4
19. How many molecules of NADH are produced in one complete process of glycolysis of one glucose molecule?
   (a) 1 (b) 2
   (c) 3 (d) 4
20. How many molecules of ATP are directly produced in one complete glycolysis of one glucose molecule?
   (a) 1 (b) 2
   (c) 3 (d) 4
21. Which among the following are correct about Glycolysis?
   (i) It is the only process that occurs in anaerobes for oxidation of glucose.
   (ii) Glucose undergoes complete oxidation to form pyruvic acid.
   (iii) At the end, there is a net gain of 4 ATP and 2 NADH.
   (a) Only ii (b) Both ii and iii
   (c) Only i (d) all of the above

22. For further complete oxidation of glucose, pyruvic acid enters to which among the following?
(a) ETS (b) Kreb's cycle
(c) EMP pathway (d) None of the above

`23. Glycolysis is originated from-
(a) Latin word (b) French word
(c) Italian word (d) Greek word

24. Meaning of glycolysis is-
(a) Splitting of water
(b) Splitting of sugar
(c) Splitting of fat
(d) Splitting of protein

25. Glycolysis is also known as ______ pathway.
(a) ETS (b) EMP
(c) ENP (d) ELP

26. The scheme of glycolysis was given by-
(a) Gustav Embden (b) Otto Meyerhof
(c) J. Parnas (d) All of the above

27. The scheme of glycolysis was given by-
(a) Gustav Embden (b) Alto Meyerhof
(c) J. Parnas (d) All of the above

28. Glycolysis occurs in which among the following?
(a) Aerobic organism
(b) Anaerobic organism
(c) Eukaryotes
(d) All of the above

29. Sucrose is converted to __(i)__ and __(ii)__ using enzyme __(iii)__

	(i)	(ii)	(iii)
(a)	Glucose	Glucose	Hexokinase
(b)	Glucose	Fructose	Hexokinase
(c)	Glucose	Glucose	Invertase
(d)	Glucose	Fructose	Invertase

30. What is the isomerised produce of glucose– 6 – phosphate in the steps of glycolysis?
(a) Fructose 1, 6 – bisphosphate
(b) Fructose –6 – phosphate
(c) Fructose –1, 3, 6 – triphosphate
(d) Fructose –3 – phosphate

31. Glycolysis is a how many steps of process?
(a) Ten (b) Eight
(c) Eleven (d) Five

32. End product of glycolysis is?
(a) Pyruvate
(b) Phenol
(c) Prusic acid
(d) Phosphoenolpyruvate

33. ATP is utilised in which steps of glycolysis.
(i) Conversion of BPGA to PGA
(ii) Conversion of glucose into glucose –6– phosphate
(iii) Conversion fructose 6 – phosphate to fructose 1, 6 – bisphosphate
(iv) Conversion of PEP to pyruvate
(a) Only i (b) Only iii
(c) Both ii and iii (d) Only ii, iii, iv

34. One molecule of glucose is converted into how many molecules of pyruvic acid?
(a) 1 (b) 2
(c) 3 (d) 4

35. When PGAL is converted into BPGA in process of respiration there is formation of?
(a) 1 molecule of ATP
(b) 1 molecule of H_2O
(c) 1 molecule of NADH + H+
(d) 1 molecule of ADP

36. PGAL get ________ and get converted to BPGA?
(a) reduced (b) hydrolysed
(c) oxidized (d) all of these

37. Conversion of 2-phosphoglycerate to phosphoenolpyruvate leads of formation of?
(a) ATP (b) $NADH_2$
(c) H_2O (d) ADP

38. What does PGAL stands for?
(a) 3 – Phosphoglyceraldehyde
(b) 5 – Phosphoglyceraldehyde

(c) 3 – Phosphoglyceric acid
(d) 5 – Phosphoglyceric acid

39. What is the full form of PEP?
(a) Pyroenol pyruvate
(b) Pyruvic pyruvate
(c) Phosphoenolpyruvate
(d) None of the above

40. Pyruvic acid is composed of how many carbon atom?
(a) Two (b) Three
(c) Four (d) Five

41. Which among the following step in glycolysis yields energy?
(i) Conversion of BPGA to PGA
(ii) Conversion of fructose – 6 –phosphate to fructose 1, 6 –bisphosphate
(iii) Conversion of PEP to pyruvic acid
(iv) Conversion of glucose – 6 –phosphate to fructose – 6 – phosphate
(a) Both ii and iv (b) i, ii, iii
(c) Both i and iii (d) All of the above

Topic 3 Fermentation

42. How many statements are correct about fermentation?
(i) Very low amount of energy is released, < 7% of energy in glucose is released in fermentation
(ii) In animal cells, when oxygen is inadequate acetic acid is formed during respiration
(iii) It is dangerous process as it leads to acid and alcohol formation.
(a) 0 (b) 1
(c) 2 (d) 3

43. The range beyond which yeasts poison themselves to death in alcohol fermentation when the concentration of alcohol reaches to?
(a) 13% (b) 15%
(c) 12% (d) 17%

44. Which among the following is the processes steps in, complete cellular respiration which don't need oxygen molecule (O2)?
(a) Glycolysis
(b) Tricarboxylic acid cycle
(c) ETC
(d) Both A and B

45. Fermentation occurs when there is
(a) Complete supply of oxygen
(b) No supply of oxygen
(c) Complete supply of water
(d) No supply of water

46. In alcoholic fermentation, pyruvate is converted to which among the following?
(a) Ethanol, CO2, NADH
(b) CO2 and Methanol
(c) CO2 and Ethanol only
(d) CO2 and Carboxylic acid

47. Which enzyme is responsible for alcoholic fermentation?
(a) Pyruvic acid decarboxylase
(b) Lactate dehydrogenase
(c) Alcohol dehydrogenase
(d) More than one option is correct

48. Which enzyme is involved in lactic acid fermentation?
(a) Pyruvic acid decarboxylase
(b) Lactate dehydrogenase
(c) Alcohol dehydrogenase
(d) More than one option is correct

49. Choose the correct option In the fermentation process:-
(i) Oxidation of ADP to ATP takes place
(ii) Reduction of ATP to ADP takes place
(iii) Reducing agent NADH + H+ is reoxidised to NAD+
(iv) Formation of NADH + H+ takes place by oxidation
(a) ii and iv (b) ii and iii
(c) Both i and iii (d) Only iii

Topic 4 Aerobic Respiration

50. Complete the following reaction- Pyruvic acid + (i) + NAD^+ $\xrightarrow[\text{Enzyme}]{\text{(ii)}}$ Acetyl COA + (iii) + NAD + H^+
(a) (i) O_2 (ii) Mg^{2+} (iii) CO_2
(b) (i) O_2 (ii) Na^+ (iii) H_2O
(c) (i) CoA (ii) Na^+ (iii) CO_2
(d) (i) CoA (ii) Mg^{2+} (iii) CO_2

51. Pyruvate enters to the mitochondrial matrix and undergoes.
(a) Reductive decarboxylation
(b) Oxidative carboxylation
(c) Reductive carboxylation
(d) Oxidative decarboxylation

52. Which enzyme catalyse the reaction going on in mitochondrial matrix in respiration?
(a) Pyruvate carboxylase
(b) Lactate dehydrogenase
(c) Alcohol dehydrogenase
(d) Pyruvate dehydrogenase

53. Who elucidated Tricarboxylic Acid cycle?
(a) Johns Elen (b) Hans Krebs
(c) Meyerhoff (d) Elena Parker

54. Formation of Acetyl coenzyme A from Pyruvate in mitochondrial matrix yields which among the following?
(a) CO_2 (b) H_2O
(c) NADPH + H^+ (d) Both A and C

55. How many molecules of NADH + H+ are produced when pyruvate converts to Acetyl CoA in TCA cycle?
(a) 0 (b) 1
(c) 2 (d) 3

Topic 5 Tricarboxylic Acid Cycle

56. One molecule of glucose synthesizes how many molecules of NADH + H^+ at the end of TCA cycle?
(a) 6 (b) 7
(c) 8 (d) 10

57. How many molecules of FADH2 are yielded from one glucose molecule at the end of TCA cycle?
(a) 1 (b) 2
(c) 3 (d) 4

58. How many net ATP molecules are directly yielded from complete oxidation of one glucose (including ATP of TCA)?
(a) 4 (b) 2
(c) 3 (d) 8

59. Which among the following is wrong?
(i) Glycolysis occurs in all living organism.
(ii) TCA cycle and ETS only occurs in aerobes.
(iii) Complete oxidation of pyruvate occurs by removal of all hydrogen atom in TCA cycle.
(a) (i) (b) (ii)
(c) (iii) (d) None of the above

60. Where does TCA cycle occurs?
(a) Cytoplasm
(b) Mitochondria cell wall
(c) Mitochondrial matrix
(d) Chloroplast

61. What is the first product of TCA cycle?
(a) Acetyl CoA (b) Citric acid
(c) Isocitric acid (d) OAA

62. What is the 1st step of TCA cycle?
(a) Formation of citrate from isocitrate
(b) Formation of citrate from the acetyl coenzyme A
(c) Formation of citrate from decarboxylation of succinic acid
(d) None of the above

63. Which enzyme catalyses the first step of TCA cycle?
(a) Citrate Synthase
(b) Citrate Reductase

(c) Citrate Oxidase
(d) None of the above

64. What is the first member of TCA cycle that accepts Acetyl CoA?
(a) Citrate
(b) CoA
(c) Oxaloacetic acid
(d) Both A and C

65. Which among the following is synthesised during the conversion of succinyl – CoA to succinic acid in TCA cycle?
(a) $FADH_2$ (b) GTP
(c) $NADH_2$ (d) ATP

66. How many total CO_2 molecule are released from Pyruvate to completion of TCA cycle?
(a) 0 (b) 1
(c) 2 (d) 3

67. How many total NADH2 are produced from pyruvate to completion of TCA cycle?
(a) 2 (b) 3
(c) 4 (d) 5

68. How many FADH2 are produced in TCA cycle?
(a) 1 (b) 2
(c) 3 (d) 4

Topic 6 Electron Transport System and Oxidative Phosphorylation

69. Which among the following is the role of O2 in whole respiration process? (Pg 233, M)
(i) Act as hydrogen removal from the system.
(ii) Act as final hydrogen acceptor.
(iii) It bond with C atom and released CO_2, one of the byproduct of respiration.
(a) ii and iii (b) iii only
(c) Both i and ii (d) All of the above

70. ETS of respiration process is called
(a) Reductive phosphorylation
(b) Oxidative phosphorylation
(c) Oxidative photophosphorylation
(d) Both B and C

71. Which among the following is wrong about ATP synthase?
(i) It is also called complex V.
(ii) This is used to synthesis ATP by utilising the energy released during ETS.
(iii) It works on the basis of proton gradient.
(iv) It consist of two major components, F1 and F0.
(a) Only ii (b) Both i and iii
(c) i and iv (d) None of the above

72. What is F1 in ATP synthase?
(a) It contain a site for protein synthesis.
(b) It contain a site for ADP synthesis from ATP.
(c) It contain a site for ATP production from ADP.
(d) It act as a channel through which proton cross the inner membrane.

73. What is the role of F0 in ATP synthase?
(a) It act as a channel through which ecrosses the inner membrane.
(b) It act as a channel through which proton crosses the inner membrane.
(c) It act as a mobile protein carrier of electron across the inner membrane.
(d) Acts as site for ATP synthesis

74. For each ATP produced, ________ passes through F0 from intermembrane space to the matrix down the electrochemical proton gradient.
(a) H^+ (b) $2H^+$
(c) $3H^+$ (d) $4H^+$

75. ETS occurs in which place?
(a) Outer membrane of mitochondria
(b) Cytoplasm
(c) Inner membrane of mitochondria
(d) Matrix of mitochondria

76. Energy stored in NADH + H^+ $FADH_2$ are released in ETS through ________.
(a) Reduction of these molecules

(b) Oxidation of these molecules
(c) Hydrolysis of these molecules
(d) Both A & B

77. ETS stands for
(a) Electrical Transport System
(b) Electron Transmission System
(c) Electron Transport System
(d) None of the above

78. When the electrons are passed onto O_2 in ETS it leads to formation of what?
(a) CO_2 (b) ATP
(c) H_2O (d) $NADH + H^+$

79. Ubiquinone is located at ________.
(a) inner membrane of mitochondria
(b) outer membrane of mitochondria
(c) inner membrane of nucleus
(d) outer membrane of nucleus

80. Ubiquinone receives electrons from which of the following?
(i) From NADH produced in mitochondrial matrix during TCA.
(ii) From FADH2 produced during oxidation of succinate in TCA.
(a) Only i (b) Only ii
(c) Both i and ii (d) None of the above

81. Electrons from NADH produced during TCA are oxidised by which enzyme?
(a) NAD+ hydrogenase
(b) NADH dehydrogenase
(c) NAD+ hydroxylase
(d) NADH dehydroxylase

82. The reduced ubiquinone are also called what?
(a) Ubiquinate (b) Ubiquinase
(c) Ubiquinal (d) Ubiquinol

83. Cytochrome c is ________?
(a) Lipid (b) Carbohydrate
(c) Protein (d) Fat

84. What is the function of cytochrome c?
(a) Act as donor of electron
(b) Passage for movement of e-
(c) Act as a receptor of e- between complex II and III
(d) Act as a mobile carrier for e- transfer between complex III and IV

85. What does cytochrome c oxidase complex contains?
(a) Cytochrome a
(b) Cytochrome a3
(c) Two copper centres
(d) All of the above

86. When e- passes from complex I to IV in ETS they are coupled to ________ for ATP production from ADP.
(a) Cytochrome c (b) Cytochrome bc1
(c) ATP synthase (d) Both A and B

87. Oxidation of one molecule of NADH2 produces how many molecules of ATP?
(a) 1 (b) 2
(c) 3 (d) 4

88. Oxidation of 2 molecule of FADH2 produces how many molecules of ATP?
(a) 1 (b) 2
(c) 3 (d) 4

Topic 7 The Respiratory Balance Sheet

89. What is the net gain of ATP molecules during aerobic respiration of one glucose molecule?
(a) 40 ATP (b) 38 ATP
(c) 36 ATP (d) 34 ATP

90. Match the following

1 Glycolysis	i Mitochondrial matrix
2 TCA	ii Cytoplasm
3 ETC	iii Inner membrane of mitochondria

(a) 1-i, 2-ii, 3-iii
(b) 1-ii, 2-i, 3-iii

(c) 1-iii, 2-ii, 3-i
(d) 1-ii, 2-iii, 3-i

91. The respiratory balance sheet is calculated on some assumptions.
Which of the following assumption is correct? (a)
The pathway is sequential, with series of glycolysis, ETS and TCA cycle in the same order for a given molecule.
(b) NADH is transferred to chloroplast where oxidative phosphorylation occurs, leading to formation of 3 ATP
(c) Only glucose is the substrate and none other substrate or intermediate enters or leaves the pathway
(d) None of these

92. In the balance sheet of fermentation, net gain is–
(a) 12 ATP molecules
(b) 38 ATP molecules
(c) 2 ATP molecules
(d) 8 ATP molecules

93. Oxidation of NADH to NAD^+ is –
(a) Slower in fermentation than aerobic respiration
(b) Faster in fermentation than aerobic respiration
(c) Equal in fermentation and aerobic respiration
(d) Cannot be compared

Topic 8 Amphibolic Pathway

94. Which among the following is wrong?
(i) Other than glucose, no other substrates can be used in respiratory process.
(ii) Respiratory pathway is an amphibolic pathway.
(iii) Different substrates enters at different stage in respiratory pathway.
(a) Only i (b) Only iii
(c) Only ii (d) Both i and ii

95. Which among the following is wrong?
(i) Other than glucose, no other substrates can be used in respiratory process.
(ii) Respiratory pathway is an amphibolic pathway.
(iii) Different substrates enters at different stage in respiratory pathway.
(a) Only i (b) Only iii
(c) Only ii (d) Both i and ii

96. Fats as a respiratory substrate converts to which compound first?
(a) Dihydroxy Aceton Phosphate
(b) Glycerol
(c) Fatty acid
(d) Both B and C

97. Match the following-

1	Amino acids	i	Pyruvic acid
2	Fatty acid	ii	Dihyroxy Acetone Phosphate
3	Glycerol	iii	Acetyl CoA

(a) 1-i, 2-iii, 3-ii (b) 1-ii, 2-i, 3-iii
(c) 1-iii, 2-i, 3-ii (d) 1-ii, 2-iii, 3-i

98. Choose the correct according to the correct sequence (from substrate to end product)
(i) Glucose 6 – phosphate
(ii) Pyruvic acid
(iii) Carbohydrate
(iv) Fructose–1, 6-bisphosphate
(v) Glucose
(vi) Dihydroxy Acetone Phosphate ↔ Glyceraldehyde 3 – phosphate
(a) i, iii, iv, v, vi, ii (b) iii, iv, v, ii, i, vi
(c) iii, v, i, ii, iv, vi (d) iii, v, i, iv, vi, ii

Topic 9 Respiratory Quotient

99. What is RQ if carbohydrates are used as a respiratory substrate?
(a) 1 (b) 0.8

(c) 0.7 (d) 0.9

100. Match the following-

1	NADH + H^+	i	1 ATP
2	$FADH_2$	ii	2 ATP
3	GTP	iii	3 ATP

(a) 1-i, 2-ii, 3-iii (b) 1-i, 2-iii, 3-ii
(c) 1-iii, 2-ii, 3-i (d) 1-ii, 2-i, 3-iii

101. In ETS O2 accept the electrons and get reduced to which of the following?
(a) Water (b) Carbon dioxide
(c) Palmitic acid (d) None of the above

102. What is the final end product of TCA cycle?
(a) 3 NADH + H^+
(b) 1 ATP
(c) 1 $FADH_2$
(d) All of the above

103. How many ATPs are produced through ETS only from 1 molecule of 3- phosphoglycerate in aerobic respiration
(a) 12 (b) 14
(c) 16 (d) 15

104. Which statement is true about RQ?
(i) It is also called respiratory ratio.
(ii) It is the volume of O_2 released over the volume of CO_2 evolved during respiration.
(iii) RQ of diff. substrate is different.
(a) Only i (b) Only iii
(c) Both i and iii (d) All of the above

105. Choose the correct.
(a) RQ = volume of CO_2 evolved/volume of O_2 consumed
(b) RQ = volume of O_2 consumed/volume of CO2 evolved
(c) RQ = volume of O_2 evolved/volume of CO_2 consumed
(d) None of the above

106. What will be the RQ for the following equation
$2(C_{31}H_{28}O6) + 145\ O_2 \rightarrow 102\ CO_2 + 98\ H_2O$ + Energy
(a) 0.9 (b) 1
(c) 0.8 (d) 0.7

107. What is RQ if proteins are used as a respiratory substrate?
(a) 1 (b) 0.8
(c) 0.9 (d) 0.7

ANSWER KEY

1. (d)	2. (a)	3. (a)	4. (d)	5. (b)	6. (c)	7. (a)	8. (c)	9. (c)	10. (c)
11. (d)	12. (b)	13. (d)	14. (c)	15. (d)	16. (c)	17. (d)	18. (a)	19. (d)	20. (d)
21. (c)	22. (b)	23. (d)	24. (b)	25. (b)	26. (d)	27. (d)	28. (d)	29. (d)	30. (b)
31. (a)	32. (a)	33. (c)	34. (b)	35. (c)	36. (c)	37. (c)	38. (a)	39. (c)	40. (b)
41. (c)	42. (c)	43. (a)	44. (d)	45. (b)	46. (c)	47. (d)	48. (c)	49. (d)	50. (d)
51. (d)	52. (d)	53. (b)	54. (a)	55. (b)	56. (c)	57. (b)	58. (a)	59. (d)	60. (c)
61. (b)	62. (b)	63. (a)	64. (c)	65. (b)	66. (d)	67. (c)	68. (a)	69. (c)	70. (b)
71. (d)	72. (c)	73. (b)	74. (b)	75. (c)	76. (b)	77. (c)	78. (c)	79. (a)	80. (c)
81. (b)	82. (d)	83. (c)	84. (d)	85. (d)	86. (c)	87. (c)	88. (d)	89. (c)	90. (b)
91. (c)	92. (c)	93. (a)	94. (a)	95. (b)	96. (d)	97. (a)	98. (d)	99. (a)	100. (c)
101. (d)	102. (d)	103. (b)	104. (c)	105. (a)	106. (d)	107. (c)			

12 Chapter Plant Growth & Development

MULTIPLE CHOICE QUESTIONS

1. Development is
   (a) Growth
   (b) Differentiation
   (c) Growth + differentiation
   (d) Growth – differentiation

Topic 1	Growth

2. Growth of living being is/are-
   (a) Irreversible
   (b) Increase in size
   (c) Increase in weight
   (d) All of these

3. Swelling of piece of wood when placed in water is
   (a) Growth but not development
   (b) Development but not growth
   (c) No growth and development
   (d) Both growth & development

Topic 2	Plant Growth generally is Indeterminate

4. Secondary growth does occur in-
   (a) All angiosperms and no gymnosperms
   (b) Some angiosperm and gymnosperms
   (c) No angiosperm and only gymnosperms
   (d) All angiosperm and all gymnosperms

5. Secondary growth does not mean-
   (a) Increase in length of plant
   (b) Increase in girth of plant
   (c) Increase in diameter in plant
   (d) Both A & C

6. 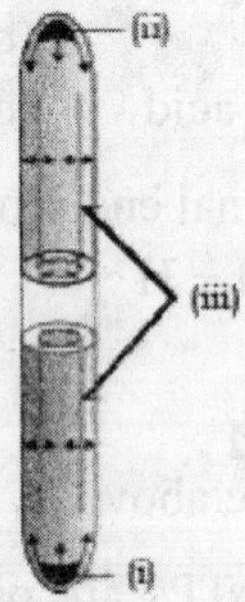

Choose the correct label-

	(i)	(ii)	(iii)
(a)	Shoot apical meristem	Root apical meristem	Vascular cambium
(b)	Root apical meristem	Shoot apical meristem	Vascular cambium
(c)	Shoot parenchym a	Root parenchy ma	Vascular bundle
(d)	Root parenchym a	Shoot parenchy ma	Vascular bundle

7. Plant grows
   (a) For a limited time in life
   (b) For a very long time in life
   (c) For a very short time in a life
   (d) For unlimited time throughout life

8. The given figure shows-

(a) Germination & development in bean
(b) Germination & development in maize
(c) Germination & development in pea
(d) Germination & development in gram

9. Plant grows throughout the life due to-
(a) Meristematic tissue
(b) Parenchyma tissue
(c) Epidermal tissue
(d) More than one option is correct

10. Open form of growth involves
(a) Continuous division in all cells forming new set of cells which divide further necessarily.
(b) Division in some cells forming new set of cells which do not divide further necessarily
(c) No division at all
(d) Division in some cells at the time of injury which forms new set of cells to heal the injury

Topic 3 Growth is Measurable

11. Growth can be measured by measuring increase in –
(a) Amount of protoplasm
(b) Dry weight
(c) Cell number
(d) Both B & C

12. Match the column in respect with the measurement growth-

	Column I		Column II
a	Maize	i	Length
b	Watermelon	ii	Surface area
c	Pollen tube	iii	Cell number
d	Dorsiventralleaf	iv	Cell size

(a) a-iii, b-iv, c-i, d-ii
(b) a-iii, b-ii, c-i, d-iv
(c) a-iv, b-iii, c-i, d-ii
(d) a-i, b-ii, c-iv, d-iii

Topic 4 Phases of Growth

13. The three phases of growth in correct order is-
(a) Meristematic, maturation, elongation
(b) Elongation, meristematic, maturation
(c) Meristematic, elongation, maturation
(d) Elongation, maturation, meristematic

14. Cells in meristematic phase of growth-
(a) Have small nuclei
(b) Have low plasmodesmatal connections
(c) Have thick cell wall
(d) Are rich in protoplasm

15. The phase of maturation lies-
(a) Proximal to cells undergoing elongation, towards the tip
(b) Proximal to cells undergoing elongation, away from the tip
(c) Proximal to cells of meristematic zone, away from the tip
(d) Proximal to cells of meristematic zone, towards the tip

16. Increased vacuolation is a characteristics of –
(a) Meristematic phase
(b) Maturation phase
(c) Elongation phase
(d) All of these

17. Thickest cell wall present in –
(a) Meristematic phase
(b) Maturation phase
(c) Elongation phase
(d) Both B & C

Topic 5 Growth Rates

18. Choose the correct set of option for size or weight of organ against time
(i) W1 = Wo ert
(ii) Lt = Lo + rt
(iii) Linear growth curve
(iv) Sigmoid growth curve

(v) Arithmetic growth

(vi) Geometric growth

(a) (i), (iii), (iv) (b) (ii), (iii), (vi)

(c) (i), (iv), (vi) (d) (ii), (iv), (v)

19. The three phases of sigmoid growth occur in order are –

(a) Log, lag, stationary

(b) Log, stationary, lag

(c) Lag, log, stationary

(d) Lag, stationary, log

20. In the formula W1 = Wo ert , r is –

(a) Relative growth rate

(b) Efficiency index

(c) Ability of plant to produce new plant material

(d) All of these

21. choose the correct option –

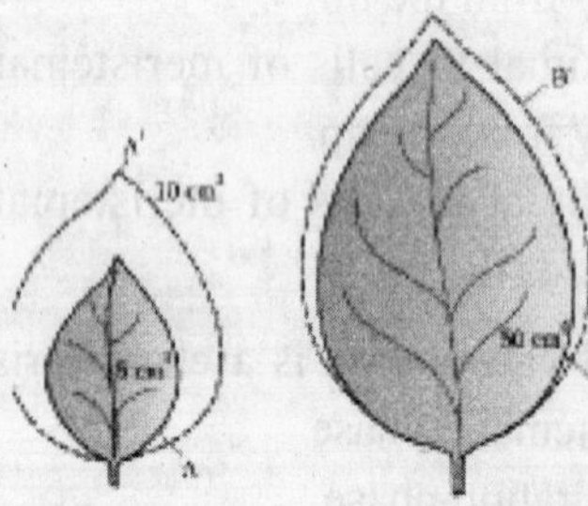

(a) Absolute growth rate of A is more than that of B

(b) Absolute growth rate of B is more than that of A

(c) Relative growth rate of A is more than that of B

(d) Relative growth rate of B is more than that of A

22. Increased growth per unit time is called

(a) Growth index

(b) Growth rate

(c) Growth efficiency index

(d) Both A & C

23. Identify the given figure

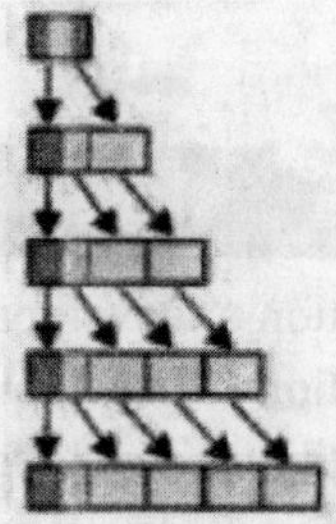

(a) Figure shows geometric growth with formula $L_t = L_o + r_t$

(b) Figure shows geometric growth with formula $W_1 = W_o$ ert

(c) Figure shows arithematic growth with formula $L_t = L_o + rt$

(d) Figure shows arithematic growth with formula $W_1 = W_o$ ert

24. Identify the given figures

(a) Figure shows geometric growth with formula $L_t = L_o + rt$

(b) Figure shows geometric growth with formula W_1 = Wo ert

(c) Figure shows arithematic growth with formula $L_t = L_o + rt$

(d) Figure shows arithematic growth with formula $W_1 = W_o$ ert

25. Choose the correct option with respect to given figures of stages of embryo development

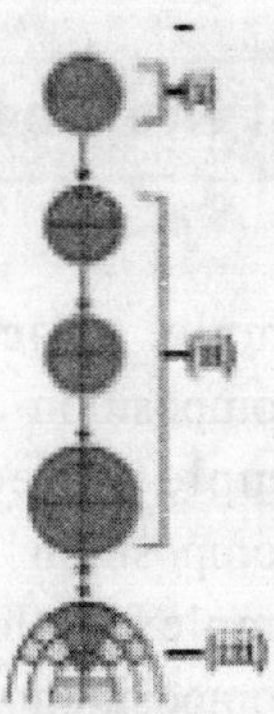

(a) (i) is arithmetic growth phase
(b) (i) is geometric growth phase
(c) (ii) is arithmetic growth phase
(d) (ii) is geometric growth phase

Topic 6 Conditions Growth

26. The factors affecting growth can be –
(a) Water, temperature, light, gravity
(b) Water, temperature, light but not gravity
(c) Water, light but not temperature & gravity
(d) Water, light, gravity but not temperature

Topic 7 Differentiation, Dedifferentiation and Redifferentiation

27. During differentiation
(a) Structural changes occur in nucleus and cytoplasm
(b) Functional changes occur in nucleus and cytoplasm
(c) Structural changes occur in cell wall and protoplasm
(d) All of these

28. Dedifferentiation can be seen in formation of–
(a) Intrafascicular parenchyma
(b) Intrafascicular cambium only
(c) Interfascicular cambium only
(d) Entire vascular cambium

29. Read the following statements –
(i) Cork cambium is a layer of meristem formed from parenchyma cells
(ii) Intrafascicular cambium is a layer of parenchyma cells formed from meristem
(iii) Vascular cambium divide and produce cells that differentiate again
(iv) Plant growth can be determinate orindeterminate
(v) Final structure of cells are never determined by location of cells
(vi) The differentiation in plant cells are closed and dependent on position

How many of the above statements are correct?
(a) 3 (b) 4
(c) 5 (d) 6

Topic 8 Development

30. identify the correct labels –

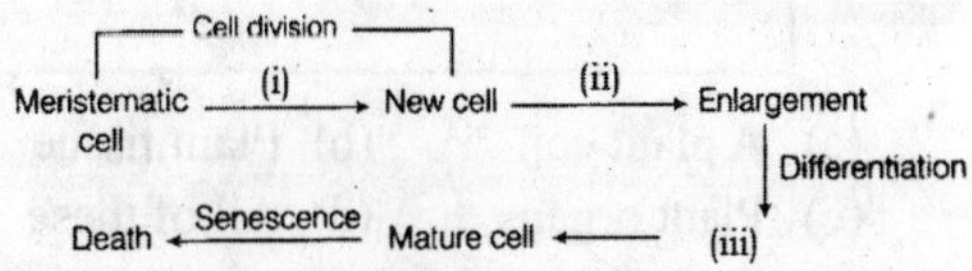

(a) (i)-Expansion, (ii)-Plasmatic growth, (iii)-Maturation
(b) (ii)-Elongation, (i)-Plasmatic growth, (iii)-Maturation
(c) (iii)- Elongation, (ii)-Plasmatic growth, (i)-Maturation
(d) (iii)-Expansion, (i)-Plasmatic growth, (ii)-Maturation

31. Plasticity can be seen in –
(a) Cotton (b) Coriander
(c) Buttercup (d) All of these

32. Environmental heterophylly can be seen in –
(a) Larkspur (b) Cotton
(c) Coriander (d) Buttercup

33. The given figure show –

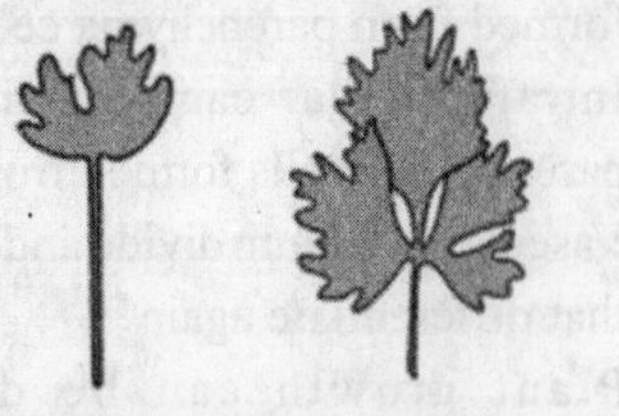

(a) Developmental heterophylly in buttercups
(b) Environmental heterophylly in larkspur
(c) Environmental heterophylly in buttercup
(d) Developmental heterophylly in Larkspur

34. The figure shows sequence of the development process in

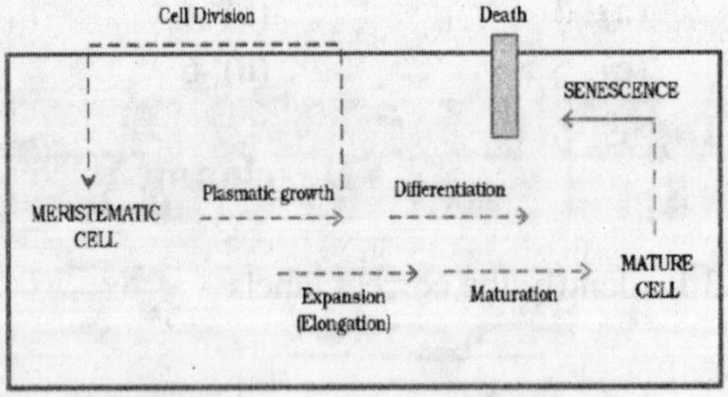

(a) A plant cell (b) Plant tissue
(c) Plant organs (d) All of these

35. The given figure show –

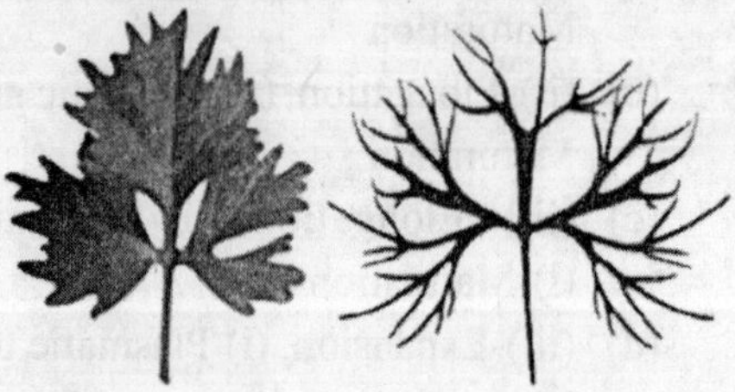

(a) Developmental heterophylly in buttercups
(b) Environmental heterophylly in larkspur
(c) Environmental heterophylly in buttercup
(d) Developmental heterophylly in larkspur

36. Intrinsic factors responsible for growth & development include –
(a) Intracellular (plant growth regulators)
(b) Intercellular (genetic)
(c) Intercellular (plant growth regulators)
(d) More than one option is correct

Topic 9 Plant Growth Regulation

37. PGRs are –
(a) Small, simple, molecules of diverse chemical composition
(b) Large, simple molecules of diverse chemical composition
(c) Small, complex molecules of diverse chemical composition
(d) Large, complex molecules of diverse chemical composition

38. An example of adenine derivative PGR
(a) IAA (b) Kinetin
(c) ABA (d) Gibberellic acid

39. Gibberellic acid is –
(a) Indole compound
(b) Adenine compound
(c) Carotenoid derivative
(d) Terpene derivative

40. Abscisic acid is –
(a) Indole compound
(b) Adenine compound
(c) Carotenoid derivative
(d) Terpene derivative

41. Kinetin is –
(a) Indole compound
(b) Adenine compound
(c) Carotenoid derivative
(d) Terpene derivative

42. A gaseous PGR is
(a) ABA (b) Ethylene
(c) GA3 (d) IAA

43. The PGRs that play important role in plant responses to wounds are –
(a) Auxin (b) Cytokinin
(c) Gibberellin (d) Abscisic acid

44. PGR Ethylene can fit into –

(a) Group of plant growth promoters
(b) Group of plant growth inhibitors
(c) Both the groups of promoter & inhibitors but mostly inhibitor
(d) Both the groups of promoter and inhibitors but mostly promoter

Topic 10 The Discovery of PGR

45. Discovery of how many out of 5 major PGRs was accidental?
(a) 2 (b) 1
(c) 3 (d) 5

46. Match the scientist with the plant they worked on –

	Column I		Column II
a	F.W.Went	i	Canary grass
b	E. Kurosawa	ii	Avena(oat)
c	Charles & Francis Darwin	iii	Tobacco
d	F.S koog	iv	Rice

(a) a-iv, b-ii, c-iii, d-i
(b) a-iii, b-i, c-ii, d-iv
(c) a-ii, b-iv, c-i, d-iii
(d) a-i, b-iii, c-iv, d-ii

47. Cousins confirmed the release of a volatile substance _(i)_, from ripened _(ii)_ that hastened the ripening of stored unripe _(iii)_.
(a) (i)-ABA, (ii)-banana, (iii)-orange
(b) (i)-ABA, (ii)-orange, (iii)-banana
(c) (i)-C2H4, (ii)-banana, (iii)-orange
(d) (i)- C2H4, (ii)- orange, (iii)-banana

48. Match the PGR with the plants which played role in their discovery –

	Column I		Column II
a	Auxin	i	Tobacco
b	Gibberellin	ii	Rice
c	Cytokinin	iii	Orange
d	Ethylene	iv	Avena

(a) a-iii, b-i, c-ii, d-iv
(b) a-iv, b-ii, c-i, d-iii
(c) a-i, b-iii, c-iv, d-ii
(d) a-ii, b-iv, c-iii, d-i

49. Abscisic acid was independent discovered by three different researchers and named as –
(a) Inhibitor-A, abscission-III, dormin
(b) Inhibitor-C, abscission-II, dormane
(c) Inhibitor-B, abscission-II, dormin
(d) Inhibitor-B, abscission-II, dormane

50. The term kinetin was given by –
(a) Muller et al (b) Kurosawa et al
(c) Cousins et al (d) None of these

51. The foolish seedling disease of rice is caused by–
(a) Nematode (b) Bacteria
(c) Fungus (d) Virus

52. Auxin was isolated by _(i)_ from _(ii)_

	(i)	(ii)
A	Francis Darwin	Wleoptile of oat seedling
B	Francis Darwin	Cleoptile of canary grass
C	F.W. Went	Cleoptile of oat seedling
D	F.W. Went	Cleoptile of canary grass

53. The figure shows –

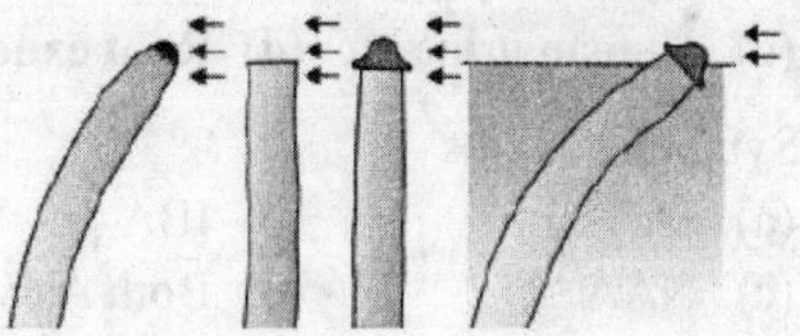

(a) Experiment to show that tip of coleoptile produces gibberelin
(b) Experiment to show that tip of coleorhiza produces gibberelin
(c) Experiment to show that tip of coleorhiza produces auxin
(d) Experiment to show that tip of coleoptile produces auxin

Topic 11 Auxin

54. Auxin –
(a) Promotes apical dominance
(b) Prevent apical dominance
(c) Both promote & prevent apical dominance based on condition
(d) Can't say

55. In tea plantation and hedge -making
(a) Lateral buds are removed
(b) Shoot tips are removed
(c) Root tips are removed
(d) None of these

56. Auxin –
(a) Induces parthenocarpy
(b) Act as herbicide
(c) Help in cell division
(d) All of these

57. 2, 4-D is used to –
(a) Kill gymnosperms usually
(b) Kill dicot usually
(c) Kill monocot usually
(d) Both A & B

58. Auxin was first isolated from –
(a) Xylem sap (b) Phloem sap
(c) Human urine (d) Root exudates

59. Synthetic auxins –
(a) IAA (b) IBA
(c) NAA (d) Both A & C

60. Auxin isolated from plant –
(a) NAA (b) IAA
(c) 2, 4-D (d) Both A & B

61. Auxins are generally produced by –
(a) Growing apex of root & shoot
(b) Epidermal tissue of shoots only
(c) Xylem secondary cells
(d) Endodermal cells of roots & shoot

62. Assertion: Auxin is used in plant propagation widely. Reason: Auxin initiate rooting.

Choose the correct option -
(a) Both Assertion and Reason are correct, and Reason is correct explanation of Assertion
(b) Both Assertion and Reason are correct, but Reason doesn't explain Assertion
(c) Assertion is correct but Reason is wrong
(d) Both Assertion and Reason are incorrect

63. Auxin promote –
(a) Rooting in stem cutting
(b) Flowering in pineapple
(c) Bolting in beet
(d) Both A & B

64. Statement-I Auxin prevent fruit and leaf drop at early stage Statement-II Auxin promote abscission of older mature leaves & fruits

Choose the appropriate option –
(a) Statement-I is correct and Statement-II is incorrect
(b) Statement-I is incorrect and Statement-II is correct
(c) Both Statement-I & Statement-II are correct
(d) Both Statement-I & Statement-II are incorrect

Topic 12 Gibberellins

65. Gibberellins is used to improve yields of –
(a) Apple (b) Grape
(c) Sugarcane (d) All of these

66. Assertion: Juvenile conifers are sprayed with GA

Reason: GA delays senescence and malting period

Which of the given options are correct?
(a) Both assertion & reason are correct and reason is correct explanation of assertion
(b) Both assertion & reason are correct but reason is not explanation of assertion
(c) Assertion is correct but reason is wrong
(d) Both assertion & reason are in correct

67. Bolting is –
(a) Yellowing of leaves
(b) Node elongation prior to flowering
(c) Early maturing and seed production
(d) None of these

68. GA promote bolting in –
(a) Beet
(b) Cabbage
(c) Plants with rosetle habit
(d) All of these

69. Gibberellins are
(a) Promotory PGR
(b) Inhibitory PGR
(c) Neither promotory nor inhibitory PGR
(d) Both promotory and inhibitory PGR

70. Which of the following statements are incorrect–
(a) There are more than 100 gibberellins reported
(b) GA3 was one of the gibberellins to be discovered
(c) All GA are basic
(d) GAs are also reported in fungi

71. Gibberellins is used in grapes for –
(a) Increase in length of stalk
(b) Increase in girth of stalk
(c) Decrease in length of stalk
(d) Decrease in girth of stalk

72. Gibberellins –
(a) Promotes senescence
(b) Delay senescence
(c) Neither promote not delay senescence
(d) Both promote & delay senescence based on situation

73. Assertion: Spraying gibberellins on fruits extend its market period

Reason: Gibberellins delays senescence Choose the best option –
(a) Both assertion & reason are correct and reason is correct explanation of assertion
(b) Both assertion & reason are correct but reason is not explanation of assertion
(c) Assertion is correct but reason is wrong
(d) Both assertion & reason are in correct

74. Which of these are correct –
(a) ABA shows the malting process in brewing industry
(b) ABA speeds the malting process in brewing industry
(c) GA3 slows the malting process in brewing industry
(d) GA3 speeds the malting process in brewing industry

Topic 13 Cytokinins

75. Cytokinins were discovered as –
(a) Kinin (b) Kinetin
(c) Kinesin (d) Zentin

76. Kinetin is a modified form of –
(a) Purine – guanine
(b) Pyrimidine – cytosine
(c) Purine – adenine
(d) Pyrimidine – thymine

77. Kinetin was discovered from –
(a) Autoclaved herring egg DNA
(b) Human urine
(c) Corn kernel
(d) None of these

78. Read the following statements regarding cytokinin –
(i) Kinetin occurs naturally in plants
(ii) Kinetin was discovered from coconut milk
(iii) Zeatin does not occur naturally in plants
(iv) Zeatin was isolated from human DNA

How many of the statements are incorrect –
(a) 1 (b) 2
(c) 3 (d) 4

79. Natural cytokinin may be synthesized in –
(a) Root apex
(b) Developing shoot buds

(c) Young fruits
(d) More than one option is correct

80. Cytokinin helps to produce –
(a) Chloroplast in leaves
(b) Elongation in sugarcane
(c) Synchronized fruit set in pineapple
(d) Flowering in pineapple

Topic 14 Ethylene

81. Ethylene –
(a) Promotes root hair formation and increase in absorption surface
(b) Demotes root hair formation and increase in absorption surface
(c) Promotes root hair formation and decrease in absorption surface
(d) Demotes root hair formation and decrease in absorption surface

82. Ethylene initiates –
(a) Flowering in pineapple
(b) Flowering in mango
(c) Synchronizing fruit-set in pineapple
(d) All of these

83. Most widely used source of ethylene –
(a) Is ethepene
(b) Hasters fruit ripening in tomato and apple
(c) In aqueous solution is absorbed on the plant roots
(d) More than one option is correct

84. Ethylene promotes –
(a) Female flowers in cucumber, increasing the yield
(b) Female flowers in cucumber, decreasing the yield
(c) male flowers in cucumber, increasing the yield
(d) male flowers in cucumber, decreasing the yield

85. Ethylene is synthesized in large amounts by –
(a) ripening fruit
(b) Tissues undergoing senescence
(c) Newly developed leaves
(d) More than one option is correct

86. Ethylene causes _(i)_ growth of seedling, swelling of axis and apical hook formation in _(ii)_ seedling

	(i)	(ii)
(a)	Horizontal	Monocot
(b)	Vertical	Dicot
(c)	Horizontal	Dicot
(d)	Vertical	Monocot

87. Ethylene promotes –
(a) Senescence and abscission of flowers
(b) Senescence but not abscission of flowers
(c) Abscission of flowers but not senescence
(d) Neither senescence nor abscission of flowers

88. Ethylene in fruits causes –
(a) Rise in rate of respiration called respiratory anti-climactic
(b) Rise in rate of respiration called respiratory climactic
(c) Fall in rate of respiration called respiratory anti-climactic
(d) Fall in rate of respiration called respiratory climactic

89. Ethylene –
(a) Promotes seed and bud dormancy
(b) Promotes only seed dormancy
(c) Promotes only bud dormancy
(d) None of these

90. Statement-I Ethylene promotes internode/petiole elongation in deep water rice plants

Statement-II ethylene helps upper part of shoot to remain above water.

Choose the appropriate option –
(a) Statement-I is correct and Statement-II is incorrect

(b) Statement-I is incorrect and Statement-II is correct
(c) Statement-I and Statement-II are correct
(d) Statement-I and Statement-II are incorrect

Topic 15 Abscisic Acid

91. ABA is –
(a) Inhibitor of plant growth and metabolism
(b) Inhibitor of plant growth but not metabolism
(c) Inhibitor of plant metabolism but not growth
(d) Inhibitor of neither plant growth nor metabolism

92. ABA –
(a) Inhibit seed germination
(b) Promote seed dormancy
(c) Inhibit seed dormancy
(d) Both A & B

93. ABA is called stress hormone because
(a) It causes stress to plant
(b) It is released during stress in plant
(c) It helps plant to fight stress conditions
(d) More than one option is correct

94. ABA stimulates –
(a) Stomata opening for more CO_2 exchange
(b) Stomata closure to decrease transpiration
(c) Stomata opening to cause transpiration & calling effect
(d) Both A & C

95. In most situation, ABA acts as an antagonist to-
(a) Auxin (b) GA
(c) Cytokinin (d) Ethylene

Topic 16 Photoperiodism

96. Identify the correct labels –

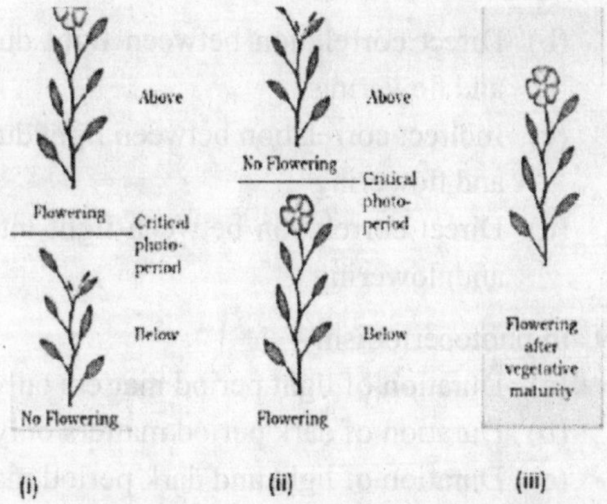

	(i)	(ii)	(iii)
(a)	Long day plant	Day neutral plant	Short day plant
(b)	Short day plant	Day neutral plant	Long day plant
(c)	Long day plant	Short day plant	Day neutral plant
(d)	Short day plant	Long day plant	Day neutral plant

97. Read the given statements –
Statement-I Some plants require a periodic exposure to light to include flowering
Statement-II Some plants are able to measure the duration of exposure to light
Choose the correct statements –
(a) Both statements are correct and are about photoperiodism
(b) One of the statements is correct about photoperiodism
(c) Both statements are correct and is about vernalization
(d) One of the statements is correct about vernalization

98. For day neutral plants, there is
(a) No correlation between light duration and flowering

(b) Direct correlation between light duration and flowering
(c) Indirect correlation between light duration and flowering
(d) Direct correlation between light intensity and flowering

99. In photoperiodism –
(a) Duration of light period matters only
(b) Duration of dark period matters only
(c) Duration of light and dark period matters
(d) Duration of light or dark period doesn't matter

100. Photoperiods are perceived by –
(a) Shoot apices (b) Flowering apices
(c) Nodal buds (d) Leaves

Topic 17 Vernalization

101. Vernalization is –
(a) Quantitative dependence of flowering on low temperature
(b) Qualitative dependence of flowering on low temperature
(c) Quantitative dependence of flowering on high temperature
(d) A and B are correct

102. Vernalization can be seen in –
(a) Spring varieties of rice
(b) Spring varieties of wheat
(c) Winter varieties of rice
(d) Winter varieties of wheat

103. Vernalization is also seen in –
(a) Biennials like – sugarbeet
(b) Biennials like maize
(c) Perennials like sugarbeet
(d) Perennials like maize

Topic 18 Seed Dormancy

104. Germination may not occur due to –
(i) Hard seed coat
(ii) Immature embryo
(iii) Chemical inhibitors
(iv) Harsh environment

How many of above are correct?
(a) 1 (b) 2
(c) 4 (d) 3

105. Seed coat dormancy is broken by –
(a) Mechanical abrasions
(b) Microbial actions in gut of animals
(c) Certain chemicals
(d) More than one option is correct

106. Effects of inhibitory substances on dormancy is removed by application of certain chemicals are–
(a) Auxin (b) Gibberellic acid
(c) Cytokinin (d) ABA

107. Some seeds which do not germinate even when external conditions are favorable, are understood to be going through –
(a) Dormancy controlled by external environment
(b) Dormancy controlled endogenously
(c) Dormancy controlled by conditions within the seed
(d) More than one option is correct

108. Chemical inhibitions of germination are –
(a) Abscisic acid (b) Phenolic acid
(c) Para-ascorbic acid (d) All of these

ANSWER KEY

1. (c)	2. (d)	3. (c)	4. (b)	5. (a)	6. (b)	7. (d)	8. (a)	9. (a)	10. (a)
11. (d)	12. (a)	13. (c)	14. (d)	15. (b)	16. (c)	17. (b)	18. (c)	19. (c)	20. (d)
21. (c)	22. (b)	23. (c)	24. (b)	25. (d)	26. (a)	27. (c)	28. (c)	29. (a)	30. (b)
31. (d)	32. (b)	33. (d)	34. (a)	35. (a)	36. (d)	37. (a)	38. (b)	39. (d)	40. (c)
41. (b)	42. (b)	43. (d)	44. (c)	45. (d)	46. (c)	47. (d)	48. (b)	49. (c)	50. (d)
51. (c)	52. (d)	53. (d)	54. (a)	55. (b)	56. (d)	57. (b)	58. (c)	59. (d)	60. (b)
61. (a)	62. (b)	63. (d)	64. (c)	65. (c)	66. (c)	67. (d)	68. (d)	69. (a)	70. (c)
71. (a)	72. (b)	73. (b)	74. (d)	75. (b)	76. (c)	77. (d)	78. (d)	79. (d)	80. (a)
81. (a)	82. (d)	83. (a)	84. (a)	85. (d)	86. (c)	87. (a)	88. (b)	89. (d)	90. (c)
91. (a)	92. (d)	93. (d)	94. (b)	95. (b)	96. (c)	97. (a)	98. (a)	99. (c)	100. (d)
101. (d)	102. (d)	103. (a)	104. (c)	105. (d)	106. (b)	107. (d)	108. (d)		

CLASS XII- BOTANY

13 Chapter

Sexual Reproduction in Flowering Plant

MULTIPLE CHOICE QUESTIONS

1. The end products of sexual reproduction in Phanerogamae are
(a) Fruit (b) Seeds
(c) Flower (d) A & B both

Topic 1 Flower

2. Floriculture deals with the cultivation of –
(a) Flower (b) Seed culture
(c) Fruit (d) Both B & C

3.

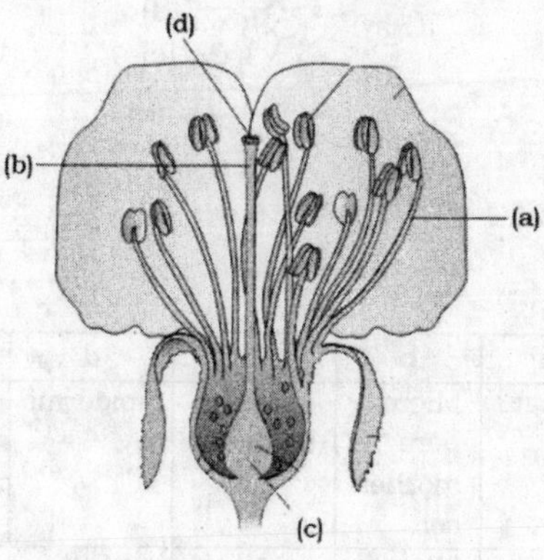

a	b	c	d	
A	Style	Filament	Stigma	Ovary
B	Filament	Style	Ovary	Stigma
C	Filament	Style	Thalamus	Anther
D	Style	Filament	Stigma	Ovule

Topic 2 The Pistil, Megasporangium and Embryosac

4. Choose incorrect statement –
(a) Several hormonal & structural changes are initiated which lead to redifferentiation and further development of the floral primordium.
(b) Inflorescences are formed which bear the floral buds and then the flower
(c) In the flower male and female reproductive structures, the androecium and the gynoecium differentiate and develop
(d) None of these

5. Reproductive organ of flower doesnot comprises
(a) Androecium (b) Stamen
(c) Gynoecium (d) Tepals

Topic 3 Stamen, Microsporangium & Pollen grains

6. When pollen grain mature –
(a) It consist of two cell that are two male gamete only.
(b) It consist of two cell that are generative & vegetative cell
(c) It consist of two cell that are two male gamete arise from vegetative cell and one generative cell
(d) It consist of three cell that are two male gamete develop meiotically from generative cell and one vegetative cell

7. Choose incorrect statement among following:

(a) In over 60% of angiosperm, pollen grains are shed at 3 – cell stage
(b) In over 60% of angiosperm, pollen grains are shed at 2 – cell stage
(c) Both A & B
(d) None of these

8. Pollen allergy is not correlated with-
(a) Cause of parthenium
(b) Cause chronic respiratory disorder
(c) Carrot grass that come into india as a contaminant with imported rice
(d) None of these

9. Pollen grain of rice is viable upto-
(a) 30 min
(b) Several month
(c) Same as in sonaceae
(d) Both B & C

10. Which temperature is correct to store semen for artificial insemination-
(a) 196°C (b) -196°C
(c) 34°C (d) 4°C

11. A typical angiosperm anther is ______ with each lobe having ______ theca i.e they are ______
(a) Bilobed, two, dithecous
(b) Dithecous, two, bilobed
(c) Bilobed , four, dithecous
(d) Dithecous, four, bilobed

12. Often theca is separated by
(a) Transverse groove (b) Longitudinal groove
(c) Diagonal groove (d) All of these

13. The dithecous consist of ______ microsporangia located at the corners, ______ in each lobe.
(a) Two, one (b) Two, two
(c) Four, two (d) Both A & C

14. Arrange microsporangial wall in sequence of outside to inside
(a) Epidermis, middle layer, endothecium, tapetum
(b) Epidermis, endothecium, middle layer, tapetum
(c) Epidermis, middle layer, tapetum, endothecium
(d) Endothecium, middle layer, tapetum, epidermis

15. How many of microsporangial wall perform function of protection and help in dehiscence of anther to release pollen
(a) 1 (b) 2 (c) 3 (d) 4

16. ______ is responsible for nourishment of pollen grain.
(a) Tapetum (b) Endothecium
(c) Epidermis (d) Middle layer

17. Which of the following undergo meiotic division to form microspore tetrad
(a) Sporogenous tissue (b) Generative tissue
(c) Microspore (d) A and B

18.

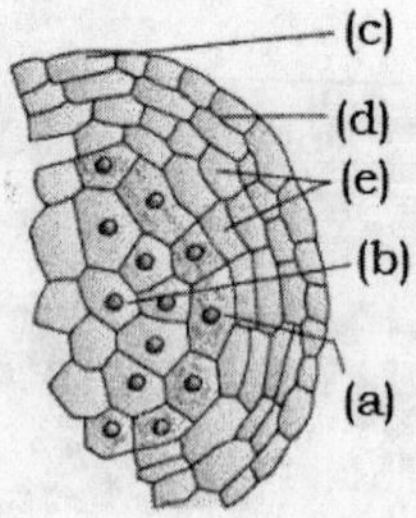

	a	b	c	d	e
A	Tapetum	Microspore mother cell	endothecium	epidermis	Middle layer
B	Tapetum	Microspore mother cell	epidermis	endothecium	Middle layer
C	Tapetum	Middle layer	Microspore mother cell	endothecium	epidermis
D	epidermis	Middle layer	Microspore mother cell	endothecium	Tapetum

19. Microsporogenesis –
(a) Process of formation of microspore
(b) Development of pollen grain from pollen mother cell
(c) It involve meiosis
(d) All of these

20. Pollen grain represents –
(a) Male gametophyte (b) Male sporophyte
(c) Female gametophyte (d) Female sporophyte

21. Choose correct about pollen grain wall
(i) It has two layered prominent wall
(ii) Hard outer layered prominent wall
(iii) Exine is composed of sporopollenin
(iv) Sporopollenin form continuous exine
(a) i, ii, iii, iv (b) i, ii, iii
(c) i, iii (d) i & iv

22. Sporopollenin is absent in –
(a) Intine (b) Germpore
(c) Exine (d) A & B both

23. Pollen grains are well preserved as fossil because of
(a) Presence of intine
(b) Presence of germpore
(c) Presence of sporopollenin
(d) All of these

24. Sporopollenin is degraded by –
(a) Engyme
(b) High temperature
(c) Strong acid & alkali
(d) None of these

25. Inner wall of pollen grain is –
(a) Intine, made up of cellulose & lignin
(b) Thin discontinuous intine
(c) Both A & B
(d) None of these

Topic 4 Pre-Fertilization: Structure and Events

26. Papaver show
(i) Multicarpellary
(ii) Apocarpous
(iii) Syncarpous
(iv) Monocarpellary
(a) i, ii (b) i, iii
(c) iv, ii (d) iv, iii

27. Given diagram is of –

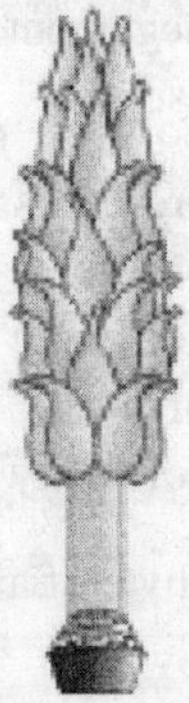

(a) Multicarpellary apocarpous gynoecium of michelia
(b) Multicarpellary synocarpous gynoecium of michelia
(c) Multicarpellary synocarpous gynoecium of papaver
(d) Multicarpellary apocarpous gynoecium of papaver

28. Which of following serves as a landing platform for pollen grain?
(a) Stigma (b) Style
(c) Anther (d) Filament

29. Choose correct statement:
(a) Inside the ovary is the ovarian cavity, also known as lodicule
(b) Megasporangia is commonly called ovules
(c) The placenta is located outside ovarian cavity
(d) A & C both

30. Choose incorrect statement –
(a) The number of ovules in an ovary is one in paddy
(b) The number of ovules in an ovary is many in papaya
(c) The number of ovules in an ovary is one in orchid

(d) Wheat mango consist of one ovule

31. Ovule is attached to placenta by –
(a) Funicle (b) Integument
(c) Hilum (d) Nucellus

32. Hilum represents the junction between
(a) Ovule & ovary
(b) Ovule & funicle
(c) Ovule & integument
(d) None of these

33. Chalaza end represent –
(a) Basal part of ovule
(b) Apical part of ovule
(c) Basal part of ovary
(d) Apical part of ovary

34. Female gametophyte of angiosperm represented by –
(a) Nucellus (b) Embryosac
(c) Integument (d) Both A & B

35. An ovile generally has ______ embryo sac formed from a megaspore through ______ division
(a) Single, equational
(b) Single, reductional
(c) Four, meiotic
(d) Four, mitotic

36. Megasporogenesis is not related to –
(a) Formation of megaspore from megaspore mother cell
(b) MMC undergoes meiotic division for megaspore
(c) Formation of microspore
(d) Both A & C

37. Ovules generally differentiate a single megaspore mother cell in –
(a) Chalazal end (b) Micropylar region
(c) Both A & B (d) Integument

38. In a majority of flowering plants
(a) One of the megaspore is functional while other three degenerate
(b) All four megaspore can developes into female gametophyte in almost all angiosperm
(c) Three megaspore is functional while other one degenerated
(d) Both A & B

39. Monosporic embryo development involve –
(a) One functional megaspore
(b) One haploid cell formed in egg apparatus
(c) Four functional megaspore
(d) None of these

40. Choose correct statement –
(a) The nucleus of the functional megaspore divides mitotically to form two nuclei which move to the opposite poles, forming the 2-nucleate embryo sac
(b) Two more sequential mitotic nuclear division in 2-nucleate embryo sac result in formation of 4- nucleate
(c) Mitotic division in embryo sac formation upto 8-celled is strictly free nuclear
(d) All of these

41. How many of eight nuclei of typical embryosac is surrounded by cell wall
(a) 2 (b) 4
(c) 6 (d) 7

42. Central cell of typical embryosac is situated –
(a) Below egg apparatus
(b) Above egg apparatus
(c) At chalazal end
(d) None of these

43. Choose the correct about egg apparatus of typical embryosac
(a) Situated at micropylar end
(b) Consist of three cells
(c) Both A & B
(d) Consist of all cells having special cellular thickening at micropylar tip

44. Typical embryo-sac of angiosperm at maturity is –
a) 8 celled, 8 nucleate
b) 7 celled, 8 nucleate

c) 8 celled, 7 nucleate
d) 7 celled, 7 nucleate

45.

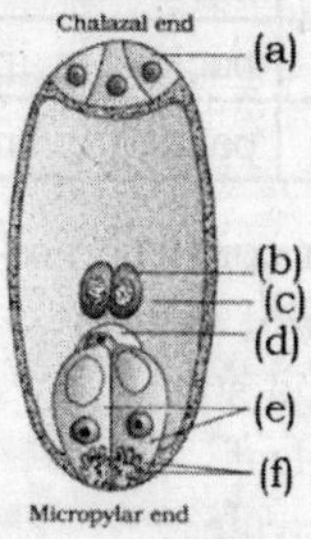

	a	b	c	d	e	f
A	Antipo dal	Polas nucle i	Centr al cell	Egg	Synergi d	Filiform apparat us
B	Antipo dal	Polas nucle i	Embr yo sac	Egg	Synergi d	Filiform apparat us
C	Antipo dal	Polas nucle i	Egg	Embr yo sac	Filiform apparat us	Synergi d
D	Antipo dal	Centr al cell	Polar nucle i	Egg	Filiform apparat us	Synergi d

Topic 5 Pollination

46. Pollination is-
(a) transfer of motile pollen to stigma
(b) transfer of non-motile anther to stigma
(c) transfer of motile anther to stigma
(d) transfer of non-motile pollen to stigma

47. Transfer of pollen from anther to stigma of same flower is called-
(a) Autogamy (b) Geitonogany
(c) Xenogeny (d) None of these

48. Read the given statements-
(i) Autogamy cannot occur in open flower.
(ii) Geitonogany cannot occur in closed flower.
Choose the appropriate answer-
(a) (i) is correct but (ii) is wrong
(b) (i) is wrong but (ii) is correct
(c) (i) and (ii) are both correct
(d) (i) and (ii) are both wrong

49. Complete autogamy is rare in-
(a) Closed flower
(b) Open flower
(c) Both open and closed flower
(d) Neither open nor closed flower

50. Flowers that do not open at all are called
(a) Chasmogamous (b) Polygamous
(c) Cleistogamous (d) Xenogamous

51. Oxalis produce-
(a) Cleistogamous flowers
(b) Chasmogamous flowers
(c) Both (a) and (b)
(d) Can't say

52. How many of the given characters are necessarily present in cleistogamous flower.
(i) Anthex and stigma lie close to each other.
(ii) There is synchrony in pollen release and stigma receptivity.
(iii) Lengths of anther and stigma are very different.
(iv) Flower is necessarily dioecious.
(v) Assured seed-set even without pollinators.
(a) 1 (b) 2
(c) 3 (d) 4

53. Geitonogamy is-
(a) Functionally self-pollination and genetically cross-pollination
(b) Genetically self-pollination and functionally cross-pollination
(c) Cross-pollination both genetically and functionally
(d) Self-pollination both genetically and functionally

54. Xenogamy is-
(a) Functionally self-pollination and genetically cross-pollination
(b) Genetically self-pollination and functionally cross-pollination

(c) Cross-pollination both genetically and functionally
(d) Self-pollination both genetically and functionally

55. Autogamy is-
(a) Functionally self-pollination and genetically cross-pollination
(b) Genetically self-pollination and functionally cross-pollination
(c) Cross-pollination both genetically and functionally
(d) Self-pollination both genetically and functionally

56. Genetically different type of pollen is brought to stigma by-
(a) Atutogamy only
(b) Geitonogamy only
(c) Xenogamy only
(d) More than one options

57. A. Pollination by abiotic agents is a chance factor.
B. Pollen is produced in enormous amount as compared to number of ovules.
Choose the best answer.
(a) A and B are correct and B is the reason for A
(b) A and B are correct and A is the reason for B
(c) A is incorrect and B is correct
(d) B is incorrect and A is correct

58. Which is more common abiotic agent for pollination-
(a) Wind (b) Insect
(c) Water (d) Animal

59. The pollen grains in wind pollinated plants should be-
(a) Heavy and sticky
(b) Heavy and non-sticky
(c) Light and sticky
(d) Light and non-sticky

60. Wind pollinated flowers often have ____ in each ovary and flowers are after ____.

	(i)	(ii)
(a)	Single	Single
(b)	Multiple	Single
(c)	Single	packed in inflorescence
(d)	Multiple	packed in inflorescence

61. The tassels in corn cob are-
(a) Filaments of anthers
(b) Stigma and style
(c) Reduced leaf
(d) Stalk of ovule

62. Match the columns.

	Column-I		Column-II
(i)	Wind pollination	(a)	Maize
(ii)	Water pollination	(b)	Hydrilla
(iii)	Biotic pollination	(c)	Monocots
(iv)	Freshwater pollination	(d)	Amorphophallus

(a) (i)-d, (ii)-b, (iii)-d, (iv)-c
(b) (i)-c, (ii)-d, (iii)-a, (iv)-b
(c) (i)-a, (ii)-c, (iii)-d, (iv)-b
(d) (i)-b, (ii)-a, (iii)-c, (iv)-d

63. (a) Distribution of some bryophytes & pteridophytes is limited. (r) Transport of male gamete in bryophytes & pteridophyte is dependent on water. Choose the correct options.
(a) a and r are correct but r is correct explanation for a
(b) a and r are correct but r is not correct explanation for a
(c) Both a and r are incorrect
(d) A is correct but r is incorrect

64. Aquatic plants pollinated by water are given, except-
(a) Zostera
(b) Hydrilla
(c) Water hyacinth
(d) More than one option

65. Pollination in water lily occurs by-
(a) Water (b) Wind

(c) Insects (d) Both B and C

66. Choose the correct statements for pollination in sea grasses-

(i)Female flower reach surface of water.
(ii)Female flower remain submerged.
(iii)Pollen released on water surface.
(iv)Pollen release inside water.
(v)Pollen grains are carried passively by water.
(vi)Pollen grains are carried actively in water.
(vii)Most of the pollen reach stigma.
(viii)Some of the pollen reach stigma.

(a) (i), (iii), (v), (vii)
(b) (ii), (iv), (vi), (vii)
(c) (ii), (iv), (v), (vii)
(d) (ii), (iv), (v), (viii)

67. Choose correct statements for pollination in vallisneria-

(i) Female flower reach surface of water.
(ii) Female flower remain submerged.
(iii) Pollen released on water surface.
(iv) Pollen release inside water.
(v) Pollen grains are carried passively by water.
(vi) Pollen grains are carried actively in water.
(vii)Most of the pollen reach stigma.
(viii)Some of the pollen reach stigma.

(a)(i), (iii), (v), (vii)
(b)(ii), (iv), (vi), (vii)
(c)(i), (iii), (v), (viii)
(d)(ii), (iv) (v), (viii)

68. Requirement for pollen of water pollinated plants are-

(i) Light pollen
(ii) Pollen with mucilagenous cover
(iii) Non-sticky pollen
(iv) Long ribbon-like pollen

(a) (i) and (iii) (b) (iii) and (iv)
(c) (i) and (ii) (d) (ii) and (iv)

69. Majority of angiosperms use ____ for pollination-

(a) Wind b) Water
(c) Animals d) Both A and B

70. Identify the given labels-

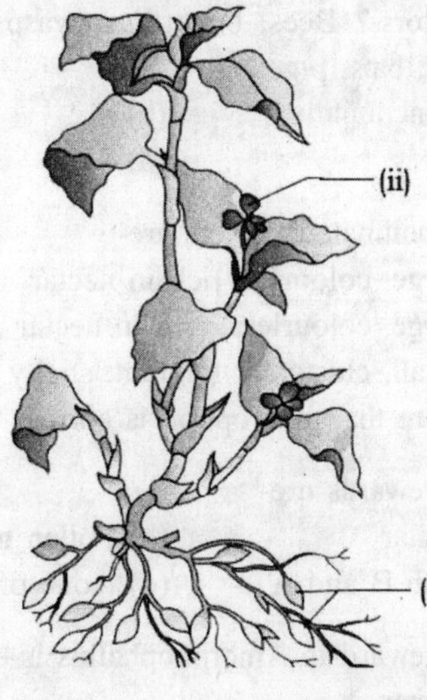

	(i)	(ii)
a)	Chasmogamous, autogamy	Cleistogamous, allogamy
b)	Chasmogamous, allogamy	Cleistogamous, autogamy
c)	Cleistogamous, autogamy	Chasmogamous, allogamy
d)	Cleistogmous, allogamy	Chasmogamous, autogamy

71. The figure shows

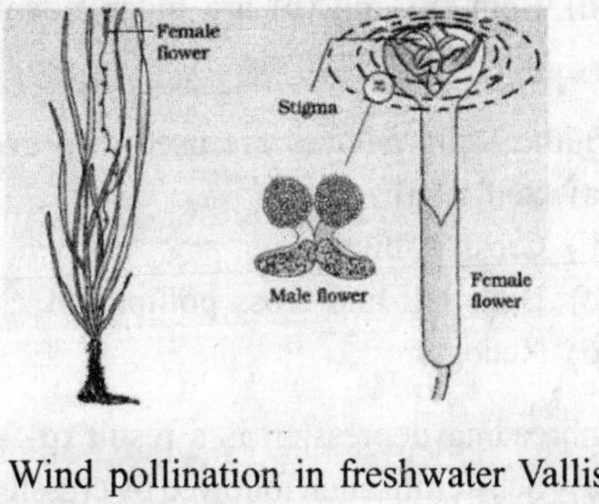

(a) Wind pollination in freshwater Vallisnaria
(b) Water pollination in marine Hydrilla
(c) Water pollination in marine Lostera
(d) Water pollination in freshwater Vallineria

72. Dominant biotic pollinating agents are-

(a) Bees (b) Birds
(c) Butterflies (d) Ants

73. How many of the following may act as pollinators-? Bees, butterflies, wasps, beetles, leopard, bats, pigeon

(a) 5 (b) 4
(c) 3 (d) 2

74. Insect-pollinated flowers are-

(a) Large, colourful, rich in nectar
(b) Large, colourless, rich in nectar
(c) Small, clustered, fragrant, sticky
(d) More than one option is correct

75. Floral rewards are-

(a) Nectar (b) Pollen grains
(c) Both B and A (d) None of these

76. Floral reward in Amorphophallus is-

(a) Nectar
(b) Safe place to lay-eggs
(c) Colourful petals
(d) Fragrance to attract insects

77. Which of the statements is true about (a) Pronuba moth and (b) Yucca plant?

(a) (a) is dependent on (b) for life cycle but the opposite is not true
(b) (b) is dependent on a for life cycle but the opposite is not true
(c) Both (a) and (b) are interdependent on each other for their life cycle
(d) Both (a) and (b) are independent of each other for life cycle

78. Outbreeding devices are used to prevent-

(a) Self-fertilization
(b) Cross-pollination
(c) Both self and cross pollination
(d) Xenogamy

79. Inbreeding depression is a result of-

(a) Self-fertilization followed by crossfertilization
(b) Cross-fertilization followed by selffertilization
(c) Continued cross-fertilization
(d) Continued self-fertilization

80. Self-pollination can be prevented by separation of anther and stigma in-

(a) time (maturity) (b) place (position)
(c) none of these (d) both of these

81. Self-incompatibility is not-

(a) Genetic mechanism
(b) Positional separation of anther and stigma
(c) Prevention for geitonogamy
(d) More than one option

82. Production of unisexual flowers on a plant assures prevention of-

(a) Autogamy only
(b) Autogamy and geitonogamy
(c) Geitonogamy only
(d) Autogamy and Xenogamy

83. Monoecious plants assures-

(a) no autogamy
(b) no autogamy and geitonogamy
(c) no autogamy and xenogamy
(d) no geitonogamy and xenogamy

84. Dioecious plants assures-

(a) no autogamy
(b) no autogamy and geitonogamy
(c) no autogamy and xenogamy
(d) no geitonogamy and xenogamy

85. If a wrong pollen (from other species or self-incompatible) lands on stigma-

(a) Pollen germinates but pollen tube cannot grow in style
(b) Pollen germinates, grows in style but cannot enter ovary
(c) Does not germinate at all
(d) Both A and C

86. When pollen grain germinates and produce pollen tubes

(a) Content of pollen grain is distributed uniformly
(b) Content of pollen grain move into pollen tube
(c) Content of pollen grain is distributed non-uniform, more in pollen grain
(d) Content of pollen grain is distributed non-uniformly, more in pollen tube

87. Filiform apparatus is present at-
(a) Micropylar part of synergid
(b) Chalazal part of synergid
(c) Micropylar part of antipodal
(d) Chalazal part of antipodal

88. Emasculation is done in-
(a) Male parent
(b) Female parent
(c) Both male and female parent
(d) Depends on the project

89. The emasculated flowers are bagged to-
(a) Protect flower from strong sunlight
(b) Protect flower from rain
(c) Protect flower from unwanted pollen
(d) Protect flower from insects

90. If female parent produces unisexual flowers, there is-
(a) no need of emasculation & bagging
(b) need of emasculation & bagging
(c) no need of emasculation but bagging is needed
(d) no need of bagging but emasculation is needed

91. Identify the filiform apparatus in given figure-

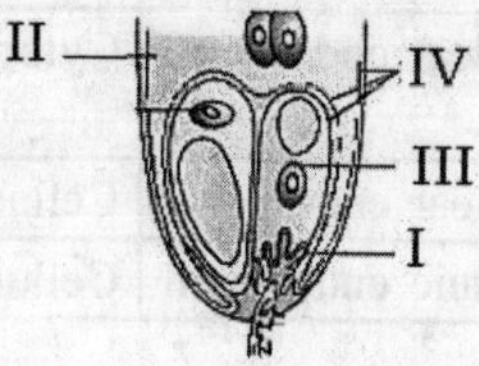

(a) I (b) II
(c) III (d) IV

Topic 6 Double Fertilization

92. Pollen tube release male gametes into-
(a) Cytoplasm of Egg cell
(b) Nucleus of Egg cell
(c) Cytoplasm of Synergids
(d) Cytoplasm of Antipodals

93. Which of the following is incorrect about double fertilization?
(a) One male gamete fuses with nucleus of egg cell
(b) Syngamy results into dyad of cells
(c) Second male gamete move toward polar nuclei
(d) Triple fusion results into PEN

94. Triple fusion is-
(a) Fusion of third male gamete with polar nuclei
(b) Fusion of three haploid cells
(c) Fusion of second male gamete with egg cell
(d) Fusion of three haploid nuclei

95. Which of these is correct?
(a) Syngamy = Triple fusion + Double fertilization
(b) Double fertilization = Syngamy + Triple fusion
(c) Triple fusion = Double fertilization – Syngamy
(d) More than one option is correct

96. Central cell after double fertilization becomes-
(a) Zygote (b) PEN
(c) PEC (d) Embryo

97. Identify the correct labels.

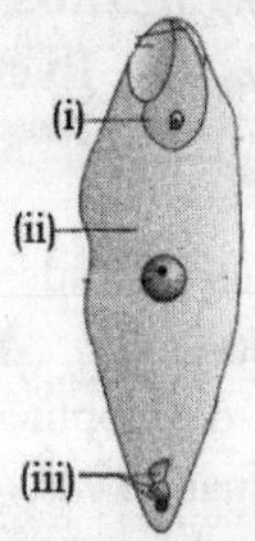

	(i)	(ii)	(iii)
(a)	Zygote	PEN	Degenerating antipodals
(b)	Zygote	PEC	Degenerating antipodals
(c)	Zygote	PEN	Degenerating synergids
(d)	Zygote	PEC	Dengenerating synergids

98. Identify the correct labels.

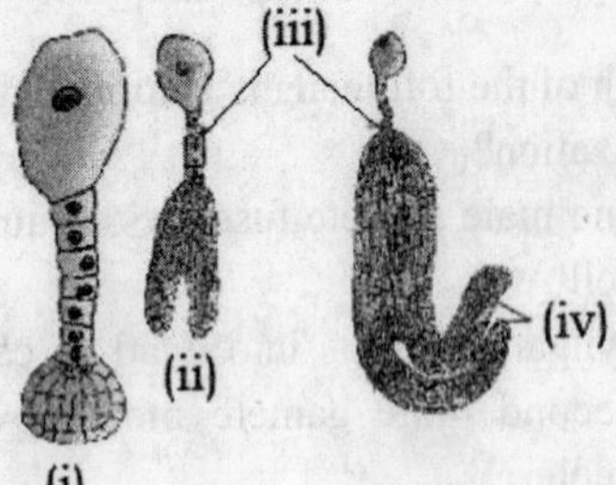

	(i)	(ii)	(iii)	(iv)
A	Globular embryo	Heart-shaped embryo	Suspensor	Cotyledon
B	Heart-shaped embryo	Globular embryo	Cotyledon	Suspensor
C	Globular embryo	Heart-shaped embryo	Cotyledon	Suspensor
D	Heart-Shaped embyro	Globular embryo	Suspensor	Cotyledon

99. The figure shows stages in-
(a) Embryo development in dicot
(b) Embryo development in monocot
(c) Embryo development in gymnosperm
(d) Both A and B

Topic 7 Post Fertilization : Structure and Events

100. Post fertilization includes how many of the following events-
(i) endosperm development
(ii) zygote formation
(iii) embryo development
(iv) seed formation
(v) fruit formation
(a) 5 (b) 4
(c) 3 (d) 2

101. Select correct statement-
(a) Endosperm development proceeds embryosac development
(b) Endosperm development precedes embryo development
(c) Embryo development precedes endosperm development
(d) More than one option is correct

102. Endosperm tissue is-
(a) Haploid (b) Diploid
(c) Triploid (d) Tetraploid

103. In free-nuclear endosperm-
(a) PEN undergoes successive nuclear divisions
(b) PEC undergoes successive cellular divisions
(c) PEN undergoes successive cellular divisions
(d) More than one option is correct

104. Cells of endosperm tissue are filled with-
(a) reserve food materials for plant cells
(b) reserve food material for embryo
(c) reserve food material for developing zygote
(d) more than one option is correct

105. Coconut water from tender coconut is ____ and white kernel is ____.

	(i)	(ii)
(a)	Cellular endosperm	Free-nuclear endosperm
(b)	Free nuclear endosperm	Cytoplasmic endosperm
(c)	Free-nuclear endosperm	Cellular endosperm
(d)	Cytoplasmic endosperm	Cellular endosperm

106. Endosperm is completely consumed by developing embryo before seed maturation in-
(a) Groundnut (b) Castor
(c) Coconut (d) All of these

107. Endosperm persists in mature seeds in-
(a) Castor
(b) Pea
(c) Beans
(d) More than one option is correct

Topic 8 Embryo

108. Embryo develops at
(a) micropylar end
(b) chalazal end
(c) either micropylar or chalazal end
(d) neither microplar nor chalazal end

109. Choose the correct order of embryo development in dicots-
(i) Zygote
(ii) Heart-shaped embryo
(iii) Mature embryo
(iv) Proembryo
(v) Globular embryo
(a) i-iv-ii-v-ii (b) i-iv-ii-v-iii
(c) i-iv-iii-ii-v (d) ii-iv-v-ii-iii

110. How many of the given parts are present in dicot embryo-Embryonal axis, Cotyledons, Scutellum, Hypocotyl, Root cap
(a) 5 (b) 4
(c) 3 (d) 2

111. How many of the given parts are present in monocot embryo-Cotyledon, scutellum, Coleoptile, Radicle, Root cap
(a) 5 (b) 4
(c) 3 (d) 2

112. Hypocotyl terminates in-
(a) Plumule
(b) Radicle
(c) Root tip
(d) More than one option is correct

113. (i) In dicot embryo, root tip is covered by root cap.
(ii) In dicot embryo, scutellum is situated towards one side of embryonal axis.
(iii) Cylindrical portion below the level of cotyledons is hypocotyl in dicots embryo.
(iv) In dicot embryo, epicotyl terminates with stem tip.
How many of the above statements is incorrect?
(a) Zero (b) One
(c) Two (d) Three

114. In grass family, the scutellum is-
(a) Cotyledon (b) Root tip
(c) Epiblast (d) Shot tip

115. Identify the given figures

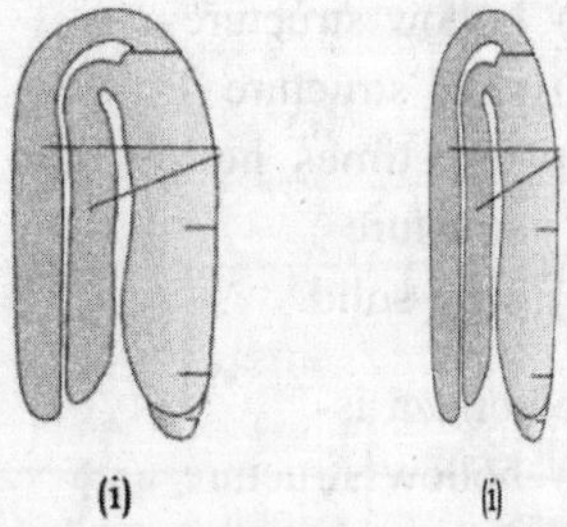

(a) (i) is embryo of grass
(b) (ii) is embryo of dicots
(c) (ii) is embryo of monocot
(d) More than one option is correct

116. Identify the correct labels-

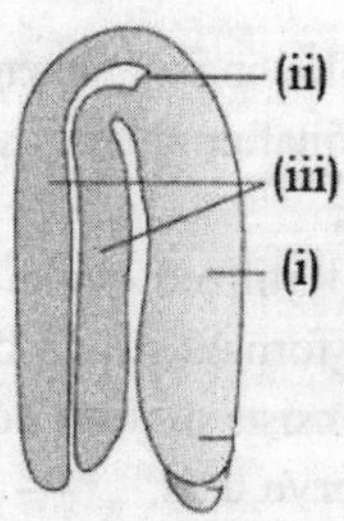

(i)	(ii)	(iii)
(a)	Cotyledon	Plumule	Hypocotyl
(b)	Radicle	Cotyledon	Plumule
(c)	Hypocotyl	Plumule	Cotyledon
(d)	Cotyledon	Plumule	Epicotyl

117. Identify the correct labels-

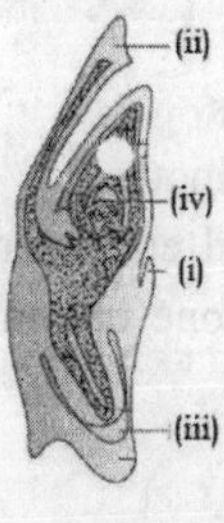

	(i)	(ii)	(iii)	(iv)
A	Epiblast	Scutellum	Coleoptile	Root cap
B	Scutellum	Epiblast	Shoot apex	Radicle
C	Epiblast	Scutellum	Root cap	Shoot apex
D	Scutellum	Epiblast	Radicle	Coleoptile

118. Coleoptile is-
(a) hollow structure
(b) solid structure
(c) sometimes hollow and sometimes solid structure
(d) semi-solid

119. Coleorhiza is-
(a) hollow structure
(b) foliar structure
(c) undifferentiated sheath
(d) more than one option is correct

Topic 9 Seed

120. Read the following statements-
(i) Seed is final product of sexual reproduction is plant.
(ii) Seed is fertilized ovule.
(iii) Seed is formed inside fruit.
(iv) Seed consists of seed coat(s), cotyledon(s) and embryo axis.
How many of the statements is incorrect?
(a) Zero (b) One
(c) Two (d) Three

121. Non-albuminous seeds-
(a) have residual endosperm
(b) retain a part of endosperm
(c) is found in castor
(d) None of these

122. Groundnut is-
(a) Albuminous
(b) Non-albuminous
(c) Has residual endosperm in mature seed
(d) More than one option is correct

123. Perisperm is-
(a) Persistent nucleus
(b) Found in beet
(c) Residual endosperm
(d) More than one option

124. Integument of ovules mature into-
(a) Ovary wall (b) Pericarp
(c) Seed coat (d) Perisperm

125. Micropyle is-
(a) Absent in seed (b) Present inside seed
(c) Present on surface of seed
(d) Present on seed coat

126. Micropyle plays role of-
(a) Stalk for seed (b) Scar of stalk
(c) Facilitating entry of water into seed
(d) Facilitating escape of seed metabolites

127. Mature seed has-
(a) More water content and more metabolism
(b) Less water content and more metabolism
(c) Less water content and less metabolism
(d) More water content and more metabolism

128. The embryo in a mature seed-
(a) Germinates essentially
(b) May enter dormancy
(c) Always enters dormancy first, followed by germination
(d) Both B and C

129. Choose the correct match regarding the maturing of flower into fruit-
(a) Wall of ovule – pericarp
(b) Nucellus – periderm
(c) Ovary – seed
(d) None of these

130. Fleshy fruit is-
(a) Mustard (b) Groundnut
(c) Guava (d) More than one

131. In false fruits, select incorrect statement-
(a) Floral parts other than ovary are involved
(b) Thalamus may contribute to fruit formation
(c) Examples include apple, cashew, groundnut
(d) Fruit does not develop from ovary

132. Which of these is incorrect about parthenocarpy-
(a) Plant formed without fertilization
(b) Banana is example
(c) Induced by application of growth harmones
(d) Such fruits are seed less

133. Identify the endosperm in the given figure–

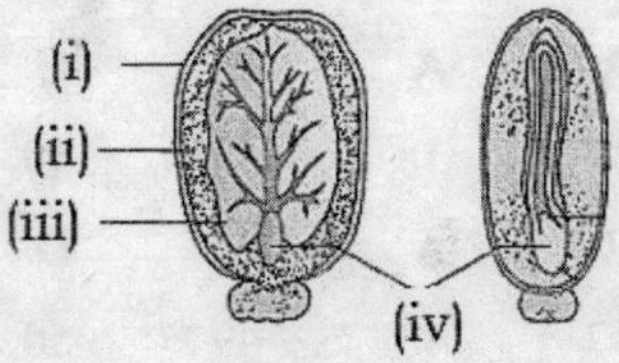

(a) I (b) II
(c) III (d) IV

134. Identify cotyledon in the given figure of seed

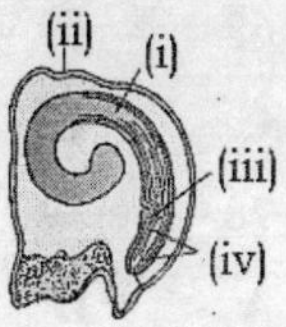

(a) I (b) II
(c) III (d) IV

135. Identify scutellum in the given figure –

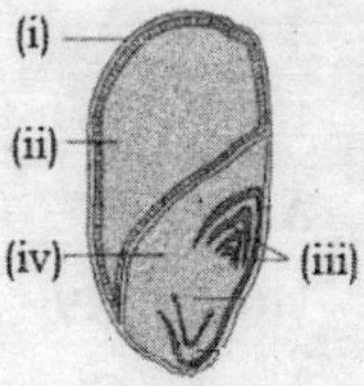

(a) I (b) II
(c) III (d) IV

136. The given figure shows –

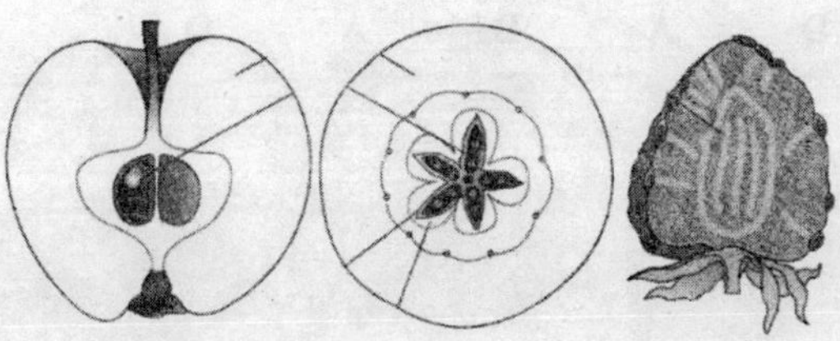

(a) Eucarp of apple and lithi
(b) Pseudocarp of apple litchi
(c) Eucarp of apple and strawberry
(d) Pseudocarp of apple and strawberry

137. In angiosperm, pollination and fertilization are –
(a) Both independent of water
(b) Both dependent of water
(c) Only pollination is essentially on water
(d) Only fertilization is dependent on water

138. For storage of seeds –
(a) Dehydration is important
(b) Dormancy is important
(c) Neither dehydration nor dormancy is needed
(d) Both dehydration and dormancy are crucial

139. The oldest yet viable seed found is –
(a) Lupinus from arctic tundra
(b) Phoenix from arctic tundra
(c) Lupinus from king herod's palace
(d) Phoenix from king herod's palace

140. Phoenix dactylifera is commonly known as–
(a) Fig (b) Coconut
(c) Cashew (d) None of these

Topic 10 | Apomixis and Polyenbryony

141. Apomixis is –
(a) Fruit without fertilization
(b) Seed without fertilization
(c) Plant without fertilization
(d) More than one option

142. Apomixis is –
(a) A form of sexual reproduction that mimics asexual reproduction
(b) A form of asexual reproduction that mimics sexual reproduction
(c) Both of the above
(d) None of these

ANSWER KEY

Q.	1	2	3	4	5	6	7	8	9	10
Ans.	**D**	**A**	**B**	**A**	**D**	**B**	**A**	**C**	**A**	**B**
Q.	11	12	13	14	15	16	17	18	19	20
Ans.	**A**	**B**	**C**	**B**	**C**	**A**	**A**	**B**	**D**	**A**
Q.	21	22	23	24	25	26	27	28	29	30
Ans.	**B**	**D**	**C**	**D**	**D**	**B**	**A**	**A**	**B**	**C**
Q.	31	32	33	34	35	36	37	38	39	40
Ans.	**A**	**B**	**A**	**B**	**B**	**C**	**B**	**A**	**A**	**D**
Q.	41	42	43	44	45	46	47	48	49	50
Ans.	**C**	**A**	**C**	**B**	**A**	**D**	**A**	**A**	**B**	**C**
Q.	51	52	53	54	55	56	57	58	59	60
Ans.	**C**	**C**	**B**	**C**	**D**	**C**	**B**	**A**	**D**	**C**
Q,	61	62	63	64	65	66	67	68	69	70
Ans.	**B**	**C**	**A**	**C**	**D**	**D**	**D**	**C**	**C**	**C**
Q.	71	72	73	74	75	76	77	78	79	80
Ans.	**D**	**A**	**A**	**D**	**C**	**B**	**C**	**A**	**D**	**D**
Q.	81	82	83	84	85	86	87	88	89	90
Ans.	**B**	**A**	**A**	**B**	**D**	**B**	**A**	**B**	**C**	**C**
Q.	91	92	93	94	95	96	97	98	99	100
Ans.	**A**	**C**	**B**	**D**	**D**	**C**	**B**	**A**	**A**	**B**
Q.	101	102	103	104	105	106	107	108	109	110
Ans.	**B**	**C**	**A**	**D**	**C**	**A**	**A**	**A**	**D**	**B**
Q.	111	112	113	114	115	116	117	118	119	120
Ans.	**B**	**D**	**B**	**A**	**C**	**C**	**C**	**A**	**C**	**A**
Q.	121	122	123	124	125	126	127	128	129	130
Ans.	**D**	**B**	**B**	**C**	**D**	**C**	**C**	**B**	**D**	**C**
Q.	131	132	133	134	135	136	137	138	139	140
Ans.	**A**	**A**	**B**	**A**	**D**	**D**	**A**	**D**	**A**	**D**
Q.	141	142								
Ans.	**B**	**B**								

14 Chapter

Principles of Inheritance and Variation

MULTIPLE CHOICE QUESTIONS

Topic 1 Mendel's Laws of Inheritance

1. Which technique was used by Mendel during his experiments on pea plant?
(a) Artificial pollination
(b) Cross pollination
(c) Self-pollination
(d) All of these

2. Choose the correct statement(s) from the following.
(I) During Mendel's investigation, statistical analysis and mathematical logic were applied to problems in Biology.
(II) Mendel investigated characters in the garden pea plant that were manifested as two opposing traits.
(III) Mendel conducted artificial pollination experiments using several true-breeding pea lines.
(IV)Mendel selected eight true-breeding pea plant varieties as pairs.
(a) I and II (b) III and IV
(c) I, II and III (d) All of these

3. The contrasting trait(s) selected by Mendel was/were
(a) smooth or wrinkled seed
(b) yellow or green seed
(c) smooth or inflated pods
(d) all of these

4. Assertion: Mendel conducted hybridization experiments on garden pea plant. Reason: He proposed laws of inheritance in living organisms.
(a) Both assertion and reason are true and reason is the correct explanation of assertion.
(b) Both assertion and reason are true but reason is not correct explanation of assertion.
(c) Assertion is true, but reason is false.
(d) Both assertion and reason are false.

5. Assertion: Mendel used contrasting traits for his studies. Reason:He used Ocimum plant for his experiments.
(a) Both assertion and reason are true and reason is the correct explanation of assertion.
(b) Both assertion and reason are true but reason is not correct explanation of assertion.
(c) Assertion is true, but reason is false.
(d) Both assertion and reason are false.

6. Assertion: Mendel used true-breeding pea lines for his experiments.

Reason: A true-breeding line is one that has undergone continuous self pollination.
(a) Both assertion and reason are true and reason is the correct explanation of assertion.
(b) Both assertion and reason are true but reason is not correct explanation of assertion.
(c) Assertion is true, but reason is false.
(d) Both assertion and reason are false. Inheritance of one Gene

7. Genetics is the subject that deals with
(a) inheritance
(b) variation of characteristics
(c) reproduction
(d) both (a) and (b)

8. The basis of heredity is
(a) variation (b) inheritance
(c) mutation (d) linkage

9. Humans knew from as early as 8000–1000 BC that one of the causes of variation was hidden in
(a) sexual reproduction
(b) asexual reproduction
(c) vegetative propagation
(d) none of these

10. Choose the incorrect statement from the following.
(a) Humans knew from very early that sexual reproduction is one of the causes of variation.
(b) They exploited the variation to obtain plants and animals of desirable characters through selective breeding.
(c) Sahiwal cows were obtained through artificial selection and domestication from ancestral wild cows.
(d) Our ancestors were very well aware about the scientific basis of inheritance of characters and variation.

11. Which one from the following is the period for Mendel's hybridization experiments?
(a) 1840–1850 (b) 1857–1869
(c) 1870–1877 (d) 1856–1863

12. Who proposed the 'Laws of Inheritance' in living organisms?
(a) Mendel (b) Morgan
(c) de Vries (d) Correns

13. Match Column-I with Column-II and choose the correct answer from the codes given below.

Column-I		Column-II	
(a)	Genetics	(1)	Process of passing characters from parent to offspring
(b)	Inheritance	(2)	Laws of inheritance
(c)	Variation	(3)	A branch of Biology
(d)	Mendel	(4)	Degree of difference of progeny from their parents

Codes-

	A	B	C	D
(a)	1	4	2	3
(b)	4	2	3	1
(c)	3	1	4	2
(d)	2	3	1	4

14. Mendel investigated characters in the garden pea plant that were manifested as two
(a) linked traits (b) opposing traits
(c) similar traits (d) none of these

15. How many pairs of contrasting characters in pea plants were studied by Mendel in his experiments?
(a) Six (b) Eight
(c) Seven (d) Four

16. Which contrasting trait was not studied by Mendel during his experiments?
(a) Seed colour (b) Leaf colour
(c) Flower colour (d) Stem height

17. Among the following, which one is not a dominating trait?
(a) Axial position of flower
(b) Green colour of pod
(c) Violet colour of flower
(d) Green colour of seed

18. A true-breeding line is one that
(a) has undergone continuous selfpollination
(b) shows stable trait inheritance
(c) shows expressions of trait for several generations
(d) all of these

19. Match Column-I with Column-II and choose the correct option from the codes given below.

Column-I	Column-II
(a) Axial flower	(1) Undergone continuous self-pollination
(b) Terminal flower	(2) Father of genetics
(c) Mendel	(3) Dominant trait
(d) True-breeding line	(4) Recessive trait

Codes-

	A	B	C	D
(a)	3	4	2	1
(b)	4	3	1	2
(c)	1	2	4	3
(d)	2	1	3	4

20. Refer to the given figures (A–d) showing traits of pea plant studied by Mendel. Among these, choose the dominant trait.

A B C D

(a) B (b) A
(c) D (d) C

Topic 2 | Inheritance of One Gene

21. Choose the incorrect statement about law of dominance.
(a) It is used to explain the expression of only one of the parental characters in a monohybrid cross in F1-generation.
(b) It does not explain the expression of both parental characters in F2-generation.
(c) It also explains the proportion of 3: 1 obtained in F2-generation.
(d) It states that characters are controlled by discrete units called factors.

22. Match Column-I with Column-II and choose the correct option from the codes given below.

Column-I		Column-II	
(a)	First law of inheritance	(1)	Law of segregation
(b)	Second law of inheritance	(2)	3: 1
(c)	Monohybrid cross	(3)	Law of dominance
(d)	Test cross	(4)	1: 1

Codes-

	A	B	C	D
(a)	3	1	2	4
(b)	1	3	4	2
(c)	2	3	1	4
(d)	4	2	3	1

23. The second law of inheritance, i.e., law of segregation is based on the fact that
(a) alleles do not show any blending.
(b) both characters are recovered as such in F2 generation.
(c) one allele dominates the other allele.
(d) Both (a) and (b)

24. The factor controlling any character is discrete and independent. It was concluded on the basis of
(a) results of F3-generation of a cross.
(b) observations of a cross made between the plants having two contrasting traits where offspring shows only one trait without any blending.
(c) self-pollination of F1-offspring.
(d) cross pollination of parental generations.

25. In Antirrhinum (Snapdragon), a red flower was crossed with a white flower and in F1 generation, pink flowers were obtained. When pink flowers were selfed, the F2 generation showed white, red and pink flowers. Choose the incorrect statement from the following.
(a) The experiment does not follow the principle of dominance.
(b) Pink colour in F1 is due to incomplete dominance.
(c) Ratio of F2 is ¼ (Red): 2/4 (Pink): ¼ (white).
(d) Law of segregation does not apply in this experiment.

26. It was being observed that sometimes, the F1 shows a phenotype that does not resemble either of the two parents and remains in between the two. It can be explained by
(a) Law of dominance

(b) Law of segregation
(c) Law of incomplete dominance
(d) None of these

27. The genotypic ratio obtained in incomplete dominance is
(a) 3 : 1 (b) 1 : 1 : 2
(c) 2 : 1 : 1 (d) 1 : 2 : 1

28. In case of co-dominance, the F1 progeny
(a) resembles either of the two parents
(b) is in between of parents
(c) resembles both the parents
(d) none of these

29. A person of AB blood group has IA and IB genes. It is an example of
(a) pleiotropy (b) segregation
(c) co-dominance (d) None of these

30. In a marriage between male with blood group A and female with blood group B, the progeny had either blood group AB or B. What could be the possible genotype of parents?
(a) IAi (Male); IBi (Female)
(b) IAi (Male); IBIB (Female)
(c) IAIA (Male); IBIB (Female)
(d) IAIA (Male); IBi (Female)

31. A person has 'O' blood group. His mother has 'A' while father has 'B' blood group. What would be the genotype of mother and father?
(a) Mother is homozygous for 'A' blood group and father is heterozygous for 'B' blood group.
(b) Mother is heterozygous for 'A' blood group and father is homozygous for 'B' blood group.
(c) Both mother and father are homozygous for 'A' and 'B' blood groups respectively.
(d) Both mother and father are heterozygous for 'A' and 'B' blood groups respectively.

32. Which of the following characteristics represent 'inheritance of blood groups' in humans?
(I) Dominance
(II) Co-dominance
(III) Multiple dominance
(IV) Incomplete dominance
(V) Polygenic inheritance
(a) II, III and V (b) I, II and III
(c) II, IV and V (d) I, III and V

33. A man with blood group 'A' marries a woman with blood 'B'. What are all possible blood groups of their offsprings?
(a) A, B and AB only
(b) A, B, AB and O
(c) O only
(d) A and B only

34. The genotypes of a husband and wife are IAIB and IAi. Among the blood types of their children, how many different genotypes and phenotypes are possible?
(a) 3 genotypes: 4 phenotypes
(b) 4 genotypes: 3 phenotypes
(c) 4 genotypes: 4 phenotypes
(d) 3 genotypes: 3 phenotypes

35. Multi alleles are present
(a) at different loci on the same chromosome
(b) at the same locus of the chromosome
(c) on non-sister chromatids
(d) on different chromosome

36. Match Column-I with Column-II and choose the correct answer from the codes given below.

	Column-I		Column-II
(a)	Dominance	(1)	ABO blood group
(b)	Codominance	(2)	Appearance of pink flowers in snapdragon in F1 generation
(c)	Incomplete dominance	(3)	Starch synthesis in pea seeds
(d)	Pleiotropy	(4)	Appearance of violet flowers in F1 generation in garden pea

Codes-

	A	B	C	D
(a)	4	1	2	3

(b)	1	4	3	2
(c)	3	2	4	1
(d)	2	3	1	4

37. ABO blood grouping is a good example of
(a) incomplete dominance
(b) mutation
(c) multiple alleles
(d) pleiotropy

38. Sometimes a single gene product may produce more than one effect. This phenomenon is known as
(a) mosaicism (b) pleiotropy
(c) multiple allelism (d) polygeny

39. Starch synthesis in pea seeds is an example of
(a) multiple allelism
(b) incomplete dominance
(c) co-dominance
(d) pleiotropy

40. Pea seeds having Bb genotype produce starch grains of
(a) large size
(b) small size
(c) intermediate size
(d) they do not produce starch.

41. Choose the incorrect statement from the following about pleiotropy.
(a) In pleiotropy, a single gene produces more than one effect.
(b) Starch synthesis in pea seeds is controlled by one genc.
(c) Pea seeds having BB genotypes, produce small starch grains.
(d) bb homozygotes of pea produce wrinkled seeds.

42. Assertion: The law of dominance is used to explain the expression of only one of the parental characters in a monohybrid cross. Reason: It also explains the proportion of 3: 1 obtained at F2 generation.
(a) Both assertion and reason are true and reason is the correct explanation of assertion.
(b) Both assertion and reason are true but reason is not correct explanation of assertion.
(c) Assertion is true, but reason is false.
(d) Both assertion and reason are false.

43. Assertion: The pink flower of dog plant show incomplete dominance. Reason: In pink flowers, both alleles are expressed equally.
(a) Both assertion and reason are true and reason is the correct explanation of assertion.
(b) Both assertion and reason are true but reason is not correct explanation of assertion.
(c) Assertion is true, but reason is false.
(d) Both assertion and reason are false.

44. Assertion: A person having IAIB genotype has AB blood group. Reason: IA and IB alleles are co-dominant
(a) Both assertion and reason are true and reason is the correct explanation of assertion.
(b) Both assertion and reason are true but reason is not correct explanation of assertion.
(c) Assertion is true, but reason is false.
(d) Both assertion and reason are false.
Inheritance of two Genes

45. The first hybrid generation of Mendel's experiment is known as
(a) Filial1 progeny
(b) F1-generation
(c) Father generation
(d) Both (a) and (b)

46. When Mendel crossed true-breeding tall and dwarf plants, in F1-generation all tall plants were obtained. On self-crossing in the F2 generation, he obtained
(a) 1/4th dwarf and 3/4th tall plants
(b) 3/4th dwarf and 1/4th tall plants
(c) 2/4th dwarf and 2/4th tall plants
(d) All dwarf plants

47. During the study of inheritance of one character in F2 generation, Mendel obtained phenotype in
(a) 2 : 1 ratio (b) 3 : 1 ratio
(c) 1 : 2 : 1 ratio (d) 1 : 1 : 1 : 1 ratio

48. The 'factors' of Mendel are today known as
(a) genome (b) gene
(c) DNA (d) allele

49. The slightly different forms of the same genes are called
(a) genome (b) DNA
(c) allele (d) cistron

50. Alleles are
(a) true-breeding homozygotes
(b) different molecular forms of a gene
(c) heterozygotes
(d) different phenotype

51. What would be the phenotype of a plant that had a genotype 'Tt'? Here 'T' represent tall trait while 't' represents dwarf trait.
(a) Tall
(b) Intermediate height
(c) Dwarf
(d) None of these

52. In homozygous condition, a particular gene has
(a) different alleles on homologous chromosomes.
(b) no alleles on homologous chromosomes.
(c) same alleles on homologous chromosomes.
(d) none of these

53. Tall and dwarf are the two alleles of gene of height. The dominant trait is
(a) dwarf
(b) tall
(c) both are equally dominant
(d) both are recessive

54. Match Column-I with Column-II and choose the correct option from the code given below.

Column-I		Column-II	
(a)	Genes	(1)	Slightly different forms of the same gene
(b)	Alleles	(2)	Genetic composition of an organism
(c)	Genotype	(3)	Physical appearance of an organism
(d)	Phenotype	(4)	Unit of inheritance

Codes–

	A	B	C	D
(a)	4	1	2	3
(b)	1	4	3	2
(c)	3	2	4	1
(d)	2	3	1	4

55. A cross that is performed for the study of a single character is
(a) dihybrid cross
(b) test cross
(c) monohybrid cross
(d) back cross

56. The given figure is the diagrammatic representation of a monohybrid cross. In the figure, some plants are mentioned as A and B. What will be the genotype of these plants?
(a) A – tt, B – Tt
(b) A – Tt, B – tt
(c) A – TT, B – TT
(d) A – Tt, B – Tt

57. Choose the incorrect statement about Mendel's monohybrid cross.

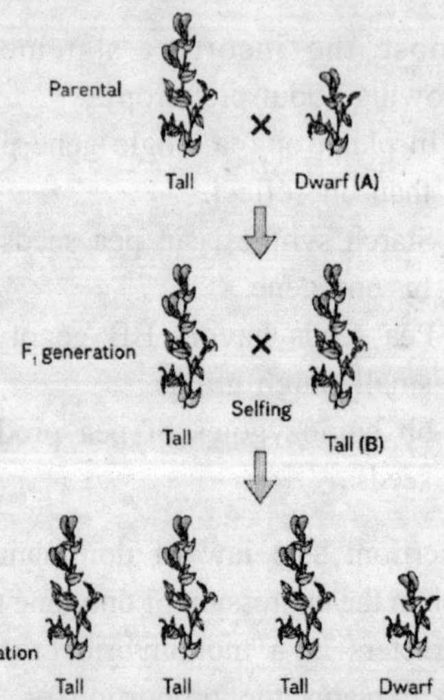

(a) The recessive parental trait is expressed without any blending in F2 generation.

(b) The alleles of parental pair segregate from each other and both alleles are transmitted to a gamete.
(c) The segregation of alleles is a random process.
(d) There is a 50% chance of a gamete containing either allele.

58. The production of gametes by the parents the formation of zygotes, the F1 and F2 plants, can be understood by using
(a) Wenn diagram
(b) Pie diagram
(c) A pyramid diagram
(d) Punnett square

59. Select the correct statement.
(a) Franklin Stahl coined the term 'linkage'.
(b) Punnett square was developed by a British scientist.
(c) Spliceosomes take part in translation.
(d) Transduction was discovered by S Altman.

60. In the test cross, organism whose genotype is to be determined, is crossed with the
(a) recessive parent
(b) dominant parent
(c) both parents one by one
(d) none of these

61. On crossing two tall plants, in F1-generation few dwarf offspring were obtained. What would be the genotype of the both the parent? a)
(a) TT and Tt (b) Tt and Tt
(c) TT and TT (d) TT and tt

62. Based on his observations of monohybrid cross, Mendel proposed which law of inheritance?
(a) Law of dominance
(b) Law of segregation
(c) Law of independent assortment
(d) Both (a) and (b)

63. According to Mendel, characters are controlled by discrete units called
(a) genes (b) factors
(c) alleles (d) allelomorph

Topic 3 | Inheritance of Two Genes

64. Crosses that are performed to study two contrasting characters at a time are called
(a) monohybrid cross
(b) dihybrid cross
(c) test cross
(d) back cross

65. The phenotypic ratio obtained by Mendel in his dihybrid cross was
(a) 1 : 2 : 1 : 2 (b) 3 : 2 : 2 : 1
(c) 9 : 3 : 3 : 1 (d) 2 : 3 : 1 : 2

66. The third law of inheritance proposed by Mendel is
(a) Law of dominance
(b) Law of independent assortment
(c) Law of incomplete dominance
(d) Law of segregation

67. The ratio 9: 3: 3: 1 of a dihybrid cross denotes that
(a) it is a multigenic inheritance.
(b) the alleles of two genes are interacting with each other.
(c) it is a case of multiple allelism.
(d) the alleles of two genes are segregating independently.

68. The numbers of phenotypes and genotypes in F2 generation of a Mendelian dihybrid cross are
(a) phenotypes 4: genotypes 16
(b) phenotypes 4: genotypes 8
(c) phenotypes 9: genotypes 4
(d) phenotypes 4: genotypes 9

69. Mendel's law of independent assortment is true for the genes situated on the
(a) same chromosome
(b) non-homologous chromosomes
(c) homologous chromosomes
(d) extra nuclear genetic element

70. Genes A and B are linked. The F1 heterozygote of a dihybrid cross involving these genes is crossed with homozygous recessive parental type (aabb). What would be the ratio of offspring in the next generation?
(a) 1: 1 (b) 1: 1: 1: 1
(c) 9: 3: 3: 1 (d) 3: 1

71. Mendel's work remained unrecognized for many years. Find out the true reason for the same.
(I) Mendel's concept of genes was not accepted by his contemporaries as an explanation for the continuous variation seen in nature.
(II) The approach of using mathematics was new and unacceptable by other biologists.
(III) He could not provide any physical proof for the existence of factors. (IV) Communication was not easy in those days and his work could not be widely published.
(a) I and II
(b) II and III
(c) III and IV
(d) All of these

72. Mendel's results on the inheritance of characters were rediscovered by:
(a) de Vries (b) Correns
(c) von Tschermak (d) all of these

73. Among the following, who noted that the behaviour of chromosomes was parallel to the behavior of genes?
(a) Walter Sutton (b) Theodore Boveri
(c) Von Tschermak (d) Both (a) and (b)

74. Refer to the given figure showing meiosis and germ cell formation in a cell with four chromosomes. Which law of Mendel can be effectively explained by this figure?
(a) Law of dominance
(b) Law of segregation
(c) Law of independent assortment
(d) All of these

75. The chromosomal theory of inheritance was proposed by
(a) Sutton (b) Boveri
(c) Morgan (d) Both (a) and (b)

76. Match Column-I with Column-II and choose the correct option from the codes given below.

	Column-I		Column-II
(a)	Mendel	(1)	Rediscovery of Mendel's law
(b)	Correns, Tschermak and Vries	(2)	Worked on Drosophila melanogaster
(c)	Sutton and Boveri	(3)	Law of independent assortment
(d)	T.H. Morgan	(4)	Chromosomal theory of inheritance

Codes-

	A	B	C	D
(a)	3	1	4	2
(b)	1	4	3	2
(c)	2	3	1	4
(d)	4	2	3	1

77. Morgan performed his experiments on
(a) Garden pea (b) Drosophila
(c) Snapdragon (d) None of these

78. When two genes are located on the same chromosome, the proportion of parental gene combination is
(a) higher than non-parental
(b) lower than non-parental
(c) equal to non-parental
(d) None of these

79. Genes which are present on the same chromosome
(a) do not form any linkage group.
(b) affect the phenotype by forming interactive groups.
(c) form a linkage group.
(d) form different groups depending upon their relative distance.

80. The term used to describe the generation of nonparental gene combination is
(a) linkage (b) recombination
(c) mutation (d) none of these

81. Which type of relationship is found between the distance of genes and percentage of recombination?
(a) Inverse (b) Parallel
(c) Direct (d) None of these

82. Among the following which will not cause variations among siblings?
(a) Linkage
(b) Independent assortment of genes
(c) Crossing over
(d) Mutation

83. Match Column-I with Column-II and choose the correct answer from the codes given below.

	Column-I		Column-II
(a)	Linkage	(1)	Non-parallel gene combination
(b)	Recombination	(2)	Genetic map
(c)	Sturtevant	(3)	Unit of distance between gene
(d)	Centimorgan	(4)	Physical association of genes

Codes-

	A	B	C	D
(a)	3	2	4	1
(b)	2	3	1	4
(c)	4	1	2	3
(d)	1	4	3	2

84. What map unit (centimorgan) is adopted in the construction of genetic maps?
(a) A unit distance between two expressed genes, representing 10% cross over.
(b) A unit distance between two expressed genes, representing 100% cross over.
(c) A unit distance between genes on chromosomes, representing 1% cross over.
(d) A unit distance between genes on chromosomes, representing 50% cross over.

85. The concept of genetic map was given by
(a) de Vries (b) Morgan
(c) Sturtevant (d) Mendel

86. Assertion: Mendel proposed the law of independent assortment on the basis of results of dihybrid cross.

Reason: When two pairs of traits are combined in a hybrid, segregation of one pair of characters is independent of the other pair of characters.
(a) Both assertion and reason are true and the reason is the correct explanation of assertion.
(b) Both assertion and reason are true but the reason is not the correct explanation of assertion.
(c) Assertion is true but reason is false.
(d) Both assertion and reason are false.

87. Assertion: The chromosomal theory of inheritance was proposed by T. H. Morgan.

Reason: Morgan worked on garden pea plants to give this theory.
(a) Both assertion and reason are true and the reason is the correct explanation of assertion.
(b) Both assertion and reason are true but the reason is not the correct explanation of assertion.
(c) Assertion is true but reason is false.
(d) Both assertion and reason are false.

88. Assertion: Morgan coined the term linkage to describe the physical association of genes on a chromosome.

Reason: Linkage shows more non-parental type combination of genes.
(a) Both assertion and reason are true and the reason is the correct explanation of assertion.
(b) Both assertion and reason are true but the reason is not the correct explanation of assertion.
(c) Assertion is true but reason is false.
(d) Both assertion and reason are false.

Topic 4 | Sex Determination

89. X-body was discovered by
(a) Mendel (b) Morgan
(c) Henking (d) de Vries

90. In XO type of sex determination, who does possess the X chromosome?
(a) Female
(b) Male
(c) Sometimes female and sometimes male
(d) None of these

91. X-chromosome is designated as
(a) autosome
(b) sex chromosome
(c) somatic chromosome
(d) none of these

92. Which type of sex determination is found in grasshoppers?
(a) XX – XY type (b) XX – XO type
(c) ZZ – ZW type (d) None of these

93. In a specific taxon of insects, some possess 17 chromosomes while others have 18 chromosomes. These 17 and 18 chromosomes bearing organisms are
(a) All males
(b) All females
(c) Females and males, respectively
(d) Males and females, respectively

94. In Drosophila, males possess
(a) XO chromosomes
(b) XX chromosomes
(c) XY chromosomes
(d) YY chromosomes

95. Match Column-I with Column-II and choose the correct option from the codes given below.

Column-I	Column-II		
(a)	X-body	(1)	Autosomes
(b)	X and Y chromosome	(2)	Henking
(c)	Somatic chromosome	(3)	Grasshopper
(d)	XO-types of sex determination	(4)	Allosomes

Codes-

	A	B	C	D
(a)	2	4	1	3
(b)	4	2	1	3
(c)	3	1	4	2
(d)	1	3	2	4

96. XY type of sex determination is found in
(a) Drosophila (b) humans
(c) grasshopper (d) both (a) and (b)

97. Choose the incorrect statement about XY type of sex determination.
(a) Both males and females have same number of chromosomes.
(b) The counter part of X chromosome is distinctly smaller and called Y chromosome.
(c) Males and females possess different number of autosomes.
(d) This type of sex determination is found in Drosophila.

98. Male heterogamety is found in
(a) grasshopper (b) Drosophila
(c) humans (d) all of these

99. In female heterogamety, females
(a) one type of gametes
(b) two types of gametes
(c) three types of gametes
(d) none of these

100. ZZ/ZW type of sex determination is the characteristics feature of
(a) platypus (b) snails
(c) peacock (d) cockroach

101. Among the following, which has a different mechanism of sex determination?
(a) Birds (b) Humans
(c) Drosophila (d) None of these

102. Refer to the given figure which is followed by few statements. Choose the incorrect statement about it.

(a) It shows male heterogamety.
(b) Both possess same types of autosomes.
(c) The sex of progeny is determined by females.
(d) This type of sex determination is different from humans.

103. In humans, sex is determined by
(a) females
(b) males
(c) environmental factors
(d) none of these

104. Match Column-I with Column-II and choose the correct option from the codes given below.

	Column-I		Column-II
(a)	XO-type	(1)	Drosophila
(b)	XY-type	(2)	Grasshopper
(c)	ZZ-ZW type	(3)	Birds
		(4)	Humans progeny from their parents

Codes-

	A	B	C	D
(a)	1	4	2	3
(b)	2	1	4	3
(c)	3	2	1	4
(d)	4	3	2	1

105. Match the items of Column I with Column II.

	Column-I		Column-II
(a)	XX-XO method of sex determination	(1)	Turner's syndrome
(b)	XX-XY method of sex determination	(2)	Female heterogametic
(c)	Karyotype-45	(3)	Grasshopper
(d)	ZW-ZZ method of sex determination	(4)	Female homogametic

Codes-

	A	B	C	D
(a)	4	2	1	3
(b)	2	4	1	3
(c)	1	4	2	3
(d)	3	4	1	2

106. Select the incorrect statement.
(a) Male fruit fly is heterogametic.
(b) In male grasshoppers, 50% of sperms have no sex chromosome.
(c) In domesticated fowls, sex of progeny depends on the type of sperm rather than egg.
(d) Human males have one of their sex chromosome much shorter than the other.

107. Assertion: Grasshoppers show male heterogamety.
Reason: Male grasshoppers produce two types of gametes.
(a) Both assertion and reason are true and the reason is the correct explanation of assertion.
(b) Both assertion and reason are true but the reason is not the correct explanation of assertion.
(c) Assertion is true but reason is false.
(d) Both assertion and reason are false.

108. Assertion: In fruitfly, sex of progeny is decided by females.
Reason: Females produce two types of gametes.
(a) Both assertion and reason are true and the reason is the correct explanation of assertion.
(b) Both assertion and reason are true but the reason is not the correct explanation of assertion.
(c) Assertion is true but reason is false.
(d) Both assertion and reason are false.

109. Assertion: Birds show female heterogamety.
Reason: In birds, the sex of progeny is determined by males.
(a) Both assertion and reason are true and the reason is the correct explanation of assertion.

(b) Both assertion and reason are true but the reason is not the correct explanation of assertion.
(c) Assertion is true but reason is false.
(d) Both assertion and reason are false.

Topic 5 Mutation and Genetic Disorders

110. Refer to the given figure. It is showing the characteristic features of

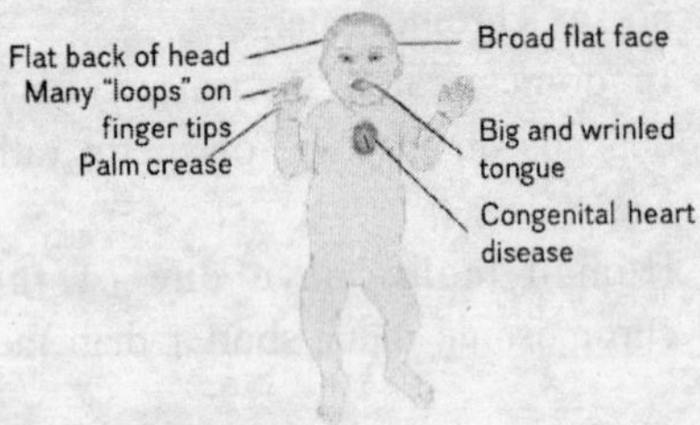

(a) Down's syndrome
(b) Turner's syndrome
(c) Klinefelter's syndrome
(d) None of these

111. The disease caused by the trisomy of chromosome number 21 is
(a) Turner's syndrome
(b) Haemophilia
(c) Klinefelter's syndrome
(d) Down's syndrome

112. An abnormal human baby with 'XXX' sex chromosomes was born due to
(a) formation of abnormal ova in the mother.
(b) fusion of two ova and one sperm.
(c) fusion of two sperms and one ovum.
(d) formation of abnormal sperms in the father.

113. What is the genetic disorder in which an individual has an overall masculine development, gynaecomastia and is sterile?
(a) Turner's syndrome
(b) Klinefelter's syndrome
(c) Edward's syndrome
(d) Down's syndrome

114. In which genetic condition, each cell in the affected person, has three sex chromosomes XXY?
(a) Turner's syndrome
(b) Thalassemia
(c) Kleinfelter's syndrome
(d) Phenylketonuria

115. A disorder caused due to the absence of one of the X chromosomes is
(a) Turner's syndrome
(b) Down's syndrome
(c) Klinefelter's syndrome
(d) Edward's syndrome

116. Assertion: The possibility of a female becoming a haemophilic is extremely rare. Reason: For being haemophilic, the mother of such a female has to be at least carrier and the father should be haemophilic.
(a) Both assertion and reason are true and the reason is the correct explanation of assertion.
(b) Both assertion and reason are true but the reason is not the correct explanation of assertion.
(c) Assertion is true but reason is false.
(d) Both assertion and reason are false.

117. Assertion: Aneuploidy is the gain or loss of chromosomes. Reason: It is caused due to the failure of cytokinesis after telophase stage of cell division.
(a) Both assertion and reason are true and the reason is the correct explanation of assertion.
(b) Both assertion and reason are true but the reason is not the correct explanation of assertion.
(c) Assertion is true but reason is false.
(d) Both assertion and reason are false.

118. Assertion: Klinefelter's syndrome is caused due to the presence of an additional copy of X-chromosome.
Reason: Such individuals are sterile.
(a) Both assertion and reason are true and the reason is the correct explanation of assertion.
(b) Both assertion and reason are true but the

reason is not the correct explanation of assertion.
(c) Assertion is true but reason is false.
(d) Both assertion and reason are false.

119. The phenomenon which results in alteration of DNA sequences is
(a) mutation (b) transpiration
(c) transcription (d) translation

120. Chromosomal aberrations are commonly observed in
(a) cardiac cells (b) cancer cells
(c) skeletal cells (d) none of these

121. A classical example of point mutation is
(a) gout
(b) night blindness
(c) sickle cell anaemia
(d) Turner's syndrome

122. The factors that cause mutations are called
(a) mutagens (b) teratogens
(c) allergens (d) none of these

123. An analysis of traits in several of generations of a family is called
(a) mutation
(b) pedigree analysis
(c) genetic map formation
(d) none of these

124. In 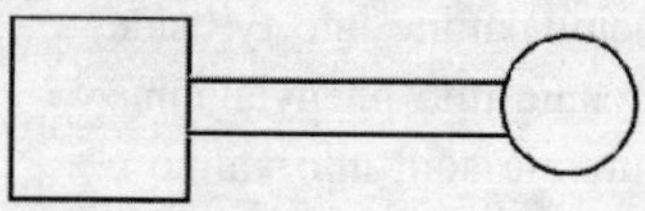 symbol repı

(a) affected individuals
(b) mating
(c) consanguineous mating
(d) unspecified sex

125. Pedigree analysis is used to study the inheritance pattern of a gene over generations. The character that is studied in the pedigree analysis is equivalent to
(a) Mendelian trait
(b) Maternal trait
(c) Polygamic trait
(d) Quantitative trait

126. Mendelian disorders are mainly determined by alteration or mutation in the
(a) chromosomes
(b) single gene
(c) array of genes
(d) none of these

127. Among the following which one is a Mendelian disorder?
(a) Haemophilia
(b) Sickle cell anaemia
(c) Cystic fibrosis
(d) All of these

128. Choose the incorrect statement about Mendelian disorders.
(a) These are usually caused by mutation in a single gene.
(b) These disorders are transmitted to the offspring according to the laws of inheritance.
(c) Mendelian disorders are always sex linked.
(d) The trait in question can be dominant or recessive.

129. A genetic disease transmitted from a carrier female that is phenotypically normal to only some male progeny is [NCERT Exemplar]
(a) sex-linked dominant
(b) sex-linked recessive
(c) autosomal dominant
(d) autosomal recessive

130. Refer to the given pedigree analysis. It is related to the analysis of

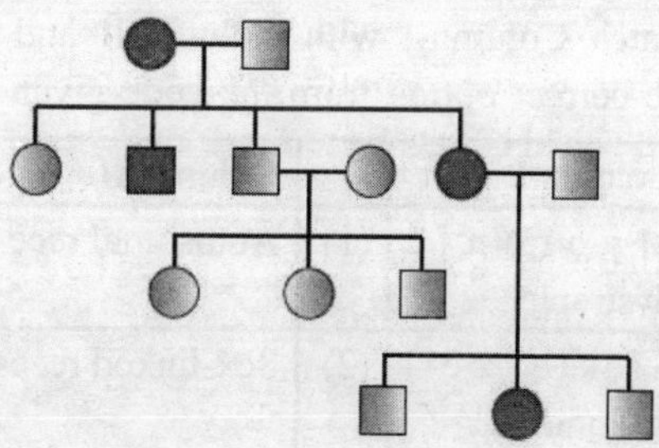

(a) autosomal dominant trait

(b) autosomal recessive trait
(c) sex-linked dominant trait
(d) sex-linked recessive trait

131. Haemophilia is a/an
(a) sex-linked recessive disease
(b) sex-linked dominant disease
(c) autosomal recessive disease
(d) autosomal dominant disease

132. The possibility of a female becoming a haemophilic is
(a) extremely high (b) extremely rare
(c) equal to a male (d) none of these

133. Haemophilia A and B are due to deficiencies of respectively clotting factor
(a) VIII and IX (b) IX and VIII
(c) VII and IX (d) X and VII

134. Sickle cell anaemia is a/an
(a) sex-linked recessive disease
(b) sex-linked dominant disease
(c) autosomal recessive disease
(d) autosomal dominant disease

135. In sickle cell anaemia, valine replaces glutamic acid. This valine is coded by the triplet
(a) AAG (b) GGG
(c) GUG (d) GAA

136. Sickle Cell Anaemia (SCa) is transferred from parents to offspring when
(a) father is affected and mother is normal.
(b) father is normal and mother is carrier.
(c) father is normal and mother is affected.
(d) both mother and father are carrier.

137. Match Column-I with Column-II and choose the correct option from the code given below.

	Column-I		Column-II
(a)	Myotonic dystrophy	(1)	Autosomal recessive
(b)	Sickle cell anaemia	(2)	Sex-linked recessive
(c)	Haemophilia	(3)	Sex-linked dominant
(d)	Rett syndrome	(4)	Autosomal dominant

Codes-

	A	B	C	D
(a)	4	2	1	3
(b)	2	4	1	3
(c)	1	4	2	3
(d)	3	4	1	2

138. Thalassemia and sickle cell anaemia are caused due to a problem in globin molecule synthesis. Select the correct statement.
(a) Both are due to a quantitative defect in globin chain synthesis.
(b) Thalassemia is due to less synthesis of globin molecules.
(c) Sickle cell anaemia is due to quantitative problem of globin molecules.
(d) Both are due to qualitative defect in globin chain synthesis.

139. The person suffering from phenylketonuria disease lacks enzyme
(a) phenylalanine hydroxylase
(b) phosphates
(c) enolase
(d) none of these

140. Phenylketonuria is an inborn error in which affected individual lacks an enzyme that converts
(a) phenylalanine into tyrosine
(b) tyrosine into phenylalanine
(c) glutamic acid into valine
(d) valine into glutamic acid

141. Phenylketonuria is a/an
(a) autosomal dominant trait
(b) autosomal recessive trait
(c) sex-linked dominant trait
(d) sex-linked recessive trait

142. If a colourblind man marries a women who is homozygous for normal colour vision, the probability of their son being colour blind is
(a) 0.75 (b) 1
(c) 0 (d) 0.5 25.

143. The chromosomal disorders are
(a) absence of one or more chromosomes
(b) excess of one or more chromosomes
(c) abnormal arrangement of chromosomes
(d) all of these

144. Condition of having 2n ± 1 or 2n ± 2 chromosomes is called
(a) polyploidy (b) aneuploidy
(c) allopolyploidy (d) monosomy

145. An increase in a whole set of chromosomes in an organism is called
(a) aneuploidy (b) linkage
(c) polyploidy (d) none of these

146. Condition (2n + 1) of chromosomes isknown as
(a) trisomy (b) monosomy
(c) polyploidy (d) haploidy

147. Match Column-I with Column-II and choose the correct option from the codes given below.

	Column-I		Column-II
(a)	Deletion	(1)	Loss of a gene or a segment of chromosome
(b)	Duplication	(2)	A segment of chromosome is turned around 180° within a chromosome
(c)	Inversion	(3)	Presence of a gene or segment of chromosome more than once
(d)	Translocation	(4)	Exchange of segments between two homologous chromosomes

Codes-

	A	B	C	D
(a)	4	2	1	3
(b)	2	4	1	3
(c)	1	4	2	3
(d)	3	4	1	2

148. Match Column-I with Column-II and choose the correct option from the codes given below.

	Column-I		Column-II
(a)	Aneuploidy	(1)	An increase in whole set of chromosomes
(b)	Polyploidy	(2)	2n + 1
(c)	Trisomy	(3)	Gain or loss of a chromosome
(d)	Monosomy	(4)	2n − 1

Codes-

	A	B	C	D
(a)	1	3	4	2
(b)	2	1	2	4
(c)	4	2	3	1
(d)	2	4	1	3

149. A disease caused by an autosomal primary nondisjunction is
(a) Klinefelter's syndrome
(b) Turner's syndrome
(c) Sickle cell anaemia
(d) Down's syndrome

ANSWER KEY

Q	01	02	03	04	05	06	07	08	09	10
Ans	**D**	**C**	**D**	**B**	**C**	**B**	**D**	**B**	**A**	**D**
Q	11	12	13	14	15	16	17	18	19	20
Ans	**D**	**A**	**C**	**B**	**C**	**B**	**D**	**D**	**A**	**A**
Q	21	22	23	24	25	26	27	28	29	30
Ans	**B**	**A**	**D**	**B**	**D**	**C**	**D**	**C**	**C**	**B**
Q	31	32	33	34	35	36	37	38	39	40
Ans	**D**	**B**	**B**	**B**	**B**	**A**	**C**	**B**	**D**	**C**
Q	41	42	43	44	45	46	47	48	49	50
Ans	**C**	**B**	**C**	**A**	**D**	**A**	**B**	**B**	**C**	**B**
Q	51	52	53	54	55	56	57	58	59	60
Ans	**A**	**C**	**B**	**A**	**C**	**A**	**B**	**D**	**B**	**A**
Q	61	62	63	64	65	66	67	68	69	70
Ans	**B**	**D**	**B**	**B**	**C**	**B**	**D**	**D**	**C**	**B**
Q	71	72	73	74	75	76	77	78	79	80
Ans	**D**	**D**	**D**	**D**	**D**	**A**	**B**	**A**	**C**	**B**
Q	81	82	83	84	85	86	87	88	89	90
Ans	**C**	**A**	**C**	**C**	**C**	**B**	**D**	**C**	**C**	**A**
Q	91	92	93	94	95	96	97	98	99	100
Ans	**B**	**B**	**D**	**C**	**A**	**D**	**C**	**D**	**B**	**C**
Q	101	102	103	104	105	106	107	108	109	110
Ans	**A**	**A**	**B**	**B**	**D**	**C**	**A**	**D**	**C**	**A**
Q	111	112	113	114	115	116	117	118	119	120
Ans	**D**	**A**	**B**	**C**	**A**	**A**	**C**	**B**	**A**	**B**
Q	121	122	123	124	125	126	127	128	129	130
Ans	**C**	**A**	**B**	**C**	**A**	**B**	**D**	**C**	**D**	**A**
Q	131	132	133	134	135	136	137	138	139	140
Ans	**A**	**B**	**A**	**C**	**C**	**D**	**A**	**B**	**A**	**A**
Q	141	142	143	144	145	146	147	148	149	
Ans	**B**	**C**	**D**	**B**	**C**	**A**	**A**	**B**	**D**	

15 Chapter

Molecular Basis of Inheritance

MULTIPLE CHOICE QUESTIONS

1. Monomer of nucleic acids are –
(a) Peptides
(b) Nucleosides
(c) Ribonucleosides
(d) None of these

2. DNA and RNA are types of –
(a) Nucleotides (b) Nucleosides
(c) Nucleic acids (d) Nucleamides

Topic 1 The DNA

3. Length of DNA is usually defined as-
(a) Number of nucleotides present in it
(b) Number of pair of nucleotides present in it
(c) Number of base pairs present in it
(d) All of these

4. Match the length of DNA with the correctorganisms –

A	B		
I	Φ×174	1	4.6×106 bp (base pairs)
II	Bacteriophage γ	2	3.3×109 bp
III	E. coli	3	48502 bp
IV	Human DNA (haploid)	4	5386 nucleotides

	I	II	III	IV
(a)	4	3	1	2
(b)	3	4	2	1
(c)	4	3	2	1
(d)	3	4	1	2

Topic 2 Structure of Polynucleotides Chain

5. A nucleotides contains –
(a) Hexose sugar + nitrogenous base + phosphate group
(b) Pentose sugar + nitrogenous base + phosphate group
(c) Hexose sugar + nitrogenous base + sulphate group
(d) Pentose sugar + nitrogenous base + sulphate group

6. Which of these is a purine –
(a) Cytosine (b) Adenine
(c) Thiamine (d) More than one is correct

7. Which of these is a pyrimidine –
(a) Adenine (b) Thymine
(c) Guanine (d) None of these

8. Which of these is a correct combination for a DNA nucleotides
(a) Oxyribose + Phosphate + Uracil
(b) Oxyribose + Phosphate + Thymine
(c) Deoxyribose + Phosphate + Uracil
(d) Deoxyribose + Phosphate + Thymine

9. All the given nucleotides exists, except
(a) Deoxy uridine (b) Thymine
(c) Both A & B (d) None of these

10. Nitrogenous base is linked to which carbon of pentose sugar
(a) 1'C (b) 2'C
(c) 3'C (d) 5'C

11. Nitrogenous base is linked to pentose sugar by which bond –
(a) N – Glycosidic bond
(b) Phosphoester bond
(c) Phosphodiester bond
(d) Peptide bond

12. Phosphate group is linked to which carbon of pentose sugar
(a) 1'C (b) 2'C
(c) 3'C (d) 5'C

13. Identify the free ends of given polynucleotides chain –

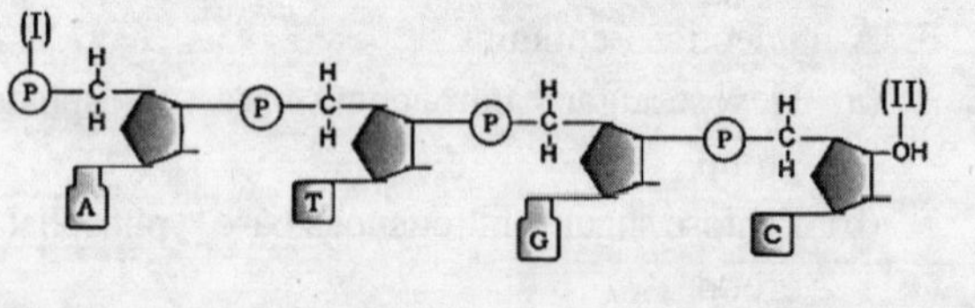

	I	II
(a)	3' phosphate	5' hydroxyl
(b)	5' hydroxyl	3' phosphate
(c)	5' phosphate	3' hydroxyl
(d)	3' hydroxyl	5' phosphate

14. Backbone of polynucleotide chain is formed due to –
(a) Sugar and N-base
(b) Sugar and phosphate
(c) Phosphate and N – Base
(d) All of these

15. Which is correct about thymine & uracil –
(a) Uracil is 5-methyl thymine
(b) Thymine is 5-methyl uracil
(c) Uracil is 5-ethyl thymine
(d) Thymine in 5-ethyl uracil

16. DNA is –
(a) Acidic and positively charged
(b) Basic and positively charged
(c) Acidic and negatively charged
(d) Basic and negatively charged

17. Name of DNA as 'Nuclein' was given by –
(a) Francis crick
(b) Erwin Chargaff
(c) Friedrich Meischer
(d) Rosalind Franklin

18. Double Helix for structure of DNA model was proposed by –
(a) Wilkins and Franklin based on their Xraydiffraction date
(b) Watson and Crick based on their X-ray diffraction date
(c) Chargaff based on their X-ray diffraction data
(d) None of these

19. The proposition of base pairing between the two stands of polynucleotide chain in double Helix model of DNA was based on observation of –
(a) Maurice Wilkins
(b) Rosalind Franklin
(c) Erwin Chargaff
(d) Both A & B

20. The two chains of double Helix DNA have –
(a) Parallel polarity
(b) Anti-parallel polarity
(c) No polarity
(d) Depends on organism

21. The bases in two stands of DNA are paired through
(a) Hydrogen bond (b) Peptide bond
(c) Glycosidic bond (d) Sulfide bond

22. Which of the following is true about base pairing in DNA –
(a) Adenine forms two hydrogen bond with Guanine
(b) Adenine forms three hydrogen bond with Guanine
(c) Adenine forms two hydrogen bond with Thymine
(d) Adenine forms three hydrogen bond with Thymine

23. Which of the following is true about base pairing in DNA –
(a) Guanine forms two H-bond with Cytosine
(b) Guanine forms three H-bond with Cytosine

(c) Guanine forms two H-bond with Adenine
(d) Guanine forms three H-bond with Adenine

24. Uniform distance between two stands of Helix is due to –
(a) Double and triple bond formed between base pairs
(b) Sugar – phosphate backbone
(c) Purine – pyrimidine base pairing
(d) None of these

25. How many of the following statements about Double – helix structure of DNA is correct –
i) Two chains are coiled in right – handed fashion
ii) Pitch of helix is 3.6 nm
iii) There are roughly 10 bp in each turn
iv) Plane of one base pair stacks over the other
(a) 1 (b) 2
(c) 3 (d) 4

26. Pitch of helix in double helix DNA is –
(a) 3.6 nm (b) 3.4 nm
(c) 3.2 nm (d) 3.8 nm

27. Central dogma in molecular biology was proposed by —
(a) Crick (b) Watson
(c) F. Meischar (d) Chragaff

28. Identify correct labels —

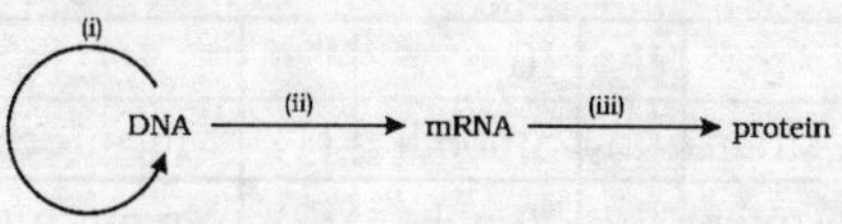

	(i)	(ii)	(iii)
(a)	Replication	Translation	Transcription
(b)	Replication	Transcription	Translation
(c)	Transcription	Replication	Translation
(d)	Translation	Replication	Transcription

Topic 3 Packaging of DNA Helix

29. If length of E. coli DNA is 1.36 mm, calculate number of base pair in E. coli? Given distance between consecutive base pairs is 0.34 × 10-9 m. –
(a) 4 × 106 (b) 4 × 109
(c) 4 × 10-6 (d) 4 × 1012

30. Assertion : In E. coli, DNA is scattered throughout the cell

Reason: In E. coli, there is no defined nucleus
(a) Both Assertion & Reason are correct & Reason is correct explanation for Assertion
(b) Both Assertion & Reason are correct but Reason is not correct explanation for Assertion
(c) Assertion is correct and Reason is incorrect
(d) Reason is correct and Assertion is incorrect

31. Histones are –
(a) Positive and acidic in eukaryotes
(b) Positive and acidic in prokaryotes
(c) Positive and basic in eukaryotes
(d) Positive and basic in prokaryotes

32. Assertion – Histones are positively charged
Reason – Histones are rich in basic amino acid residues lysine and arginine
(a) Assertion & Reason are correct and Reason is correct explanation for Assertion
(b) Assertion & Reason are correct and Reason is not the correct explanation for Assertion
(c) Assertion is correct and Reason is wrong
(d) Both Assertion and Reason are wrong

33. Histones are organized into –
(a) Hexamer (b) Octamer
(c) Tetramer (d) Dimer

34. A typical nucleosome contain _____ bp of DNA has
(a) 200 (b) 400
(c) 600 (d) 800

35. Repeating unit of chromatin –
(a) Are nucleosomes
(b) Are seen as 'beads-on-string' under electron microscope
(c) Are packed to form fibres
(d) All of these

36. The figure show –

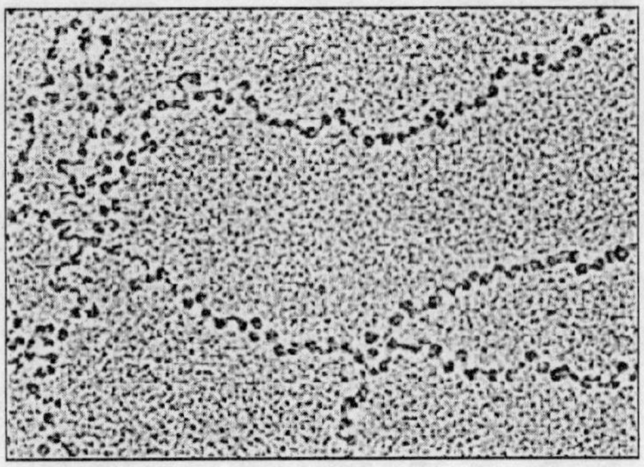

(a) Beads-on-string
(b) A nucleosome
(c) Chromatin
(d) More than one option is correct

37. Identify the correct label for given figure

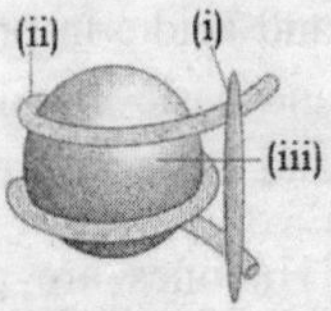

	(i)	(ii)	(iii)
(a)	H2 histone	DNA	Histone octamer
(b)	H1 histone	Histone octamer	DNA
(c)	H2 histone	Histon octamer	DNA
(d)	H1 histone	DNA	Histone octamer

38. Chromosomes are connected chromatin fibres present –
(a) At all times in cell
(b) Only during cell division – formed at prophase
(c) Only during cell division – formed at metaphase
(d) Only during cell division – formed at Interphase

39. In a typical nucleus, euchromatin & hetero chromatin are present. Choose the correct set of characters for heterochromatin –
(i) Loosely packed
(ii) Densely packed
(iii) Light stain
(iv) Dark stain
(v) Inactive chromatin
(vi) Active chromatin
(a) i, iii, v (b) ii, iv, vi
(c) i, iii, vi (d) ii, iv, v

40. Choose correct set of characters for euchromatin
(i) Loosely packed
(ii) Densely packed
(iii) Light stain
(iv) Dark stain
(v) Inactive chromatin
(vi) Active chromatin
(a) i, iii, v (b) ii, iv, vi
(c) i, iii, vi (d) ii, iv, v

Topic 4 The Search for Genetic Material

41. Griffith's experiments were conducted in-
(a) 1928 (b) 1958
(c) 1978 (d) 1968

42. The experiment of Griffith was performed in-
(a) Diplococcus pneumoniae, bacteria
(b) Haemophilus influenzas, fungi
(c) Streptococcus pneumoniae, fungi
(d) None of these

43. Match the given columns- **(Pg. 100, M)**

	I	II	III		
i.	R-strain	(a)	Smooth	(1)	Mucous coat
ii.	S-strain	(b)	Rough colonies	(2)	No mucous coat

(a) (i)-(a)-(1) (b) (i)-(b)-(1)
(c) (ii)-(a)-(1) (d) (ii)-(a)-(2)

44. Which strain of the microbe used Griffith is virulent-
(a) S-strain (b) R-strain
(c) Both (d) None

45. Griffith observed that the mice died surprisingly the following combination of strains was used, which was unusual-
(a) S-strain heat killed

(b) Heat killed S-strain
(c) Heat killed R-strain + Live S-strain
(d) Heat killed S-strain + Live R-strain

46. In Griffith experiment
(a) R-strain transformed to S-strain and became virulent
(b) R-strain transformed to S-strain and lost virulence
(c) S-strain transformed to R-strain and became virulent
(d) S-strain transformed to R-strain and lost virulence

47. Griffith claimed that-
(a) Some protein was transferred among bacteria
(b) Some DNA was transferred among bacteria
(c) Some carbohydrates was transferred among bacteria
(d) None of these Biochemical Characterization of Transforming Principle

48. Prior to work of Avery, Macleod and McCarty, genetic material was thought to be
(a) Protein (b) DNA
(c) RNA (d) None

49. Avery, Macleod & McCarty discovered that-
(a) DNA caused transformation
(b) RNA caused transformation
(c) Protein caused transformation
(d) Lipid caused transformation

50. Which enzyme inhibited the transformation-
(a) Protease (b) RNase
(c) DNase (d) All

Topic 5 | The Genetic Material is DNA

51. Unequivocal proof that DNA is genetic material came from experiments of-
(a) Avery, Macleod & McCarty
(b) Hershey and Chase
(c) de Vries, Correns and Tschermak
(d) Sutton and Boveri

52. The scientists of Q-11 worked with-
(a) a virus (b) a bacteria
(c) a fungi (d) a nematode

53. In the experiment performed for proving DNA as genetic material, the bacteriophages were grown on medium containing-
(a) radioactive sulfur
(b) radioactive nitrogen
(c) radioactive phosphorous
(d) More than one option

54. The bacteriophages growing in presence of radioactive phosphorous __(i)__ contained radioactive __(ii)__.

	(i)	(ii)
(a)	P32	DNA
(b)	P35	Protein
(c)	P32	Protein
(d)	P35	DNA

55. Bacteriophages grown on radioactive Sulphur __(i)__ contained radioactive __(ii)__.

	(i)	(ii)
(a)	S32	DNA
(b)	S32	Protein
(c)	S35	DNA
(d)	P35	Protein

56. The bacteria involved in Hershey & chase experiment of 1952 was-
(a) Bacteriophage
(b) E. coli
(c) S. pneumoniae
(d) C. butyliwm

57. Bacteria infected with virus that showed radioactivity had-
(a) radioactive DNA (S32)
(b) radioactive DNA (S35)
(c) radioactive DNA (P32)
(d) radioactive DNA (P35)

58.

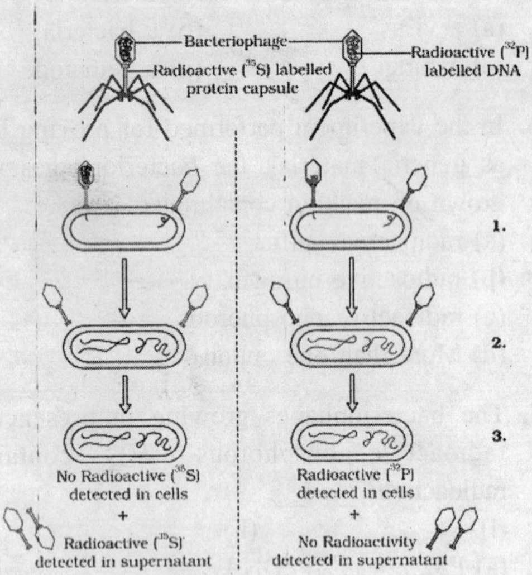

Identify the correct label.

	1	2	3
(a)	Blending	Infection	Centrifugation
(b)	Infection	Blending	Centrifugation
(c)	Centrifugation	Infection	Blending
(d)	Blending	Centrifugation	Infection

Topic 6 Properties of Genetic

59. RNA is genetic material in-
(a) TMV
(b) QB Bacteriophage
(c) Both A and B
(d) None of these

60. Properties of genetic material include-
(a) Stable (b) Mutable
(c) Replicable (d) All of these

61. A - Stability as a property of genetic material was very evident in Griffith's transforming principle.

R - Heat can kill the bacteria and completely destroy the properties of genetic material (Pg. 103, M)
(a) Both A and R are true and R is correct explanation for A
(b) Both A and R are true but R is not correct explanation for R
(c) A is true but R is false
(d) Both A and R are false

62. Which is more structurally and chemically stable?
(a) DNA (b) RNA
(c) Protein (d) All

63. RNA viruses show-
(a) Less mutation
(b) Faster evolution
(c) Slower evolution
(d) More than one option is correct

64. DNA is preferred by nature over RNA for-
(a) Storage of genetic information
(b) Transmission of genetic information
(c) Expression of genetic information
(d) More than one

65. Choose incorrect statement RNA-
(a) was first genetic material
(b) acts as catalyst too
(c) is more stable than DNA
(d) has protein synthesizing mechanism built around it

Topic 7 Replication

66. Scheme for replication of DNA was proposed by-
(a) Watson & Crick
(b) Meselson & Stahl
(c) Taylor
(d) Hershey & Chase

67. The replication of DNA is-
(a) Conservative
(b) Non-conservative
(c) Semi-conservative

(d) All of these depending on organism

68.

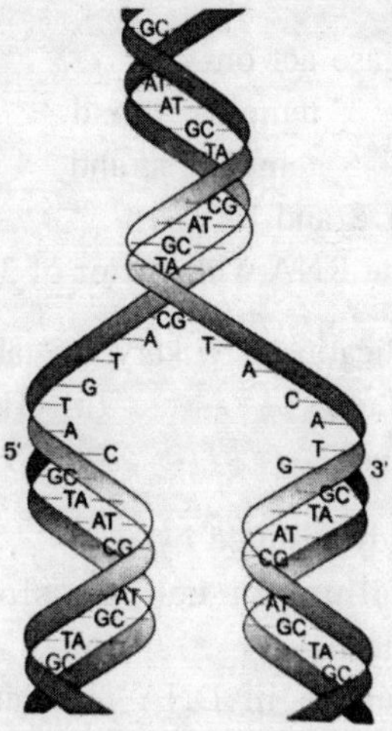

The figure shows-

(a) Conservative DNA replication model
(b) Semi-conservative DNA replication model
(c) Non-conservative DNA replication model
(d) Can't say

69. The DNA replication model experimental proof was first shown in-

(a) Human cells (b) E. coli
(c) Plant cell (d) Vicia faba

70. Meselson and stahl performed experiment for proving DNA replication scheme in-

(a) 1952 (b) 1953
(c) 1958 (d) 1961

71. The bacteria were grown in medium containing-

(a) 15NH4Cl – 15N is heavy isotope of nitrogen
(b) 14NH4Cl – 14N is heavy isotope of nitrogen
(c) 15NH4Cl – 15N is normal isotope of nitrogen
(d) 14NH4Cl – 14N is normal isotope of nitrogen

72. The heavy DNA molecule containing heavy isotope of N is distinguished from normal DNA by-

(a) UV rays
(b) Ethidium bromide solution
(c) Centrifugation is CsCl density gradient
(d) PCR technique

73. In Meselson & Stahl experiment, first they-

(a) grew bacteria on heavy isotope of N medium followed by normal one
(b) grew bacteria on normal isotope of N medium followed by heavy one
(c) grew bacteria on radioactive N followed by Keavy one
(d) grew bacteria on heavy isotope of N followed by radioactive one

74.

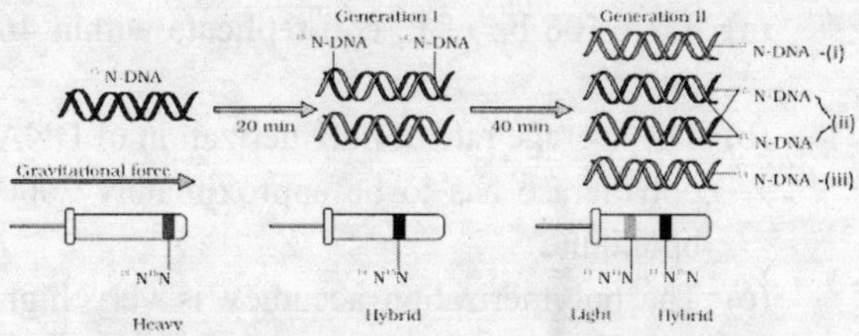

Identify the correct label

	(i)	(i)	(iii)
(a)	Light	Heavy	Hybrid
(b)	Heavy	Hybrid	Light
(c)	Light	Hybrid	Light
(d)	Heavy	Hybrid	Heavy

75. In Meselson & Stahl expt a bacteria after dividing in 20 minutes had a hybrid DNA. What will be the ratio of Hybrid to Light after 80 minutes?

(a) 2 : 14 (b) 14 : 2
(c) 16 : 2 (d) 2 : 16

76. Similar experiment on Vicia faba was conducted by ____ to detect distribution of newly synthesized DNA in chromosomes

(a) Taylor (b) Stahl
(c) Gamow (d) Nirenberg

77. Experiment on Vicia faba involved use of-

(a) Radioactive uridine
(b) Radioactive thymidine
(c) Radioactive adenosine
(d) Radioactive cytidine

Topic 8 The Machinery and the Enzymes

78. The main enzyme of replication is-
(a) RNA dependent RNA polymerase
(b) RNA dependent DNA polymerase
(c) DNA dependent DNA polymerase
(d) DNA dependent RNA polymerase

79. Choose correct statement with regard with efficiency of DNA polymerase.
(a) 4.6 × 106 bp of E. coli replicate within 46 minutes
(b) The average rate of polymerization of DNA polymerase has to be approximately 2000 bp/minute
(c) The polymerization accuracy is very high and very fast
(d) All of these

80. What is function of deoxyribonucleoside triphosphate-
(a) It act as substrate
(b) Provide energy for polymerization
(c) A and B both
d) It is product formed after polymerization

81. Assertion: The two strands of DNA cannot be separated in their length.

Reason: Separation required very high energy.
(a) Both Assertion & Reason are correct and reason is correct explanation of assertion
(b) Both Assertion & Reason are correct and reason is not correct explanation of assertion
(c) Assertion is correct, Reason is false
(d) Assertion & Reason are false

82. Polymerization by DNA polymerase is in-
(a) 3' → 5' direction only
(b) 5' → 3' direction only
(c) A and B both
(d) Random

83. The template of replication fork with polarity 5' → 3' is ____ while 3' → 5' is ____.
(a) continuous, continuous
(b) continuous, discontinuous
(c) discontinuous, continuous
(d) discontinuous, discontinuous

84. DNA ligase act on-
(a) 5' → 3' template strand
(b) 3' → 5' template strand
(c) Both A and B
(d) Ligate RNA with vector of 3' → 5' polarity

85. The replication is eukaryotes take place in-
(a) M-phase (b) G1 phase
(c) S-phase (d) G2 phase

86. Polyploidy resulted by-
(a) A failure in cell division after DNA replication
(b) A failure in DNA replication after cell division
(c) A failure in cell division before DNA replication
(d) A and C both

87.

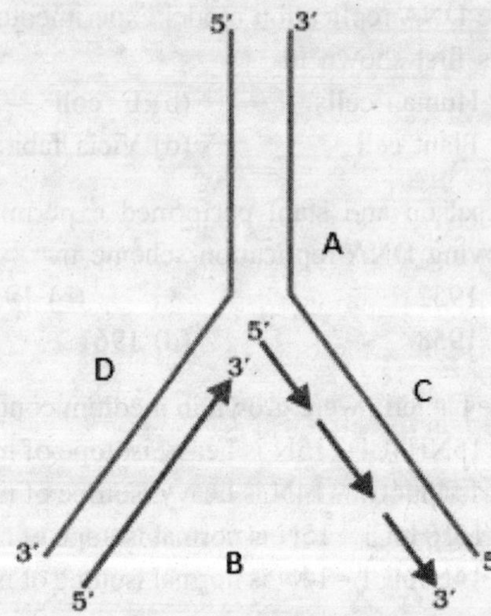

Correct label of A, B, C, D is-
(i) A = Template parental strand
(ii) B = Newly synthesized strand
(iii) D = Continuous strand
(iv) C = Discontinuous strand
(a) i, ii only (b) iii, iv only
(c) i, ii, iii, iv (d) None of these

6.5 Transcription

88. Transcription is-
(a) The process of copying genetic information from both strand of DNA into RNA
(b) The process of copying genetic information from one strand of DNA into RNA
(c) The process of copying genetic information from RNA into DNA
(d) A and B both

89. In transcription, adenosine bind with
(a) Thymine (b) Uracil
(c) Cytosine (d) A and B both

90. Why both the strand of DNA are not copied during transcription-
(a) If both strands act a template, they would code for RNA molecules with same sequence
(b) RNA formed by transcription of both strand, when code for protein, the sequence of amino acid in protein are same
(c) The two RNA molecules if produced simultaneously would be complementary to each other
(d) All of these

91. Translation of RNA would be prevented if-
(a) RNA is single strand
(b) RNA is double-stranded
(c) RNA is produced by both strand of DNA
(d) B and C both

6.5.1 Transcription Unit

92. Transcription unit primarily consist of-
(a) 1 region (b) 2 regions
(c) 3 regions (d) None of these

93. Transcription primarily required-
(a) RNA dependent RNA polymerase
(b) DNA dependent RNA polymerase
(c) DNA dependent DNA polymerase
(d) RNA dependents DNA polymerase

94. Template strand of transcription unit is/are-
(a) 5' → 3' strand of DNA
(b) 3' → 5' strand of DNA
(c) Site of catalysis of enzyme required for transcription
(d) B and C both

95. What is coding strand of given template strand 3' – AGCATGCA – 5'
(a) 5' – TACGTACGT – 3'
(b) 5' – UACGUACGU – 3'
(c) 3' – UACGUACGU – 5'
(d) 3' – TACGTACGT – 5

96. Label A, B, C, D, E of given diagram.

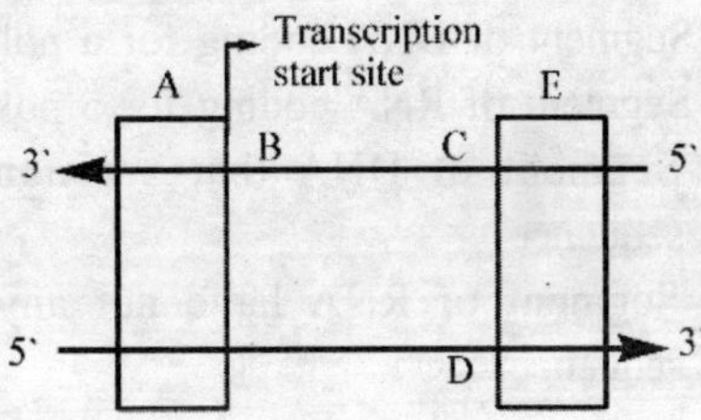

	a	b	c	d	e
A)	Promoter	Structural gene	Template strand	Coding strand	Terminator
B)	Terminator	Structural gene	Coding strand	Template strand	Promoter
C)	Promoter	Template strand	Coding strand	Structural gene	Terminator
D)	None of these				

97. Promoter is located-
(i) 3' end
(ii) 5' end
(iii) upstream of structural gene
(iv) downstream of structural gene
(a) i, iii (b) ii, iii
(c) i, iv (d) ii, iv

98. Terminator is located at
(i) 3' end
(ii) 5' end

(iii) upstream of structural gene
(iv) downstream of structural gene
(a) i, iii (b) ii, iii
(c) i, iv (d) ii, iv

6.5.2 Transcription unit and the gene

99. A gene is defined as-
(a) Functional unit of inheritance
(b) Non-functional region of DNA that haven't any information
(c) A and B both
(d) None of these

100. Cistron is-
(a) Segment of DNA coding for a polypeptide
(b) Segment of RNA coding for a polypeptide
(c) Segment of DNA that are non-coding sequence
(d) Segment of RNA have not any coding sequence

101. Choose the correct statement.
(a) Monocistronic eukaryotic structural gene have interrupted coding sequence.
(b) Polycistronic prokaryotic structural gene have interrupted coding sequence.
(c) Monocistronic prokaryotic structural gene have interrupted coding sequence.
(d) A and B both

102. Exons are-
(a) Coding sequence
(b) Non-coding sequence
(c) Expressed sequence
(d) A and C both

103. Intron-
(a) appear is mature or processed RNA
(b) do not appear in mature or processed RNA
(c) appear is prokaryotes
(d) B and C both

Topic 9	Transpiration

104. Which of following play role is protein synthesis of prokaryote?
(a) r-RNA (b) t-RNA
(c) m-RNA (d) All of these

105. The function of some RNA are given below choose the incorrect one.
(a) mRNA provide template strand
(b) mRNA provide non-template strand
(c) tRNA bring amino acid
(d) rRNA play structural & catalytic role

106. How many polymerase required is bacteria for transcription of all type of RNA?
(a) One (b) Two
(c) Three (d) Five

107. Choose incorrect step about transcription.
(a) RNA polymerase binds to promoter and initiate transcription.
(b) Nucleotide triphosphate act as substrate and polymerization in a template.
(c) A short stretch of RNA remains bound to enzyme.
(d) Last step is termination.

108. Initiation factor and termination factor are-
(a) Sigma and Rho factor respectively.
(b) Rho and Sigma factor respectively.
(c) Rho and Rho factor respectively.
(d) Sigma and Sigma factor respectively.

109. Translation & transcription in eukaryote occur in
(a) Cytoplasm & nucleus respectively
(b) Nucleus & cytoplasm respectively
(c) Cytosol
(d) Nucleus

110. Which of following can be coupled in bacteria?
(a) Replication & transcription
(b) Transcription & translation
(c) Replication & translation
(d) None of these

111. Transcription of 18 s rRNA is done by ____ in eukaryote.
(a) RNA pol. I (b) RNA pol. II
(c) RNA pol. III (d) All of these

112. Choose incorrect statement.
(a) 5.8 s r-RNA and 5 s-RNA transcribes by same RNA polymerase in eukaryote.
(b) hnRNA & mRNA transcribe by same RNA polymerase in eukaryote.
(c) tRNA & snRNA transcribes by sameRNA polymerase in eukaryote.
(d) None of these

113. Splicing is required to-
(a) remove intron in eukaryote
(b) remove exon in eukaryote
(c) remove exon in prokaryote
(d) remove intron in prokaryote

114. Capping is-
(a) Addition of methyl guanosine triphosphate at 5' end
(b) addition of adenylate residue at 3' end
(c) addition of methyl guanosine triphosphate at 3' end
(d) addition of adenylate residue at 5' end

115. The fully processed hnRNA is-
(a) tRNA (b) mRNA
(c) rRNA (d) None of these

Topic 10 Genetic Code

116. George Gamow argued-
(a) There are only 5 bases and if they have code for 20 amino acid the code should constitute a combination of bases
(b) There are only 4 bases and if they have code for 20 amino acid the code should constitute a combination of bases
(c) Genetic code is triplet
(d) B and C

117. Which of following have maximum codon in genetic code-
(a) Leu (b) Met
(c) Cal (d) Phe

118. Which of following is/are showing dual function-
(a) UUU (b) AUG
(c) UGA (d) GUA

119. Least number of codon is for-
(a) Met (b) Phe
(c) Gls (d) Gly

Topic 11 Mutations and Genetic Code

120. Sickle cell anaemia is classical example of-
(a) point mutation
(b) frameshift mutation
(c) deletion mutation
(d) addition mutation

121. In sickle cell anaemia, there are changes in gene for-
(a) alpha globin chain
(b) beta globin chain
(c) gamma globin chain
(d) delta globin chain

122. In sickle cell anaemia, resultant effect of mutation is change of amino acid residue-
(a) Valine to alanine
(b) Valine to glutamic acid
(c) Alanine to valine
(d) Glutamic acid to valine

123. The following is an example of- BIG RED CAP - BIG REM DCA P
(a) Deletion mutation
(b) Point mutation
(c) Addition mutation
(d) More than one option

124. RAM HAS CAP - RAM HAS BIG CAP
The given example shows-
(a) Addition mutation
(b) Deletion mutation
(c) Substitution mutation
(d) More than one option

Topic 12 tRNA - The Adapter Molecule

125. tRNA has-
(a) Codon loop (b) Anticodon loop
(c) Both (d) Neither

126. The presence of adapter molecule to read the code on DNA and bind to amino acids was postulated by-
(a) James Watson
(b) Francis Crick
(c) Friedrich Meisher
(d) Both A and B

127. tRNA was also called-
(a) s RNA (soluble RNa)
(b) s RNA (single RNa)
(c) s RNA (smart RNa)
(d) s RNA (simple RNa)

128. The amino acid acceptor end of tRNA is-
(a) 5'
(b) 3'
(c) Can be any of these
(d) Free end

129. For initiation translation,
(a) Only tRNA carries initiator amino acid to the site.
(b) Specific rRNA carries initiator amino acid to the site.
(c) Any rRNA carries initiator amino acid to the site.
(d) Specific tRNA carries initiator amino acid to the site.

130. For stop codon-
(a) There are specific tRNAs with aminoacids.
(b) There are specific tRNAs which do not bind to any amino acids.
(c) There are no tRNA.
(d) There are tRNAs which may or may not bind to amino acids.

131.

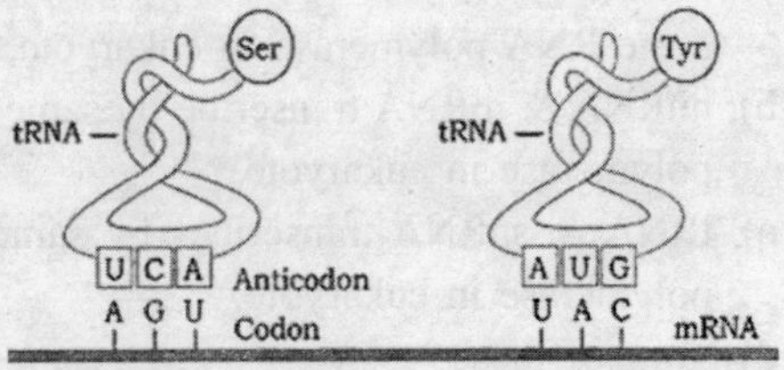

The given figure shows-
(a) Secondary structure of tRNA – Cloverleaf
(b) Primary structure of tRNA – clover-leaf
(c) Secondary structure of tRNA – inverted-L
(d) Primary structure of tRNA – inverted-L

Topic 13 Translation

132. Translation refers to process of-
(a) Making RNA from DNA
(b) Making DNA from RNA
(c) Polymerization of nucleotide to form a DNA
(d) Polymerization of amino acid to form a polypeptide

133. The order and sequence of amino acid during translation are defined by-
(a) The sequences of bases in r-RNA
(b) The sequences of bases in t-RNA
(c) The sequences of bases in m-RNA
(d) All of these

134. Which of following bond is formed during translation?
(a) Glycosidic bond
(b) Phosphodiester bond
(c) Peptide bond
(d) All of these

135. First phase of translation does not involve-
(a) Charging of RNA
(b) Amino acids are activated in presence of ATP
(c) Activated amino acid are linked to their cognate tRNA
(d) None of these

136. Initiation or first phase of translation is-
(a) Amino acylation of tRNA
(b) Amino acylation of mRNA
(c) Both A and B
(d) Deamino acylation of mRNA

137. The cellular factory responsible for synthesizing protein is-
(a) Ribosome (b) Lysosome
(c) Peroxisome (d) None of these

138. In inactive state, protein factory of cell exist in
(a) Two state
(b) 4 state in prokaryote
(c) 6 state in eukaryote
(d) B and C both

139. Which of following is sign as beginning of translation?
(a) When the large subunit of protein factory of cell encounters an mRNA.
(b) When the small subunit of protein factory of cell encounters an mRNA.
(c) When the small subunit of protein factory of cell encounter a tRNA.
(d) When the large subunit of protein factory of cell encounters a tRNA.

140. The bond formation (peptide) between charged tRNA is accomplished due to-
(a) Presence of ATP and catalyst
(b) Two such charged tRNA are brought close by two site in large subunit of ribosome
(c) Two charged tRNA are brought close by two site in small subunit of ribosome & presence of ATP along with catalyst
(d) A and B both

141. The ribosome act as catalyst during bond formation (peptide) as in-
(a) 28 s rRNA in bacteria
(b) 23 s rRNA in bacteria
(c) 23 s rRNA in eukaryote
(d) 28 s tRNA in bacteria

142. Choose the correct statement-
(a) A translational unit in mRNA is sequence of RNA that is flanked by start codon and stop codon and codes for polypeptide.
(b) A translational unit is sequence of DNA that is flanked by start codon & codes for polypeptide.
(c) A transcriptional unit in tRNA is the sequence of RNA that is flanked by start codon and stop codon and codes for polypeptide.
(d) A transcriptional unit in rRNA is the sequence of RNA that is flanked by start codon (AUG) and stop codon and codes for polypeptide.

143. UTR is/are-
(i) Untranslated region of mRNA
(ii) It present at both 5' end (start codon) cand 3' end (before stop codon)
(iii) They are required for efficient translation process
(iv) It present at both 3' end (before start codon) and 3' end (after stop codon)
(a) i, ii, iii are correct
(b) i, ii, iii and iv are correct
(c) i, iii, iv are correct
(d) i, iii are correct

144. Initiator tRNA binds with -
(a) AUG codon of mRNA
(b) at initiation of protein synthesis
(c) ATG codon of dsDNA
(d) A and B both

145. Choose the correct about elongation of translation-
(a) Complexes composed of an amino acid linked to tRNA, sequentially bind to appropriate codon in mRNA by forming complementary base pairs with the tRNA anticodon
(b) The ribosome moves from codon to codon along the mRNA in (3' → 5').
(c) Complexes composed of an amino acid linked to tRNA, sequentially bind to appropriate anticodon in mRNA by forming complementary base pair with tRNA codon.
(d) A and B both

146. Termination of translation complex is done when-
(a) Release factor binds with stop codon (AUG)
(b) Release factor binds with UGA like codon
(c) Complete translation of DNA including UTR occurs in eukaryotes
(d) B and C both

Topic 14 | Regulation of Gene Expression

147. Gene regulation is eukaryote exerted at-
(a) Formation of primary transcript
(b) Transport of mRNA from nucleus to cytoplasm
(c) A and B both
(d) Regulation of splicing of tRNA

148. ***B***-glactosidase is used to catalyze the hydrosis of
(a) Lactose into galactose and glucose
(b) Lactose into fructose & glucose
(c) Lactose into fructose & fructose
(d) None of these

149. E.coli do not have lactose around them to utilized for energy source, they would-
(a) No longer require the synthesis of enzyme α-galactosidase
(b) Synthesized enzyme β-galactosidase
(c) Die due to lack of carbon source and energy source
(d) None of these

150. In prokaryote, predominant site for control of gene expression is-
(a) Control of rate of transcriptionalc initiation
(b) Control of rate of translational
(c) Control of rate of transcriptional elongation
(d) B and C both

151. Given below are statement. Choose the incorrect statement.
(a) The development and differentiation of embryo into adult organisms are result of coordinated regulation of expression of several sets of genes.
(b) Regulatory proteins act positively in activator.
(c) In a transcriptional unit the activity of RNA polymerase at a given promoter is in turn regulated by interaction with accessory protein.
(d) None of these

152. Operator-
(a) Region adjacent to sequence by which repressor mRNA formed
(b) Bind with repressor protein
(c) Bind with inducer
(d) A and B both

153. Each operon has-
(a) Same operator and same repressor
(b) Same operator but specific repressor
(c) Specific operator but same repressor
(d) Specific operator and specific repressor

154. Lac operon was studied first by-
(a) Francois Jacob
(b) Jacque Monod
(c) Geneticist and Biochemist
(d) None of these

155. Lac operon is/are-
(a) Monocistronic structural gene is regulated by a common promoter & regulators genes
(b) Polycistronic structural gene is regulated by a common promoter & regulatory genes.
(c) Consist of one regulatory gene, monocistronic structural gene having five gene along with promotor & operator.
(d) A and C both

156. Regulatory gene of lac-operon is-
(a) p-gene (b) i-gene
(c) o-gene (d) z-gene

157. i in i-gene stand for-
(a) inducer
(b) inhibitor
(c) A and B both
(d) Inactive repressor

158. Match Column-I & Column-II.

Column-I (Gene)	Column-II (Product)

(a)	Z-gene	(i)	Repressor mRNA
(b)	a-gene	(ii)	β-gal
(c)	Y-gene	(iii)	Permease
(d)	i-gene	(iv)	Transacetylase

	a	b	c	d
(a)	(iii)	(ii)	(iv)	(i)
(b)	(iv)	(iii)	(ii)	(i)
(c)	(i)	(ii)	(iv)	(iii)
(d)	(ii)	(iv)	(iii)	(i)

159. The monomeric product of lactose is chiefly hydrolyzed by-
(a) i-gene (b) z-gene
(c) a-gene (d) y-gene.

160. Lac in lac-operon is for -
(a) Monosaccharide
(b) Disaccharide
(c) Polysaccharide
(d) Insect

161. In absence of preferred carbon source, if lactose is provided in growth medium of bacteria, the lactose is transported into cell through by action of product formed by-
(a) i-gene (b) z-gene
(c) a-gene (d) y-gene

162. Allolactose is-
(a) Inducer of lac-operon
(b) Inductive repressor
(c) Form of lactose that bind with product of repressor mRNA and inhibit transcription of structural gene
(d) All of these

163. Lac operon is-
(a) Negative regulation operon
(b) Positive regulation operon
(c) A and B both
(d) None of these

164. Inducer of lac-operon is-
(a) Glucose (b) Galactose
(c) Lactose (d) Fructose

165.

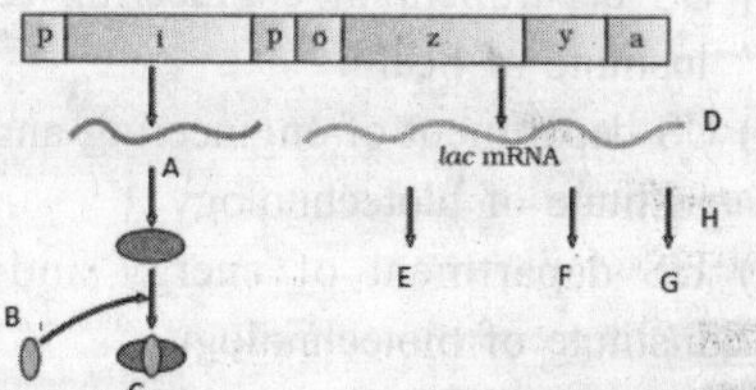

(a) The given diagram is in presence of lactose
(b) The given diagram is in absence of lactose
(c) The given diagram is of gene off
(d) D and H is same process

Topic 15 Human Genome Project (HGP)

166. HGP was launched in-
(a) 1980 (b) 1970
(c) 1990 (d) 2000

167. HGP was called a-
(a) Minor project
(b) Hexagonal project
(c) Mega project
(d) None of these

168. Human genome has approx. ____ bp.
(a) 3 × 109 (b) 3 × 106
(c) 6 × 109 (d) 6 × 106

169. If cost of sequencing required is US $ 3 per bp, then total cost of sequencing human genome as per 8-3 will be:
(a) US $ 18 billion
(b) US $ 9 billion
(c) US $ 18 million
(d) US $ 9 million

170. There were approx. ____ genes in human DNA, as per the goals of HGP
(a) 20,000 – 25,000
(b) 40,000 – 45,000
(c) 10,000 – 15,000
(d) 50,000 – 60,000

171. HGP was a ____ year project -
(a) 15 (b) 12
(c) 13 (d) 14

172. HGP was coordination by-
(a) US department of engineering & national institute of health
(b) US department of engineering and national institute of biotechnology
(c) US department of energy and national institute of biotechnology
(d) US department energy and National Institute of Health

173. The _____ of U.K was a major partner of HGP
(a) Wellcome trust (b) Health trust
(c) Social trust (d) Welcome trust

174. Project was completed in-
(a) 2005 (b) 2004
(c) 2003 (d) 2002

175. Additional contributes to HGP was-
(a) Japan (b) China
(c) Germany (d) All of these

176. Caenorhabditis elegans is a-
(a) Fungi (b) Nematode
(c) Bacteria (d) Virus

177. Caenorhabditis elegans is -
(a) Free living , non-pathogenic
(b) Parasitic , pathogenic
(c) Free living , pathogenic
(d) Parasitic , non-pathogenic

178. Methods / approaches of HGP include-
(a) Excess sequence tags
(b) Expressed sequence tags
(c) Exercise sequence tags
(d) Exerted sequence tags

179. Sequence annotations refer to-
(a) Identifying all genes expressed as RNA and then sequencing then
(b) Sequencing the whole set of genome and then assigning different regions with functions
(c) Identifying and sequencing the genome simultaneously
(d) More than one option is correct

180. For sequencing, the DNA is-
(a) Partially extracted from cell
(b) Totally isolated from cell
(c) Not needed to isolated from cell
(d) None of these

181. The DNA for sequencing is converted to fragments of small size. The fragments are made-
(a) On a pre – decided basis
(b) On a pre – defined basis
(c) Randomly
(d) Depending upon organism

182. The step in DNA sequencing after fragmentation of DNA is-
(a) Cloning in host using vectors
(b) Cloning in vectors using host
(c) Amplification of DNA fragments
(d) More than one option

183. Commonly used hosts for DNA cloning include-
(a) Bacteria (b) BAC
(c) YAC (d) Both A and C

184. BAC stands for-
(a) Bacterial artificial colour
(b) Binominal artificial characterization
(c) Bacterial artificial chromosome
(d) Bacterial articular chromosome

185. Fragments were sequenced using automated DNA sequence that worked on principle of a method developed by-
(a) Erwin Chargaff
(b) Marshal Nirenberg
(c) Frederick Sanger
(d) George Gamow

186. Method for determination of amines acid sequence in protein was developed by-
(a) Erwin Chargaff
(b) Marshal Nirenberg
(c) Frederick Sanger
(d) George Gamow

187. The last of the 24 human chromosomes to be sequenced was-
(a) Chromosome 1 (b) Chromosome X
(c) Chromosome 22 (d) Chromosome Y

Topic 16 Features of Human Genome

188. According to HGP, human genome contains-
(a) ~ 3000 million bp
(b) ~ 6000 million bp
(c) ~ 9000 million bp
(d) ~ 1000 million bp

189. Dystrophin was found to be-
(a) Largest known human gene with 2.4 million bases
(b) Smallest known human gene with 2.4 million bases
(c) Largest known human gene with 4.8 million bases
(d) Smallest known human gene with 4.8 million bases

190. Which chromosome was found to have most genes-
(a) Chr 22 (b) Chr 1
(c) Chr 5 (d) Chr Y

191. Which chromosome was found to have fewest genes-
(a) Chr X (b) Chr Y
(c) Chr 1 (d) Chr 5

Topic 17 DNA Finger Printing

192. The DNA sequence in which small stretch of DNA is repeated many times is called-
(a) SNP (b) Repetitive DNA
(c) Polymorphic DNA (d) More than one option

193. Satellite DNA classified into different categories like micro - satellite , mini - satellite , etc based on-
(a) Length of segment (b) Number of repetitive
(c) Base composition (d) All of these

194. Polymorphism arises due to-
(a) Mutation – inheritable
(b) Stability of genetic material
(c) Mutation - non-heritable
(d) All of these

195. DNA polymorphism is observed more in-
(a) non-coding DNA sequence as its mutation affects reproduction
(b) coding DNA sequence as its mutation affects reproduction
(c) non-coding DNA sequence as it mutation may not affect reproduction ability
(d) Coding DNA sequence as its mutation may not affect reproduction ability

196. Technique of DNA fingerprinting was initially developed by-
(a) James Watson (b) Jansley
(c) Alec Jeffreys (d) Maheshwari

197. VNTR stands for -
(a) Various number of Tendon Repeats
(b) Variable Number of Tendon Repeats
(c) Various Number of Tandem Repeats
(d) Variable Number of Tandem Repeats

198. VNTR belongs to-
(a) Micro-satellite
(b) Macro-satellite
(c) Mini-satellite
(d) All of these

199.

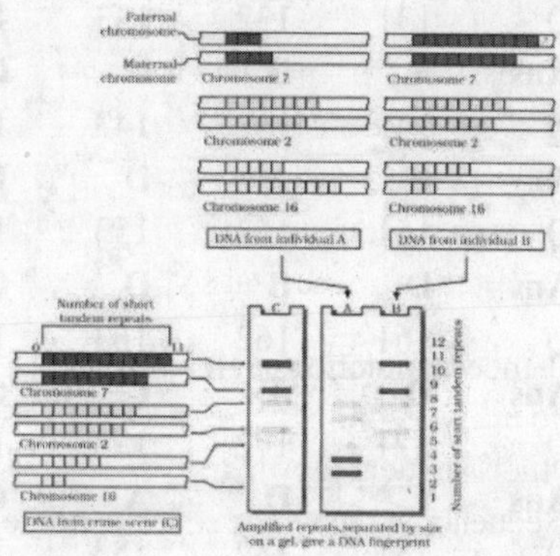

In the given figure if 'C' is the DNA collected from crime site and 'A' & 'B' are samples from suspects, than who is the criminal?
(a) B (b) A
(c) Both A and B (d) None of these

ANSWER KEY

Q	01	02	03	04	05	06	07	08	09	10
Ans	**D**	**C**	**D**	**A**	**B**	**B**	**B**	**D**	**D**	**A**
Q	11	12	13	14	15	16	17	18	19	20
Ans	**A**	**D**	**C**	**B**	**B**	**C**	**C**	**A**	**C**	**B**
Q	21	22	23	24	25	26	27	28	29	30
Ans	**A**	**C**	**B**	**C**	**C**	**B**	**A**	**B**	**A**	**B**
Q	31	32	33	34	35	36	37	38	39	40
Ans	**C**	**A**	**B**	**A**	**D**	**D**	**D**	**C**	**D**	**C**
Q	41	42	43	44	45	46	47	48	49	50
Ans	**A**	**D**	**C**	**A**	**D**	**A**	**D**	**A**	**A**	**A**
Q	51	52	53	54	55	56	57	58	59	60
Ans	**B**	**A**	**D**	**A**	**D**	**B**	**C**	**B**	**C**	**D**
Q	61	62	63	64	65	66	67	68	69	70
Ans	**C**	**A**	**B**	**A**	**C**	**A**	**C**	**B**	**B**	**C**
Q	71	72	73	74	75	76	77	78	79	80
Ans	**A**	**C**	**A**	**C**	**A**	**A**	**B**	**C**	**C**	**C**
Q	81	82	83	84	85	86	87	88	89	90
Ans	**A**	**B**	**C**	**A**	**C**	**A**	**C**	**B**	**B**	**C**
Q	91	92	93	94	95	96	97	98	99	100
Ans	**D**	**C**	**B**	**D**	**A**	**A**	**B**	**C**	**A**	**A**
Q	101	102	103	104	105	106	107	108	109	110
Ans	**C**	**D**	**B**	**D**	**B**	**A**	**B**	**A**	**A**	**A**
Q	111	112	113	114	115	116	117	118	119	120
Ans	**A**	**A**	**A**	**A**	**B**	**B**	**A**	**B**	**A**	**A**
Q	121	122	123	124	125	126	127	128	129	130
Ans	**B**	**D**	**D**	**A**	**B**	**B**	**A**	**B**	**D**	**C**
Q	131	132	133	134	135	136	137	138	139	140
Ans	**A**	**D**	**C**	**C**	**D**	**A**	**A**	**A**	**B**	**D**
Q	141	142	143	144	145	146	147	148	149	150
Ans	**B**	**A**	**D**	**D**	**A**	**B**	**C**	**A**	**D**	**A**
Q	151	152	153	154	155	156	157	158	159	160
Ans	**D**	**B**	**D**	**C**	**B**	**B**	**B**	**D**	**B**	**B**
Q	161	162	163	164	165	166	167	168	169	170
Ans	**D**	**A**	**C**	**C**	**A**	**C**	**C**	**A**	**B**	**A**
Q	171	172	173	174	175	176	177	178	179	180
Ans	**C**	**D**	**A**	**C**	**D**	**B**	**A**	**B**	**B**	**B**
Q	181	182	183	184	185	186	187	188	189	190
Ans	**C**	**D**	**A**	**C**	**C**	**C**	**A**	**A**	**A**	**B**
Q	191	192	193	194	195	196	197	198	199	
Ans	**B**	**B**	**D**	**A**	**C**	**C**	**D**	**C**	**A**	

16 Chapter Microbes in Human Welfair

MULTIPLE CHOICE QUESTIONS

Topic 1 | Microbes in Household Products and Beverages

1. Choose the incorrect statement among the following.

(a) Different varieties of cheese are known by their characteristic texture, flavour and taste, the specificity coming from the microbes used.

(b) Large holes in 'Swiss cheese' are due to production of a large amount of SO_2 by a bacterium named Propionibacterium sharmanii.

(c) The 'Roquefort cheese' are ripened by growing a specific fungi named Penicillium roqueforti which gives them a particular flavour.

(d) Adenovirus is diamond like in structure and causes respiratory infections.

2. Yeast is used in the production of

(a) citric acid and lactic acid

(b) lipase and pectinase

(c) bread and beer

(d) cheese and butter

3. Big holes in Swiss cheese are made by a

(a) a machine

(b) methanogens

(c) the bacterium Propionibacterium sharmanii producing a large amount of carbon dioxide

(d) Lactobacillus

4. How many of the following beverages are produced by distillation of the fermented broth? Whisky, wine, rum, brandy, beer

(a) 5 (b) 4

(c) 3 (d) 1

5. Which of the following statements is incorrect?

(a) Besides microscopic plants and animals, microbes are the major components of biological systems on this earth.

(b) Microbes are present everywhere even in the harsh environments such as deep inside the geysers (thermal vents) where the temperature may be as high as 1000C.

(c) Microbes are diverse–protozoa, bacteria, fungi and microscopic plant viruses, viroids and also prions.

(d) Prions are proteinaceous noninfectious agents.

6. Which of the following statements is correct?

(a) Microbes like protozoa can be grown on nutritive media to form colonies.

(b) Microbes can cause diseases in animals not plants.

(c) Microbes are not found in highly acidic environments.

(d) All microbes are not harmful; several microbes are useful to human beings in diverse ways.

7. Following is the image of a bacteriophage.

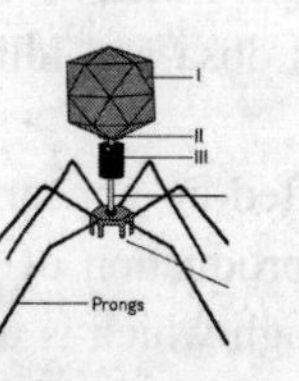

Which of the following options correctly labels its various parts?

(a) I: Head, II: Neck, III: Collar
(b) I: Collar, II: Head, III: Tail
(c) I: Collar, II: Tail, III: Head
(d) I: Tail, II: Collar, III: Head

8. The organisms responsible for converting milk into curd is-
(a) Lactobacillus
(b) Propionibacterium sharmanii
(c) LAB
(d) Both (a) and (c)

9. Which of the following statements is not incorrect?
(a) LAB produces acids that coagulate and completely digest the milk proteins.
(b) A small amount of cheese is added to the fresh milk as inoculum or starter contain millions of LAB,
(c) LAB multiply at a suitable temperature, thus converting curd to milk, which also improves its nutritional quality by increasing vitamin B12 in our stomach.
(d) LAB plays very beneficial role in checking disease causing microbes.

10. The vitamin whose content increases during curd formation by lactic acid bacteria is
(a) Vitamin C (b) Vitamin D
(c) Vitamin B_{12} (d) Vitamin E

11. The small amount of curd added to the fresh milk to convert it in**to curd is called**
(a) starter (b) inoculum
(c) implant (d) both (a) and (b)

12. How many of the following statements are correct?
I. The dough which is used for making food such as dosa and idli is also fermented by yeast.
II. The puffed-up appearance of dough is due to the production of CO2 gas.
III. The dough which is used for making bread, is fermented using brewer's yeast
IV. 'Toddy', a traditional drink of some parts of Southern India is made by fermenting fruits from palms
V. Microbes are also used to ferment fishes, soyabean and bamboo-roots to make foods.
(a) 4 (b) 3
(c) 2 (d) 1

Topic 2 Microbes in Industrial Products

13. Which of the following statements is incorrect?
(a) Very large vessels in which microbes are grown for the purpose of production of beverages and antibiotics on industrial scale are termed as fermenters.
(b) Microbes especially yeasts have been used from time immemorial for the production of beverages like wine, beer, whisky, brandy or rum.
(c) For this purpose the same yeast Saccharomyces cerevisiae used for bread-making and commonly called brewer's yeast, is used for fermenting malted cereals and fruit juices, to produce ethanol.
(d) Antibiotics produced by microbes are regarded as one of the most significant discoveries of the 19th century and have greatly contributed towards the welfare of the human society

14. How many of the following statements are not incorrect?
I. Anti is a Greek word that means 'against', and bio means 'life', together they mean 'against life'.
II. Antibiotics are chemical substances, which are produced by some microbes and can kill or retard the growth of other (disease-causing) microbes.
III. Penicillin is the first antibiotic to be discovered.
IV Alexander Fleming discovered Penicillin while working on a Staphylococci bacteria.
V. The full potential as an effective antibiotic was established much later by Ernest Chain and Howard Florey.

(a) 5 (b) 4
(c) 3 (d) 1

15. Which of the following statements is correct?
(a) Penicillin was used to treat American soldiers wounded in World War I.
(b) Fleming, Chain and Florey were awarded the Nobel Prize in 1954, for this discovery.
(c) Before Penicillin, antibiotics were also purified from other microbes.
(d) Antibiotics have greatly improved our capacity to treat deadly diseases such as plague, whooping cough, diphtheria and leprosy.

16. Microbes are used for commercial and industrial production of certain chemicals like organic acids, alcohols and enzymes. Which of the following is incorrect regarding this?
(a) Aspergillus – fungus – citric acid
(b) Acetobacter aceti – bacteria – vinegar
(c) Clostridium butylicum – protozoan - butyric acid
(d) Lactobacillus - bacteria - lactic acid.

17. The substance which is used in detergent formulations and helpful in removing oily stains are –
(a) Pectinases (b) Proteases
(c) Lipases (d) Statins

18. Bottled juices are clarified by the enzyme-
(a) Pectinase and lipases
(b) Pectinase and peptidase
(c) Pectinase and protease
(d) Pectinase only

19. Match the following columns:

Column-I		Column-II	
(a)	Cyclosporin A	1	Blood Cholesterol lowering agen
(b)	Streptokinase	2	Immunosuppressive agen
(c)	Statins	3	Clot buster for removing clots
(d)	Saccharomyces	4	Production of ethanol

Which of the following is the correct option?

	a	b	c	d
(a)	1	3	4	2
(b)	3	2	1	4
(c)	2	3	1	4
(d)	4	3	2	1

20. How many of the following statements are correct?
I. A bioactive molecule, cyclosporin A, is produced by the fungus Trichoderma polysporum.
II. Statins are produced by the fungus Monascus purpureus.
III. Statins act by competitively inhibiting the enzyme responsible for synthesis of cholesterol.
IV. Streptokinase is used to remove clots from the blood vessels of patients who have undergone myocardial infarction leading to heart attack.
(a) 3 (b) 4
(c) 2 (d) 1

21. A good producer of citric acid is
(a) Pseudomonas (b) Clostridium
(c) Saccharomyces (d) Aspergillus

22. Which of the following is correctly matched for the product produced by them?
(a) Acetobacter aceti: Antibiotics
(b) Methanobacterium: Lactic acid
(c) Penicillium notatum: Acetic acid
(d) Saccharomyces cerevisiae: Ethanol

23. Which of the following is wrongly matched in the given table?

	Microbe		Product		Application
A	Trichoderma polysporum	I	Cyclospor in A	1	Immunosuppressive

B	Monascus purpureus	ii	Statins	2	Blood Cholesterol lowering agent
C	Streptococcus	iii	Streptokinase	3	Removal of clot from blood vessels
D	Clostridium acetobutylicum	iv	Lipase	4	Removal of oil stains

24. Match Column-I with Column-II and select the correct option using the codes.

	Column-I		Column-II
(a)	Citric acid	(1)	Trichoderma
(b)	Cyclosporin	(2)	Clostridium
(c)	Statins	(3)	Aspergillus
(d)	Butyric acid	(4)	Monascus

Which of the following is the correct option?

	a	b	c	d
(a)	3	1	4	2
(b)	4	3	1	2
(c)	2	1	4	3
(d)	2	3	1	4

25. Match the following list of microbes and their importance:

	Column-I		Column-II
(a)	Saccharomyces	(1)	Production of immunosuppressive agents
(b)	Monascus purpureus	(2)	Ripening of Swiss cheese
(c)	Trichoderma polysporum	(3)	Commercial production of ethanol
(d)	Propionibacterium sharmanii	(4)	Production of blood cholesterol-lowering agent

Which of the following is the correct option?

	a	b	c	d
(a)	2	4	1	3
(b)	4	3	1	2
(c)	2	1	4	3
(d)	3	4	1	2

Topic 3 MIcrobes in Sewage Treatment

26. How many of the following are not incorrect with respect to wastewater treatment?

I. A major component of waste water is human excreta.

II. The municipal waste-water is called sewage which contains large amounts of inorganic matter and microbes. Many of which are pathogenic.

III. Sewage is treated in sewage treatment plants (STPs) to make it non- polluting.

IV. Treatment of waste water is done by the aerobic microbes naturally present in the sewage.

(a) 4 (b) 3
(c) 2 (d) 1

27. Which of the following statements is correct with respect to wastewater treatment?

(a) Treatment of wastewater is carried out in three stages- Primary, Secondary and Biological Treatment.

(b) Primary treatment basically involves physical removal of particles – large and small – from the sewage through sedimentation then filtration.

(c) All solids that settle form the primary sludge, and the supernatant forms the effluent.

(d) The sludge from the primary settling tank is taken for secondary treatment.

28. Which of the following statements is incorrect with respect to wastewater treatment?

(a) The primary effluent is passed into large aeration tanks where it is constantly agitated mechanically and air is pumped into it. This allows vigorous growth of useful aerobic microbes into flocs.

(b) Flocs are masses of bacteria associated with fungal filaments to form mesh like structures.
(c) While growing, microbes consume the major part of the organic matter in the effluent. This significantly elevates the BOD (biochemical oxygen demand) of the effluent.
(d) BOD refers to the amount of the oxygen that would be consumed if all the organic matter in one liter of water were oxidised by bacteria.

29. Which of the following statements is wrong with respect to wastewater treatment?
(a) The BOD test measures the rate of uptake of oxygen by micro-organisms in the sample of water.
(b) BOD is a direct measure of the organic matter present in the water.
(c) The greater the BOD of waste water, more is its polluting potential.
(d) Once the BOD of sewage or waste water is reduced significantly, the effluent is then passed into a settling tank where the bacterial 'flocs' are allowed to sediment.

30. The gases which evolve from the anaerobic sludge digester constituting biogas are-
(a) Methane, sulphur dioxide, carbon sulphide
(b) Carbon dioxide, hydrogen chloride, methane
(c) Methane, hydrogen chloride and carbon dioxide
(d) Methane, carbon dioxide, hydrogen sulphide

31. Which of the following statement is wrong?
(a) In settling tank, the sedimented bacterial flocs are called activated sludge.
(b) A small part of the activated sludge is pumped back into the anaerobic sludge digester to serve as the inoculum.
(c) In anaerobic sludge digesters, other kinds of bacteria which grow anaerobically, digest the bacteria and the fungi in the sludge.
(d) The biogas can be used as source of energy as it is inflammable.

32. Which of the following statements is incorrect?
(a) The effluent from the secondary treatment plant is generally released into natural water bodies like rivers and streams.
(b) Microbes play a major role in treating millions of gallons of waste water everyday across the globe. This methodology has been practiced for more than a decade now, in almost all parts of the world.
(c) Till date, no manmade technology has been able to rival the microbial treatment of sewage.
(d) The Ministry of Environment and Forests has initiated Ganga Action Plan and Yamuna Action Plan to save these major rivers of our country from pollution.

33. The primary treatment of wastewater removes
(a) dissolved impurities
(b) stable particles
(c) methane
(d) pathogens

34. Which of the following in sewage treatment removes suspended solids?
(a) Tertiary treatment
(b) Secondary treatment
(c) Primary treatment
(d) Sludge treatment

35. Secondary sewage treatment is mainly a
(a) physical process
(b) mechanical process
(c) chemical process
(d) biological process

36. BOD of wastewater is represented as
(a) total inorganic matter
(b) biodegradable matter
(c) carbon dioxide evolution
(d) oxygen consumption

37. The high value of BOD (Biochemical Oxygen Demand) indicates that
(a) water is pure.
(b) water is highly polluted.

(c) water is less polluted.
(d) consumption of organic matter in the water is higher by microbes.

38. If the activated sludge flocs do not get enough oxygen supply,
(a) it will increase the rate of the treatment.
(b) the center of flocs will become anoxic, which would cause the death of bacteria and eventually breakage of flocs.
(c) it will increase the size of flocs.
(d) protozoa would grow in large numbers.

39. Methanogenic bacteria are not present in
(a) gobar gas plant
(b) stomach of ruminants
(c) bottom of water-logged paddy fields
(d) activated sludge

40. Activated sludge should settle quickly so that
(a) it is rapidly pumped back from sedimentation tank to aeration tank
(b) absorb pathogenic bacteria present in wastewater
(c) it is anaerobically digested
(d) absorbs inorganic matter

41. The sludge generated by wastewater treatment is treated by
(a) anaerobic digesters
(b) floc
(c) chemicals
(d) oxidation pond

42. What gases are produced in anaerobic sludge digesters?
(a) Methane and CO2 only.
(b) Methane, Hydrogen sulfide and CO2.
(c) Hydrogen Sulfide and CO2.
(d) Methane and CO2.

43. During sewage treatment, biogases are produced which include
(a) hydrogen sulfide, nitrogen, methane
(b) methane, hydrogen sulfide, carbon dioxide
(c) methane, oxygen and hydrogen sulfide
(d) hydrogen sulfide, methane and sulfur dioxide

Topic 4 | Microbes in Production of Biogas

44. The predominant gas in biogas is-
(a) Methane
(b) Carbon dioxide
(c) Hydrogen sulphide
(d) ethane

45. Which of the following is incorrect with respect to production of biogas?
(a) The type of the gas produced depends upon the microbes and the organic substrates they utilise.
(b) Certain bacteria, which grow anaerobically on cellulosic material, produce large amount of methane along with CO_2 and N_2.
(c) Methanogens are commonly found in the anaerobic sludge during sewage treatment, the common example is Methanobacterium.
(d) Methanogens are also present in the rumen of cattle where they help in the breakdown of cellulose and play an important role in the nutrition of cattle.

46. The depth of concrete tank of biogas plant in which bio-wastes are collected and a slurry of dung is fed is
(a) 10-12 meters (b) 10-12 feet
(c) 10-15 meters (d) 10-15 feet

47. How many of the following statements are incorrect?
I. The gobar gas is a special type of biogas generated from the dung of cattle.
II. A floating cover is placed over the slurry, which keeps on rising as the gas is produced in the tank due to the microbial activity.
III. A typical biogas plant has two main outlets, one for biogas which is used by nearby houses and the other for spent which may be used as fertiliser
IV. The technology of biogas production was developed in India mainly due to the efforts of IARI and KVIC.
V. Biogas plants are more often built in rural

areas due to availability of large quantities of cattle dung and thus the biogas produced is used for cooking and lighting.

(a) 5 (b) 4
(c) 3 (d) none

48. IARI stands for –
(a) Indian Academy of Research and Intelligence
(b) International Agricultural Research Institute
(c) Indian Agricultural Research Institute
(d) International Academy of Research and Intelligence

49. KVIC stands for –
(a) Khadi and Village Industries Company
(b) Kisan and Village Industries Company
(c) Khadi and Village Industries Commission
(d) Khadi and Village Interest Commission

50. Select the correct option among the following

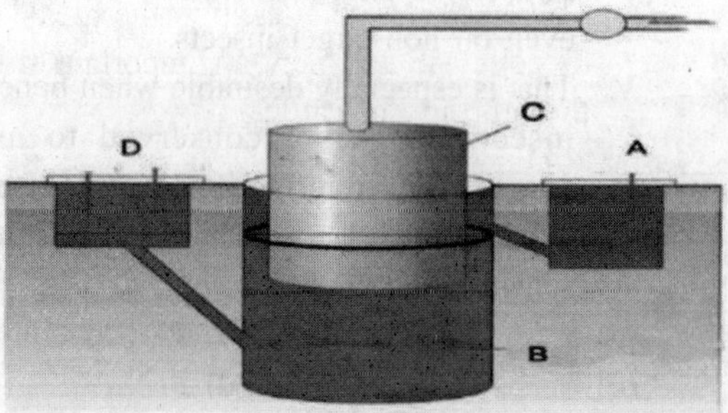

(a) A – slurry, B – Digester, C – Gas Holder, D - Sludge
(b) A – sludge, C – Gas mixture, B – Slurry, D – Water + Dung
(c) A – sludge, B – Digester, C – Gas Holder, D – Slurry
(d) A – slurry, C – Gas mixture, B – Sludge, D – Water + Dung

51. Which of the following is mainly produced by the activity of anaerobic bacteria on sewage?
(a) Propane (b) Mustard gas
(c) Marsh gas (d) Laughing gas

52. Which of the following statements about methanogens is not correct?
(a) They produce methane gas.
(b) They can be used to produce biogas.
(c) They are found in the rumen of cattle and their excreta.
(d) They grow aerobically and breakdown cellulose-rich food.

53. Select the correct statement from the following:
(a) Activated sludge-sediment in settlement tanks of the sewage treatment plant is a rich source of aerobic bacteria.
(b) Methanobacterium is an aerobic bacterium found in the rumen of cattle.
(c) Biogas is produced by the activity of aerobic bacteria on animal waste.
(d) Biogas, commonly called gobar gas, is pure methane.

Topic 5 Microbes as Biocontrol Agent

54. The use of biological methods for controlling plant disease and pest is known as –
(a) Bioinvestment (b) Bioremediation
(c) Biofortification (d) Biocontrol

55. How many of the following statements are incorrect?
I. In agriculture, there is a method of controlling pests that relies on natural predation rather than introduced chemicals.
II. A key belief of the organic farmer is that biodiversity furthers health.
III. The conventional farming practices rarely used chemical methods to kill both useful and harmful life forms indiscriminately.
IV. The organic farmer holds the view that the eradication of the creatures that are often described as pests is not only possible, but also undesirable.
(a) 4 (b) 3
(c) 2 (d) 1

56. In a system of checks and balances-
(a) The insects that are sometimes called pests

are completely eradicated.
(b) All insects are eradicated.
(c) When pests are beyond manageable levels, they are completely eradicated.
(d) Pests are not eradicated but they are kept at manageable levels.

57. The biopesticide used to get rid of aphids is –
(a) Dragonflies
(b) Baculoviruses
(c) Ladybird
(d) Bacillus thuringiensis

58. Microbial biocontrol agent that can be introduced in order to control butterfly caterpillar is-
(a) Fungus Trichoderma
(b) Bacillus tracin
(c) Baculoviruses
(d) Bacillus thuringiensis

59. The bacteria bacillus thuringiensis is available in the market in the form of –
(a) Tablets (b) Powder
(c) Spores (d) Gel form

60. Which of the following statements is incorrect?
(a) Eradication of the pests is undesirable as without them the beneficial predatory and parasitic insects which depend upon them as food or hosts would not be able to survive.
(b) An important part of the biological farming approach is to become familiar with the various life forms that inhabit the field, predators as well as pests, and also their life cycles, patterns of feeding and the habitats that they prefer.
(c) The very familiar beetle with red and black markings is the Ladybird
(d) Bt spores are mixed with water and sprayed onto valuable plants such as brassicas and fruit trees, where these are eaten by the insect adults.

61 How many of the following are correct with respect to Trichoderma?
I. Developed for use in the treatment of plant diseases
II. It's species are free living fungi that are very common in root ecosystems.
III. It belongs to the division Ascomycetes
IV. They are very effective biocontrol agents for several plant pathogens.
(a) 4 (b) 3
(c) 2 (d) 1

62. How many of the following are correct with respect to baculoviruses?
I. They are pathogens that attack insects and other arthropods.
II. The majority of baculoviruses used as biological control agents are in the family nucleopolyhedrovirus.
III. These viruses are excellent candidates for species-specific, narrow spectrum insecticidal applications.
IV. They have been shown to have no negative impacts on plants, mammals, birds, fish or even on non-target insects.
V. This is especially desirable when beneficial insects are being conserved to aid in an overall integrated pest management programme, or when an ecologically sensitive area is being treated.
(a) 5 (b) 4
(c) 3 (d) 2

63. Which one of the following is not an example of carrying out biological control of pests/diseases using microbes?
(a) Ladybird beetle against aphids in mustard.
(b) Bt-cotton to increase cotton yield.
(c) Nucleopolyhedrovirus against white rust in Brassica.
(d) Trichoderma sp. against certain plant pathogens.

64. Match the items in Column-I and Column- II and choose the correct answer.

	Column-I		Column-II
(a)	Ladybird	(1)	Methanobacterium

(b)	Mycorrhiza	(2)	Trichoderma
(c)	Biological control	(3)	Aphids
(d)	Biogas	(4)	Glomus

Which of the following is the correct option?

	a	b	c	d
(a)	1	4	3	2
(b)	3	4	2	1
(c)	4	1	2	3
(d)	3	2	1	4

Topic 6 Microbes as Biofertilizer

65. The bacteria which can fix atmospheric nitrogen in it's free-living form in the soil are –
(a) Azospirillum and Azorhizobium
(b) Azospirillum and Azotobacter
(c) Azotobacter and Rhizobium
(d) Frankia and Azospirillum

66. Which one of the following microbes form a symbiotic association with plants and helps them in their nutrition?
(a) Azotobacter (b) Aspergillus
(c) Glomus (d) Trichoderma

67. Which one of the following helps in the absorption of phosphorus from the soil by plants?
(a) Anabaena (b) Glomus
(c) Rhizobium (d) Frankia

68. Which of the following is not a biofertilizer?
(a) Rhizobium (b) Nostoc
(c) Mycorrhiza (d) Agrobacterium

69. Which of the following is incorrect?
(a) Biofertilizers are organisms that enrich the nutrient quality of the soil.
(b) Realising the problems associated with the overuse of chemical fertilisers, there is a large pressure to switch to organic farming – the use of biofertilisers.
(c) The main sources of biofertilisers are bacteria, fungi, protozoans and cyanobacteria.
(d) In the root nodules of leguminous plants, symbiotic association is formed by Rhizobium.

70. The fungal symbiont in mycorrhiza association absorbs ______ from soil and passes it to the plant.
(a) N (b) N and K
(c) P (d) Ca

71. How many of the following are correct with respect to mycorrhiza?
I. Fungi are also known to form symbiotic associations with plants (mycorrhiza).
II. Many members of the genus Glomus form mycorrhiza.
III. It provide resistance to root-borne pathogens.
IV. It provide tolerance to salinity and drought.
V. It helps in overall increase in plant growth and development.
(a) 5 (b) 4
(c) 3 (d) 1

72. Which of the following is incorrect?
(a) Rhizobium fixes atmospheric nitrogen into inorganic forms which are used by plants as nutrients.
(b) Cyanobacteria are autotrophic microbes widely distributed in aquatic and terrestrial environments many of which can fix atmospheric nitrogen.
(c) Blue green algae add organic matter to the soil and increase its fertility.
(d) In paddy fields, cyanobacteria serve as an important biofertiliser.

ANSWER KEY

Q	1	2	3	4	5	6	7	8	9	10
Ans	**B**	**C**	**C**	**C**	**D**	**D**	**A**	**D**	**D**	**C**
Q	11	12	13	14	15	16	17	18	19	20
Ans	**D**	**A**	**D**	**A**	**B**	**C**	**C**	**D**	**C**	**B**
Q	21	22	23	24	25	26	27	28	29	30
Ans	**D**	**D**	**D**	**A**	**D**	**B**	**A**	**C**	**B**	**D**
Q	31	32	33	34	35	36	37	38	39	40
Ans	**B**	**B**	**B**	**C**	**D**	**D**	**B**	**C**	**D**	**A**
Q	41	42	43	44	45	46	47	48	49	50
Ans	**A**	**B**	**B**	**A**	**B**	**D**	**D**	**C**	**C**	**C**
Q	**51**	**52**	**53**	**54**	**55**	**56**	**57**	**58**	**59**	**60**
Ans	**C**	**D**	**D**	**D**	**D**	**D**	**C**	**D**	**C**	**D**
Q	61	62	63	64	65	66	67	68	69	70
Ans	**B**	**A**	**B**	**B**	**B**	**C**	**B**	**A**	**C**	**C**
Q	71	72								
Ans	**A**	**A**								

17 Chapter

Organisms and Populations

MULTIPLE CHOICE QUESTIONS

1. Which one of the following is the Indian father of Ecology?
(a) E. hackle (b) Ramdeo Mishra
(c) P. odum (d) Tansley

Topic 1 Organism and its Environment

2. Which one of the following interaction takes place among organisms?
(a) Recreation (b) Reproduction
(c) Survival (d) Both B and C

3. The basic unit of ecological hierarchy is-
a) Population b) Community
c) Ecosystem d) Organism

4. Identify the following which is not correctly matched -

	Biome	Mean annual temp.(.c)	Mean annual precipitatio n (em)
1	Tropical forest	20-25	130-430
2	Artic and alpine	-12-2	10-125
3	Coniferous forest	-5-5	100-200
4	Temperate forest	8-22	5-225

(a) 3 (b) 1
(c) 2 (d) 4

5. Different biomes are formed due to annual variations in ________ over the earth's surface. -
(a) Temperature
(b) Precipitation
(c) Incident of solar radiation
(d) All of these

6. Which one of the following are the examples of Deserts, rainforests, tundra, etc.-
(a) Community (b) Niche
(c) Biomes (d) Ecosystem

7. The key element that determines difference in environment conditions of different habitats include. -
(a) Temperature (b) Light
(c) Soil (d) All of these

Topic 2 Major Abiotic Factors

8. Which one of the following statement are correct?
(i) Organisms capable to tolerate a wide range of temperature are called stenothermal organisms.
(ii) Thermal tolerance of different species determines their geographical distribution to a large extent.
(iii) Average temperature in tropical desert in summer is <50ºC.
(iv) Thermal spring cannot sustain life due to very high avg. temperature i.e. >100ºC.
(a) ii (b) i, iii, iv
(c) i, ii, iv (d) iv

9. Organism which tolerate narrow range of temperature--

(a) Stenothermal (b) Eurithermal
(c) Eurihaline (d) None of these

10. Mango tree do not grow in
(a) Temperate country
(b) Tropical country
(c) Sub-tropical country
(d) None of these

11. Mango do not and cannot grow in the above region. The most important environmental factor responsible for it is-
(a) Soil (b) Temperature
(c) Water (d) Light

12. Snow leopard are not found in ________ and tuna fish rarely caught beyond ________ latitudes in the ocean.
(a) Tropical, Kerala
(b) Kerala, tropical
(c) Kerala, temperate
(d) Kerala, sub-tropical

13. Organism which tolerate wide range of temperature?
(a) Furithermal (b) Stenothermal
(c) Stenohaline (d) None of these

14. Match the following salinity.

1	Sea	a	5
2	Hypersaline Lagoons	b	30-35
3	Inland water	c	>100

	1	2	3
(a)	a	c	b
(b)	b	c	a
(c)	c	a	b
(d)	b	a	c

15. A fresh water organisms cannot survive in a water body that has greater ________ than its original habitat-
(a) Nutrients (b) Depth
(c) Salt concentration (d) Water clarity

Topic 3 Responses to Abiotic factor

16. The organism try to maintain the constancy of its internal environment and the process is called
(a) Hibernation (b) Aestivation
(c) Homeostasis (d) None of these

17.

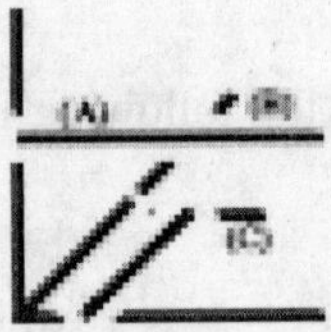

(a) A-Regulators, B-Conformers, C-Partial Regulator
(b) A-Conformers, B-Regulators, C-Partial Regulator
(c) A-Partial Regulator, B-Regulators, CConformers
(d) A-Conformers, B-Partial Regulators, CRegulator

18. Regulators maintain homeostatic by which means-
(a) Chemical (b) Physiological
(c) Behavioural (d) Both B and C

19. We maintain constant body temperature of-
(a) 39° C (b) 37° C
(c) 33° C (d) 34° C

20. The organism in which body temperature changes according to the ambient temperature is known as
(a) Conformers (b) Regulator
(c) Partial Regulators (d) Endothermal

21. Thermoregulation is energetically expensive process for
(a) Shrews (b) Mammals
(c) Humming bird (d) Both A and C

22. Thermoregulation is energetically expensive process for small animals due to their-
(a) Small surface area relative to their size

(b) Large size relative to surface area
(c) Both B and A
(d) Large surface area relative to size

23. The organism which moves away temporarily from stressful situation is known as-
(a) Migrators (b) Conformers
(c) Regulators (d) Endothermals

24. Keolado National Park is situated in
(a) Rajasthan (b) Raipur
(c) Gujarat (d) Madhya Pradesh

25. Match the following

A	B		
1	Regulators	i	Humming birds
2	Conformers	ii	Shrenes
3	Migrators	iii	Mammals
4	Suspendors	iv	Siberian birds
		v	Bacteria, fungi and lower plants

	1	2	3	4
(a)	i, ii	iii	iv	v
(b)	iii	i, ii	v	iv
(c)	iii	i, ii	iv	v
(d)	iv	iii	i,ii	v

26. A stage of suspended development is called
(a) Diapause (b) Aestivation
(c) Hibernation (d) Migration

27. Winter sleep is known as ________ and summer sleep is known as ________.
(a) Hibernation, Aestivation
(b) Migration, Aestivation
(c) Aestivation, Hibernation
(d) Aestivation, Migration

28. Match the following (Pg. 225, M)

	A		B
1	Bear	i	Aestivation
2	Zooplankton	ii	Migration
3	Snail	iii	Hibernation
4	Siberian clane	iv	Diapause

	1	2	3	4
(a)	ii	iv	iii	i
(b)	iii	iv	i	ii
(c)	iii	i	iv	ii
(d)	iv	iii	ii	i

Topic 4 Adaptation

29. Any attributes of the organism that enable them to survive and reproduce its habitat is known as
(a) Migration (b) Diapause
(c) Adaptation (d) Dormancy

30. Kangaroo rat in ________ American deserts is capable to meet all its water requirement through ________.
(a) East, internal fat oxidation
(b) North, internal fat oxidation
(c) North, internal protein oxidation
(d) West, internal fat oxidation

31. Desert plants have special photosynthetic pathway which is known as-
(a) C3 cycle (b) C4 cycle
(c) CAM pathway (d) None of these

32. Desert plant do not have following one characteristics-
(a) Bread leaf (b) Flattened stem
(c) Sunken stomata (d) Thick cuticle

33. Mammals from Colder climates generally have shorter ears and limbs to minimize the heat loss. This Rule was give by-
(a) Charles Darwin (b) Jansely
(c) P. Odum (d) Allen

34. Desert lizard manage to keep their body temperature constant by ________ means.
(a) Physiological (b) Behavioural
(c) Chemical (d) Both A and B

35. Many marine aquatics lives in very high pressure. Which type of adaptation shown by them?
(a) Biochemical (b) Behaviourals
(c) Physical (d) None of these

36. Altitude sickness in high altitude is due to-
(a) Low atmospheric pressure
(b) Low oxygen
(c) High atmospheric pressure
(d) Both A and B

Topic 5 Population Attributes

37. What is a group of individual belonging to the same species called-
(a) Population (b) Biomes
(c) Community (d) Family

38. ______ links ecology to population genetics and evolution
(a) Ecosystem
(b) Biomes
(c) Population ecology
(d) Population attributes

39. Population has certain attributes which are-
(a) birth rates (b) death rates
(c) sex ratio (d) All of these

40. If the age distribution is plotted for the population, the resulting structure is called-
(a) Population attributes
(b) Population ecology
(c) Age pyramids
(d) None of these

41. What type of human population is represented by the following are pyramid?

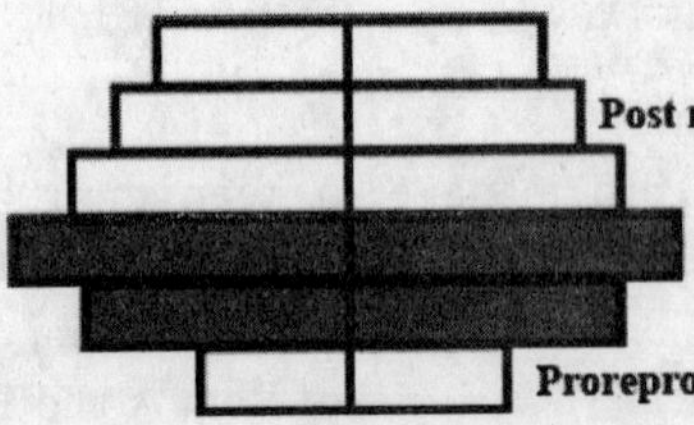

(a) Stable population
(b) Declining population
(c) Expanding population
(d) Vanishing population

42. The tiger census in our national parks and tiger reserves is after based on-
(a) Pug marks
(b) Fecal pellets
(c) Counting no. of tigers
(d) Both A and B

43. ________ is more meaningful measures of the population size of parthenium
(a) Total no. (b) Biomass
(c) Age (d) None of these

44. The age distribution of a population is determined by:
(a) Timing of birth
(b) Timing of death
(c) The rate at which the population is growing
(d) All are correct

Topic 6 Population Growth

45. What four factors define population growth?
(a) Birth, deaths, immigration, emigration
(b) Survivorship, age-specific mortally, fecundity, death rate
(c) Mark-capture, cenrus, sampling transects
(d) Age-specific birth rates, Metapopulation structure, quad rate, ectone

46. ________ contribute to an increase in population density
(a) Natality and emigration
(b) Mortality and emigration
(c) Mortality and Immigration
(d) Natality and Immigration

47. ________ refers to the no. of deaths in the population during a period.
(a) Natality (b) Immigration
(c) Mortality (d) Birth rate

48. If N is the population density at time t, then its

density at time t + 1 is
(a) Nt+1 = Nt + (B + E) – (D + I)
(b) Nt+1 = Nt + (B + d) – (E + I)
(c) Nt+1 = Nt – (B + I) – (D + E)
(d) Nt+1 = Nt + (B + I) – (D + E)

49. Fill up A and B Boxes in the given diagram with correct options:

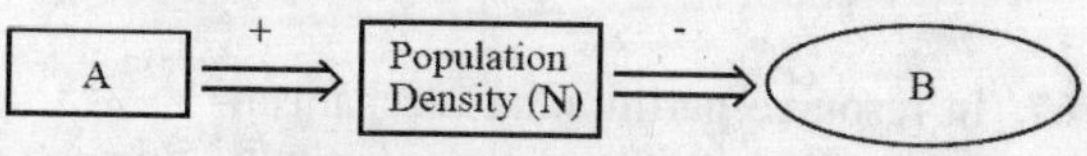

(a) A = Natality + Immigration, B = Mortality + Emigration
(b) A = Natality + Mortality, B = Immigration + Emigration
(c) A = Birth rate + Death rate, B = Migration + Emigration
(d) A = Natality + Emigration, B = Mortality + Immigration

50. A biologist studied the population of rates in a born. He found that average Natality was 260, average Mortality 250, Immigration 30 and emmigration 40. The net increase in population is-
(a) 10 (b) 0
(c) 15 (d) 20

51. The formula for exponential population growth is
(a) dt/dN = rN (b) dN/rN = dt
(c) rN/dN = dt (d) dN/dt = rN

52. Which of the following is not a factor that would limit the growth of population?
(a) Food shortage (b) Immigration
(c) Disease (d) Famine

53. Birth rate = B, Death rate = D, Emigration = E and Immigration = I

	Column I		Column II
a	Population is increasing	i	B + I = D + E
b	Population is decreasing	ii	B + I < D + E
c	Population is stable	iii	B + I > D + E

	a	b	c
(a)	iii	ii	i
(b)	ii	iii	i
(c)	i	ii	iii
(d)	ii	i	iii

54. Darwinian fitness is represented by
(a) Low r value (b) High r value
(c) High k value (d) Low k value

55. What are Labelled phase A, B and C in given sigmoid growth curve?

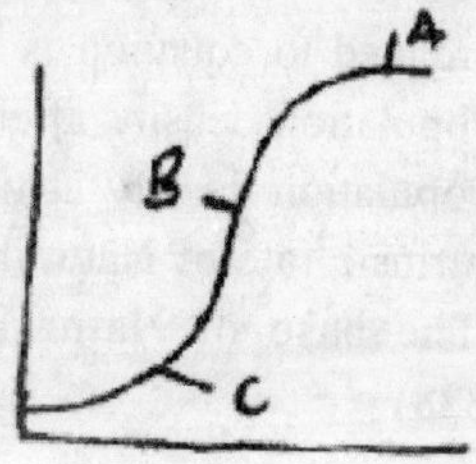

(a) A-Lag, B-Log, C-Stationary
(b) A-Stationary, B-Log, C-Lag
(c) A-Lag, B-Stationary, C-Log
(d) A-Stationary, B-Lag, C-Log

56. Carrying capacity is denoted as
(a) r (b) N
(c) K (d) I

57. Figure For calculation of the r value, which of the following is required?
(a) Birth rates (b) Death rates
(c) Both a and b (d) None

58. Which of the following equation is correct for Logistic growth?
(a) $N_t = N_{08rt}$
(b) dN/dt = rN
(c) dt/dN = rN (K-N/K)
(d) dN/dt = rN (K-N/K)

59. ________ refers to the no. of births during a given period in the population that are added to the initial density.
(a) Natality (b) Mortality
(c) Immigration (d) Survival

60. Logistic curve is ________.
(a) L-shaped (b) J-shaped
(c) Sigmoid curve (d) None of these

61. A plot of N in relation to time (t) results in sigmoid curve. This type of population growth is called-
(a) J shaped Curve
(b) U shaped Curve
(c) Verhulst-Pearl Logistic Growth
(d) Constant Growth

62. $N_t = N_{0ert}$ is the integral form of the exponential-growth equation Which of the following statement related to equation is not correct:
(a) N_t = Population density after time t
(b) N_0 = Population density at time zero
(c) r = Intrinsic rate of Natural decrease
(d) e = The base of Natural logarithmics (2.71828)

63. Which growth model is considered as more realistic one?
(a) Exponential growth
(b) Constant growth
(c) Logistic growth
(d) None of these

64. A population growing in a habitat with limited resources show initially a __(a)__ followed by phase of __(b)__ and finally __(c)__.
(a) A-Lag phase, B-Acceleration and deceleration, C-an asymptote
(b) A-Log phase, B-Acceleration and deceleration, C-an asymptote
(c) A-Log phase, B-Acceleration and deceleration, C-a symptote
(d) A-Log phase, B-Acceleration and deceleration, C-a symptote

65. To calculate the current r value for human population we need to know about
(a) Birth rate (b) Death rate
(c) Carrying capacity (d) Both A and B

66. Human population follows the ________ as the carrying capacity increase or we do not meet yet our population carrying capacity.
(a) J-shaped growth curve
(b) Z-shaped growth curve
(c) S-shaped growth curve
(d) All of the above

Topic 7 | Life History Variation

67. In resource partitioning mechanism-
(a) Species divide a niche to avoid competition for resources
(b) Two different species eat the same thing at the same time of a day
(c) Individuals of the same species that compete with each other
(d) Two species that share the same niche

68. In accordance with their lifestyles, parasites evolved special adaptations such as ________.
(a) the loss of unnecessary sense organs
(b) presence of adhesive organs or suckers to ding on to the host
(c) loss of digestive system and high reproductive capacity
(d) All of the above

69. The human liver fluke (a nematode parasite) depends on two intermediate host to complete its life cycle that is-
(a) insect and cow
(b) insect and human
(c) a snail and fish
(d) None of these

70. Mosquito is-
(a) Not a parasite (b) Parasite
(c) Endoparasite (d) Holoparasite

71. Parasites that feed on the external surface of the host organism are called-
(a) endoparasite (b) ectoparasite
(c) Holoparasite (d) Hemiparasite

72. Which one is/are the example of ectoparasite?
(a) Lice on human
(b) tick one dogs

(c) sea anemone and clown fish
(d) both A and B

73. Match the following:

a	Marine fish	i	Brood parasitism
b	Cuscutta	ii	Copepods
c	Cattle egret	iii	Grazing cattle
d	Koel	iv	Parasite

	a	b	c	d
(a)	ii	iv	iii	i
(b)	iii	iv	i	ii
(c)	iv	ii	iii	i
(d)	i	iii	iv	ii

74. ________ are those that live inside the host body at different sites (Liver, kidney, lungs, red blood cells, etc)
(a) endoparasite (b) ectoparasite
(c) Hemiparasite (d) Both B and C

75. The life cycle of endoparasite are more complex why?
(a) Because of their complex morphology
(b) Because of their food habit
(c) Because of their extreme specialization
(d) All of these

76. Which of the following group do not comes under commensalism?
(a) Orchid growing on Mango branch
(b) Lichens and fungi
(c) Cattle egret and grazing cattle
(d) Sea Anemone and clown fish

77. Orchid grows as a ________ on a mango branch.
(a) Acrophyte (b) Parasite
(c) Epiphyte (d) Both A and B

78. Lichens represent an intimate mutualistic relationship between ________.
(a) Fungus and algae
(b) Cyanobacteria and fungus
(c) Archaebacteria and fungus
(d) Both A and B

79. Mycorrhiza are associations between
(a) fungi and higher root plants
(b) fungi and algae
(c) Algae and lichen
(d) Both B and C

80. Who showed that 5 closely related species of Warblers living on same tree were able to avoid competition and co-exist by behavioural difference?
(a) C. Darwins (b) Connell
(c) Mac Arther (d) Gause

81. (A) The female wasp uses the fig fruit only for oviposition.
(B) Female wasp shows commensalism.
(a) A and B both are correct
(b) A and B both are wrong
(c) A is correct B is wrong
(d) A is wrong and B is correct

82. The Mediterranean orchid employs ________ to get pollination done by a species of bee.
(a) Shelter (b) Food
(c) Sexual deceit (d) Egg-laying sites

83. Pseudocopulation is an example of-
(a) Wasp and fig (b) Ophrys and bees
(c) Ophrys and warp (d) None of these

84. If the female bee's colour pattern changes, the orchid flower co-evolves to maintain the resemblance of its petals for the successful pollination is known as
(a) Commensalism (b) Protocooperation
(c) Co-evolution (d) None of these

85. Which of the following organism breeds only once on their life time?
(a) Pacific salmon fish
(b) Oyester
(c) Bamboo
(d) Both A and C

86. Which of the following organism produces a large no. of small sized offspring?

(a) Pacific salmon fish
(b) Oyester and pelogic fishes
(c) Oyester and pacific salmon fish
(d) Birds and mammals

87. Match the following

1	Both the species benefited	a	Amensalism
2	Both the species get harmed	b	Mutualism
3	One species get benefits	c	Pairasitism
4	One is harmed other species unaffected	d	Predation
		e	Competition

(a) 1-c, 2-b, 3-a, 4-d
(b) 1-b, 2-e, 3-c, d, 4-a
(c) 1-a, 2-e, 3-c, d, 4-b
(d) 1-b, 2-a, 3-c, d, 4-c

88. Prickly pear cactus caused havoc in the early 1920's in-
(a) Canada (b) Austria
(c) India (d) Australia

89. Which predator brought the control over the inversive growth of prickly pear cactus
(a) Moth (b) Bollworm
(c) Caterpillar (d) Grasshopper

90. (a) Chemical control methods are adopted in agricultural pest control, based on ability of predator to regulate prey population.
(b) Penicilium and stryptomyees show Amensalism.
(a) a) Statement A is correct
(b) Statement B is correct
(c) Statement A and B both are correct
(d) Statement A and B are wrong

91. (A) Predators helps in maintaining species diversity in community.
(B) It reduces the intensity of competition among competing prey species.
(a) Statement A is correct
(b) Statement B is correct
(c) Both statement is wrong
(d) Option A and B

92. Starfish Pisaster is
(a) Parasite (b) Hemi-parasite
(c) Predator (d) Prey of invertebrate

93. More than 10 species of vertebrates disappeared a year after removing the starfish from habitat is due to-
(a) interspecific competition
(b) Brood parasitism
(c) Intra specific competition
(d) None of these

94. Prey species defenses themselves through there behaviour-
(a) Camouflaged (b) Highly distasteful
(c) Poisonous (d) All of these

95. Which Butterfly is highly distasteful?
(a) Monarch Butterfly (b) Viceroy Butterfly
(c) Queen Butterfly (d) All of these

96. Butterfly is highly distasteful which is acquire by them by tending on poisonous weed during
(a) Caterpillar Stage (b) Adult Butterfly
(c) Pupa State (d) All of these

97. "Camouflage" means
(a) Cryptically coloured
(b) Feeding on young ones of other species
(c) Poisonous
(d) Feeding on own species

98. Darwin has given the statement of
(a) Survival of fittest
(b) Struggle for existence
(c) Both A and B
(d) None of these

99. Who has convinced that interspecific competition is a patent force in organic evolution?
(a) Darwin (b) P. odum
(c) Jansely (d) None of these

100. (a) It is generally believed that competition occurs when closely related species complete for same resources that are limiting.

(b) Totally unrelated species could also complete for the same resource.
(a) A is true B is false
(b) Both A and B is false
(c) A and B both are true
(d) A is false and B is true

101. The feeding efficiency of one species might be reduced due to the interfering and inhibitory presence of the other species even if resource are abundant in known as-
(a) Interspecific predation
(b) Interfering competition
(c) Both A and B
(d) commensalism

102. When certain exotic species are introduced in to a geographical area they become invasive mainly because:
(a) The invaded land has unlimited resources for the introduced species.
(b) The population of the introduced species in the invaded land is very low.
(c) Introduced species do not face any competition in the introduced land.
(d) The invaded land does not have its natural predator.

103. What was the result, when all pisaster starfish were removed from an enclosed intertidal area, in a field experiment?
(a) Extinction of many invertebrate species
(b) Increase in diversity of invertebrates
(c) Inability of the pisaster to enter the area again
(d) Replacement of pisaster by other starfish

104. A species whose distribution is restricted to a small geographical area because of the presence of a competitively superior species is found to expand its distributional range dramatically when the competing species is experimentally removed. This is called-
(a) Competitive Exclusion
(b) Competitive Supermacy
(c) Competitive Inclusion
(d) Competitive Release

105. Which of the following is not a function of predators?
(a) They decrease the species competition in a community
(b) They act as conduits for energy transfer access trophic levels
(c) They help in stabilization of the ecosystem
(d) They decrease the species diversity in a community

106. Connell's field experiment on the rockey sea coast of scotland the larger and competitively superior bernacle Balances dominates the intertidal area and excludes the smaller barnacle chathamalus from that zone. This happened due to:
(a) Mutualism (b) Predation
(c) Competition (d) Parasitism

107. The principle of competitive exclusion was stated by:
(a) Gause (b) C. Darwin
(c) Mac Arther (d) Connelli

108. Gause's principle of competitive exclusion states that:
(a) More abundant species will exclude the less abundant through competition
(b) Larger organism will exclude smaller one
(c) No two closely related species can occupy same niche indefinitely for the same limiting resources
(d) Both A and B

ANSWER KEY

Q	1	2	3	4	5	6	7	8	9	10
Ans	**B**	**D**	**D**	**A**	**D**	**C**	**D**	**A**	**A**	**A**
Q	11	12	13	14	15	16	17	18	19	20
Ans	**B**	**B**	**A**	**B**	**C**	**C**	**A**	**D**	**B**	**A**
Q	21	22	23	24	25	26	27	28	29	30
Ans	**D**	**D**	**A**	**A**	**C**	**A**	**A**	**B**	**C**	**B**
Q	31	32	33	34	35	36	37	38	39	40
Ans	**C**	**A**	**D**	**B**	**A**	**D**	**A**	**C**	**D**	**C**
Q	41	42	43	44	45	46	47	48	49	50
Ans	**B**	**D**	**B**	**D**	**A**	**D**	**C**	**D**	**A**	**B**
Q	51	52	53	54	55	56	57	58	59	60
Ans	**D**	**B**	**A**	**B**	**B**	**C**	**C**	**D**	**A**	**C**
Q	61	62	63	64	65	66	67	68	69	70
Ans	**C**	**C**	**C**	**A**	**D**	**A**	**A**	**D**	**C**	**A**
Q	71	72	73	74	75	76	77	78	79	80
Ans	**B**	**D**	**A**	**A**	**C**	**B**	**C**	**D**	**A**	**C**
Q	81	82	83	84	85	86	87	88	89	90
Ans	**D**	**C**	**C**	**D**	**B**	**B**	**D**	**A**	**B**	
Q	91	92	93	94	95	96	97	98	99	100
Ans	**D**	**C**	**A**	**D**	**A**	**A**	**A**	**C**	**A**	**C**
Q	101	102	103	104	105	106	107	108		
Ans	**B**	**D**	**A**	**D**	**D**	**C**	**A**	**C**		

18 Chapter

Ecosystem

MULTIPLE CHOICE QUESTIONS

Topic 1 Ecosystem: Structure and Function

1. The autotrophic components include
(a) Phytoplankton (b) Some algae
(c) Marginal plants (d) All of these

2. The decomposers is/are the
(a) Fungi (b) Bacteria
(c) Flagellates (d) All of these

3. The consumers is/are
(a) Zooplankton (b) Phytoplanktons
(c) Marginal plants (d) All of these

4. Vertical distribution of different species occupying different levels is called as
(a) Standing crop (b) Standing state
(c) Stratification (d) Decomposition

5. Identification and enumeration of plant and animal species of an ecosystem gives its
(a) Productivity
(b) Species composition
(c) Physical structure
(d) Vertical distribution

6. Which one of the following is odd one out from others
(a) Decomposition
(b) Energy flow
(c) Nutrient cycling
(d) None

Topic 2 Productivity

7. Net primary productivity (NPP) equals to
(a) NPP = R – GPP (b) GPP – R = NPP
(c) NPP = GPP + R (d) GPP = R – NPP

8. The rate of formation of new organic matter by consumers is called as
(a) primary productivity
(b) Gross primary productivity
(c) Secondary productivity
(d) Respiratory loss

9. Primary productivity depends on
(a) Variety of environmental factors
(b) Availability of nutrients
(c) Photosynthetic capacity of plant
(d) All of these

10. The annual net primary productivity of the whole biosphere is approximately
(a) 190 million tons (b) 170 million tons
(c) 170 billion tons (d) None of these

11. What is the basic requirement for any ecosystem to function and sustain.
(a) Primary production
(b) Decomposers
(c) Constant input of solar energy
(d) Nutrient cycling

12. ________ is defined as the amount of biomass or organic matter produced per unit area over a time period by plants during photosynthesis
(a) Gross primary productivity

(b) Primary production
(c) Secondary production
(d) None of these

13. Primary production is expressed as-
(a) K *Clam*^{2} (b) K Clam/m^2
(c) *g/m*^{2} (d) both B & C

14. ______ of an ecosystem is the rate of production of organic matter during photosynthesis
(a) Net primary productivity
(b) Secondary production
(c) Gross primary productivity
(d) None of these

Topic 3 Decomposition

15. Which one of the following is called as "farmer's friend"?
(a) Cow (b) Bacteria
(c) Earthworm (d) Crops

16. Who breaks down complex organic matter into inorganic substances like CO2, water etc.
(a) Crop roots (b) Decomposers
(c) Grazing Cattle (d) None of these

17. The process of breaks down complex organic matter into inorganic substances is called as
(a) Fragmentation (b) Humification
(c) Decomposition (d) Leaching

18. Detritus is/are
(a) Dead plant (b) Dead animals
(c) Fecal matter (d) All of these

19. The correct way of decomposition
(a) Fragmentation → leaching → humification → catabolism → mineralization
(b) Fragmentation → leaching → catabolism → humification → mineralization
(c) Fragmentation → catabolism → leaching → mineralization → humification
(d) Fragmentation → mineralization → catabolism → leaching → humification

20. ________ break down detritus into smaller particles
(a) Earthworm
(b) Detritivores
(c) Phytoplanktons
(d) Both A & B are correct

21. Bacteria and fungal enzymes degrade detritus into simpler inorganic substances. This process is called as
(a) Leaching (b) Fragmentation
(c) Catabolism (d) Humification

22. Humification leads to accumulation of a dark coloured amorphous substance called ______
(a) Pectin (b) Humus
(c) Lignin (d) None of these

23. Decomposition rate is slower if
(a) Detritus rich in lignin & chitin
(b) Rich in nitrogen & sugars
(c) Low in nitrogen & chitin
(d) Low in lignin

24. ________ favours decompositions
(a) Warm & dry environment
(b) Warm & moist environment
(c) cold & dry environment
(d) cold & moist environment

Topic 4 Energy flow

25. PAR stands for
(a) Percent active radiation
(b) Photosynthetically active radiation
(c) Power angel regulation
(d) None of these

26. Plant capture only ______ of the PAR and this amount of energy sustains the entire living world
(a) 50 – 60 % (b) 40 – 80 %
(c) 2 – 10 % (d) 20 – 40 %

27. The green plant in the ecosystem are called
(a) Primary consumer
(b) Producer
(c) Secondary consumer
(d) None of these

28. Producers in an aquatic ecosystem
(a) Phytoplankton (b) Algae
(c) Zooplanktons (d) Both A & B

29. Generally, primary consumers will be
(a) Carnivores (b) Producers
(c) Herbivores (d) All of these

30. In ecosystem, GFC stands for
(a) Generic flow control
(b) Global fund for children
(c) Grazing food chain
(d) None of these

31. Decomposers are also known as
(a) Autotrophs (b) Standing crops
(c) Saprotrophs (d) None of these

32. Based on the source of their nutrition or food, organisms occupy a specific place in the food chain that is known as their
(a) Food web (b) Trophic level
(c) Niche (d) Eco level

33. Match the following

Column I		Column II	
i	Plants	a	Lion
ii	Carnivores	b	Phytoplanktons
iii	Herbivores	c	Wolf
iv	Top Carnivores	d	Cow

(a) i-b, ii-c, iii-d, iv-a (b) i-c, ii-d, iii-b, iv-a
(c) i-b, ii-d, iii-a, iv-c (d) i-d, ii-b, iii-a, iv-c

34. Each tropical level has a certain mass of living material at a particular time called as the
(a) Biomass (b) Standing crop
(c) Standing state (d) None of these

35. The standing crop is measured as the
(a) Mass of living organisms
(b) Biomass
(c) The no. in a unit area
(d) All of these

36. Choose the correct sequence –
(a) Producer → herbivore → primary carnivore → secondary carnivore
(b) Producer → primary carnivore →herbivore → secondary carnivore
(c) Primary carnivore → secondary carnivore → herbivore → Producer
(d) None of these

Topic 5 Ecological Pyramids

37. Ecological pyramids are
(a) Pyramid of number
(b) Pyramid of energy
(c) Pyramid of biomass
(d) All of these

38. The pyramid of biomass in sea is
(a) Always upright
(b) Generally inverted
(c) Both A & B
(d) None of these

39. Pyramid of energy is
(a) Always inverted
(b) Sometimc upright
(c) Always upright
(d) Sometimes inverted

40. Identify the pyramid

(a) Pyramid of number
(b) Pyramid of biomass
(c) Pyramid of energy
(d) None of these

41. Identify the pyramid

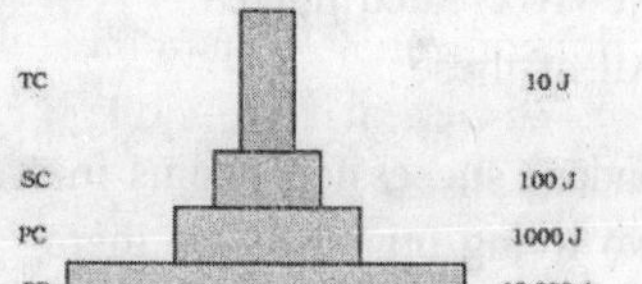

(a) Pyramid of number
(b) Pyramid of biomass
(c) Pyramid of energy
(d) None of these

42. Identify the pyramid

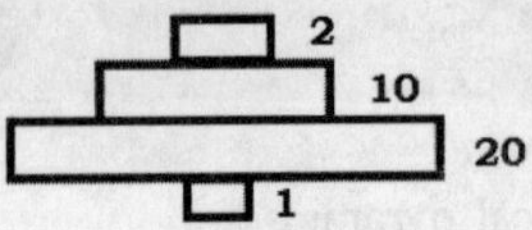

(a) Pyramid of number
(b) Pyramid of biomass
(c) Pyramid of energy
(d) None of these

Topic 6 | Ecological succession

43. A community that is in near equilibrium with the environment is called as
(a) Pioneer community
(b) Middle community
(c) Climax community
(d) Sere

44. The gradual and fairly predictable change in the species composition of a given area is called
(a) Hydrarch succession
(b) Ecological succession
(c) Pioneer succession
(d) None of these

45. The entire sequence of communities that successively change in a given area are called
(a) Ecosystem (b) Pioneer
(c) Sere (d) All of these

46. Areas where primary succession occurs
(a) Bare rock
(b) Newly cold lava
(c) Newly created pond
(d) All of these

47. Secondary succession begins in areas where
(a) No living organism are there
(b) Lost all the living organism
(c) Natural biotic communities have been destroyed
(d) Both B & C

48. Areas where secondary succession occurs
(a) Burned and cut forests areas
(b) Land that have been flooded
(c) Abandoned farm lands
(d) All are correct

49. Select the correct statement
(a) secondary succession is faster than primary succession
(b) primary succession is faster
(c) Both are a equal speed
(d) None of these

50. The individual transitional communities are termed as
(a) Seral stages
(b) Pioneer
(c) Seral communities
(d) Both A & C are correct

Topic 7 | Succession of plants

51. Which type of succession takes place in wet areas
(a) Hydrarch succession
(b) Xerarch succession
(c) Mesarch succession
(d) None of these

52. Xerarch succession occurs in
(a) Wet areas (b) Cold areas
(c) Dry areas (d) All of these

53. The species that invade a bare area called
(a) Sere (b) Pioneer species
(c) Climax species (d) None of these

54. In hydrarch succession the successional series progress from
(a) Mesic to hydric condition
(b) Hydric to mesic condition
(c) Hydric to xeric condition

(d) Xeric to mesic condition

55. In xerarch succession, the succession series progress from
(a) Xeric to hydric condition
(b) Xeric to mesic condition
(c) Mesic to xeric condition
(d) None of these

56. Which one of the following occur as a pioneer species on rocks
(a) Bryophytes
(b) Phytoplankton
(c) Lichens
(d) Blue algae

57. Choose the correct sequence of succession in water
(a) Phytoplanktons → rooted-submerged plants → rooted floating angiosperms → free floating plants → reed swamp → marsh-meadow → scrub → the trees → forest
(b) Phytoplanktons → free floating plants → rooted-submerged plants → rooted floating angiosperms → reed swamp → scrub → marsh-meadow → the trees → forest
(c) Phytoplanktons → rooted-submerged plants → reed swamp → rooted floating angiosperms → free floating plants → marsh-meadow → scrub → the trees → forest
(d) None of these

58. Choose the correct statement-
(a) All succession whether taking place in water or on land, proceeds to a different climax community
(b) All succession whether taking place in water or on land, proceeds to a similar climax community the mesic
(c) All succession whether taking place in water or on land, proceeds to a similar climax community the xeric
(d) All of these

59. Which one of the following is not the part of hydrarch succession
(a) Scrub stage
(b) Tree
(c) Zooplankton
(d) Submerged plant stage

60. During succession some species colonise an area and their population become more numerous whereas population of other species
(a) Increases
(b) Decline and even disappear
(c) Migrate
(d) None of these

61. Why does secondary succession is faster?
(a) Because soil is already there
(b) They have special power
(c) Growth of plants is faster
(d) All of these

62. The climax community remains ________ as long as the environment remains ________.
(a) Unstable, unchanged
(b) Stable, unchanged
(c) Stable, changed
(d) Stable, changed

63. The word (term) use for medium water conditions
(a) Xeric (b) Hydric
(c) Mesic (d) None of these

64. Choose the correct sequence
i) lichens
ii) Grasses
iii) Bryophytes
iv) Higher plants
v) Forest
(a) i → ii → iii → iv →
(b) i → iii → ii → iv → v
(c) i → iv → ii → iii → v
(d) v → iv → i → ii → iii

65. in hydrarch succession, after climax with time the water body is converted into
(a) River (b) Ocean
(c) Land (d) None of these

66. In hydrarch succession, the pioneer and climax

community are respectively
(a) Forest, Phytoplanktons
(b) Phytoplanktons, Forest
(c) Mess, Trees
(d) Lichen, Trees

Topic 8 Nutrient Cycling

67. The amount of nutrients such as carbon, nitrogen, phosphorus, calcium etc present in the soil at any given time is referred to as the
(a) Nutrients cycle (b) Standing crop
(c) Standing state (d) None of these

68. Standing state varies in
(a) Different kinds of ecosystem
(b) On a season basis
(c) Different kinds of nutrients
(d) Both A & B

69. The movement of nutrients elements through the various components of an ecosystem can be called
(a) Gaseous cycle
(b) Nutrient cycling
(c) Sedimentary cycle
(d) All of these

70. Another name of nutrient cycling is
(a) Gaseous cycle
(b) Biological cycle
(c) Biogeochemical cycle
(d) Biophysical

71. Reservoir for gaseous type of nutrient cycle
(a) Earth's crust (b) Rock
(c) The atmosphere (d) Water bodies

72. Reservoir for sedimentary type of Nutrient cycle
(a) Ocean (b) Earth's crust
(c) Rock (d) Atmosphere

73. Environmental factor to regulate the rate of release of nutrients into the atmosphere.
(a) Soil
(b) Moisture
(c) Temperature & pH
(d) All of the above

74. Reason behind nutrients never lost from ecosystem.
(a) Because nutrients present in large amount
(b) Because they are recycled
(c) Because they have no use
(d) All of the above

Topic 9 Ecosystem-Carbon Cycle

75. Percent of carbon constitutes in dry weight of organism
(a) 60% (b) 39%
(c) 49% (d) 71%

76. Which is the first & second most abundant constituent of an organism?
(a) Water, phosphorus
(b) Water, carbon
(c) Carbon, water
(d) Carbon, phosphorus

77. How much of total quantity of global carbon is dissolved in the oceans?
(a) 88% (b) 75%
(c) 81% (d) 71%

78. Carbon cycling occurs through
(a) Atmosphere
(b) Living & dead organism
(c) Ocean
(d) All of the above

79. How much at carbon in fixed annually in the biosphere through photosynthesis?
(a) 8 × 1012 kg (b) 4 × 1012 kg
(c) 4 × 1013 kg (d) 4.9 × 10 kg

80. Additional sources for releasing CO_2 in the atmosphere is/are-
(a) Burning of wood (b) Forest fire
(c) Fossil fuel (d) All of the above

81. Human activities have significantly increased the rate of released of CO_2 into the atmosphere by
(a) Rapid deforestation

(b) Massive burning of fossil
(c) Both A and B
(d) None of the above

Topic 10 | Ecosystem-Phosphorus Cycle

82. Phosphorus is a major constituent of
(a) Biological membranes
(b) Nucleic acids
(c) Cellular energy transfer unit
(d) All of the above

83. Rock is the natural reservoir of
(a) Carbon
(b) Nitrogen
(c) Phosphorus
(d) None at these

84. Herbivores & other animals obtain Phosphorus from
(a) Rock (b) Plants
(c) Ocean (d) Lake

85. The waste products and the dead organism are decomposed by _______ releasing phosphorus.
(a) Fungi
(b) Phosphate-solubilising bacteria
(c) Phosphate-unsolubising bacteria
(d) None of the above

86. Choose the correct statement.
(a) Atmospheric inputs of phosphorus through rainfall are much smaller than carbon inputs.
(b) Atmospheric inputs of phosphorus through rainfall are larger than carbon inputs.
(c) Atmospheric inputs of phosphorus through rainfall are equal to the carbon inputs.
(d) None of the above

87. Choose the more correct statement.
(a) Gaseous exchange at phosphorus b/w organism & environment are very high.
(b) Gaseous exchange of phosphorus b/w organism & environment are low.
(c) Gaseous exchange at phosphorus b/w organism & environment are negligible.
(d) None at these

88. In natural resevoirs, phosphorus present in the form of
(a) Phosphite (b) Pyrophosphate
(c) Phosphates (d) None of the above

89. Identify the blanks

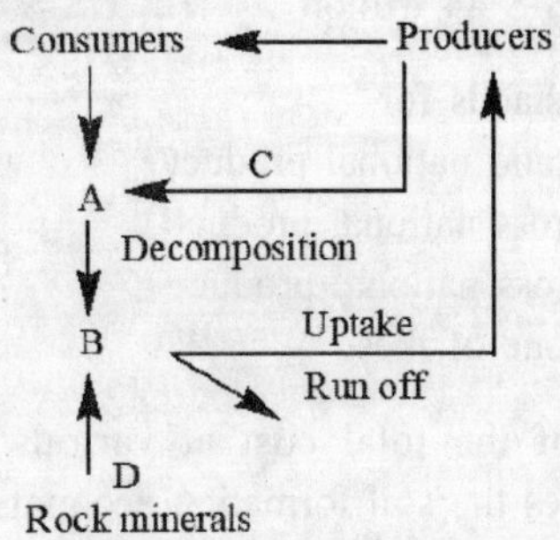

	A	B	C	D
(a)	Detritus	Weathering	Soil solution	Litter fall
(b)	Litter fall	Weathering	Detritus	Soil solution
(c)	Weathering	Litter fall	Soil solution	Detritus
(d)	Detritus	Soil solution	Litter fall	Weathering

90. Which one of the following is not a Gaseous nutrient cycle?
(a) Oxygen cycle (b) Nitrogen cycle
(c) Sulphur cycle (d) None of the above

91. Animals need large quantities of phosphorus to make
(a) Shells (b) Teeth
(c) Bones (d) All of the above

Topic 11 | Ecosystem Services

92. _______ & his colleagues have very recently tried to put price tags on nature's life-support services.
(a) Robert frost
(b) Robert Constanza

(c) Robert hook
(d) Robert Clive

93. Researchers have put an average price tag of ________ a year on fundamental ecosystem services.
(a) US $ 33 billion (b) US $ 44 billion
(c) US $ 44 trillion (d) US $ 33 trillion

94. GNP stands for
(a) Grand national product
(b) Gross national product
(c) Gross national produce
(d) None of these

95. Out of the total cost at various ecosystem services the soil formation accounts for about.
(a) 40% (b) 60%
(c) 50% (d) 30%

96. The cost of climate regulation & habitat for wildlife are
(a) 8% each (b) 6% at overall
(c) 6% each (d) None at the above

97. The value of the global GNP
(a) US $ 28 trillion (b) US $ 18 Billion
(c) US $ 33 trillion (d) US $ 18 trillion

98. Choose the correct statement.
(a) Value of Ecosystem services at biodiversity is difficult to determine.
(b) Value of Ecosystem services of biodiversity is very easy to determine.
(c) No need to determine the value of Ecosystem services.
(d) None of these

99. The products of ecosystem processes are named as
(a) Environmental services
(b) Ecosystem goods
(c) Ecosystem services
(d) All of the above

100. Healthy ecosystems are the base for a
(a) Wide range of economic
(b) Environmental
(c) Aesthetic goods & services
(d) All of the above

101. Examples of Ecosystem services
(a) Healthy forest ecosystem purify air & water
(b) Generate fertile soil
(c) Provide storage site for carbon
(d) All of the above

ANSWER KEY

Q	1	2	3	4	5	6	7	8	9	10
Ans	**D**	**B**	**A**	**C**	**B**	**D**	**B**	**C**	**D**	**C**
Q	11	12	13	14	15	16	17	18	19	20
Ans	**C**	**B**	**D**	**C**	**C**	**B**	**C**	**D**	**B**	**D**
Q	21	22	23	24	25	26	27	28	29	30
Ans	**B**	**B**	**A**	**B**	**B**	**C**	**B**	**D**	**C**	**C**
Q	31	32	33	34	35	36	37	38	39	40
Ans	**C**	**C**	**A**	**B**	**D**	**A**	**D**	**B**	**C**	**B**
Q	41	42	43	44	45	46	47	48	49	50
Ans	**C**	**A**	**C**	**B**	**C**	**D**	**D**	**D**	**A**	**D**
Q	51	52	53	54	55	56	57	58	59	60
Ans	**A**	**C**	**B**	**B**	**B**	**A**	**A**	**B**	**C**	**B**
Q	61	62	63	64	65	66	67	68	69	70
Ans	**A**	**B**	**B**	**B**	**C**	**B**	**C**	**D**	**D**	**C**
Q	71	72	73	74	75	76	77	78	79	80
Ans	**C**	**B**	**D**	**B**	**C**	**B**	**D**	**D**	**C**	**D**
Q	81	82	83	84	85	86	87	88	89	90
Ans	**C**	**D**	**C**	**B**	**B**	**A**	**C**	**C**	**D**	**C**
Q	91	92	93	94	95	96	97	98	99	100
Ans	**D**	**B**	**D**	**B**	**C**	**C**	**D**	**A**	**C**	**D**
Q	101									
Ans	**D**									

19 Chapter

Biodiversity and Conservation

MULTIPLE CHOICE QUESTIONS

Topic 1	Biodiversity

1. Highest variation of species in plants is in-
(a) Fungi (b) Mosses
(c) Algae (d) Lichens

2. In vertebrates maximum No. of species found in
(a) Birds (b) Mammals
(c) Fishes (d) Reptiles

3. Problem behind no given figures about prokaryotes is
(a) Conventional taxonomic methods are not suitable for identifying microbial species
(b) Many species are simply not culturable under laboratory conditions
(c) Because prokaryotes are non-living
(d) A & B both are correct

4. India has only …….. percent at the world's land area. Its share of the global species diversity in an impressive….percent.
(a) 2.4, 8.1 (b) 8.1, 2.4
(c) 4.1, 8.2 (d) 8.4, 2.1

5. India is one of the ….mega diversity countries of the world.
(a) 13 (b) 14
(c) 12 (d) 11

6. How may plant species found in India roughly
(a) 1,00,000 (b) 45,000
(c) 3,00,00 (d) None of the above

7. According to May's global estimate there are probably…..plant species & more than …..animal species.
(a) 3,00,000 & 1,00,000
(b) 30,000 & 10,000
(c) 1,00,000 & 3,00,000
(d) 10,000 & 30,000

8. Term 'Biodiversity' popularized by
(a) Robert hook (b) Ernst Haeckel
(c) Edward Wilson (d) G. Tansley

9. Edward Wilson was a -
(a) Mathematician (b) Sociobiologist
(c) Psychologist (d) None of the above

10. 'Biodiversity' was popularized
(a) To describe the combined diversity
(b) to describe individual diversity
(c) To describe plant diversity
(d) None of the above

11. Types of Biodiversity is/are-
(a) Genetic diversity
(b) Species diversity
(c) Ecological diversity
(d) All of these

12. The genetic variation shown in
(a) Rouwolfia vomitoria
(b) Rice
(c) Mango
(d) All of the above

13. Active chemical in Rauwolfia vomitoria
(a) Terpine

(b) Reserpine
(c) Asprine
(d) None of the above

14. Which one is correct about Genetic diversity
a) Genetic variation shown by Rauwolfia vomitoria
(b) Genetic variation shown in western ghats.
(c) India has less than 5000 genetically different strain of rice
(d) India has 1,000 varieties of rice

15. India has more than…….. genetically different strains of rice, and ……varieties of Mango.
(a) 50,000, 1,000
(b) 1,000 & 50,000
(c) 5,000, 1,000
(d) 1,000 & 5,000

16. The diversity at the species level is c/a
(a) Ecological diversity
(b) Genetic diversity
(c) Species diversity
(d) None of these

17. The ……..Ghats have a greater amphibian species diversity than the ….. Ghats.
(a) Eastern, Western
(b) Western, Eastern
(c) Western, Southern,
(d) Eastern, Southern

18. Ecological diversity is in
(a) Deserts
(b) Rain forests, Mangroves
(c) Coral reefs, Wetlands
(d) All of the above

19. Biodiversity and its conservation is of
(a) national concern
(b) concern in some states of india
(c) international concern
(d) all of these

20. IUCN stands for-
(a) International unity for conservation of Nature
(b) Indian union for conservation of Nature
(c) Italian union for conservation at Nature
(d) International Union for conservation of Nature and Natural Resources

21. Robert may estimates the global species diversity at about
(a) 7 billion (b) 70 million
(c) 7 million (d) 70 billion

22. Most species rich taxonomic group are
(a) Crustaceans (b) Insects
(c) Molluscs (d) None of the above

23. In all available species more than 70% of all the species recorded are ______ which camprise no more than 22% of the total
(a) Invertebrates, Plants
(b) Plants, Animal
(c) Animal, Plants
(d) Plant, Invertebrates

Topic 2 | Patterns of Biodiversity

24. Species diversity decreases as
(a) We move away from the equator towards the poles
(b) We move away from the poles towards equator
(c) We move hill to plane area
(d) None of the above

25. How many species of birds found in India?
(a) 1500 (b) 1400
(c) 1200 (d) 1700

26. Greatest biodiversity on earth is found in
(a) Sundarbans
(b) Amazon rain forest
(c) Congo rain forest
(d) None of the above

27. Amazon Rainforest situated in
(a) North America (b) North India
(c) South America (d) West America

28. Choose the correct option

1	Plant	i	378

2	Fishes	ii	40,000
3	Birds	iii	427
4	Mammals	iv	3,000
5	Reptiles	v	1,300

	1	2	3	4	5
a)	ii	iv	v	iii	i
b)	iii	ii	v	I	iv
c)	ii	iv	v	I	iii

(d) None of the above

29. How many insect species waiting to be discovered & named.
(a) 2 million (b) 20,000
(c) 2,000 (d) 2,00,000

30. Which one is wrong about species found in Amazon rainforest.
(a) Reptiles-378
(b) Fishes-3,000
(c) Birds-1300
(d) All of the above are correct

31. More......Energy available in tropics, which contributes to higher productivity
(a) Solar
(b) Thermal
(c) Nuclear
(d) None of the above

32. The relation b/w species richness & area for a wide variety of taxa turns out to be a
(a) Linear
(b) Sigmoid
(c) Rectangular Hyperbola
(d) None of the above

33. Correct equation of species-Area relationship
(a) log A = log S + Z log C
(b) log S = log C + Z log A
(c) log S = log Z + C log A
(d) log S = log C – Z log A

34. In species area relationship equation, Z stands for-
(a) Slope of the line
(b) Regression coefficient
(c) Species richness
(d) Both a & B are correct

35. Value of Z is
(a) 20-30 (b) 1 – 2
(c) 0.1 – 0.2 (d) None of the above

36. The value of Z in the entire continents
(a) 0.8 – 1.8 (b) 0.6 – 1.2
(c) 0.1 - 0.2 (d) None of the above

37. The value of Z for frugivorous birds & mammals in the tropical forests of different continents
(a) 2.15 (b) 0.5
(c) 1.15 (d) 0.8

38. Concept of species-Area relationship given by
(a) Robert Frost
(b) Paul Ehrlich
(c) Alexander hook
(d) Alexander Von Humboldt

39. Alexander von Humboldt is a
(a) German naturalist
(b) Philosopher
(c) Geographer
(d) Both A & C are correct

40. According to Alexander species richness
(a) Increased with increasing explored area, but only up to a limit
(b) Decreased with increasing explored area, but only up to a limit
(c) Increased with decreasing explored area, but there are no limitations
(d) None of the above

41. Find out A and B and also select the correct answer

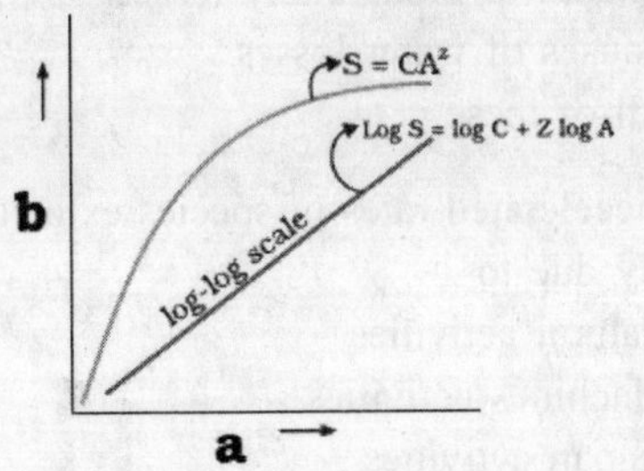

(a) a = species richness b = Area
(b) a = area b = species richness
(c) a = Area b = Regression coefficient
(d) none of these

42. In species-area relationship equation
(a) S = species richness
(b) A = area
(c) C = Y – intercept
(d) All of these

43. Which one is correct about a stable community?
(a) Should not show too much variation in productivity from year to year
(b) It must be either resistant or resilient on surface disturbance
(c) It must also be resistant to invasions by alien species
(d) All of these are correct

44. Find out correct option about David Tilman's long term ecosystem experiments
(a) Plots with more species showed less year-to-year va riation in total biomass
(b) Plots with more species show too much variation in productivity
(c) Increased diversity contributed to higher productivity
(d) Both A & C are correct

45. 'Rivet popper hypothesis' is given by
(a) Alexander von Humboldt
(b) Paul Ehrlich
(c) Tilman
(d) Robert hook

46. Paul Ehrlich is a
(a) Stanford ecologist
(b) Stanford psychologist
(c) Stanford micrologist
(d) Stanford geologist

47. In Rivet popper hypothesis, rivets depict as
(a) Ecosystem (b) Extinct species
(c) Species (d) Plants

48. According to Ehrlich's rivet popper hypothesis loss of rivets on the wings affects in ecosystem as-
(a) Loss of key species that drive major ecosystem function
(b) Proper functioning of ecosystem
(c) Species to become extinct
(d) None of these

49. According to rivet popper hypothesis, which one is not correct
(a) Rivet – species
(b) Airplane – Ecosystem
(c) Rivets to take home – species to become extinct
(d) All of these are correct

Topic 3 Loss of Biodiversity

50. The colonization of tropical pacific islands by humans is said to have led to the extinction of more than ______ of native birds
(a) 2000 species (b) 1200 species
(c) 20000 species (d) 200 species

51. According to red list (2004) No. of total extinct species in last 500 years
(a) 504 species (b) 387 species
(c) 478 species (d) 784 species

52. How many species of invertebrates become extinct in last 500 years
(a) 87 species (b) 784 species
(c) 359 species (d) None of these

53. According to recent extinction which one is incorrect option –
(a) Dodo – Mauritius

(b) Steller's sea cow – USA
(c) Quagga – Africa
(d) Thylacine – Australia

54. Subspecies of tiger which become recently extinct
(a) Bali (b) Caspian
(c) Javan (d) All of these

55. The last 20 years alone have witnessed the disappearance of ______
(a) 300 species (b) 10 species
(c) 27 species (d) 1000 species

56. Match the following-

Species		% of threat of extinction	
a	Birds	1	23%
b	Mammals	2	32%
c	Amphibians	3	12%
d	Gymnosperms	4	31%

(a) a-1, b-3, c-2, d-4
(b) a-3, b-1, c-2, d-4
(c) a-3, b-1, c-4, d-2
(d) a-3, b-4, c-1, d-2

57. Select the correct statement about 'Sixth Extinction'
(a) The extinction rates are estimated to be 100 - 1000 times faster than in the prehuman times.
(b) Human activities are responsible for the faster rates.
(c) Half of all the species on Earth might be wiped out within the next 1000 years.
(d) Both A and B option are correct.

58. Loss of biodiversity in a region may lead to
(a) decline in plant production,
(b) lowered resistance to environmental. perturbations such as drought.
(c) Increased variability in ecosystem process.
(d) All of these.

59. 'The Evil Quartet' is the term used to describe
(a) Causes of habitat losses
(b) Causes of biodiversity losses.
(c) Causes of water losses
(d) All of these.

60. The accelerated rates of species extinctions are largely due to
(a) Natural activities
(b) Machines activities
(c) human activities
(d) none of these

61. Which one is not involved in the evil quartet?
(a) Habitat loss and fragmentation
(b) Alien species invasions
(c) over - exploitation
(d) all are involved

62. The most important cause of animals and plants to extinction
(a) Over exploitation
(b) Alien species invasions
(c) Co-existence
(d) Habitat loss and fragmentation.

63. Lungs of the planet term used for
(a) Tropical rain forest,
(b) Amazon rainforest,
(c) temperate reason,
(d) none of these.

64. The Amazon rainforest harboring probably millions of species to is being cut and cleared for.
(a) For cultivating soya beans.
(b) Conversion to grassland for raising beef cattle.
(c) Cultivating new forest.
(d) Both A and B are correct.

65. When large habitats are broken up into small fragments due to various ______, mammals and birds requiring large territories and certain animals with migratory habits are badly affected, leading to __b__.
(a) a – Nature activities, b – Population declines
(b) a – Human activities, b – Population declines
(c) a – Human activities, b–Population increasing

(d) none of these

66. Once covering more than ______ percent of Earth's land surface, these rain forests now cover no more than ______ percent.

(a) 6, 14 (b) 14, 8
(c) 14, 6 (d) 12, 8

67 Species extinct in the last 500 years due to over exploitation by humans.

(a) Steller's sea cow
(b) Cichlid fish
(c) Passenger pigeon
(d) A & C both are correct

68. How many species of cichild fish is extinct due to introduction of Nile perch in lake victoria?

(a) 2 (b) 20
(c) 200 (d) 2000

69. The environmental damage caused and threat posed to our native species by invasive weed species like.

(a) Carrot grass (b) water hycinth
(c) lantana (d) all of these.

70. African catfish ______ is posing as threat to the indigenous catfishes.

(a) Parthenium
(b) Clarias garlepinus
(c) Clarias branchysoma
(d) None of these

71. Which one is correct statement about coextinction?

(a) When a species become extinct, the plant and animal species associated with it in an obligatory way also become extinct.
(b) When a host becomes extinct, it's unique assemblage of parasites also meets the same fate
(c) In the plant pollinator· mutualism, where extinction of one invariably leads to the extinction of the other.
(d) All these statements are correct.

72. Which one is not an example of Alien species invasions?

(a) Parthenium
(b) Steller's sea cow
(c) Catfish
(d) Lantana

73. Purpose of introducing the African cat fish in the Indian River.

(a) For Aquaculture
(b) For ornamental purpose
(c) For agriculture
(d) None of these.

Topic 4 | Biodiversity Conservation

74. Sacred groves are that-

(a) Place which are protected for animals.
(b) Place where all people are worship.
(c) Place where religion and cultural traditions that emphasized protection of nature
(d) None of these

75. Sacred groves, Khasi and Jaintia hills located in-

(a) Meghalaya (b) Manipur
(c) Mizoram (d) Madhya Pradesh

76. What is endangered or threatened?

(a) Organisms facing a very high risk of death due to environment.
(b) Organisms facing a very high risk of extinction, in the near future.
(c) Organism facing a very high risk of danger from other organisms.
(d) None of these.

77. Which one of the following is not an example of Ex-situ conservation?

(a) Zoological parks
(b) Botanical Gardens
(c) Biosphere reserves
(d) none

78. Gametes of threatened species can be preserved in variable and fertile condition for long period using

(a) Heat (b) Cryopreservation

(c) Both (d) None of these

79. Threatened plant species can be propagated by-

(a) Tissue Culture (b) Aquaculture

(c) Cryopreservation (d) None of these

80. The Earth's summit held in Rio de Janeiro in.

(a) 2000 (b) 1990

(c) 1992 (d) 2002

81. The world's summit on sustainable development held in 2002 in.

(a) Johannesburg, South Africa.

(b) Cape town South Africa.

(c) Rio de Janeiro.

(d) East America

82. How many countries pledged in world summit held in 2002 in Johannesburg?

(a) 200 (b) 180

(c) 170 (d) 190

83. In which type of conservation threatened animals and plants are taken out from their natural habitat and placed in special selting place where they can be protected and given a special care?

(a) Cryopreservation

(b) Ex-situ conservation

(c) In-situ conservation

(d) None of these

84. Seeds of different genetic strains of commercially important plants can be kept for long period in

(a) Cryopreservation

(b) Tissue culture

(c) Seed bank

(d) None of these

85. Which are involved in In-situ conservation?

(i) Biosphere reserve

(ii) Cryopreservation

(iii) Tissue culture

(iv) Seed bank

(v) National park

(vi) Zoological park

(vii) Sacred groves

(viii) Safari parks

(a) iii, vii, v (b) ii, iii, i

(c) i, v, vii (d) iv, vi, i

86. Pledge of Earth Summit held in Rio de Janeiro.

(a) All nations to take appropriate measure for conservation of biodiversity and sustainable utilization of its benefit.

(b) All nations are free to utilize natural resources and also harm the ecosystem.

(c) Significant reduction in the current rate of biodiversity loss at global, regional and local levels.

(d) None of these

87. What is correct about the historic convention on biological diversity?

(i) It held in Rio de Janeiro

(ii) In 1992

(iii) Also called the earth summit

(iv) 190 states Pledged in this summit

(v) Commitment to achieve by 2020

(a) i, ii, iv (b) i, ii, iii

(c) iii, iv, I (d) all

88. In recent years, which type of conservation has advanced beyond keeping threatened species in enclosures?

(a) In-situ conservation

(b) Ex-situ conservation

(c) None of these

(d) Both A & B

89. Reason behind conserving the biodiversity.

(a) Narrowly utilitarian.

(b) Broadly utilitarian.

(c) Ethical

(d) All of these

90. More than 25% of the drugs currently sold in the market worldwide are derived from __i__ and __ii__ species of plants contribute to the traditional medicines used by native peoples

around the world.
(a) i- plants, ii- 25,000
(b) i- Animals, ii- 25,000
(c) i- Animals, ii- 25,00
(d) i- plants, ii- 25,00

91. How much oxygen is produced by Amazon Forest through Photosynthesis?
(a) 20% of the total oxygen.
(b) 30% of the total oxygen.
(c) 25% of the total oxygen.
(d) None of these.

92. Select the correct option about ethical argument.
(a) In ethical argument of conserving biodiversity relates to what we are to million of plant, animals and micro species with whom we share this planet.
(b) We need to realize that every species has an intrinsic value.
(c) We have a moral duty to care for their well being and pass on our biological legacy in good order to future generation.
(d) All these statements are correct.

93. Species confined to that region and not found anywhere else is called as-
(a) In situ (b) Exotic
(c) Endemic (d) None of these

94. Initially ___i___ biodiversity hotspots were identified, but after some time ___ii___ more have been added and the total number of biodiversity hotspot in the world to __iii__ .

(a) i- 25, ii- 9
(b) ii- 8, iii- 30
(c) i- 23, iii- 34
(d) none of these

95. Match the following

	Column I		Column II
a	Biosphere reserve	1	448
b	National parts	14	
c	Wildlife centuries	90	

(a) a-2, b-3, c-1 (b) a-1, b-2, c-3
(c) a-2, b-1, c-3 (d) a-1, b-3, c-2

96. Sacred groves are/is found in
(a) Khasi and Jaintia hills in Meghalaya
(b) Aravalli hills of Rajasthan
(c) Western ghat region of Karnataka
(d) All of these are correct

97. In which sacred growth the last refugees for a large number of rare and threatened plants are there?
(a) In Rajasthan
(b) In western ghat
(c) In Meghalaya
(d) In Karnataka.

98. 'Biodiversity hotspots' is a region where-
(a) Very high level of species richness and high degree of endemism
(b) Very high level of species richness and low degree of endemism
(c) Very low level of species richness and also low degree of endemism
(d) None of these

99. Which one is not a characteristic of biodiversity hotspots?
(a) High level of species richness.
(b) Endemism
(c) 38 in numbers
(d) Accelerated habitat loss protection

100. Which one is not included under in situ conservation?
(a) Biosphere reserves
(b) National parks
(c) Zoological parks
(d) Sacred groves

101. Which one is not the hotspot in India?
(a) Western ghats (b) Indo Burma
(c) Eastern Ghats (d) Himalaya

ANSWER KEY

Q	1	2	3	4	5	6	7	8	9	10
Ans	**A**	**C**	**D**	**A**	**C**	**A**	**C**	**C**	**B**	**A**
Q	11	12	13	14	15	16	17	18	19	20
Ans	**D**	**D**	**B**	**A**	**A**	**C**	**B**	**D**	**C**	**D**
Q	21	22	23	24	25	26	27	28	29	30
Ans	**C**	**B**	**C**	**A**	**C**	**B**	**C**	**A**	**A**	**D**
Q	31	32	33	34	35	36	37	38	39	40
Ans	**A**	**C**	**B**	**D**	**C**	**B**	**C**	**D**	**D**	**A**
Q	41	42	43	44	45	46	47	48	49	50
Ans	**B**	**D**	**D**	**D**	**B**	**A**	**C**	**A**	**D**	**A**
Q	51	52	53	54	55	56	57	58	59	60
Ans	**D**	**C**	**B**	**D**	**C**	**B**	**D**	**D**	**B**	**C**
Q	61	62	63	64	65	66	67	68	69	70
Ans	**D**	**D**	**B**	**D**	**B**	**C**	**D**	**C**	**D**	**B**
Q	71	72	73	74	75	76	77	78	79	80
Ans	**D**	**B**	**A**	**D**	**A**	**B**	**C**	**B**	**A**	**C**
Q	81	82	83	84	85	86	87	88	89	90
Ans	**C**	**D**	**B**	**C**	**C**	**A**	**D**	**B**	**D**	**A**
Q	91	92	93	94	95	96	97	98	99	100
Ans	**A**	**D**	**C**	**A**	**A**	**D**	**C**	**A**	**C**	**C**
Q	101									
Ans	**A**									

20 Chapter

Environmental Issues

MULTIPLE CHOICE QUESTIONS

1. Environment (protection) Act was passed in ______

(a) 1990 (b) 1996

(c) 1986 (d) 1948

2. Environment (protection) Act was passed by ______

(a) WHO

(b) Forest and environment minister of USA

(c) Water conservation committee of world

(d) Government of india

3. ________ are the agents that bring undesirable changes in physical, chemical or biological characteristics of air, land, water or soil.

(a) Pollution (b) Pollutants

(c) Contagion (d) Particulate matter

Topic 1 Air Pollution and Its Control

4. In India, the air (prevention and control of pollution) Act come into force in ______

(a) 1980 (b) 1982

(c) 1988 (d) 1981

5. Label the figure correctly using given correct option –

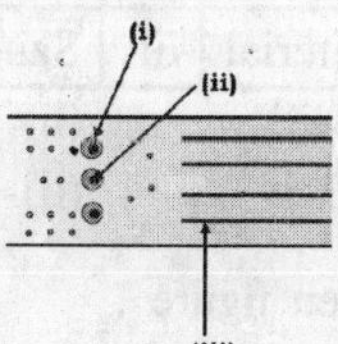

	(i)	(ii)	(iii)
A	Discharge corona	Negatively charged wire	Collection plate grounded
B	Electron	Nucleus	Collection plate
C	Negatively charged wire	Discharge corona	Collection plate
D	Dust particles	Negatively charged wire	Collection plate grounded

6. Which among the following is used in scrubber to clean the exhaust air?

(a) Water or lime (b) Chlorine

(c) Platinum (d) Molybdenum

7. ______ is the most widely used method of removing particulate matter from air

(a) Wet scrubbers

(b) Venturi scrubbers

(c) Fume scrubber

(d) Electrostatic precipitator

8. In electrostatic precipitator, ______ attach to dust particle giving them a net negative charge

(a) Proton

(b) Negatively charged wire

(c) Electron

(d) Another dust particles

9. CPCB stands for

(a) Control pollution central board

(b) Central population control board

(c) Central pollution control board
(d) Control population central board

10. According to CPCB, particulate size ________ are responsible for causing the greatest harm to human health
(a) 2.0 micrometers or more in diameter
(b) 2.0 micrometers or less in diameter
(c) 2.5 micrometers or less in diameter
(d) 2.5 micrometers or more in diameter

11. Catalytic converters, having expensive metals namely ______ and ______ as the catalysts are fitted into auto mobiles for reducing emission of poisonous gases.
(a) platinum – palladium and rhodium
(b) osmium – platinum and rhodium
(c) silver – platinum and indium
(d) ruthenium – palladium and rhodium

Topic 2 Air Pollution and Its Control

12. Use of ______ and ________ reduces the vehicular pollution.
(a) Leaded petrol and low sulphur petrol
(b) Unleaded petrol and high sulphur petrol
(c) Unleaded petrol and low sulphur petrol and diesel.
(d) Leaded petrol and high sulphur petrol

13. Which among the following is correct option – ACC. to Euro III norms –
(a) Sulphur be controlled at 350 ppm in diesel
(b) Sulphur be controlled at 150 ppm in diesel
(c) Sulphur be controlled at 370 ppm in diesel
(d) Sulphur be controlled at 360 ppm in diesel

14. Bharat stage II is equivalent to which among the following –
(a) Euro – III norms (b) Euro – I norms
(c) Euro – II norms (d) Euro – IV norms

15. (i) Bharat stage III – norms is applicable to 4 wheelers, 3 wheelers, 2 wheelers (ii) Bharat stage III norm is implemente throughout the country since May 2017
(a) Both (i) and (ii) are correct
(b) (i) is correct and (ii) is wrong
(c) (i) is wrong and (ii) is correct
(d) Both (i) and (ii) are wrong

16. What does PIL stands for
(a) People Interest Limitation
(b) Public Interest Litigation
(c) Population Interest Litigation
(d) Personal Interest Litigation

17. Which among the following is best to use as a fuel for public transport?
(a) Petrol
(b) Diesel
(c) Compressed Natural Gas (CNG)
(d) Gasoline

18. Statement P: CNG is better than petrol and diesel
Statement Q: CNG burns most efficiently and left with very little of unburnt residue
(a) P is correct and Q is wrong
(b) Both statements are correct
(c) P is wrong and Q is correct
(d) Both statements are wrong

Topic 3 Water Pollution and It's Control

19. Water (Prevention and Control of pollution) Act was passed in which year?
(a) 1981 (b) 1974
(c) 1989 (d) 1986

20. Choose the correct option – Impurities in waste water

1	Suspended solids	i	Fecal matter, bacteria
2	Colloidal material	ii	Nitrate, Calcium
3	Dissolved material	iii	Sand, Silt, Clay

(a) 1-i, 2-ii, 3-iii (b) 1-ii, 2-i, 3-iii
(c) 1-iii, 2-i, 3-ii (d) 1-ii, 2-iii, 3-i

21. Label the given figure

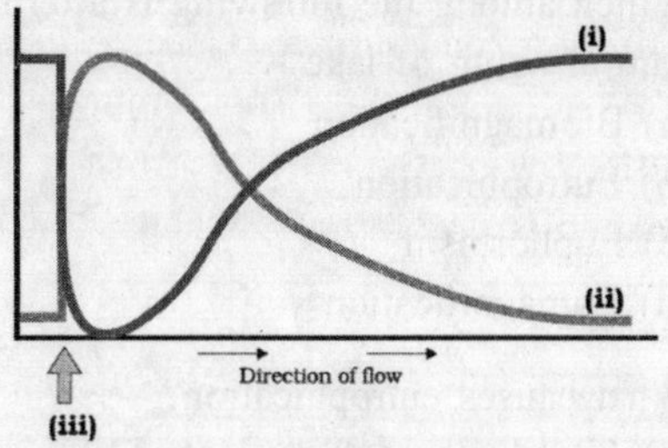

	(i)	(ii)	(iii)
A	BOD	Dissolved O_2	Clean water
B	BOD	Dissolved O_2	Sewage discharge
C	Dissolved O_2	BOD	Sewage water
D	Dissolved O_2	BOD	Clean water

22. BOD stands for –
(a) Biological Oxygen Demand
(b) Biochemical Oxygen Demand
(c) Biodegradable Oxygen Demand
(d) Biological Oxygen Dissolved

23. (i) Dissolved salts like nitrates, phosphates, toxic metal ions and organic compounds are most easy to remove from sewage water
(ii) A mere 0.1 % impurities make domestic sewage unfit for human use
(iii) Most of domestic sweage are not readily decomposed

How many among the above statements are correct –
(a) 0 (b) 1
(c) 2 (d) 3

24. Assertion : It is possible to estimate the amount of non – biodegradable organic matter in sewage water by measuring BOD.
Reason : Bacteria and other microorganism multiply using the organic substances as substrate and utilizes some of the components of sewage –
(a) Assertion is wrong Reason is correct
(b) Both are wrong
(c) Assertion is correct but Reason is wrong
(d) Both are correct

25. Free-Floating algae are also called ______
(a) Euglenoids
(b) Planktonic-algae
(c) Hydro algae
(d) Dinoflagellates algae

26. What is the cause of algal bloom?
(a) Excessive growth of planktonic algae due to presence of large amount of nutrient in water
(b) Excess of fish mortality in water
(c) Discharge of harmful elements like lead, mercury, etc in water bodies through industries
(d) Nitrogen – fixing bacteria present in water bodies fixes nitrogen and results in the spread of vegetation

27. Which among the following are not consequences of algal bloom –
(a) Imparting distinct colors to the water bodies
(b) Deterioration of water quality
(c) Fish natality
(d) Toxicity to human being and animals

28. Scientific name of water hyacinth –
(a) Eichhornia crassipes
(b) Elodeo canadensis
(c) Epomoea quamoclit
(d) Etroplus surotensis

29. Which among the following is called ' Terror of Bengal ' ? –
(a) Nelumbo nucifera
(b) Ceratrophyllum submeresum
(c) Eichhornia crassipes
(d) Lemma minor

30. Which among the following is not true about water hyacynth? –
(a) There are the world's most problematic aquatic weed.
(b) These are introduced in the India for their rapid growing ability
(c) They have very beautiful flowers

(d) They cause imbalance in the ecosystem dynamics of the water bodies

31. Match the following –

1 Water	i DDT 0.5 ppm
2 Zooplankton	ii DDT 2 ppm
3 Small fish	iii DDT 0.003 ppm
4 Large fish	iv DDT 25 ppm
5 Fish – Eating birds	v DDT 0.04 ppm

(a) 1-iii, 2-v, 3-i, 4-ii, 5-iv
(b) 1-i, 2-ii, 3-v, 4-iv, 5-iii
(c) 1-ii, 2-v, 3-iv, 4-i, 5-iii
(d) 1-iii, 2-iv, 3-i, 4-v, 5-ii

32. Heavy metals like mercury, lead, etc have density greater than ______? –
(a) 5 g/cm^3
(b) 5 mg/cm^3
(c) 25 g/cm^3
(d) 25 mg/cm^3

33. Which among the following refers to the increase in concentration of the toxicant at successive trophic level –
(a) BOD
(b) Biomagnification
(c) Biofortification
(d) Eutrophication

34. Assertion : Toxic substances get passed on to the next higher trophic level

Reason: Toxic substances get accumulated by an organism cannot be metabolished or excreted –
(a) Both are correct
(b) Assertion is correct but Reason is wrong
(c) Assertion is wrong but Reason is correct
(d) Both are wrong

35. High concentration of DDT disturb calcium metabolism in bird which causes –
(a) Thinning of eggshell
(b) Premature breaking of eggshell
(c) Decline in bird population
(d) All of these

36. Which among the following is a term used for natural aging of lake –
(a) Biomagnification
(b) Eutrophication
(c) Euphorbism
(d) Eumagnification

37. What causes eutrophication? –
(a) Introduction of nutrients like nitrogen and phosphorus
(b) Introduction of harmful elements like lead
(c) Introduction of plastic waste
(d) None of these

38. Which among the following statements are correct? –
i) Eutrophication is long process which may even take thousands of years
ii) Eutrophication encourages growth of aquatic organisms
iii) Eutrophication leads to the conversion of fresh lake into polluted and deteriorate lake
(a) (ii) and (iii) (b) Only (i)
(c) (i) and (ii) (d) All

39. Eutrophication finally leads to the conversion of lake into which among the following? –
(a) Deep swampy pond
(b) Land
(c) Polluted pond
(d) Lake with silt and organic content at base

40. Which among the following can accelerate the process of eutrophication –
(a) Organic waste from household
(b) Plastic waste
(c) Sewage and agricultural waste
(d) Both A & C

41. Accelerated aging process of natural lake by pollutants from mans activities are called ______ –
(a) Cultural eutrophication
(b) Man-made eutrophication
(c) Accelerated eutrophication
(d) More than one option is correct

42. Prime contaminants, that causes severe eutrophication and also act as plant nutrients are ____ –
(a) Calcium (b) Nitrate
(c) Phosphate (d) Both B & C

43. Which of below option is correct? Pollutants from thermal power plants results in —

i) Elimination of organism sensitive to high temperature
ii) Elimination of organisms sensitive to low temperature
iii) Enhance the growth of organisms in extremely cold areas
iv) Certainly has no effect on the organisms
(a) Only I (b) ii and iii
(c) i and iii (d) only iv

Topic 4 | Integrated waste water Treatment

44. Arcata is situated along the –
(a) North coast of California
(b) Eastern part of Europe
(c) Southern part of Germany
(d) Middle zone of France

45. The series of six connected marshes spread over how much area in the integrated waste water treatment? –
(a) 600 hectares (b) 60 hectares
(c) 600 Acre (d) 60 Acre

46. Which among the following is true about the second stage of the Arcata integrated waste water treatment? –
(a) It comprises of marshes which contain algae, fungi and bacteria, which neutrises absorb and assimilate the pollutants
(b) Marshes go with conventional sedimentation filtering and chlorine treatment
(c) Marshes constitute of a sanctuary with a high level of biodiversity
(d) Both (i) and (iii)

47. Who are responsible for the safeguarding of inegrated waste water treatment in Arcata? –
(a) FOAM (b) POAM
(c) COAM (d) DOAM

48. What does FOAM stands for? –
(a) Family of Arcata Marsh
(b) Friends of Arcata Marsh
(c) Family of Arcata Mission
(d) Friends of Arcata Mission

49. Which among the following is a sustainable system for handling human excreta using dry composting toilets? –
(a) Recycle sanitation
(b) Biological sanitation
(c) Ecological sanitation
(d) None of these

50. Name of dry composting toilets used in many areas of Kerala and Sri Lanka–
(a) Bio san (b) Com san
(c) Chem san (d) Eco san

Topic 5 | Solid Wastes

51. Municipal solid waste include waste from
(a) Homes and schools
(b) Hospitals
(c) Stores and office
(d) All of these

52. ____ were adopted as the substitute for open - burning dumps –
(a) sanitary landfills
(b) decomposing pit
(c) shallow pit
(d) none of these

53. Problems with landfills to dump wastes are—
(a) Large amount of garbage results in rapidly filling of landfills

(b) Seepage of chemicals causing soil contamination
(c) Contamination of groundwater
(d) All of these

54. Waste that is generated is categorized into the following type–
(i) bio-degradable
(ii) nonbiodegradable
(iii) contaminated waste
(iv) recyclable waste
(a) only i and ii (b) i, ii and iii
(c) i, ii and iv (d) All of these

55. What is incinerators? –
(a) Chemicals used for degradation of waste
(b) Micro-organisms that degrade waste
(c) Area where waste products are dumped
(d) Furnace wear waste are burned

56. Crucial step for safe disposing hospital waste is –
(a) disinfecting the waste products
(b) using of incinerators
(c) disposal in water bodies
(d) none of these

57. What are irreplarable computers and electronic wastes are called? –
(a) E – waste (b) C – waste
(c) A – waste (d) B – waste

58. How E – waste are treated without damage to environment and human life? –
(a) Recycling
(b) Buried in pit / landfills
(c) Incinerated
(d) All of these

Topic 6 Remedy for Plastic Waste

59. Which among the following is produced by Ahmed Khan's company? –
(a) Bitumen (b) Polyblend
(c) Karlit (d) Qualinite

60. Polyblend is obtained by which of the following?
(a) By making a paste of recycled modified plastic
(b) By making a powder of recycled modified plastic
(c) By making a paste of non-recycled unmodified plastic
(d) By making a powder of non-recycled unmodified plastic

61. Polyblend is mixed with which among the following to lay roads? –
(a) Karolinite (b) Allinite
(c) Bitumen (d) Dalinite

62. What are the factors by which khan prove that his innovation is better way to lay road–
(a) Enhanced water repellent properties
(b) This increases the road life by a factor of 3
(c) Accumulation of plastic waste reduced
(d) All of these

63. Ahmed khan collaborated with –
(a) R.V. College of Engineering
(b) Jawaharlal Medical College
(c) Bangalore City Corporation
(d) Both A and C

Topic 7 Agro – Chemicals and their Effects

64. What are the effects of Green Revolution? –
(a) Increased crop production
(b) widely used inorganic fertilizer and pesticides
(c) increased toxicity to soils ecosystem
(d) all of these

65. Which among the following is correct? –
i) Increase the use of chemical fertilizer can be biomagnified in terrestrial ecosystem and can be toxic
ii) Increased used of chemical fertilizer is useful for aquatic ecosystem as it undergoes Eutrophication in faster rate
(a) only (i) (b) only (ii)
(c) (i) and (ii) (d) None of these

Topic 8 Case Study of Organic Farming

66. What is the key point of integrated organic farming? –
(a) Preference of mixed farming
(b) Waste product from one process are cycled in as nutrient for other process
(c) Minimizing the use of chemical fertilizer
(d) All of these

67. What are the results of integrated organic family?
(a) Maximum utilization of resources
(b) Increased efficiency of production
(c) Higher profitability to farmer
(d) All of these

68. How many among the following are true for Dagar's idea of integrated organic farming –
(i) he included Bee - keeping dairy, water harvesting, composting and agriculture in his farm
(ii) he uses cattle dung as manure to crops
(iii) he uses crop waste only to generate natural gas but not for compost formation
(iv) he uses chemical fertilizer for crops
(a) 1 (b) 2
(c) 3 (d) 4

69. Who was the founder of Haryana kisan welfare club? –
(a) Suresh Dagariya
(b) Raju Chandra Bhushan
(c) Ahmad Khan
(d) Ramesh Chandra Dagar

70. Three Mile Island and Chernobyl incidents are related to which among the following?
(a) Leakage of chemical gas
(b) Leakage of radioactive gas
(c) release of contaminated wastewater
(d) none of these

71. How radiation given off by nuclear waste is damaging to organisms? –
(a) They cause mutation at very high rate
(b) They cause disorders by altering the gene sequence
(c) they cause cancer
(d) all the above

72. What is the best method to dispose nuclear waste? –
(a) Bury then deep down in Atlantic Ocean
(b) after pre-treatment bury within the rock in shielded container 500-meterdeep below the earth
(c) taking these wastes to the other planet and disposing there.
(d) all of these

Topic 9 Greenhouse effect and Global Warming

73. From where does term 'Greenhouse effect' has been derived? –
(a) From Green Terrestrial Surface
(b) Phenomena occurring in green plants
(c) phenomena that occurs in greenhouse
(d) scientist named Greenhouse

74. Why greenhouse is actually used for plants? –
(a) It is used to maintain the soil health for growing plant
(b) It is used to grow plants especially in winter as it's warm from inside providing plants good temperature to grow
(c) It provide plants cool temperature and environment to grow
(d) all of these

75. Which among the following phenomena is responsible for natural heating of Earth's surface and atmosphere? –
(a) Absorption of heat (sun) rays
(b) Greenhouse effect
(c) Emission of radioactive raise from nuclear plant
(d) Doppler's phenomena

76. Earth receives incoming solar radiation in the form of ________. –
(a) Long wave radiation
(b) Short wave radiation

(c) Electronic wave radiation
(d) Magnetic wave radiation

77. In which form earth's surface re-emits heat, that does not escape into space? –
(a) Infrared radiation (b) UV rays
(c) PAR (d) All of these

78. Match the following- Relative contribution of various greenhouse gasses to total global warming

1	CO_2	I	20%
2	Methane	II	14%
3	CFC	III	60%
4	N_2O	IV	6%

(a) 1-II, 2-III, 3-IV, 4-I
(b) 1-III, 2-II, 3-IV, 4-I
(c) 1-IV, 2-I, 3-II, 4-III
(d) 1-III, 2-I, 3-II, 4-IV

79. Assertion: greenhouse gasses are responsible for greenhouse effect.

Reason: Earth receives incoming solar radiation in short wave radiation some radiation get reflected back, some get fall in Earth's surface heating it, earth again re-emits heat in form of infrared rays and get absorbed by greenhouse gases these gases again radiate them and cycle get repeated many times and keep earth surface warm – (Pg. 281, M)
(a) Assertion is correct and Reason is wrong
(b) Both are correct
(c) Assertion is wrong and Reason is correct
(d) Both are wrong

80. What is the result of increased level of greenhouse gasses? –
(a) Global warming
(b) Melting of ice glaciers in Polar region
(c) Abnormal weather phenomena (El Nino effect)
(d) All of these

81. What portion of incoming solar radiation get reflected by clouds and gases without reaching to the earth's surface? –
(a) 1/3 rd (b) 1/4 th
(c) 2/3 rd (d) 1/2 nd

82. What would be the average temperature of earth if there would be no greenhouse effect? –
(a) 0°C (b) 15°C
(c) -18°C (d) -105°C

Topic 10 Ozone Depletion in the Stratosphere

83. Bad ozone is found in which layer? –
(a) Upper Stratosphere (b) Thermosphere
(c) Troposphere (d) Ionosphere

84. Good ozone is found in which layer?
(a) Upper Stratosphere (b) Thermosphere
(c) Troposphere (d) Ionosphere

85. The thickness of ozone layer is measured by
(a) Fukin unit (b) Delberg unit
(c) Dobson unit (d) Colleit unit

86. The area where, ozone layer is thin is known as
(a) Ozone reduced area
(b) Ozone hole
(c) Spare ozone area
(d) Zero ozone

87. At which area, ozone hole is prominent?
(a) Above antarctica
(b) Above north Pacific ocean
(c) Above Atlantic ocean
(d) Above whole Asia

88. What is the time period at which ozone hole area Antarctica develops each year?
(a) Late August – early October
(b) January – Feburary
(c) Early March – Late March
(d) Mid December – Mid January

89. How ozone gas are formed?
(a) By action of UV rays on molecular oxygen
(b) By action of infrared rays on molecular oxygen

(c) By action of PAR on molecular oxygen
(d) none of the above

90. Full Form of CFCs is
(a) Chlorofluoro cation
(b) Chlorofluroscent carbon
(c) Chlorofluoro carbon
(d) Cationicfluoroin carbon

91. Which among the following is true about CFCs?
(i) It degrades the ozone layer
(ii) By the action of UV ray CFCs releases Cl atom
(iii) Cl bind with ozone and releases molecular oxygen and act as merely catalysts
(iv) Cl atom are consumed in the reaction
(a) Only i and ii (b) Only i and iii
(c) i, ii and iii (d) all of these

92. what are the effects of UV – B rays?
(a) Damages DNA by causing mutation
(b) Causes inflammation of Cornea, snowblindness,
(c) causes cataract
(d) All of these.

93. Which among the following international treaty was signed to control emission of ozone depleting substances?
(a) Kyoto Protocol
(b) Nagoya Protocol
(c) Earth summit
(d) Montreal Protocol

94. In which year Montreal Protocol was signed?
(a) 1990 (b) 1985
(c) 1986 (d) 1987

Topic 11 Degradation By Improper Resources Utilization And Maintainance

95. What causc soil erosion?
(a) Un-restricted grazing
(b) Deforestation
(c) Poor-irrigation practices
(d) All of these

96. What are the effect of water logging?
(a) It result in soil salinity as it draws salt from underground and deposit on land surface as thin crust.
(b) It results in accumulation of acid on soil surface.
(c) It results in desertification.
(d) All of these.

Topic 12 Air Pollution and Conservation of Forests

97. Amrita devi was a lady from which community?
(a) Garwhal (b) Bajpah
(c) Bishnoi (d) Barwal

98. Chipko movement was performed in?
(a) Diwanans Gujarat (b) Bishnoi Gir forest
(c) Garhwal Himalayas (d) Deoni Delhi

99. When was Chipko movement observed?
(a) 1972 (b) 1971
(c) 1973 (d) 1974

100. What does JFM stand for?
(a) Junior Federation of Management
(b) Junior Forest Manager
(c) Joint Fellowship Management
(d) Joint Forest Management

101. Which among the following award is given to the individual community for their extraordinary courage and dedication in protecting wildlife?
(a) Chipcombent award.
(b) Amrita Devi Bisnoe wildlife protection award.
(c) Garval wildlife protection award
(d) none of these.

102. When was National Forest Policy came?

(a) 1987 (b) 1985
(c) 1988 (d) 1986

103. Another name of slash and burn agriculture is _______.

(a) Sling cultivation
(b) Donum cultivation
(c) Jhum cultivation
(d) Fiala cultivation

104. Assertion: Jhum cultivation also causes deforestation

Reason: It include cutting down of trees in forest and burned the remaining plant. Ashes used as fertilizer and land is use for cultivation, then same process is repeated to another area due to increased population time-gap given is limited for recovery and results in deforestation.

(a) Assertion and Reason both are correct
(b) Assertion is correct but Reason is wrong
(c) Both are wrong
(d) Assertion is wrong but reason is correct

105. The process of restoration of forest that once existed, but was removed at some point of time in past is called _______ .

(a) Deforestation (b) Repeated forestation
(c) Reforestation (d) All of these

ANSWER KEY

Q	1	2	3	4	5	6	7	8	9	10
Ans	**C**	**D**	**B**	**D**	**A**	**A**	**D**	**C**	**C**	**C**
Q	11	12	13	14	15	16	17	18	19	20
Ans	**A**	**C**	**A**	**C**	**B**	**B**	**C**	**B**	**B**	**C**
Q	21	22	23	24	25	26	27	28	29	30
Ans	**C**	**B**	**B**	**A**	**B**	**A**	**C**	**A**	**C**	**B**
Q	31	32	33	34	35	36	37	38	39	40
Ans	**A**	**A**	**B**	**A**	**D**	**B**	**A**	**C**	**B**	**C**
Q	41	42	43	44	45	46	47	48	49	50
Ans	**C**	**D**	**C**	**A**	**B**	**D**	**A**	**B**	**C**	**D**
Q	51	52	53	54	55	56	57	58	59	60
Ans	**D**	**A**	**D**	**C**	**D**	**B**	**A**	**D**	**B**	**A**
Q	61	62	63	64	65	66	67	68	69	70
Ans	**C**	**D**	**A**	**D**	**A**	**D**	**D**	**B**	**D**	**B**
Q	71	72	73	74	75	76	77	78	79	80
Ans	**D**	**B**	**C**	**B**	**B**	**B**	**A**	**D**	**B**	**D**
Q	81	82	83	84	85	86	87	88	89	90
Ans	**B**	**C**	**C**	**A**	**C**	**B**	**A**	**A**	**A**	**C**
Q	91	92	93	94	95	96	97	98	99	100
Ans	**C**	**D**	**D**	**D**	**D**	**D**	**A**	**C**	**D**	**D**
Q	101	102	103	104	105					
Ans	**B**	**C**	**C**	**A**	**B**					

1 Chapter

Animal Kingdom

MULTIPLE CHOICE QUESTIONS

Topic 1 Basis of classification

1. The diagram below shows the diploblastic and triploblastic germ layers in the animals. Identify the correct option in which they are found.

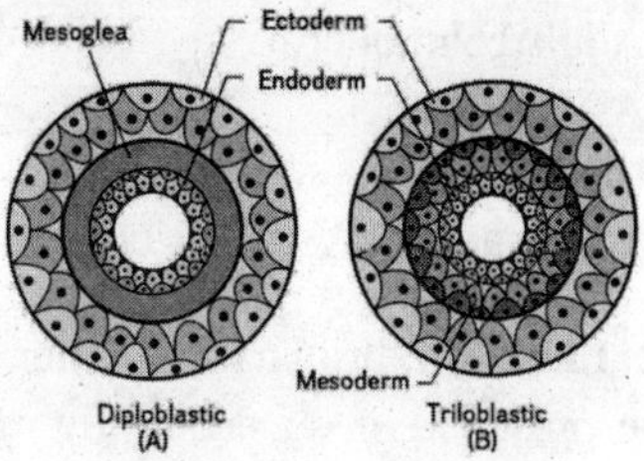

(a) A – Molluscs, B – Chordates
(b) A – Annelida, B – Porifera
(c) A – Coelenterates, B – Platyhelminthes
(d) A – Molluscs, B – Porifera

2. Choose the incorrect option
(a) True coelom is a body cavity which arises as a cavity in the embryonic mesoderm.
(b) Digestive cavity is found in acoelomates, pseudocoelomates as well as coelomates.
(c) The body cavity of arthropods and non-cephalopod molluscs is called haemocoel.
(d) There is no cavity between the body wall and gut wall in echinoderms.

3. Metamerism is present in
(a) annelids (b) arthropods
(c) chordates (d) all of these

4. Choose the incorrect match:
(a) Coelenterates – Radial symmetry
(b) Molluscs – Radial symmetry in adults
(c) Platyhelminthes – Triploblastic
(d) Ctenophores – Triploblastic

5. Choose the correct label for A, B, C and D in the broad Classification of Kingdom Animalia based on common fundamental features as given below.

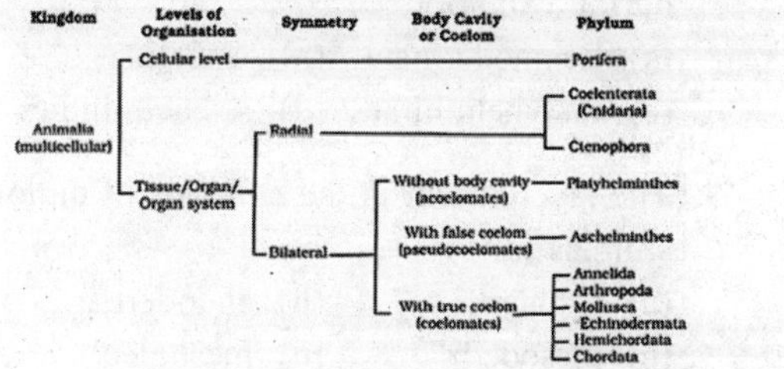

	A	B	C	D
(a)	Bilateral	Porifera	Ctenophora	Coelomate
(b)	Radial	Porifera	Ctenophora	Acoelomate
(c)	Bilateral	Porifera	Ctenophora	Coelomate
(d)	Radial	Ctenophora	Porifera	Acoelomate

6. Choose the incorrect statement.
(I) Notochord is ectodermally derived rod-like structure.
(II) Notochord is formed on the dorsal side during embryonic development.
(III) The animals from porifera to Echinoderms are without notochord.
(IV) In some chordates, notochord is replaced by the vertebral columnand these chordates are called vertebrates.
(a) I and II (b) I, II, and III
(c) II, III and IV (d) I only

7. Identify the correct labels A and B.

Notochord	Nerve Cord
A	Part of nervous system
Found in chordates only	Found in chordates as well as nonchordates
Dorsal side in chordates	B

(a) A = Exoskeleton, B = Dorsal in chordates as well as in non-chordates
(b) A = Endoskeleton; B = Ventral in chordates as well as in non-chordates
(c) A = Exoskeleton; B = Ventral in chordates and dorsal in nonchordates
(d) A = Endoskeleton; B = Dorsal in chordates and ventral in nonchordates

8. True coelom appear in which of the following during evolution?
(a) Echinodermata (b) Annelida
(c) Platyhelminthes (d) Aschelminthes

9. The layer absent in the embryos of diploblastic animals is
(a) ectoderm (b) endoderm
(c) mesoderm (d) mesoglea

10. Nerve cells and tissue level of organization first appeared in
(a) coelenterates (b) ctenophora
(c) chordate (d) porifera

11. In some animal groups, the body is found divided into compartments with at least some organs. This characteristic feature is called
(a) segmentation (b) metamerism
(c) metagenesis (d) metamorphosis

12. Body cavity is the cavity present between body wall and gut wall. In some animals the body cavity is not lined by mesoderm. Such animals are called
(a) acoelomate (b) pseudocoelomate
(c) coelomate (d) haemocoelomate

13. Match the following Columns:

Column-I (Phylum)	Column-II (Characteristic Features)
(A) Porifera	(1) Canal system
(B) Aschelminthes	(2) Water vascular system
(C) Annelida	(3) Muscular pharynx
(D) Arthropoda	(4) Joined appendages
(E) Echinodermata	(5) Metameres

Select the correct option:

	(A)	(B)	(C)	(D)	(E)
(a)	1	3	5	4	2
(b)	1	2	3	4	5
(c)	5	4	3	2	1
(d)	4	3	1	2	5

14. Which of the following animals are true coelomates with bilateral symmetry?
(a) Adult echinoderms
(b) Aschelminthes
(c) Platyhelminthes
(d) Annelids

15. **Assertion:** The primary character of chordates is the presence of dorsal hollow nerve cord.
Reason: Vertebral column is derived from the notochord.
(a) Both Assertion and Reason are true and Reason is correct explanation of Assertion.
(b) Both Assertion and Reason are true, but Reason is not the correct explanation of Assertion.
(c) Assertion is true, but Reason is false.
(d) Assertion is false, but Reason is true.

16. **Assertion:** Animals with radial symmetry has more advantage in detecting food and danger.
Reason: It allows animal to be able to respond to stimulus from any direction.
(a) Both Assertion and Reason are true and Reason is correct explanation of Assertion.
(b) Both Assertion and Reason are true, but Reason is not the correct explanation of Assertion.
(c) Assertion is true, but Reason is false.
(d) Assertion is false, but Reason is true.

17. **Assertion:** Aschelminthes represent pseudocoelomates.

Reason: In aschelminthes, mesoderm is present as scattered pouches in between ectoderm and endoderm.

(a) Both Assertion and Reason are true and Reason is correct explanation of Assertion.
(b) Both Assertion and Reason are true, but Reason is not the correct explanation of Assertion.
(c) Assertion is true, but Reason is false.
(d) Assertion is false, but Reason is true.

18. Which of the following is a basic feature of all the organisms of Animalia?

(a) Multicellular structure
(b) Sensory and neuromotor system
(c) Terrestrial habitat
(d) Locomotion

19. Which of the following lack tissue grade organization?

(a) Metazoans (b) Eumetazoans
(c) Parazoans (d) None of these

20. Match the columns.

Column-I	Column-II
(A) Organ level	(1) Pheretima
(B) Cellular aggregate level	(2) Fasciola
(C) Tissue level	(3) Spongilla
(D) Organ system level	(4) Obelia

Codes:

	(A)	(B)	(C)	(D)
(a)	2	4	3	1
(b)	2	3	4	1
(c)	4	1	2	3

(d) If both assertion and reason are false.

21. Choose the correct option

(a) Ctenophores and platyhelminthes possess complete digestive system.
(b) Aschelminthes to chordates, all possess organ system level of organization along with complete digestive system.
(c) Coelenterates and aschelminthes possess organ system level of organization along with complete digestive system.
(d) Poriferans may possess complete digestive system.

22. The entry of food and exit of waste takes place from separate openings in

(a) organisms having incomplete digestive system
(b) coelenterates, ctenophores and Platyhelminthes
(c) organisms having complete digestive system
(d) organisms having cellular level of organization

23. Which of the following is incorrect?

(a) Some division of labour (activities) occur among the cells in the members of phylum porifera.
(b) Division of labour (activities) is completely absent among the cells in poriferans.
(c) Open circulatory system is found in Tunicates, hemichordates, and noncephalopod molluscs.
(d) All of these

24. Choose the incorrect option.

(a) Complete digestive system - Two openings, mouth and anus
(b) Incomplete digestive system - Single opening system
(c) Open circulatory system - Blood is circulated through tubes
(d) Closed circulatory system - Arteries veins and capillaries present

25. Choose the correct body symmetry shown in the diagram.

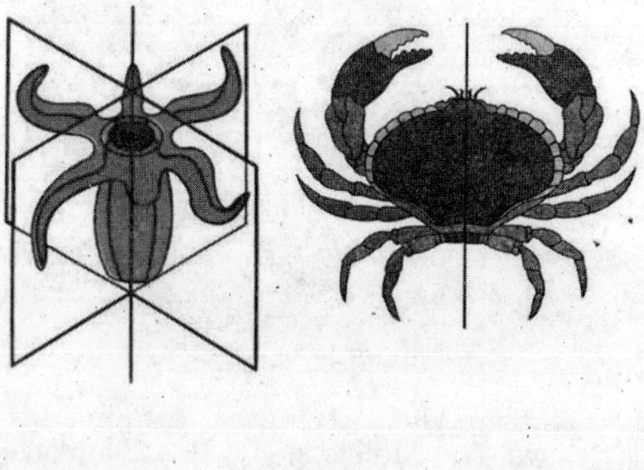

(a) A – Radial, B – Bilateral
(b) A – Bilateral, B – Pentamerous
(c) A – Radial, B – Pentamerous
(d) A – Bilateral, B – Radial

26. The diagram below shows the diploblastic and triploblastic germ layers in the animals. Identify the correct option in which they are found.

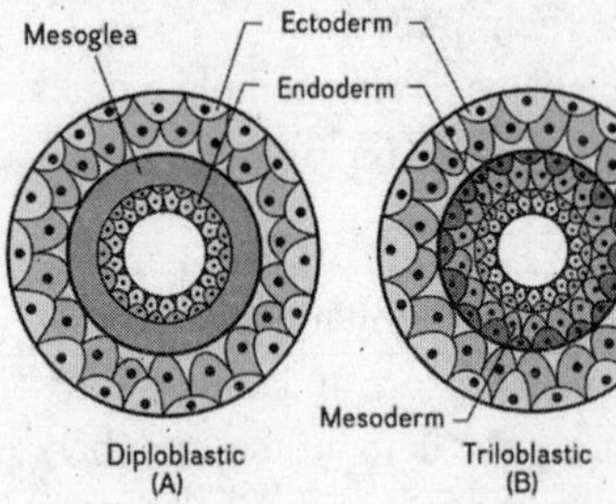

(a) A – Radial, B – Bilateral
(b) A – Bilateral, B – Pentamerous
(c) A – Radial, B – Pentamerous
(d) A – Bilateral, B – Radial

27. Choose the true statement:
(a) Animals like annelids, arthropods, aschelminthes, molluscs, hemichordates and chordates possess bilateral symmetry.
(b) Most of the animals possess bilateral symmetry.
(c) Platyhelminthes was the first phylum during evolution to exhibit bilateral symmetry.
(d) All of these

28. Study the types of animals with respect to the presence or absence of body cavities:

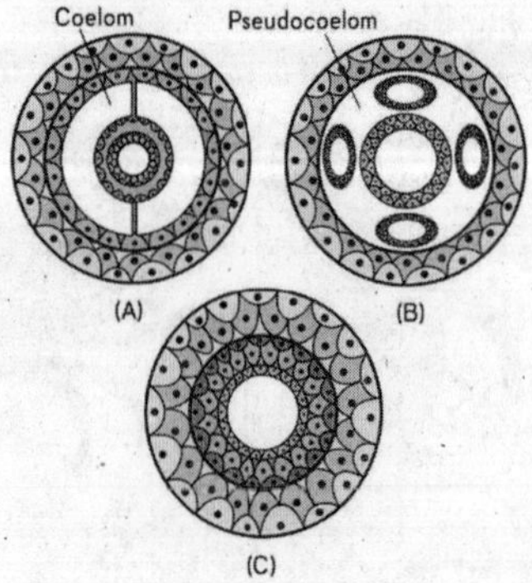

(a) A – Molluscs, B – Chordates
(b) A – Annelida, B – Porifera
(c) A – Coelenterates, B – Platyhelminthes
(d) A – Molluscs, B – Porifera

29. Which of the following option is wrong?
(a) Coelenterates and ctenophores are diploblastic.
(b) Animals from platyhelminthes to chordates are triploblastic.
(c) Radially symmetric animals remain attached to a surface by their aboral surface.
(d) Mesoglea is an undifferentiated layer which do not form any tissue or organ.

30. Choose the incorrect match
(a) Tube-within-tube body plan: Nemathelminthes, Annelida, Arthropoda, Mollusca, Echinodermata, Chordata
(b) Cell-aggregate type body plan: Coelenterates
(c) Blind-sac type body plan: Platyhelminthes and coelenterates
(d) None of these

31. Which of the following is/are the function of coelom?
(a) Absorb shock or provide hydrostatic skeleton
(b) Support shock or provide hydrostatic skeleton
(c) Allow muscles to grow independently of the body wall
(d) All of these

Topic 2 Porifera

32. Sponges are
(a) with water canal system
(b) sexually reproducing by formation of gametes
(c) both (a) and (b)
(d) sessile or free-swimming

33. In case of poriferans, the spongocoel is lined with flagellated cells called

(a) ostia (b) oscula
(c) choanocytes (d) mesenchymal cells

34. Body having meshwork of cells, internal cavities lined with food filtering flagellated cells and indirect development are the characteristics of phylum
(a) coelenterate (b) porifera
(c) Mollusca (d) protozoa

35. In most simple type of canal system of porifera, water flows through which one of the following ways?
(a) Ostia → Spongocoel → Osculum → Exterior
(b) Spongocoel → Ostia → Osculum → Exterior
(c) Osculum → Spongocoel → Ostia → Exterior
(d) Osculum → Ostia → Spongocoel → Exterior

36. Examine the figures A, B, and C.

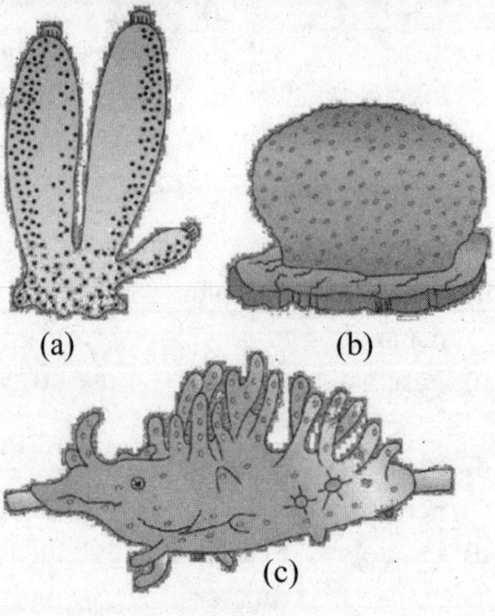

In which one of the four options all the animals (Poriferans) are correct?
(a) A - Sycon, B - Euspongia, C - Spongilla
(b) A - Euspongia, B - Spongilla, C - Sycon
(c) A - Spongilla, B - Sycon, C - Euspongia
(d) A - Euspongia, B - Sycon, C - Spongilla

37. Which of the following is a freshwater sponge?
(a) Euspongia (b) Euplectella
(c) Spongilla (d) Sycon

38. In poriferans, the rudimentary division of labour is found between the
(a) tissue (b) cells
(c) organs (d) organ-system

39. Which of the following is not a characteristic of class Porifera?
(I) Development is indirect (larval stage is present).
(II) Mostly asymmetrical and usually marine
(III) Primitive multicellular animals with cellular level of organization.
(IV) Choanocytes line the spongocoel and the canals.
(V) Sexes are separate
(a) I and IV (b) II only
(c) V only (d) III and IV

40. Choose the correct characteristic for sponges.
(a) They are highly regenerative
(b) They are universally radially symmetrical
(c) The contain clarions spicules but lack the siliceous one
(d) They are found only in fresh water

Topic 3 Coelenterata

41. Cnidocytes are
(a) also called cnidoblast or nematocyte
(b) explosive cells each of which contain giant secretory organelle called nematocyst
(c) stinging cells
(d) with all the above features

42. Consider the following statements about cnidarians:
(I) They have tissue level of organization and triploblastic.
(II) Digestion is extracellular and intracellular.
(III) Corals secrete calcium bicarbonate form a skeleton.
(IV) Corals may harbour some photosynthetic dinoflagellates for taking nutrition.
(V) They possess a central gastrovascular cavity with a single opening mouth in hypostome.
(a) Statements I and III are correct

(b) Statements II, IV and V are correct
(c) Statements I, II and III are correct
(d) Statements III and IV are incorrect

43. Here two basic body forms of cnidarians are given.

(a) A and B are false swimming forms
(b) A and B are sessile form
(c) A produce B asexually and B form the 'A' sexually
(d) B produce A sexually and A form the 'B' sexually

44. Match the columns.

	Column-I		Column-II
(a)	Gorgonia	(1)	Sea fan
(b)	Adamsia	(2)	Sea pen
(c)	Physalia	(3)	Portuguese man of war
(d)	Pennatula	(4)	Sea anemone

Select the correct option

	(A)	(B)	(C)	(D)
(a)	1	3	4	2
(b)	1	2	3	4
(c)	4	3	2	1
(d)	3	4	1	2

45. 'Stinging capsules' or nematocytes are found in
(a) sea anemone (b) sea pen
(c) sea fan (d) all of these

46. **Assertion:** Choanocytes or collar cells line the spongocoel and the canals in poriferans.
Reason: Poriferans possess spicules or spongin fibers.
(a) Both Assertion and Reason are true and Reason is correct explanation of Assertion.
(b) Both Assertion and Reason are true, but Reason is not the correct explanation of Assertion.
(c) Assertion is true, but Reason is false.
(d) Assertion is false, but Reason is true.

Topic 4 Ctenophora

47. Ctenophores
(a) perform external fertilization
(b) have indirect development
(c) both (a) and (b)
(d) have separate sexes

48. Identify the correct option specifying the names of the animals A, B, C and D.

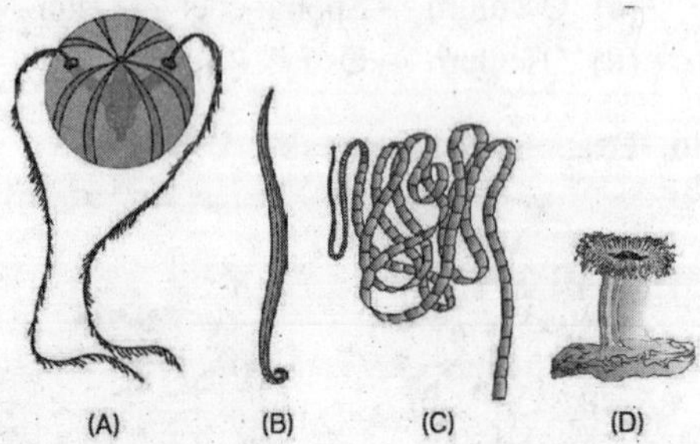

	(A)	(B)	(C)	(D)
(a)	Pleurobranchia	Tapeworm	Taenia	Aurelia
(b)	Fasciola	Tapeworm	Liverfluke	Aurelia
(c)	Pleurobranchia	Roundworm	Taenia	Adamsia
(d)	Fasciola	Roundworm	Liverfluke	Adamsia

Topic 5 Platyhelminthes

49. In tapeworms
(a) flame cells are absent
(b) both exoskeleton and endoskeleton present
(c) hooks and suckers present
(d) body is radially symmetrical

50. Which of the following is not a Platyhelminthes
(a) Wuchereria (b) Taenia
(c) Faxiola (d) Planaria

51. Ascaris is characterized by
(a) the absence of true coelom but presence of metamerism
(b) the presence of neither true coelom nor metamerism
(c) the presence of true coelom but the absence of metamerism
(d) the presence of true coelom and metamerism

52. Which of the option is correct for the statements given below.
(I) Commonly called sea walnuts or comb jellies.
(II) Bioluminescence is well marked.
(III) Body bear eight external rows of ciliated comb plates.
(IV) They have flame cells for osmoregulation and excretion.
(V) Alimentary canal is complete with a well-developed muscular pharynx.

	Ctenophores	Platyhelminthes	Aschelminthes
(a)	I, II, III	IV	V
(b)	IV	I, II	III, V
(c)	I, II	III, IV	V
(d)	IV, V	II, III	I

53. Phylum Platyhelminthes members are
(a) dorsoventrally flattened, thus called flatworms
(b) bilaterally symmetrical, triploblastic and acoelomates
(c) with organ system level of organization
(d) with all the above features

54. Assertion: Taenia Solium and Dugesia belong to Platyhelminthes.
Reason: Platyhelminthes are coelomates.
(a) Both Assertion and Reason are true and Reason is correct explanation of Assertion.
(b) Both Assertion and Reason are true, but Reason is not the correct explanation of Assertion.
(c) Assertion is true, but Reason is false.
(d) Assertion is false, but Reason is true.

55. Assertion: The organisms of Platyhelminthes are usually hermaphrodite.
Reason: These organisms possess internal as well as external fertilization.
(a) Both Assertion and Reason are true and Reason is correct explanation of Assertion.
(b) Both Assertion and Reason are true, but Reason is not the correct explanation of Assertion.
(c) Assertion is true, but Reason is false.
(d) Assertion is false, but Reason is true.

Topic 6 Aschelminthes

56. Consider the following statements about aschelminthes:
(I) Their body is circular in crosssection, so are called round worms.
(II) Alimentary canal is incomplete
(III) Muscular pharynx is present
(IV) They are hermaphrodites Which of the following is correct?
(a) I and III (b) II and IV
(c) I, II and IV (d) IV only

57. Choose the incorrect option.
(a) Mesoglea is present in between ectoderm and endoderm in Obelia.
(b) Asterias exhibits radial symmetry.
(c) Fasciola is pseudocoelomate animal.
(d) Taenia is a triploblastic animal

58. Out of the given cells, which of them can differentiate and perform different functions?
(a) Choanocytes
(b) Interstitial cells
(c) Gastrodermal cells
(d) Nematocysts

59. Blood sucking leech is
(a) Nereis (b) Hirudinaria
(c) Pheretima (d) All of these

60. Which one of the following endoparasites of humans does show viviparity?
(a) Ancylostoma duodenale
(b) Enterobius spiralis
(c) Trichinella spiralis
(d) Ascaris lumbricoides

61. Assertion: Aschelminthes and Annelids possess bilateral symmetry.
Reason: Both Aschelminthes and Annelids are coelomates.
(a) Both Assertion and Reason are true and Reason is correct explanation of Assertion.
(b) Both Assertion and Reason are true, but Reason is not the correct explanation of Assertion.
(c) Assertion is true, but Reason is false.
(d) Assertion is false, but Reason is true.

Topic 7 Annelida

62. Which of the following animals are true coelomates with bilateral symmetry?
(a) Adult echinoderms
(b) Aschelminthes
(c) Platyhelminthes
(d) Annelids

63. The name 'Annelida' is given to animal phylum having
(a) parapodia (b) metameric segments
(c) nephridia (d) all of these

64. In Annelids
(a) neural system consists of paired ganglia connected by lateral nerves to a double ventral nerve cord
(b) reproduction occur both asexually and sexually
(c) like Nereis, Pheretima and Hirudinaria have monoecious condition
(d) Aquatic forms are completely absent.

65. Match the columns

Column-I	Column-II
(A) Gills	(1) King crab
(B) Tracheal system	(2) Crab, prawn
(C) Book gills	(3) Butterfly, cockroach
(D) Book lungs	(4) Scorpion, spider

	(A)	(B)	(C)	(D)
(a)	1	2	3	4
(b)	2	3	1	4
(c)	4	3	2	1
(d)	3	1	4	2

66. Choose the incorrect statement.
(a) In cockroaches and prawns, excretion of waste material occurs through malphigian tubules.
(b) In ctenophores, locomotion is mediated by comb plates.
(c) In fasciola, flame cells take part in excretion.
(d) Earthworms are hermaphrodites and yet cross fertilization takes place among them.

67. Consider the following statements:
(I) Triploblastic, bilateral symmetry
(II) Metamerically segmental and coelomate animals
(III) Dioecious
(IV) Closed circulatory system
(V) Lateral appendages
(VI) Annelida

Which of the following information's belongs to the given animal.
(a) I, II, IV, VI (b) I, III, IV, V
(c) II, III, IV, V (d) III, IV, V, VI

Topic 8 Arthropoda

68. Choose the incorrect set with respect to arthropods.
(a) Limulus, locusta, culex
(b) Bombyx, Apis, lacifer
(c) Pinctada, Aplysia, Dentalium
(d) Aedes, Anopheles, Apis

69. Balancing organ of aquatic arthropods is
(a) Cnidoblasts (b) choanocytes
(c) scleroblasts (d) statocysts

70. Which one of the following features is not present in the phylum-Arthropods?
(a) Metameric segmentation
(b) Parapodia
(c) Jointed appendages
(d) Chitinous exoskeleton

71. Which one of the following characteristics is mainly responsible for diversification of insects on land?
(a) Segmentation (b) Bilateral symmetry
(c) Exoskeleton (d) Eyes

72. Which of the following statement is correct?
(a) Insect hemolymph has no role in oxygen transport in most cases.
(b) Insects hemolymph is mostly colourless.
(c) Both (a) and (B)
(d) None of these

73. Consider the following statements about Arthropods.

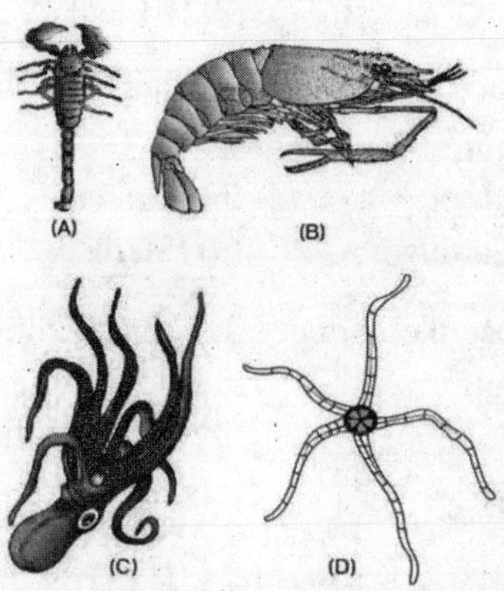

(I) Open circulatory system is found in most arthropods.
(II) Arthropods contain Haemolymph which directly bathes in internal tissues and organs.
(a) I is true but II is false
(b) I is false but II is true
(c) Both I and II are true
(d) Both I and II are false

74. Moulting
(a) is also called ecdysis
(b) occurs to shed chitin at regular intervals by many arthropods
(c) is the shedding of cuticle in many invertebrates
(d) all of these

75. Maggot is the larva of
(a) housefly (b) crab
(c) moth (d) butterfly

Topic 9 Mollusca

76. Which of the following is incorrect statement for Hemichordata?
(a) They are bilaterally symmetrical, triploblastic and coelomate.
(b) Circulation is of open type.
(c) Sexes are separate, fertilization is external and development is indirect.
(d) None of these

77. Select the feature which is/are not present in Hemichordates.
(a) Stomochord (b) Worm-like body
(c) Gills (d) All of these

78. The correct classification of given animal is
(a) Chordata – Vertebrata – Craniata
(b) Chordata – Craniata
(c) Chordata – Acraniata
(d) Non- chordata – Hemichordata

79. The body of Balanoglossus is divisible into
(a) proboscis, tunic and trunk
(b) collar, trunk and tunic
(c) proboscis, collar and trunk
(d) proboscis, stomochord and trunk

80. An important characteristics that hemichordates share with chordates is
(a) absence of notochord
(b) ventral tubular nerve cord
(c) pharynx with gill slits
(d) pharynx without gill slits

81. Choose the incorrect statement for phylum Mollusca.

(a) Body is covered by a calcareous shell and unsegmented.
(b) Feather like gills present for excretion and respiration.
(c) The anterior head region has sensory tentacles.
(d) Mostly terrestrial, triploblastic and acoelomates.

82. Choose the correct names for the following.

	A	B	C	D
(a)	Scorpion	Prawn	Loligo	Asterias
(b)	Scorpion	Prawn	Octopus	Ophiura
(c)	Locust	Butterfly	Loligo	Asterias
(d)	Locust	Prawn	Squid	Ophiura

83. Assertion: In many gastropods, the arms and the mental cavity are placed anteriorly above the head.
Reason: During embryonic development in many gastropods, one side of the visceral mass grows faster than the other side. This uneven growth rotates the visceral organs up to 180° in many gastropods.

(a) Both Assertion and Reason are true and Reason is correct explanation of Assertion.
(b) Both Assertion and Reason are true, but Reason is not the correct explanation of Assertion.
(c) Assertion is true, but Reason is false.
(d) Assertion is false, but Reason is true.

84. Assertion: In molluscs, feathers like gills are present in the mantle cavity.
Reason: These gills have respiration and excretory function.

(a) Both Assertion and Reason are true and Reason is correct explanation of Assertion.
(b) Both Assertion and Reason are true, but Reason is not the correct explanation of Assertion.
(c) Assertion is true, but Reason is false.
(d) Assertion is false, but Reason is true.

85. Which of the following statements represents the incorrect feature of Echinodermata?

(a) They are triploblastic and coelomate animals.
(b) All are marine with cellular level of organization.
(c) Endoskeleton of calcareous ossicle.
(d) None of these

86. Which of the following is the feature of water vascular system in Echinoderms?

(a) Locomotion
(b) Respiration
(c) Capture and transport of food
(d) All of these

87. Choose the correct statement for star fish.

(I) Sexes are separate and reproduction is sexual
(II) Development is indirect with freeswimming larva
(III) Mouth is present on the upper (dorsal) side and anus on the lower (ventral) side.
(IV) Their body bear jaw-like structure which is called oral arms.

(a) I and III (b) I, II and IV
(c) I, II and III (d) III and IV

88. Which one for the following animals does not undergo metamorphosis?

(a) Moth (b) Tunicate
(c) Earthworm (d) Starfish

89. Choose the correct statement for the following animals.

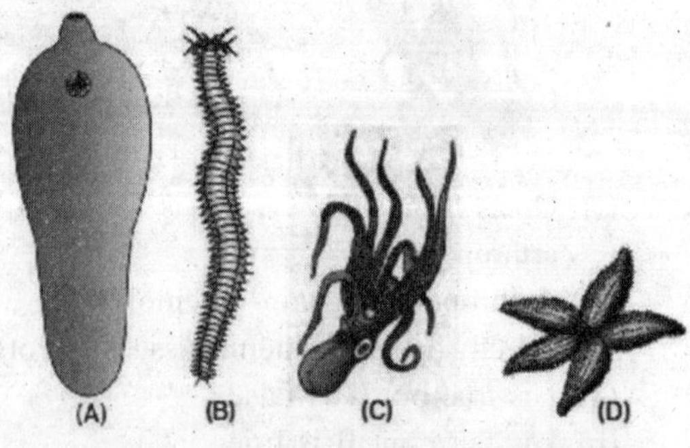

(a) All the animals are aquatic, free living
(b) All are true coelomates

(c) 'A' has radial symmetry but remaining have bilateral symmetry
(d) 'A' is monoecious but remaining are dioecious

Topic 10 Chordata

90. Syrinx present in birds
(a) helps in producing sound
(b) lie near the junction of trachea and bronchi
(c) both (a) and (b)
(d) helps in excretion of urea

91. The most unique mammalia character is
(a) the presence of two pairs of limbs
(b) reproducing young ones
(c) the presence of mammary glands
(d) the presence of skin

92. The skin of the mammals is unique in possessing
(a) glands
(b) epidermal layer
(c) hair
(d) both (a) and (c)

93. Heart is always four chambered in
(a) mammals (b) aves
(c) reptiles (d) both (a) and (b)

94. Choose the correct statements from the following:
(a) Mammals, birds, reptiles and amphibians possess 12 pairs of cranial nerves.
(b) In aquatic mammalian males, testes lie outside the body cavity in scrotal sacs
(c) The neck of mammals generally possess 5 cervical vertebrae
(d) Archeopteryx is a fossil animal.

95. Choose the odd one out.
(a) Prototheria – Ornithorlynchus
(b) Marsupilia – Macropus
(c) Metatheria – Maceaea
(d) Eutheria – Homo

96. Choose the correct option for A, B, C and D.

Prototherians	Metatherians	Eutherians
A	Viviparous	Viviparous
Nipples absent on mammary glands	B	Nipples present
C	Vagina and uterus present	Vagina and uterus present
Ear is devoid of pinna	Pinna is present	D
Scrotum absent	Scrotum present	Scrotum present

(a) A = Oviparous
B = Nipples present
C = Vagina and uterus absent
D = Pinna is absent in aquatic forms
(b) A = Oviparous
B = Nipples present
C = Vagina and uterus absent
D = Pinna is universally forms
(c) A = Viviparous
B = Nipples present
C = Vagina and uterus absent
D = Pinna is absent in aquatic forms
(d) A = Oviparous
B = Nipples absent
C = Vagina and uterus absent
D = Pinna is present only in aquatic forms.

97. Tetrapods
(a) lack paired appendages and pentadactyl limbs
(b) universally possess gills
(c) possess sense organ functional in air
(d) dwell only in terrestrial zones

98. Identify A, B, C and D in the table given below.

Amphibins	Reptiles	Birds	Mammals
Scales usually absent	A	Present on hind limbs	Absent
Cloaca present	Cloaca present	Cloaca absent	B

Erthrocytes oval, biconvex and nucleated	RBC oval, biconvex and nucleate d	Erythro cytes oval, biconvex x and nucleated	RBC circular biconcave and non-nucleated
Three chambere d heart	Three chambered heart	C	Four chambere d heart with left systemic arch
External ear absent	External ear may be present	D	External ear with pinna present

(a) A = Scales absent
B = Cloaca mainly present
C = Four chambered heart with left systemic arch
D = External ear absent
(b) A = Scales present
B = Cloaca mainly absent
C = Three chambered heart
D = External ear absent
(c) A = Scales absent
B = Cloaca mainly present
C = Three chambered heart with right systemic arch
D = External ear present
(d) A = Scales present
B = Cloaca mainly absent
C = Four chambered heart with right systemic arch
D = External ear present

99. Assertion: All vertebrates are chordates.
Reason: Vertebrates possess notochord during embryonic period.
(a) Both Assertion and Reason are true and Reason is correct explanation of Assertion.
(b) Both Assertion and Reason are true, but Reason is not the correct explanation of Assertion.
(c) Assertion is true, but Reason is false.
(d) Assertion is false, but Reason is true.

100. Assertion: All metatherians are placental mammals.
Reason: All placental mammals have menstrual cycle.
(a) Both Assertion and Reason are true and Reason is correct explanation of Assertion.
(b) Both Assertion and Reason are true, but Reason is not the correct explanation of Assertion.
(c) Assertion is true, but Reason is false.
(d) Assertion is false, but Reason is true.

101. Assertion: Duck bill platypus is not a true mammal.
Reason: True mammals are all viviparous while platypus are egg laying.
(a) Both Assertion and Reason are true and Reason is correct explanation of Assertion.
(b) Both Assertion and Reason are true, but Reason is not the correct explanation of Assertion.
(c) Assertion is true, but Reason is false.
(d) Assertion is false, but Reason is true.

102. Assertion: Bats and whales are classified as mammals.
Reason: Bats and whales have four chambered heart.
(a) Both Assertion and Reason are true and Reason is correct explanation of Assertion.
(b) Both Assertion and Reason are true, but Reason is not the correct explanation of Assertion.
(c) Assertion is true, but Reason is false.
(d) Assertion is false, but Reason is true.

103. Assertion: Mammalian teeth are heterodont.
Reason: Mammals possess more than a single tooth
(a) Both Assertion and Reason are true and Reason is correct explanation of Assertion.
(b) Both Assertion and Reason are true, but Reason is not the correct explanation of Assertion.

(c) Assertion is true, but Reason is false.
(d) Assertion is false, but Reason is true.

104. Assertion: Reptiles are referred to as poikilotherms.
Reason: Reptiles possess eggs with shells which help them to adapt in land environment.
(a) Both Assertion and Reason are true and Reason is correct explanation of Assertion.
(b) Both Assertion and Reason are true, but Reason is not the correct explanation of Assertion.
(c) Assertion is true, but Reason is false.
(d) Assertion is false, but Reason is true.

105. Assertion: Birds possess moist skin.
Reason: Birds possess oil glands throughout their body.
(a) Both Assertion and Reason are true and Reason is correct explanation of Assertion.
(b) Both Assertion and Reason are true, but Reason is not the correct explanation of Assertion.
(c) Assertion is true, but Reason is false.
(d) Assertion is false, but Reason is true

106. Which of the following is not found in the phylum chordate
(a) A dorsal hollow nerve chord
(b) Lateral paired gill slits during development
(c) A notochord at some stage of development
(d) An external skeleton

107. Animals belonging to phylum-chordata are fundamentally characterized by the presence of structure noted as A, B, C and D. Identify A, B, C and D.

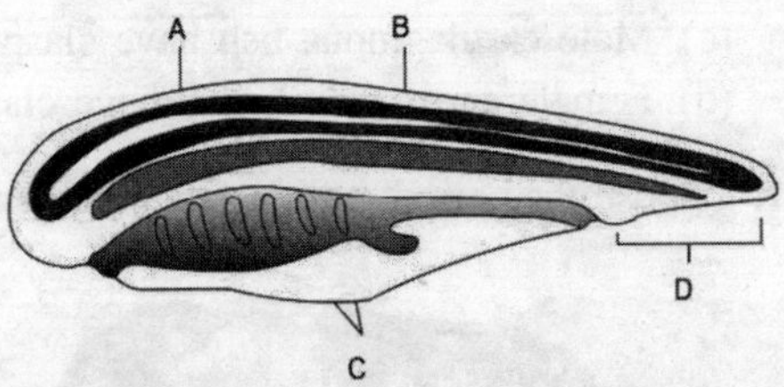

(a) A - Notochord, B - Nerve cord, C - Gill slits, D - post anal part
(b) A - Nerve cord, B - Notochord, C - Gill slits, D - Post anal part
(c) A - Nerve cord, B - Notochord, C - Post anal part, D - Gill Slits
(d) A - nerve cord, B - Gill slits, C - Notochord, D - post anal part

108. Choose the incorrect vertebrate character.
(a) Ventral muscular heart
(b) Kidneys for excretion and osmoregulation
(c) Paired appendages which may be fins or limbs
(d) None of these

109. The following flow chart for division of sub phylum vertebrata fill in the parts A, B, C and D and choose the correct option.

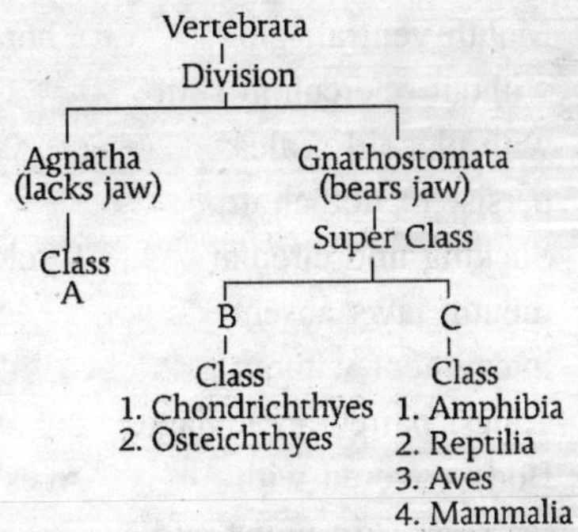

	A	B	C
(a)	Ostracodermi	Pisces	Tetrapoda
(b)	Cyclostomata	Pisces	Tetrapoda
(c)	Ostracodermi	Cyclostomata	Pisces
(d)	Pisces	Tetrapoda	Cyclostomata

110. Select the correct difference between the notochord in the following:

	Urochordata	**Cephalochordata**
(a)	Present only in larval tail	Extend from head to tail throughout life
(b)	present only in adult	Present only in larval tail
(c)	Persistent throughout their life	Present only in adult
(d)	Extend from head to tail throughout life	Present only in larval tail

111. Consider the following statements.
(I) Lancelets are jawless, primitive fishlike vertebrates.
(II) In lancelets notochord, tubular nerve cord and pharyngeal gills slits are present throughout their life.
(a) I is true, but II is false
(b) I is false, but II is true
(c) Both I and II are true
(d) Both I and II are false

112. Which of the following represents the correct combination without any exception?

	Characteristic	Class
(a)	Mammary gland; hair on body; pinnae; two pairs of limbs	Mammalia
(b)	Mouth ventral; gills without operculum skin with placoid scales; persistent notochord	Chondrichthyes
(c)	Sucking and circular mouth, jaws absent integument without scales; paired appendages	Cyclostomata
(d)	Body covered with feathers; skin moist and glandular; lungs with air sacs forelimbs from wings	Aves

113. Among the following edible fishes, which one is a marine fish having rich source of omega-3 fatty acids?
(a) Mystus (b) Mangur
(c) Mrigala (d) Mackerel

114. Which one is not cartilaginous fish?
(a) Carcharoden (great white shark), Trygon (sting ray)
(b) Exocoetus (flying fish), catla (katla), clarias (Mangur)
(c) Scolidon (dog fish)
(d) Pristis (saw fish)

115. Following are few examples of bony fishes. Choose the odd one out as marine bony fish.
(a) Flying fish
(b) Hippocampus (sea horse)
(c) Both (a) and (b)
(d) Labeo (rohu), catla, clarias

116. Which of the following is not a characteristic of class chondrichthyes?
(a) Gill slits are separated and without operculum.
(b) Predaceons with powerful jaws.
(c) Notochord is persistent throughout life.
(d) Airbladder present.

117. Which of the following characteristic features always holds true for the corresponding group of animals?
(a) Viviparous Mammalia
(b) Possess a mouth with an upper and a lower jaw Chordata
(c) Three-chambered heart with one incompletely divided ventricle Reptilia
(d) Cartilaginous endoskeleton Chondrichthyes

118. Bony fishes are
(a) having external fertilization
(b) mostly oviparous
(c) with direct development
(d) all of these

119. Bony fishes stay at any particular depth in water without spending energy due to
(a) Operculum (b) Neuromuscles
(c) Pneumatic bones (d) Swim bladder

120. Choose the incorrect statement?
(a) Both cartilaginous and bony fishes are dioecious
(b) Cartilaginous fishes show sexual dimorphism
(c) Male cartilaginous fish have claspers
(d) Female cartilaginous fish have claspers

121. Choose the correct option for the given figures.

(a) Animal A is salamandra and B is chameleon.
(b) Both A and B belongs to class Reptilia.
(c) Fertilization is external in both.
(d) Animal A has 2-chambered heart and B has 3-chambered heart.

122. Choose the incorrect option for the following animal.

(a) Cloaca present
(b) Dioecious, external fertilization, oviparous, indirect development
(c) Body divisible into head and trunk
(d) Eyes are without eyelids.

123. Which one of these animals is not a homeotherm?
(a) Camelus (b) Chelone
(c) Macropus (d) Psittacula

124. Identify the vertebrate group of animals characterized by crop and gizzard in its digestive system.
(a) Aves (b) Reptilia
(c) Amphibia (d) Osteichthyes

125. Which among these is the correct combination of aquatic mammals?
(a) Seals, dolphin, sharks
(b) Dolphins, seals, trygon
(c) Whales, dolphins, seals
(d) Trygon, whales, seals

126. Which one of the following characteristic is not shared by birds and mammals?
(a) Breathing using lungs
(b) Viviparity
(c) Warm-blooded nature
(d) Ossified endoskeleton

127. Which of the following animals is not viviparous?
(a) Flying fox (bat) (b) Elephant
(c) Platypus (d) Whale

128. Choose the correct option having animals with four chambered heart?
(a) Amphibian, reptiles, birds
(b) Crocodiles, birds, mammals
(c) Lizards, crocodiles, turtles
(d) Lizards, mammals, birds

129. The animal pair with non-glandular skin are
(a) snake and frog
(b) crocodile and tiger
(c) frog and pigeon
(d) chameleon and turtle

130. Which of the following characteristic is shared by both birds and mammals?
(a) Pigmented skin
(b) Pneumatic bones
(c) Viviparity
(d) Warm-blooded body

131. Which of the following sets of animals belongs to a single Taxonomic group?
(a) Man, monkey, chimpanzee
(b) Cuttlefish, jellyfish, silver fish, dog fish, starfish
(c) Bat, pigeon, butterfly
(d) Silkworm, tapeworm, earthworm

132. Match the following columns.

Column-I	Column-II
(A) Cyclostomes	(1) Hemichordata
(B) Aves	(2) Urochoradata
(C) Tunicates	(3) Agnatha
(D) Balanoglossus	(4) Pisces
(E) Osteichthyes	(5) Tetrapod

Codes

	A	B	C	D	E
(a)	1	2	3	4	5
(b)	2	3	4	1	5
(c)	3	5	2	1	4
(d)	3	1	5	2	4

133. Which of the following is incorrect for Petromyzon?
(a) Cranium and vertebral column are cartilaginous

(b) They are freshwater organisms but migrate for spawning to sea water
(c) After spawning within few days, they die
(d) Their larvae, after metamorphosis, return to ocean

134. Match the name of the animal in Column I with one characteristic in Column II and the phylum/class in column III to which it belongs.

Column-I	Column-II	Column-III
(a) Petromyzon	Ectoparasite	Cyclostomata
(b) Ichthyophis	Terrestrial	Reptilia
(c) Limulus	Body covered by chitinous exoskeleton	Pisces
(d) Adamsia	Radially symmetrical	Porifera

135. Choose the incorrect subphylum of Phylum Chordata
(a) Hemichordata (b) Vertebrata
(c) Cephalochordata
(d) Urochordata

136. Protochordates
(a) include Urochordata and cephalochordata
(b) are exclusively marine
(c) have notochord throughout life
(d) All of these

137. Choose the incorrect option for chordates.
(a) Paired pharyngeal gill slits
(b) Coelomate diploblastic
(c) Post anal tail
(d) Closed circulatory system

138. Choose the correct option for the animals shown below.

(a) The organism belongs to cephalochordate and genus Ascidia.
(b) Circulatory system is open type.
(c) Development is always direct.
(d) Fresh water dwelling.

139. The skin of amphibians
(a) can be smooth or rough
(b) are usually with Scales
(c) possess eutaneous glands
(d) both (a) and (b)

140 In amphibians
(a) mole copulatory organs are absent
(b) metamorphosis is usually absent
(c) tadpole stage is universally present
(d) cranial nerves are absent

141. Read the following statements.
(I) Retention of larval trait is called neoteny.
(II) The largest amphibian is Cryptobrances.
(III) Seymousia is a connecting link between amphibian and reptiles.
(IV) Larva of Ambystoma is called axolotl.
(V) Axolotls are amphibians formed without undergoing metamorphosis.
Choose the correct statement
(a) I and IV
(b) II and III
(c) I, II and III
(d) I, II, III and IV

142. Match the columns.

Column-I	Column-II
(A) Chameleon	(1) Tortoise
(B) Testudo	(2) Tree lizard
(C) Calotes	(3) Garden lizard
(D) Chelone	(4) Turtle

	A	B	C	D
(a)	1	2	3	4
(b)	4	3	2	1
(c)	2	1	3	4
(d)	3	1	4	2

143. Choose the incorrect statement for class Reptilia.
(a) Sexes are separate.
(b) Kidneys are metanephric.
(c) Limbs are always present and are two pairs.
(d) Possess creeping or crawling mode of locomotion.

144. Birds
(a) are poikilotherms
(b) have respiration performed only by the air sacs
(c) are bipeds
(d) endoskeleton is ossified partially

145. Find the incorrect match
(a) Crow – Corvus
(b) Pigeon – Columba
(c) Parrot – Psittacula
(d) Penguin – Pavo

146. Which of the following is incorrect?
(a) Aves possess poor olfactory system.
(b) Aves are partially homeotherms.
(c) Aves bones are hollow with air cavities.
(d) Aves have sexes separate, fertilization is internal, oviparous with direct development.

ANSWER KEY

1. (a)	2. (d)	3. (a)	4. (d)	5. (b)	6. (d)	7. (d)	8. (d)	9. (c)	10. (a)
11. (b)	12. (b)	13. (a)	14. (d)	15. (b)	16. (a)	17. (a)	18. (a)	19. (d)	20. (d)
21. (b)	22. (c)	23. (b)	24. (c)	25. (a)	26. (c)	27. (d)	28. (a)	29. (d)	30. (b)
31. (d)	32. (d)	33. (c)	34. (b)	35. (a)	36. (a)	37. (c)	38. (b)	39. (c)	40. (a)
41. (d)	42. (d)	43. (c)	44. (a)	45. (d)	46. (b)	47. (c)	48. (c)	49. (c)	50. (a)
51. (b)	52. (d)	53. (a)	54. (c)	55. (c)	56. (a)	57. (c)	58. (a)	59. (b)	60. (d)
61. (c)	62. (d)	63. (d)	64. (a)	65. (b)	66. (d)	67. (a)	68. (a)	69. (d)	70. (b)
71. (c)	72. (d)	73. (c)	74. (d)	75. (a)	76. (d)	77. (a)	78. (d)	79. (c)	80. (c)
81. (a)	82. (b)	83. (a)	84. (b)	85. (a)	86. (d)	87. (d)	88. (a)	89. (a)	90. (a)
91. (d)	92. (d)	93. (d)	94. (c)	95. (a)	96. (d)	97. (d)	98. (d)	99. (d)	100. (d)
101. (d)	102. (b)	103. (a)	104. (c)	105. (d)	106. (d)	107. (d)	108. (d)	109. (d)	110. (a)
111. (b)	112. (c)	113. (d)	114. (b)	115. (c)	116. (b)	117. (d)	118. (d)	119. (a)	120. (d)
121. (a)	122. (b)	123. (c)	124. (a)	125. (c)	126. (a)	127. (c)	128. (d)	129. (d)	130. (d)
131. (a)	132. (c)	133. (b)	134. (a)	135. (a)	136. (c)	137. (b)	138. (d)	139. (a)	140. (a)
141. (d)	142. (c)	143. (b)	144. (d)	145. (d)	146. (b)				

2 Chapter Structural Organisation in Animals

MULTIPLE CHOICE QUESTIONS

Topic 1 Epithelial Tissues

1. Select the incorrect statement.
   (a) Multicellular glandular epithelium is formed of clusters of cells.
   (b) Compound epithelium is actively involved in secretion and absorption of substances.
   (c) Pancreatic and salivary ducts are internally lined by compound of epithelium.
   (d) None of these

2. Cell junctions
   (a) Are formed in epithelial tissues
   (b) Provide structural and functional link between adjacent cells of tissues
   (c) Are alternatively called gap junctions

   Select the most appropriate option.
   (a) a, b, c are correct
   (b) Only a is correct
   (c) b and c are correct
   (d) a and b are correct

3. Match the following cell structure with its characteristic feature:

	Column-I		Column-II
a	Tight junctions	(1)	Cement neighbouring cells together to form sheet
b	Adhering junctions	(2)	Transmit information through chemical to another cells
c	Gap junctions	(3)	Establish a barrier to prevent leakage of fluid across epithelial cells
d	Synaptic junctions	(4)	Cytoplasmic channels to facilitate communications between adjacent cells

Select the most appropriate option.

	a	b	c	d
(a)	4	3	1	2
(b)	2	4	1	3
(c)	4	2	1	3
(d)	3	1	4	2

4. The function of the gap junction is to
   (a) stop substance from leaking across a tissue.
   (b) perform cementing to keep neighbouring cells together.
   (c) facilitate communication between adjoining cells by connecting the cytoplasm for rapid transfer of ions, small molecules and some large molecules.
   (d) separate two cells from each other.

5. **Assertion:** Squamous epithelium helps in the diffusion of gases in lungs.
   Reason: Squamous epithelium bears microvilli.
   (a) Both assertion and reason are true, assertion is incorrect explanation of assertion
   (b) Both assertion and reason are true, reason is not the correct explanation of assertion
   (c) Assertion is true, reason is false
   (d) Both assertion and reason are false

6. Assertion: Compound epithelium is composed of two or more layers of cells.

Reason: Compound epithelium has protective functions.

(a) Both assertion and reason are true, assertion is incorrect explanation of assertion
(b) Both assertion and reason are true, reason is not the correct explanation of assertion
(c) Assertion is true, reason is false
(d) Both assertion and reason are false

7. Consider the following statements:

(a) A tissue is composed of similar cells which perform specific functions.
(b) Epithelial tissues are characterized by a free surface toward body fluid or outside environment.

Select the correct option.

(a) a is true, b is false
(b) Both a and b are true
(c) a is false, b is true
(d) Both a and b are false

8. Simple epithelium consists of

(a) large intercellular spaces
(b) single layer of cells
(c) flat cells without nucleus
(d) all of these

9. Human skin is composed of [Pg-101,E]

(a) compound epithelium
(b) squamous epithelium
(c) columnar epithelium
(d) ciliated epithelium

10. Match the following columns.

	Column-I		Column-II
a	Squamous epithelium	(1)	Stomach and intestine
b	Cuboidal epithelium	(2)	Lungs and blood vessels
c	Columnar epithelium	(3)	Tubular parts of nephrons

Select the correct option

	a	b	c
(a)	3	1	2
(b)	1	2	3
(c)	2	3	1
(d)	3	2	1

11. The inner walls of large blood vessels are formed by

(a) pseudostratified epithelium
(b) squamous epithelium
(c) ciliated epithelium
(d) columnar epithelium

12. What is the similarity between cuboidal epithelium and columnar epithelium?

(a) They are composed of two layers of cells.
(b) They are composed of phagocytic cells.
(c) They perform the functions of secretion and absorption.
(d) All of these

13. Identify the tissues A and B shown in the following diagram:

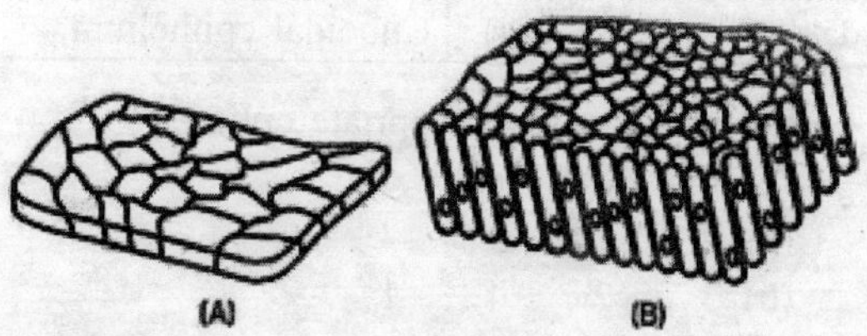

Select the correct option

	Column-I	Column-II
(a)	Squamous epithelium	Columnar epithelium
(b)	Cuboidal epithelium	Squamous epithelium
(c)	Columnar epithelium	Cuboidal epithelium
(d)	Compound epithelium	Pseudostratified

14. Efficient gaseous exchange in the air sacs of the lungs occur due to the presence of

(a) numerous microvilli
(b) ciliated epithelium
(c) flat cells
(d) columnar epithelium

15. The inner surface of hollow organs are lined by
(a) columnar epithelium
(b) compound epithelium
(c) squamous epithelium
(d) ciliated epithelium

16. The ciliated epithelial cells are required to move particles or mucus in a specific direction. In humans, these cells are mainly present in
(a) fallopian tubes and pancreatic duct
(b) Eustachian tube and salivary duct
(c) bronchioles and fallopian tubes
(d) bile duct and bronchioles

17. Match the following columns.

	Column-I		Column-II
a	Goblet cells	(1)	Multicellular glandular epithelium
b	Salivary glands	(2)	Unicellular glandular epithelium
c	Buccal cavity	(3)	Compound epithelium
d	PCT	(4)	Cuboidal epithelium

Select the most appropriate option.

	a	b	c	d
(a)	2	1	3	4
(b)	3	2	4	1
(c)	4	3	1	2
(d)	1	4	2	3

18. All the listed glands pour their secretions into ducts except
(a) salivary gland
(b) digestive glands
(c) pineal gland
(d) mammary glands

Topic 2 Connective Tissues

19. The intracellular material of cartilage is
(a) solid and pliable
(b) solid and non-pliable
(c) hollow and soft
(d) hollow and jelly-like

20. Select the incorrect statement:
(a) Most of cartilage in vertebrate embryo gets replaced by bones in adults.
(b) Chondrocytes are cartilage cells which are found in between collagen fibres.
(c) Cartilage form the human nose except its tips.
(d) Bones of vertebral column are composed of cartilage.

21. Bones are hard and non-pliable due to the presence of
(a) calcium salts (b) elastin fibres
(c) chondrocytes (d) all of these

22. The spaces in which osteocytes are present are called
(a) osteoclast (b) sinuses
(c) lacunae (d) canaliculi

23. The fluid connective tissue contains all of the following cells, except
(a) platelets (b) fibroblasts
(c) WBCs (d) RBCs

24. Match the following columns:

	Column-I		Column-II
a	Adipose tissue	(1)	Blood
b	Hyaline cartilage	(2)	Macrophages and mast cells
c	Fluid connective tissue	(3)	Fat storage
d	Areolar tissue	(4)	None

Select the correct option

	a	b	c	d
(a)	4	3	2	1
(b)	2	4	1	3
(c)	1	2	3	4
(d)	3	4	1	2

25. Select the incorrect statement regarding connective tissues:
(a) It helps to connect and support other tissues of the body.
(b) Blood is a fluid connective tissue.

(c) It is composed of structural protein fibres, viz, collagen or elastin.
(d) Their ground substance is composed of polypeptides and its derivatives.

26. Areolar tissue is a type of
(a) loose connective tissue
(b) compound epithelium
(c) dense connective tissue
(d) specialized connective tissue

27. Areolar tissues contain
(a) Tlymphocytes tissue and B lymphocytes
(b) fibroblast, macrophages, mast cells
(c) fibroblast cells only
(d) fibroblasts and fat globules

28. Fat-storing adipose tissue is
(a) loose connective tissue
(b) dense regular connective tissue
(c) dense irregular connective tissue
(d) specialized connective tissue

29. Consider the following statements:
(a) In dense connective tissues, fibroblasts are compactly packed.
(b) In dense regular connective tissues, collagen fibers are arranged in parallel rows.

Select the correct option
(a) a is true, b is false
(b) a is false, b is true
(c) Both a and b are false
(d) Both a and b are true

30. In the below diagram of areolar connective tissue, the different cells and parts have been indicated by alphabets. Choose the answer in which these alphabets correctly match with the parts and cells they indicate –

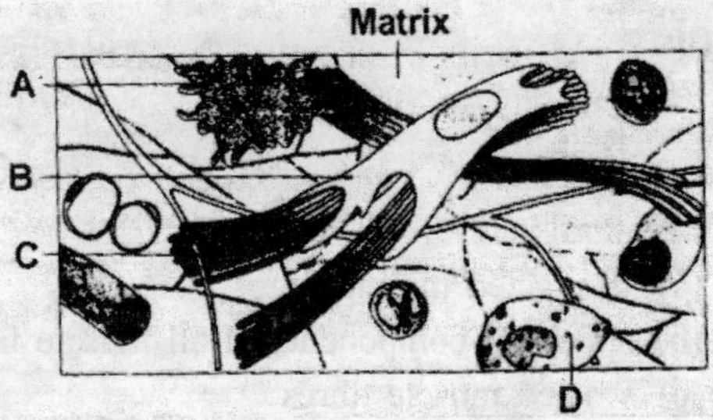

	A	B	C	D
(a)	Adipocyte	Collagen fibres	Microfilament	Mast cells
(b)	Macrophage	Collagen fibres	Microfilament	Mast cells
(c)	Macrophage	Collagen fibres	Microtubule	RBC
(d)	Macrophage	Fibroblast	Collagen fibres	Mast cells

31. Ligaments contain
(a) loosc bundles of fibres
(b) large fat storage areas
(c) parallely arranged collagen fibres
(d) irregularly placed elastin fibres

32. Match the following columns:

	Column-I		Column-II
a	Skin	(1)	Loose connective tissue
b	Tendon	(2)	Specialized connective tissue
c	Adipose tissue	(3)	Dense regular connective Tissue
d	Cartilage	(4)	Dense irregular connective tissue

Select the correct option

	a	b	c	d
(a)	3	1	2	4
(b)	4	3	1	2
(c)	2	4	3	1
(d)	1	2	4	3

Topic 3 Muscle Tissue and Neural Tissue

33. Myofibrils are
(a) contracted muscle fibres
(b) structural components of all muscle fibres
(c) striated muscle fibres
(d) skeletal muscle fibres

34. Skeletal muscles are
(a) striated in appearance
(b) smooth in appearance
(c) involuntary muscles
(d) both A and C

35. Go through the following figures.

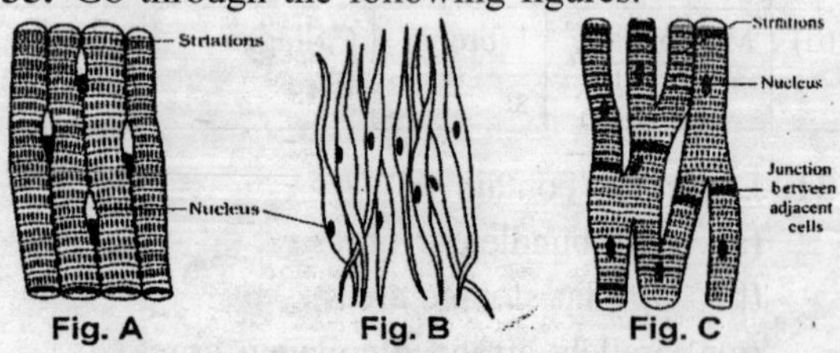

Identify these muscles (A, B and C).

	A	B	C
(a)	Smooth muscles	Striated muscles	Cardiac muscles
(b)	Cardiac muscles	Smooth muscles	Striated muscles
(c)	Striated muscles	Smooth muscles	Cardiac muscles
(d)	Involuntary muscles	Voluntary muscle	Heart muscle

36. Involuntary muscle fibers in human body are found in
(a) heart (b) blood vessels
(c) intestine (d) all of these

37. Consider the following statements:
(a) All involuntary muscles are smooth in appearance.
(b) All striated muscles are voluntary muscles.

Select the correct option
(a) (a) is true, (b) is false
(b) Both a) and (b) are true
(c) (a) is False, (b) is true
(d) Both (a) and (b) are false

38. Which type of tissue correctly matches with its locations?

	Tissue	Location
(a)	Areolar tissue	Tendons
(b)	Transitional epithelium	Tip of nose
(c)	Cuboidal epithelium	Lining of stomach
(d)	Smooth muscle	Wall of intestine

39. The muscular layer lining the stomach and intestine is
(a) striated in appearance
(b) smooth in appearance
(c) multinucleated
(d) characterized by intercalated discs

40. Heart cells have the ability to contract as a unit to the presence of
(a) multinucleate condition
(b) fusiform shape
(c) intercalated discs
(d) striations

41. The structural and functional unit of nervous system is
(a) neuron (b) neuroglia
(c) oligodendrocyte (d) ganglia

42. The neurons in the neural system are protected by
(a) dendrites (b) neuroglial cells
(c) axons (d) Nissl's granules

43. Assertion: Both skeletal muscles and cardiac muscles are striated appearance.
Reason: Cardiac muscles are involuntary in nature.
(a) Both assertion and reason are true, assertion is incorrect explanation of assertion
(b) Both assertion and reason are true, reason is not the correct explanation of assertion

(c) Assertion is true, reason is false
(d) Both assertion and reason are false

44. Assertion: Neurons are excitable cells.
Reason: Neurons are found abundantly throughout the body.
(a) Both assertion and reason are true, assertion is incorrect explanation of assertion
(b) Both assertion and reason are true, reason is not the correct explanation of assertion
(c) Assertion is true, reason is false
(d) Both assertion and reason are false

Topic 4 Cockroach

45. Identify A to E.

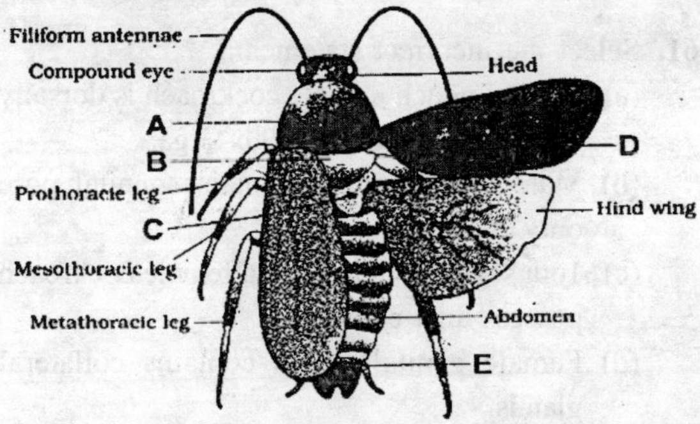

Figure — External features of cockroach

	A	B	C	D	E
(a)	Pronotum	Mesothorax	Metathorax	Tegmina	Pleura
(b)	Pronotum	Mesothorax	Metathorax	Tegmina	Sterna
(c)	Pronotum	Mesothorax	Metathorax	Tegmina	Analcerci
(d)	Pronotum	Mesothorax	Metathorax	Tegmina	Anal style

46. The body of cockroach can be morphologically distincted as
(a) head and abdomen
(b) head, thorax and abdomen
(c) head, abdomen and tail
(d) head and trunk

47. Consider the following statements:
(a) Body of cockroach is covered by hard exoskeleton made up of chitin.
(b) In male cockroach, the wings extend beyond the abdomen.

Select the correct option
(a) a is true, b is false
(b) Both a and b are false
(c) a is false, b is true
(d) Both a and b are true

48. The chitinous exoskeleton of arthropods is formed by the polymerization of
(a) lipoglycans
(b) keratin sulphate and chondroitin sulphate
(c) D – glucosamine
(d) N – acetyl glucosamine

49. Hardened plates of exoskeleton in cockroaches are known as
(a) capsids (b) spicules
(c) sclerites (d) metamere

50. Dorsal and ventral sclerites in cockroach are
(a) tergites and sternites, respectively
(b) sternites and pleurites, respectively
(c) pleurites and tergites, respectively
(d) pleurites and sternites, respectively

51. The triangular head of cockroach
(1) bear compound eyes
(2) is formed by the fusion of two segments
(3) bear chewing and lapping mouthparts

Select the correct option
(a) 1, 2, 3 are correct
(b) Only 1 is correct
(c) 2 and 3 are correct
(d) 1 and 2 are correct

52. The mouth parts of cockroach contain
(a) two mandibles, two maxillae, labrum, labium and hypopharynx
(b) two mandibles and maxillae each, two labrum and labium each
(c) one mandible and maxilla each, labarum and hypopharynx
(d) one mandible, labarum, labium and hypopharynx

53. Select the incorrect statement regarding cockroach:
(a) Hypopharynx acts as a tongue.

(b) Head is mobile in all directions due to flexible neck.
(c) Antennae possess sensory receptors to monitor the environment.
(d) Thorax is greatly reduced and nondivisible.

54. Paired walking legs in cockroaches are found on
(a) each thoracic segments
(b) mesothorax and metathorax
(c) prothorax and metathorax
(d) metathorax only

55. Forewings and hindwings in cockroaches arises from
(a) mesothorax and metathorax
(b) prothorax
(c) metathorax
(d) prothorax and metathorax

56. Match the following columns.

	Column-I		Column-II
a	Scleride	(1)	Forewings
b	Tegmina	(2)	Simple eye
c	Ocellus	(3)	Exoskeleton plate

Select the correct option

	a	b	c
(a)	3	2	1
(b)	2	1	3
(c)	3	1	2
(d)	2	3	1

57. How the forewings of cockroaches are distinguished from the hindwings?
(a) Forewings are leathery while hindwings are membranous.
(b) Forewings are much longer while hindwings are vestigial.
(c) Forewings are much reduced and hindwings are highly evolved.
(d) Forewings are transparent while hindwings are opaque.

58. In female cockroach
(a) forewings help in flight
(b) abdomen consists of 8 segments
(c) seventh sternum is boat shaped
(d) all of these

59. Which of the following sterna form the brood pouch in female cockroach?
(a) 8th, 9th (b) 7th, 8th, 9th
(c) 6th, 7th, 8th (d) 8th, 9th, 10th

60. Which of the following features is used to identify a male cockroach from a female cockroach?
(a) Presence of a boat shaped sternum on the 9th abdominal segment
(b) Presence of caudal styles
(c) Forewings with darker tegmina
(d) Presencc of anal cerci

61. Select the incorrect statement:
(a) Genital pouch in male cockroach is dorsally bounded by 9th and 10th terga.
(b) Male genital pouch contains genital pore only.
(c) 10th segment in male and female cockroach possess anal cerci.
(d) Female genital pouch contains collateral glands.

62. Which among the following structures are found in male cockroaches only?
(a) Collateral glands and anal cerci
(b) Anal cerci and spermathecal pores
(c) Spermathecal pores and collateral glands
(d) Gonapophysis and anal styles

63. Match the following columns:

	Column-I		Column-II
a	Anal cerci	(1)	7th in females
b	Boat-shaped sternum	(2)	9th segment in males
c	Anal styles	(3)	10th segment
d	Genital pouch	(4)	Bounded by 9th and

Select the correct option

	a	b	c	d
(a)	4	2	1	3
(b)	2	3	4	1
(c)	1	4	3	2
(d)	3	1	2	4

64. What is the correct arrangement of structures in Cockroach's alimentary canal?
(a) Oesophagus, Gizzard, Crop
(b) Gizzard, Oesophagus, Crop
(c) Crop, Oesophagus, Gizzard
(d) Oesophagus, Crop, Gizzard

65. Match the following.

	Column-I		Column-II
a	Proventriculus	(1)	Food storage
b	Crop	(2)	Grinding food particles
c	Hepatic Caeca	(3)	Secretion of digestive juices
d	Malpighian tubules	(4)	Removal of excretory products

Select the correct option

	a	b	c	d
(a)	4	3	2	1
(b)	2	1	3	4
(c)	2	4	1	3
D)	3	2	4	1

66. How many chitinous teeth are present in the gizzard of cockroach?
(a) Two (b) Six
(c) Ten (d) Fourteen

67. Refer to the given diagram of cockroach's alimentary canal:

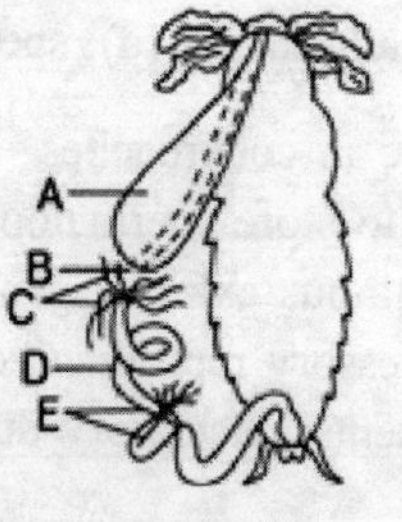

Which of the following options represent correct name and characteristic of labelled structures?
(a) A – Gizzard – Food grinding structure
(b) C – Hepatic ceca – Secretary digestive juices
(c) D – Ileum – Food absorption
(d) E – Malpighian tubules – 6–8 blind tubules

68. Select the correct sequence of organs in the alimentary canal of cockroach starting from mouth.
(a) Pharynx → Oesophagus → Gizzard → Crop → Ileum → Rectum
(b) Pharnyx → Oesophagus → Gizzard → Ileum → Crop → Colon → Rectum
(c) Pharnyx → Oesophagus → Ileum → Crop → Gizzard → Colon → Rectum
(d) Pharnyx → Oesophagus → Crop → Gizzard → Ileum → Colon → Rectum

69. During anatomical studies of cockroach, how would you differentiate malpighian tubules (A) from gastric caeca (B)?
(a) A – 6–8 in number, B – 200–300 in number
(b) A – Present at the junction of midgut and hindgut B – Present at the junction of foregut and midgut
(c) A – Blind, black-coloured tubules B – Hollow, yellow-coloured tubules
(d) All of these

70. Select the incorrect statement regarding cockroach:
(a) Cockroaches possess open circulating system.
(b) Blood vessels are highly developed and open into heart.
(c) Visceral organs found in hemocoel are bathed in hemolymph.
(d) Alary muscles associated with heart are contractile muscles.

71. The heart of cockroach possess
(a) 10 chambers (b) 8 chambers
(c) 13 chambers (d) 12 chambers

72. Match the following columns.

	Column-I (Structures in cockroach)		Column-II (Number)
a	Spiracles	(1)	6–8
b	Heart chambers	(2)	100–150
c	Hepatic ceca	(3)	13
d	Malpighian tubule	(4)	10

Select the correct option

	a	b	c	d
(a)	2	4	3	1
(b)	3	2	4	1
(c)	1	4	2	3
(d)	4	3	1	2

73. Select the incorrect statement regarding cockroach
(a) Exchange of gases occurs by diffusion at the tracheoles.
(b) Malpighian tubules remain lined by glandular and ciliated cells.
(c) Uricose glands are principal reproductive glands in female cockroach
(d) Fat bodies and nephrocytes help in excretion of nitrogenous waste

74. The principal nitrogenous waste in cockroach is
(a) ammonia (b) vasa
(c) guanine (d) uric acid

75. The body cells in cockroach discharge their nitrogenous waste in the hemolymph mainly in the form of
(a) calcium carbonate
(b) ammonia
(c) potassium urate
(d) urea

76. How many ganglia are found in the thorax and abdomen of male cockroach?
(a) 3 and 6 (b) 6 and 4
(c) 5 and 5 (d) 6 and 3

77. Consider the following statements:
(a) Cockroaches remain alive for several hours even after its head is cut off.
(b) Nervous system of cockroach is dorsally placed along the whole body.

Select the correct option
(a) a is true, b is false
(b) Both a and b are true
(c) a is false, b is true
(d) Both a and b are false

78. Which of the following statements is incorrect?
(a) Female cockroach possesses sixteen ovarioles in the ovaries.
(b) Cockroaches exhibit mosaic vision with less sensitivity and more resolution.
(c) A mushroom-shaped gland is present in the 6th-7th abdominal of male cockroach.
(d) A pair of spermatheca is present in the 6th segment of female cockroach.

79. Match the following columns

	Column-I		Column-II
a	Testes	(1)	2nd –6th segment
b	Ovaries	(2)	4th –6th segment
c	Spermatheca	(3)	6th segment
d	Mushroom glands	(4)	6th–7th segment

Select the correct option

	a	b	c	d
(a)	3	2	4	1
(b)	2	1	3	4
(c)	4	3	1	2
(d)	1	4	2	3

80. The bundles of sperms are called
(a) phallomere (b) gonapophysis
(c) spermatheca1 (d) spermatophores

81. Phallomere in cockroaches
(a) helps to store spermatophores
(b) is chitinous external genitalia
(c) is accessory reproductive gland
(d) represents ejaculatory duct

82. Ovarioles are
(a) bundles of ova
(b) ovarian tubules
(c) immature ovaries
(d) capsule containing fertilized ova

83. Match the following columns.

	Column-I		Column-II
a	Ovarioles	(1)	Opening of ejaculatory duct
b	Gonopore	(2)	Chain of developing ova
c	Phallomere	(3)	Bundles of sperms
d	Spermatophore	(4)	External genitalia

Select the correct option

	a	b	c	d
(a)	3	2	1	4
(b)	4	3	2	1
(c)	1	4	3	2
(d)	2	1	4	3

84. Which of the following structure encasc the fertilized eggs of cockroaches?
(a) Spermatheca (b) Ovariole
(c) Cocoon (d) Ootheca

85. Select the incorrect statement.
(a) Female cockroaches produce one ootheca at a time.
(b) The nymphs of cockroach resemble adults.
(c) The nymphs of cockroach moults about 13 times to reach adult form.
(d) Only adult cockroaches have wings.

86. The following figure is related to head region of cockroach. Identify A to F.

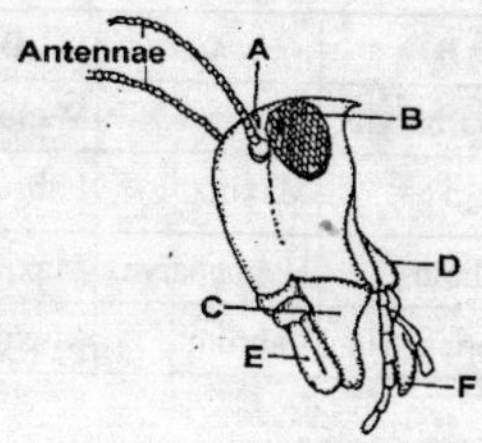

	A	B	C	D	E	F
(a)	Compound eye	Ocellus	Maxilla	Mandible	Labrum	Labium
(b)	Ocellus	Compound eye	Mandible	Maxilla	Labrum	Labium
(c)	Ocellus	Compound eye	Mandible	Maxilla	Labium	Labrum
(d)	Ocellus	Compound eye	Maxilla	Mandible	Labium	Labrum

87. Match the following columns.

	Column-I		**Column-II**
a	Collateral gland	(1)	Stimulatory organ in male
b	Titillator	(2)	Anal appendage which helps in oviposition
c	Gonapophysis	(3)	Helps in the formation of egg cases

Select the correct option

	a	b	c
(a)	2	3	1
(b)	1	2	3
(c)	3	1	2
(d)	2	1	3

88.

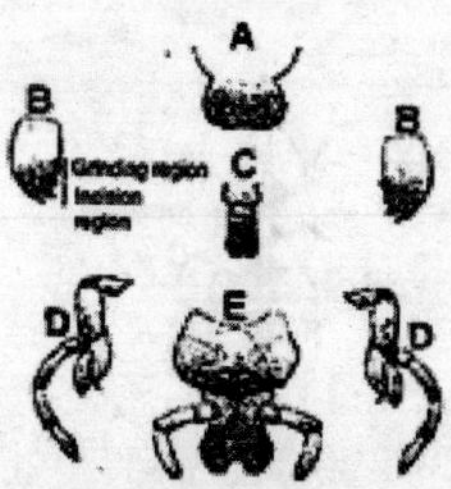

Fig. Mouthparts of cockroach

The above figure is related with mouth parts of cockroach. Identify A to E -

	A	B	C	D	E
(a)	Maxilla	Hypop harynx	Labium	Mandible	Labrum
(b)	Mandible	Labium	Maxilla	Labrum	Hypopharynx
(c)	Labrum	Mandible	Hypopharynx	Maxilla	Labium
(d)	Labium	Hypopharynx	Labrum	Maxilla	Mandible

89.

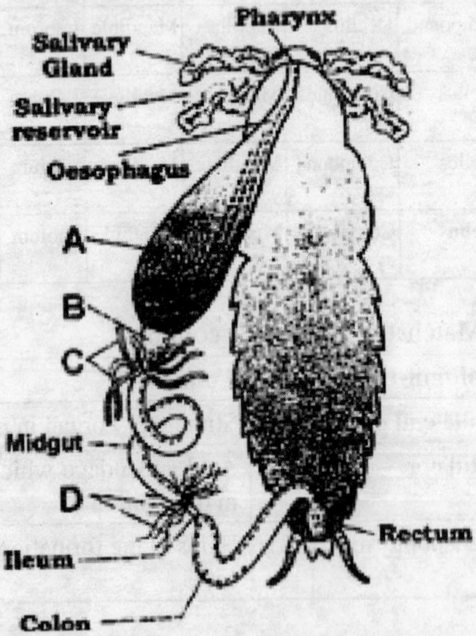

Identify structures A to D –

	A	B	C	D
(a)	Gizzard	Crop	Hepatic caecae	Malpighian tubules
(b)	Crop	Gizzard	Hepatic caecae Malpighian tubules	Labrum
(c)	Crop	Gizzard	Malpighian tubules	Hepatic caecae
(d)	Gizzard	Crop	Malpighian tubules	Hepatic caecae

90.

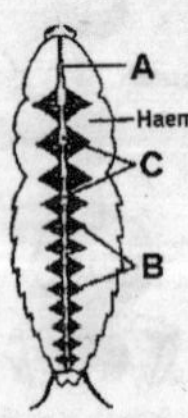

The above figure shows open circulatory system of cockroach. Identify A, B and C.

	A	B	C
(a)	Posterior aorta	Alary muscles	Chambers of hear
(b)	Anterior aorta	Ciliary muscles	Chambers of hear
(c)	Anterior aorta	Alary muscles	Chambers of hear
(d)	Anterior aorta	Ciliary muscles	Chambers of hear

91.

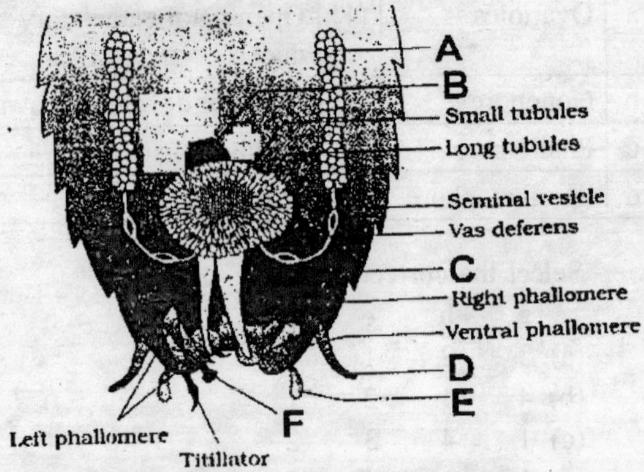

Identify A to F in above diagram –

	A	B	C	D
A	Testis	Testis	Testis	Testis
B	Collateral gland	Collateral gland	Phallic gland	Phallic gland
C	Ejaculatory duct	Ejaculatory duct	Ejaculatory duct	Ejaculatory duct
D	Anal cercus	Terga	Anal cerci	Caudal style
E	Caudal style	Caudal style	Caudal style	Caudal style
F	Pseudopenis	Pseudopenis	Pseudopenis	Pseudopenis

92. Figure refers to reproductive system of female cockroach. The correct labellings indicated by alphabets are respectively-

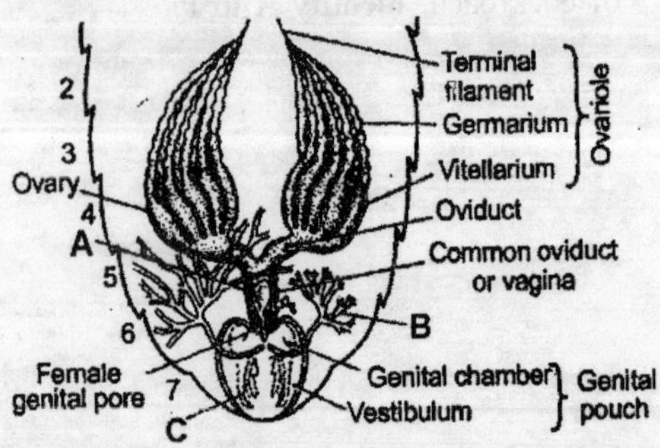

	A	B	C
(a)	Spermatheca	Collateral glands	Gonapophyses
(b)	Phallic gland	Collateral glands	Gonapophyses
(c)	Spermatheca	Seminal vesicles	Gonapophyses
(d)	Spermatheca	Collateral glands	Tegmina

Topic 5 Earchwork and Frog

93.

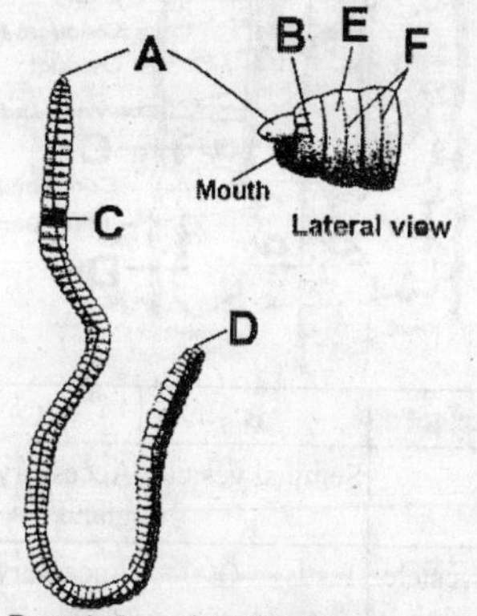

Dorsal view of earthworm

Go through the above figure. Identify A to F.

	A	B	C	D
A	Peristomium	Prostomium	Prostomium	Prostomium
B	Prostomium	Peristomium	Peristomium	Peristomium
C	Clitellum	Clitellum	Endosteum	Endosteum
D	Anus	Anus	Anus	Cloaca
E	Metameres	Metameres	Metameres	Metameres
F	Ring of setae	Ring of setae	Ring of setae	Ring of setae

94.

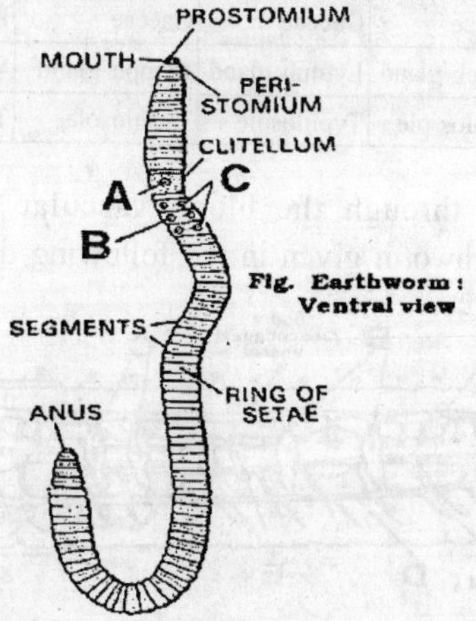

Fig. Earthworm : Ventral view

	A	B	C
(a)	Excretory pore	Female genital pore	Male genital pore
(b)	Male genital pore	Female genital pore	Genital papilla
(c)	Female genital pore	Genital papilla Male genital pore	
(d)	Female genital pore	Male genital pore	Genital papilla

95.

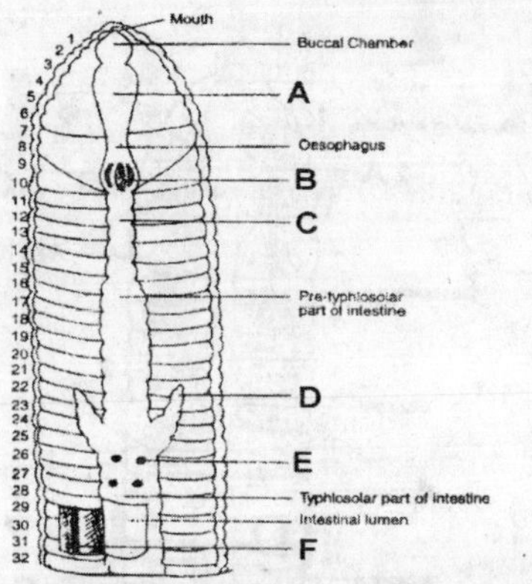

Choose the correct option of labeling from the options given-

	A	B	C	D
A	Pharynx	Gizzard	Pharynx	Pharynx
B	Stomach	Pharynx	Gizzard	Gizzard
C	Gizzard	Stomach	Stomach	Stomach

D	Caecae	Caecae	Caecae	Liver
E	Lymph gland	Lymph gland	Lymph gland	Villi
F	Typhlos ole	Typhlosole	Typhlosole	Typhlosole

96. Go through the blood vascular system of earthworm given in the following diagram -

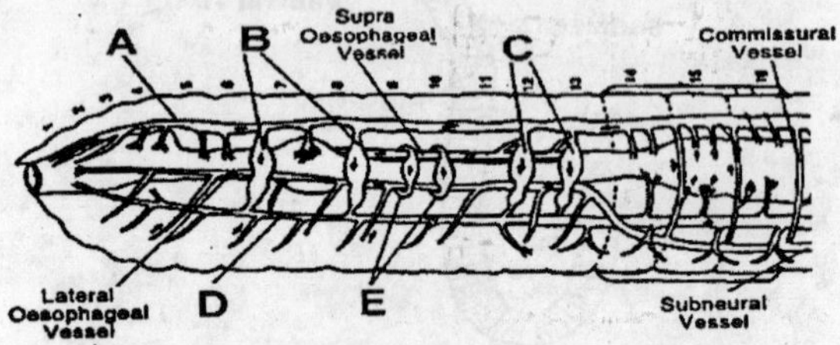

	A	B	C	D
A	Dorsal vessel	Ventral vessel	Dorsal vessel	Ventral vessel
B	Lateral hearts	Lateral hearts	Lateral hearts	Lateral hearts
C	Lateraloeso-pharyngeal heart	Lateraloesopha-ryngeal heart	Anterior loop	Anterior loop
D	Ventral vessel	Dorsal vessel	Ventral vessel	Dorsal vessel
E	Anterior loop	Anterior loop	Lateraloeso-pharyngeal heart	Lateraloeso-pharyngeal heart

97.

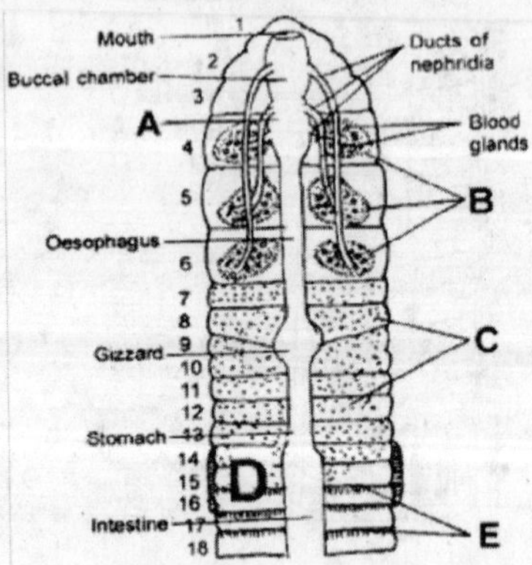

I. Septal nephridia
II. Pharynx
III. Forest of integumentary nephridia
IV. Integumentary nephridia
V. Tufts of pharyngeal nephridia

Identify the structures labeled A to E in the diagram given above from the list I to V -

	A	B	C	D	E
(a)	II	I	III	IV	V
(b)	II	V	IV	III	I
(c)	II	IV	V	I	III
(d)	II	III	IV	I	V

98. Identify A to D in the figure –

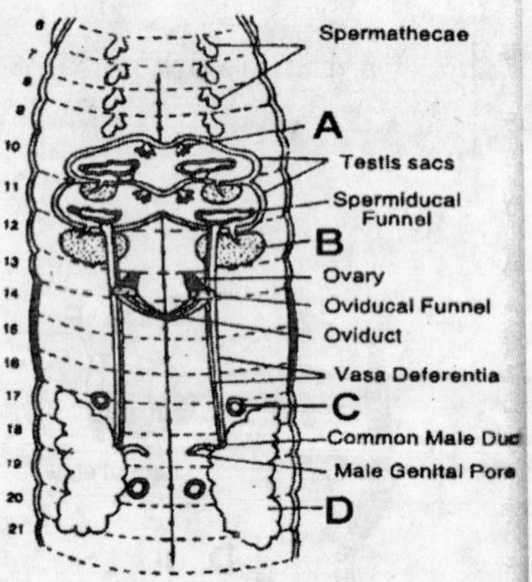

	A	B	C	D
A)	Testis	Seminal vesicle	Accessory gland	Prostate gland
B)	Seminal vesicle	Testis	Accessory gland	Prostate gland
C)	Testis	Seminal vesicle	Prostate gland	Accessory gland
D)	Seminal vesicle	Testis	Prostate gland	Accessory gland

99. Refer to the diagram of nephridial system in earthworm.

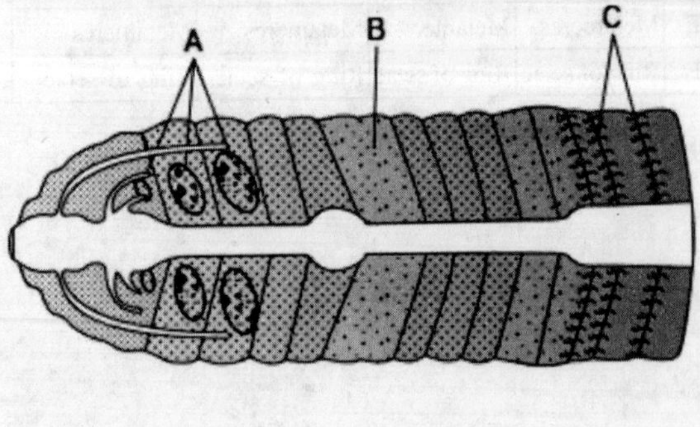

Select the option representing correct characteristic of the labelled structure:

(a) C – Septal nephridia – Open into intestine

(b) A – Pharyngeal nephridia – Open to outside

(c) B – Integumentary nephridia – Densely found on first two segments

(d) All of these

100. The above figure is associated with diagrammatic representation of internal organs of frog. Identify A to E.

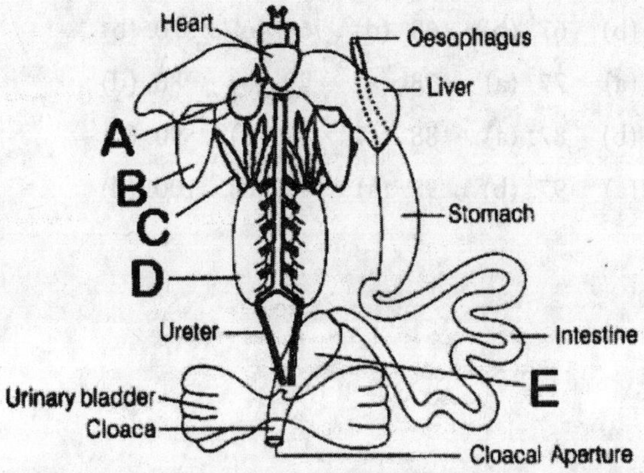

	A	B	C	D
A	Gall bladder	Gall bladder	Gall bladder	Gall bladder
B	Lung	Lung	Lung	Lung
C	Ovary	Fat bodies	Testis	Fat bodies
D	Testis	Testis	Kidney	Kidney
E	Rectum	Rectum	Rectum	Rectum

101.

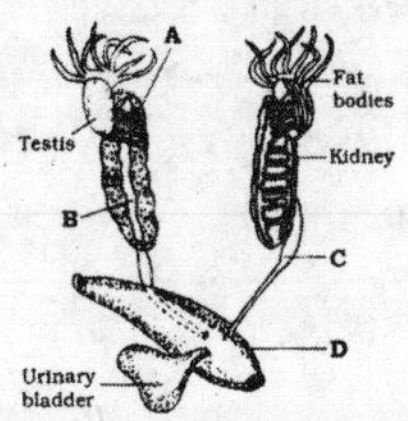

Go through the following figure indicating the male reproductive system of frog. Identify A to D

	A	B	C	D
A.	Bidder's canal	Adrenal gland	Urinogenital duct	Rectum
B.	Bidder's canal	Adrenal gland	Urinogenital duct	Cloaca
C.	Vasa efferentia	Adrenal gland	Urinogenital duct	Cloaca
D.	Vasa efferentia	Thyroid gland	Urinogenital duct	Cloaca

102. The above figure is related with female reproductive system of frog. Identify A to D.

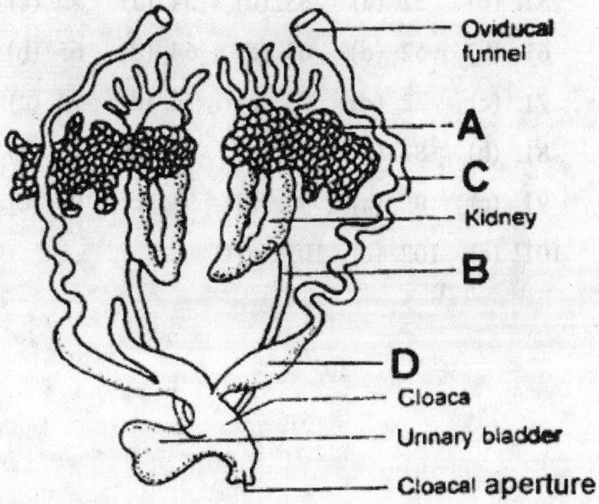

	A	B	C	D
A)	Ovary	Ureter	Oviduct	Ovisac
B)	Ovary	Urinogenital duc	Bidder's canal	Ovisac
C)	Ovary	Urinogenital duc	Ovisac Oviduct	Fat bodies
D)	Ovary	Urinogenital duc	Bidder's canal	Oviduct

103. Identify A, B and C respectively –

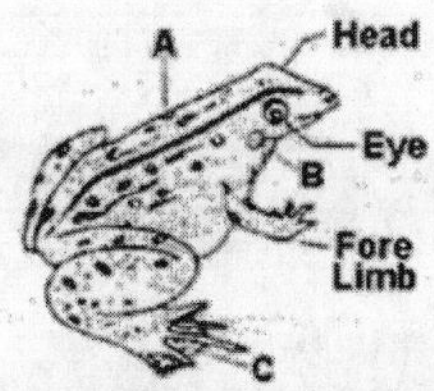

(a) Trunk, Tympanum, Web

(b) Neck, Brown eye spot, Web

(c) Trunk, Tympanum, Hind limb

(d) Neck, Tympanum, Hind limb

ANSWER KEY

1. (c)	2. (a)	3. (d)	4. (c)	5. (c)	6. (b)	7. (b)	8. (b)	9. (a)	10. (c)
11. (b)	12. (c)	13. (a)	14. (c)	15. (d)	16. (c)	17. (a)	18. (c)	19. (a)	20. (c)
21. (a)	22. (c)	23. (b)	24. (d)	25. (d)	26. (a)	27. (b)	28. (a)	29. (d)	30. (d)
31. (c)	32. (b)	33. (b)	34. (a)	35. (c)	36. (d)	37. (d)	38. (d)	39. (b)	40. (c)
41. (a)	42. (b)	43. (d)	44. (b)	45. (c)	46. (b)	47. (d)	48. (d)	49. (c)	50. (a)
51. (b)	52. (a)	53. (d)	54. (a)	55. (a)	56. (c)	57. (a)	58. (d)	59. (a)	60. (b)
61. (b)	62. (d)	63. (d)	64. (d)	65. (b)	66. (b)	67. (b)	68. (d)	69. (c)	70. (b)
71. (c)	72. (d)	73. (c)	74. (d)	75. (c)	76. (a)	77. (a)	78. (a)	79. (b)	80. (d)
81. (b)	82. (a)	83. (d)	84. (d)	85. (a)	86. (b)	87. (a)	88. (c)	89. (b)	90. (c)
91. (c)	92. (a)	93. (b)	94. (c)	95. (c)	96. (a)	97. (b)	98. (a)	99. (d)	100. (d)
101. (a)	102. (c)	103. (a)							

3 Chapter

Biomolecules

MULTIPLE CHOICE QUESTIONS

1. Elemental analysis on a plant tissue, Animal tissue or a microbial paste reveals:
(a) list of elements like C; H; O & several others
(b) Respective content per unit mass of a living tissue
(c) Both
(d) Diversity of living organism in our Biosphere.

2. Elemental list could be _____ in _____ terms of study on living tissues & earth's crust:-
(a) Same; absolute
(b) Different; absolute
(c) Different; same
(d) Same; relative

Topic 1	How to Analyse Chemical Composition?

3. Palmitic acid has _______ number of carbons including carboxyl carbon.
(a) 16 (b) 15
(c) 14 (d) 12

4. Arachidonic acid has _______ number of carbon atoms including the carboxyl
(a) 16 (b) 20
(c) 21 (d) 19

5. Fatty acids could be _______ (with double bonds) or _______ (without double bonds).
(a) Saturated; Unsaturated
(b) Unsaturated; Saturated
(c) Saturated; Saturated
(d) Unsaturated; Unsaturated

6. How many of the following is an esterified glycerol:-
Monoglyceride;
Diglyceride;
Triglyceride:
Muramic acid
Lignin;
Suberin
(a) 4 (b) 5
(c) 6 (d) 3

7. The oil have lower melting point
(a) All fats (b) triglycerides
(c) Gingelly oil (d) All

8. A phospholipid have
(a) a phosphorous
(b) a phosphorylated group
(c) Both
(d) None

9. The neural tissues have lipids with _______ structure
(a) More complex (b) Less complex
(c) More simple (d) simple

10. Carbon compounds in living organism having heterocyclic rings could be
(a) Monoglyceride (b) Adenine
(c) Cytosine (d) Both (B) & (C)

11. Adenine esterified with sugar is known as
(a) Adenylic acid (b) Adenosine
(c) Adenotine (d) None of the above

12. Nucleic acids like DNA & RNA consist of

(a) Nucleotide & nucleoside
(b) Nucleoside only
(c) Nucleotide only
(d) Nucleotide & phosphate groups.

13. With respect to other elements which element is relatively abundant in living organism than in earth's crust:-
(a) C & Ca (b) C & H
(c) S & N (d) N & Ca

14. For the chemical composition analysis____ is used:-
(a) Ch_3COOH (b) CH_3COOH -Cl
(c) Cl_3-CCOOH (d) Cl_3- COOH

15. Filtrate obtained after grinding of living tissue is also known as:-
(a) Slurry
(b) Acid - soluble
(c) Acid insoluble pool
(d) All

16. Acid – insoluble pool is also known as:-
(a) Slurry (b) Retentate
(c) filtrate (d) All

17. Analytical techniques applied to the compound gives us an idea of:-
(a) Probable structure of compounds
(b) Molecular formula of compounds.
(c) Both
(d) None

18. All the carbon compounds that we get from the living tissue can be called:-
(a) Biomolecules (b) Slurry
(c) Retentate (d) All

19. If the tissue is fully burnt:-
(a) All the carbon compounds are oxidised to gaseous forms (CO2 & water vapour).
(b) Remaining's are known as ash.
(c) Ash contains inorganic elements & inorganic compounds.
(d) All

20. Inorganic elements like sulphate and phosphates are present in
(a) Ash of burnt tissue
(b) Oxidised gaseous form
(c) Both
(d) None

21. α – Amino acids are organic compounds containing
(a) Amino group and acidic group substituted on different carbon.
(b) Keto – group & Hydrogen on different carbon.
(c) Amino group & acidic group substituted on same carbon.
(d) Keto – group & alcohol group substituted on same carbon.

22. How many substituted groups are present in an α – amino acid
(a) 1 (b) 2
(c) 3 (d) 4

23. The R – group in a proteinaceous amino acid could be
(a) Hydrogen (b) Methyl group
(c) Hydroxy methyl (d) Any of the above

24. The chemical and physical properties of amino acids are essentially of the
(a) Amino group (b) Carboxyl group
(c) The R - group (d) All of the above

25. If the R – group of amino acid is methyl
(a) Glycine (b) Serine
(c) Alanine (d) Any of the above

26. A hydrogen substituted carbon containing amino acid is:-
(a) Glycine (b) Alanine
(c) Both (A) & (B) (d) Serine

27. Number of Amino; Carboxyl & the R – functional group determines:-
(a) Acidic nature of Amino acid.
(b) Basic nature of Amino acid

(c) Neutral nature of Amino acid
(d) Any of the above

28. Which of the following group of amino acid is aromatic in nature:-
(a) tyrosine; phenylalanine
(b) tyrosine; tryptophan glutamic acid
(c) Glutamic acid; lysine; valine
(d) none of the above

29. Which of the following is neutral in nature:-
(a) Valine (b) Serine
(c) Alanine (d) All

30. A particular property of amino acid is the ionizable nature of
(a) -H (b) $-NH_2$
(c) CH_3 (d) All

31. Which of the following determines the particular property of amino acid is the Ionizable nature & structure of amino acid:-
(a) $-NH_2$ & $-COOH$
(b) $-COOH$ only
(c) $-NH_2$ only
(d) none of the above

32. In different solution; of different ____ the ______ **of amino acid changes. [Pg-144,E]**
(a) pH; pH
(b) pH; structure
(c) Structure; Structure
(d) structure; pH

33. Which of the following is a zwitterionic form.
(A) $H_3^+N-\underset{R}{CH}-COOH$ (B) $H_3^+N-\underset{R}{CH}-COO^-$
(C) $H_2N-\underset{R}{CH}-COO^-$ (D) All of the above

34. Lipids are generally ______ insoluble:-
(a) fat (b) water
(c) Lipid (d) All

35. Lipids could be a ______ fatty acids or has a ______ group attached to an R – group.
(a) Carboxyl; fatty acid
(b) Fatty acid; simple
(c) Carboxyl; simple
(d) Simple; carboxyl

36. The R – group attached to the carboxyl group in a lipid could be a
(a) $-CH_3$ (b) $-C_2H_5$
(c) Higher number of $-CH_2$
(d) All of the above

Topic 2 Primary and Secondary Metabolites

37. Alkaloids; Flavonoids; Rubber; Essential oils; antibiotics; coloured pigments; scents; Gums spices

How many of the above are primary metabolites
(a) 7 (b) 9
(c) 5 (d) None

38. Few ______ metabolites have ecological importance's:-
(a) Primary & secondary
(b) Secondary & Primary
(c) Only Primary
(d) Only Secondary

39. The diagram represent:-

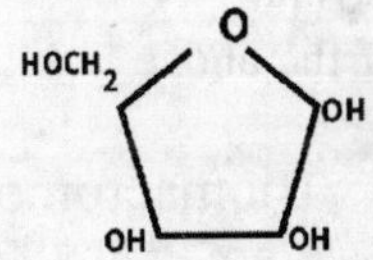

(a) Ribose (b) Glucose
(c) Both (d) None

40. $CH_3-(CH_2)_{14}-COOH$
(a) A glycerol molecule
(b) A fatty acid
(c) An amino acid
(d) A carbohydrate

41. Which of the following is the compound represents the shown figure :-

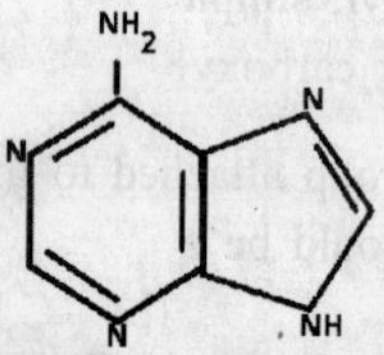

(a) A purine (Adenine)
(b) A pyrimidine (Uracil)
(c) A purine (Uracil)
(d) A pyrimidine (Adenine)

42. Which of the following is a Nucleoside:-
(a) Adenylic acid (b) Uridine
(c) Thymidylic acid (d) All

43. How many of the following are nitrogen bases:-

i) 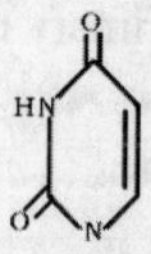 ii) 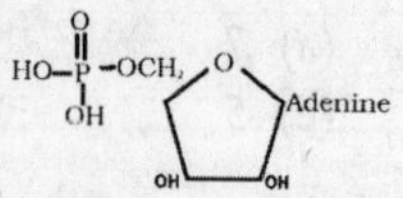

iii) Guanine iv) Uracil
(a) All four (b) Only three
(c) Only two (d) Only one

44. Which of the following group represents Lectins
(a) Abrin; Ricin
(b) Monoterpenes; Diterpenes
(c) Concanavalin – A
(d) None of the above

Topic 3 Biomacromolecules

45. They have molecular weight ranging from 18 to around 800 Da.

The above written statement represents:-
(a) About Biomacromolecules
(b) One feature common to all those compounds found in the acid insoluble fraction.
(c) Both
(d) None

46. How many of the following statements are incorrect:-
i) Acid insoluble fraction has only four types of organic compounds.
ii) All the compound in acid insoluble fraction have molecular weight in range of 10,000 Da and above.
iii) Molecular weight less than one thousand Dalton are usually referred to as Micromolecules.
iv) Biomacromolecules are simply known as Biomolecules.
(a) 1 (b) 2
(c) 3 (d) 4

47. How many of the following statement is true regarding lipids in Biomacromolecules:-
i) Lipids are polymeric substances.
ii) Have molecular weight less than 10,000 Da.
iii) Molecular weight do not exceed 800 Da.
(a) only i) & ii) (b) only iii)
(c) All i); ii) & iii) (d) only ii) & iii)

48. Which of the following statement is not correct:-
(a) After grinding cell membrane forms the vesicles.
(b) Vesicles are water soluble.
(c) Lipids are not strictly Biomacromolecules
(d) None of the above

49. The acid soluble pool roughly represents ______ composition.
(a) Cytoplasmic (b) Nuclear
(c) Mitochondrial (d) None

50. The macromolecules from the cytoplasm and organelles become the
(a) Retentate (b) Slurry
(c) Filtrate (d) All

51.

Component	% of the total cellular mass
Water	70 – 90
i)	10 – 15
ii)	3
Lipids	iii)
iv)	5 – 7
Ions	1

(a) i) protein

ii) carbohydrate
iii) 2
iv) Nucleic acid

(b) i) carbohydrate
ii) Nucleic acid
iii) 2
(iv) Protein

(c) i) Nucleic acid
ii) Protein
iii) 2
iv) Carbohydrate

(d) i) Nucleic acid
ii) carbohydrate
iii) 2
iv) Protein

Topic 4 Proteins

52. Proteins are:-
(a) Polypeptides
(b) Linear chains of amino acid linked by peptide bonds
(c) Polymer of amino acids
(d) All of them.

53. A protein if a heteropolymer:-
(a) It contains only one types of amino acids.
(b) it contains different types of amino acids.
(c) both
(d) None

54. Which statement is incorrect:-
(a) homopolymers have only one type of monomer repeating 'n' number of times
(b) Dietary proteins are source of essential amino acids.
(c) Amino acids could be essential or non – essential
(d) essential amino acids are synthesized in our body.

55. What are functions of proteins:-
i) Carry out many functions in living organism
ii) Transporter of nutrients
iii) Fight infections
iv) Regulates in the form of hormones & enzymes

(a) only two (b) only three
(c) Only four (d) None

56. The most abundant enzyme in animal world is ___i)___ while in whole of the biosphere is ___ii)___
(a) (i) Collagen (ii) PEPcase
(b) (i) RuBisCo (ii) PEPcase
(c) (i) Collagen (ii) RuBisCO
(d) None of them

Topic 5 Polysaccharide

57. Polysaccharide is the part of ____
(a) In – soluble fraction
(b) Insoluble pellet
(c) Retentate
(d) All

58. A polysaccharide contains
(a) Different Monosacharides
(b) Same type of monosaccharide
(c) like cellulose
(d) All of these

59. Cellulose and starch is a homopolymer of
(a) Glucose (b) Fructose
(c) Galactose (d) None

60. Which of the following statement is incorrect:
(a) starch is a a polysaccharide homopolymer.
(b) Inulin is a polymer of fructose
(c) In a polysaccharide chain, Right end is reducing while left end is non – reducing.
(d) Starch forms helical secondary structures.

61. (I) Starch produces blue colour after binding with I_2
(II) Cellulose cannot hold I_2
(a) Both are wrong

(b) Both are correct
(c) (I) is correct (II) is incorrect
(d) (II) is correct (I) is incorrect

62. Paper made from plant pulp and cotton fibre is
(a) Starch only
(b) Cellulose
(c) Complex polysaccharide
(d) Both (B) & (C)

63. What are examples of homopolymers:-
(a) N – acetyl galactosamine; Glucosamine
(b) Amino acids; sugars
(c) Chitin
(d) None

Topic 6 — Nucleic Acids

64. Nucleic acids are:-
(a) Polynucleosides (b) Polynucleotides
(c) Both (d) None

65. A nucleotide has ____ chemical distinct compounds:-
(a) Only one (b) Two
(c) Three (d) Four

66. A heterocyclic compound in Nucleic acid is:
(a) N_2 – Base (b) Sugar
(c) Fatty acid (d) All

67. Adenine and ____ are ______ purines
(a) Cytosine; Substituted
(b) Guanine; Substituted
(c) Uracil; Substituted
(d) Guanine; Unsubstituted

68. The sugar found in polynucleotides is either ribose (________) or ________
(a) 2' deoxyribose; monosaccharide
(b) Monosaccharide; 2' deoxyribose
(c) Disaccharide; 2' deoxyribose
(d) Disaccharide; Monosaccheride

Topic 7 — Structure of Proteins

69. In a protein the left end represents:-
(a) First amino acid & C – terminal
(b) Last amino acid &N – terminal
(c) First amino acid & N – terminal
(d) Last amino acid & C – terminal

70. In a protein the right end represents
(a) First amino acid & C – terminal
(b) Last amino acid &N – terminal
(c) First amino acid & N – terminal
(d) Last amino acid & C – terminal

71. Which of the following statement is Untrue:-
(a) A protein thread is folded in the form of a helix.
(b) Only some portion of the protein thread are arranged in the form of a helix
(c) In proteins only left handed helices are observed.
(d) Both (B) & (C)

72. The long protein chain is also folded upon itself like a hollow woolen ball known as:-
(a) Primary structure
(b) Secondary Structure
(c) Tertiary structure
(d) None of the above

73. Identify the diagram given below

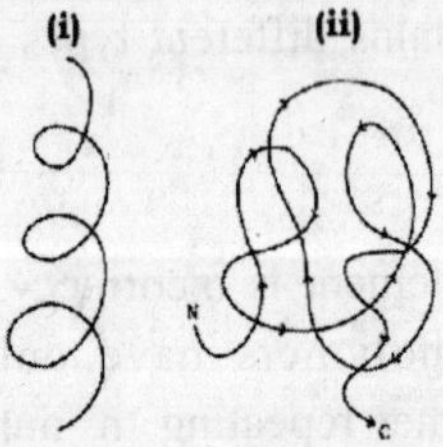

(a) (i) Primary (ii) Secondary
(b) (i) Secondary (ii) Tertiary
(c) (i) Tertiary (ii) Primary
(d) None of the above

74. Protein polypeptides or subunits arranged with respect to each other of a protein otherwise called the
(a) Primary structure

(b) Tertiary structure
(c) Quaternary structure
(d) Secondary structure

75. A adult human Hb (Haemoglobin) consists of ________ subunits.
(a) 1 (b) 2
(c) 3 (d) 4

76. ________ subunits of α – type and ____ of β – type together constitute the human haemoglobin(Hb):-
(a) 2; 4 (b) 2; 2
(c) 4; 2 (d) 4; 4

Topic 8 Nature of Bond Linking Monomers in a Polymer.

77. In polypeptide amino acids are linked by
(a) H – bond
(b) Glycosidic Bond
(c) Peptide bond
(d) Peptide and H – bond both

78. Choose the correct statement about peptide bond
(a) It is formed when carboxyl(-COOH) group of one amino acids react with carboxyl ($-NH_2$) group of other amino acid.
(b) It is formed when amino ($-NH_2$) group of one amino acid react with carboxyl (-COOH) group of other amino acid.
(c) It is formed when carboxyl group (-COOH) of one amino acid react with amino ($-NH_2$) group of other amino acid.
(d) It is formed when amino ($-NH_2$) group of one amino acid react with amino ($-NH_2$) group of other amino acid.

79. Peptide bond is formed by-
(a) Elimination of water moiety i.e. rehydration
(b) Addition of water moiety i.e. rehydration
(c) Addition of water moiety i.e. dehydration
(d) Elimination of water moiety i.e. dehydration

80. Polysaccharide is formed by linking of monosaccharide by-
(a) H – bond (b) S – bond
(c) Peptide Bond (d) Glycoside bond

81. Dehydration is cause of formation of –
(a) Peptide bond (b) Glycosidic bond
(c) Both A & B (d) None of these

82. Glycosidic bond is formed between monosaccharide while linking-
(a) Carbon & Carbon
(b) Carboxyl & amino group
(c) Carbon & Hydrogen
(d) Carbon & Oxygen

83. Match the Column- I & column – II

	Bond (Column- I)		Occurrence Column – II
a.	Peptide bond	(i)	Between Nitrogenous bases of nucleic acid
b.	Glycosidic bond	(ii)	Between adjacent amino acid
c.	Ester bond	(iii)	Between phosphate & hydroxyl group of sugar
d.	H – Bond	(iv)	Between adjacent carbon of monosaccharide

(a) a – i, b – ii, c – iii, d – iv
(b) a – ii , b – iv , c – i, d – iii
(c) a – iii, b – iv, c – i, d – ii
(d) a – ii, b – iv, c – iii, d – i

84. In nucleic acid phosphate links –
(a) 3' carbon of both sugar of succeeding sugar
(b) 3' carbon of one sugar & 5' carbon of the other sugar of succeeding nucleotide
(c) 5' carbon of one sugar of succeeding sugar.
(d) 5' carbon of one sugar & 3' carbon of other group of succeeding nucleotide.

85. What is / are number of ester bond & phosphodiester bond either side of nucleic acid respectively-
(a) 1, 2 (b) 1, 1
(c) 2, 1 (d) 2, 2

86. The famous Watson – crick model is related to-

(a) Nucleic acid (DNA)
(b) Protein
(c) Carbohydrate
(d) Enzymes

87. How many of following is / are correct with respect to Watson – crick model.
i) DNA exist as a double helix
ii) The strands of polynucleotides are antiparallel.
iii) Backbone is formed by sugar only.
iv) Nitrogen bases faces inside
v) A of one strand bound with U on other strand
(a) 2 (b) 3
(c) 4 (d) All fives

88. Choose the correct nitrogen base pairing of DNA
(a) A ≡ T (b) A = U
(c) A = T D) A ≡ U

89. Each step of ascent is represented by how many pairs of bases according to Watson – crick model.
(a) 1 (b) 2
(c) Zero (d) None of these

90. At each of ascent, the strand turn ___
(a) 63° (b) 36°
(c) 34° (d) 3.4°

91. One full turn of helix strand of B – DNA involves how many nitrogen bases
(a) 10 (b) 20
(c) 2 (d) none of these

92. Choose correct statement regarding B – DNA
(a) Pitch would be 36 A°
(b) The rise per base pair would be 3.4 A°
(c) Pitch would be 3.4A°
(d) The rise per base pair would be 36 A°

93. Cytosine (C) bond with ______ by ______ H – Bond.
(a) Guanine (G); 2 (b) Thymine; 2
(c) Guanine (G); 3 (d) Thymine; 3

Topic 9 Dynamic State of Body Constituent's Concept of Metabolism

94. What is 'turn over'?
(a) Biomolecules are never being changed into some other biomolecules and also made from some other biomolecules.
(b) Biomolecules are constantly being changed into some other biomolecules but never made from some other biomolecules.
(c) Biomolecules are never being changed into some other biomolecules nor being made from some other biomolecules.
(d) Biomolecules are constantly being changed into some other biomolecules and also made from some other biomolecules.

95. The breaking & making through chemical reaction which occur constantly in living organism are called
(a) Metabolism (b) Anabolism
(c) Catabolism (d) none of these

96. Amine are formed by-
(a) removal of (-COOH) from amino acid
(b) removal of (CO_2) from amino acid
(c) addition of (CO2) to amino acid
(d) addition of (COOH) to amino acid

97. Metabolites are converted into each other in a series of linked reactions called ______.
(a) Catabolic pathway only
(b) Anabolic pathway only
(c) Metabolic pathway
(d) None of these

98. Metabolic pathway are-
(a) Linear only
(b) Circular only
(c) May be linear or circular
(d) None of them

99. How many uncatalysed metabolic conversion is / are found in living system
(a) 1
(b) More than 1 but less than 100

(c) Zero
(d) Thousand

Topic 10 Metabolic Basis for Living

100. Metabolic pathway that lead to a more complex structure from a simples structure is / are
(a) Anabolic pathway
(b) Catabolic pathway
(c) Both A & B
(d) None of these

101. Choose the correct about catabolic pathway
i) Metabolic pathway that lead to simpler structure from a complex structure.
ii) Glucose becomes lactic acid in our skeletal muscles
iii) Acetic acid becomes cholesterol.
iv) Metabolic pathway that lead to more complex structure from a simpler structure.
(a) i & iii (b) i & ii
(c) iv & ii (d) iv & iii

102. Which of following expect to consume energy?
i) When glucose is degraded to lactic acid
ii) Assembly of protein from amino acid
iii) Anabolic pathway
iv) Catabolic pathway
(a) i & iii (b) i & iv
(c) ii & iii (d) ii & iv

103. How many of following is /are correct about glycolysis
i) Formation of glucose from lactic acid
ii) Occur in ten(10) metabolic step.
iii) Energy liberated during degradation is store in form of chemical bond.
iv) Formation of lactic acid from glucose
(a) i, ii, iii (b) ii, iii, iv
(c) i & ii (d) i & iv

104. Energy currency in living system is –
(a) Adenosine triphosphate
(b) Glucose
(c) Protein
(d) Enzyme

105. Bioenergetics deals with-
(a) How do living organism derive their energy
(b) How do living organism store energy & in what form.
(c) How do living organism convert energy into work.
(d) All of these

Topic 11 The Living State

106. The blood concentration of glucose in normal healthy individuals is
(a) Less than 2.4 mmol/L
(b) More than 10 mmol/L
(c) 4.2 mmol/L – 5.0 mmol/L
(d) None of these

107. Living state is –
(a) Equilibrium steady – state to be not to perform work.
(b) Non – equilibrium steady – state to be not to perform work.
(c) Equilibrium steady – state to be able to perform work.
(d) Non – equilibrium steady – state to be able to perform work.

108. Living process is a constant effort to prevent falling into equilibrium. This is achieved by –
(a) Energy output (b) energy input
(c) Both of these (d) None of these

Topic 12 Enzymes

109. Enzymes are chemically –
(a) Protein (b) Carbohydrate
(c) Lipid (d) Nucleic acid

110. Ribozymes are chemically

(a) Protein (b) Lipid
(c) Carbohydrate (d) Nucleic acid

111. What is / are difference between inorganic catalyst and enzyme catalyst.
(a) inorganic catalysts work efficiently at low temperature but enzyme of only thermophilic organism work efficiently at low temperature
(b) Inorganic catalyst work efficiently at high temperature but enzyme get damaged at high temperature except of microbes that are live in sulphur springs
(c) Inorganic catalyst are not efficient at high temperature but enzymes of all living organism work efficiently at high temperature.
(d) None of these

112. Choose correct regarding "active site"
1) Substrate fits
2) Enzymes catalyst through active site show low rate
3) It forms by crevices or pocket made by primary protein only.
4) It form by crevices or pocket made by tertiary protein structure
(a) 1, 2, 3 (b) 1, 2, 4
(c) 1, 3 (d) 1, 4

Topic 12.1 Chemical Reaction

113. Physical change refers to – [Pg- 154,E]
(a) Change in shape without breaking bonds.
(b) Change in state of matter
(c) Ice, water, water vapour.
(d) All of these

114. Chemical change differ from physical change in
(a) Dissociation of bond
(b) Formation of new bond
(c) A & B bond
(d) There is no difference in both

115. Hydrolysis of starch into glucose is :-
(a) Inorganic chemical reaction
(b) Organic chemical reaction
(c) Physical changes
(d) A & B both

116. Rate of physical or chemical process refer to –
(a) Amount of reactant formed per unit time
(b) Amount to product dissociate per unit time
(c) Differential of time with respect to produce
(d) Differential of product with respect to time

117. Choose the correct
(a) Rate can be called velocity if the direction is not specific.
(b) Rate of physical & chemical processes are not influenced by temperature
(c) Catalysed reaction proceeds at rates vastly lower than that of uncatalysed ones.
(d) Catalysed reaction proceeds at rates vastly higher than that of uncatalysed ones.

118. Choose the correct response
(a) For every increase by 10°C, rate is double
(b) Rate decrease by one – fourth by decrease in temperature by 10°C.
(c) When enzymes catalysed reaction are observed the rate would be vastly lower than the same but uncatalysed reaction.
(d) None of these

119. Choose correct response with respect to given equation:-
Carbon dioxide + water $\rightleftharpoons$ carbonic acid
(a) Carbonic anhydrase is enzyme required for accelerated reaction.
(b) In absence of enzyme, still this reaction is fast enough
(c) 200 molecules of H_2CO_3 being per hour formed by enzyme accelerated reaction.
(d) 600,000 molecules of H_2CO_3 being formed every second in absence of any enzyme.

120. Which of the following is correct chemical formula for pyruvic acid?
(a) $C_2H_3O_4$ (b) $C_3H_3O_3$
(c) $C_3H_4O_3$ (d) $C_6H_{12}O_6$

121. Match column – I and column – II

	Column – I (Metabolic pathway)		Column – II (Occurrence)
A.	Formation of alcohol	(i)	Anaerobic condition of skeletal muscle
B.	Formation of pyruvic acid	(ii)	Yeast
C.	Formation of lactic acid	(iii)	Aerobic condition of normal human cell

(a) A - I, B -iii, C -ii
(b) A-iii, B -ii, C -i
(c) A -ii, B -I, C -iii
(d) None of these

Topic 12.2 How do Enzymes being about such High Rates of Chemical Conversions?

122. Which of the following is correct about enzymes
(a) It is 2 – D structure
(b) Convert product into substrate
(c) They have active site
(d) All of these

123. Transition state structure is formed when –
(a) Enzyme is free
(b) Enzyme bound with product
(c) 'ES' complex
(d) Substrate structure do not change until product formed.

124. Which of following are unstable
(a) Enzyme
(b) Product
(c) Reactant
(d) Intermediate structural states.

125.

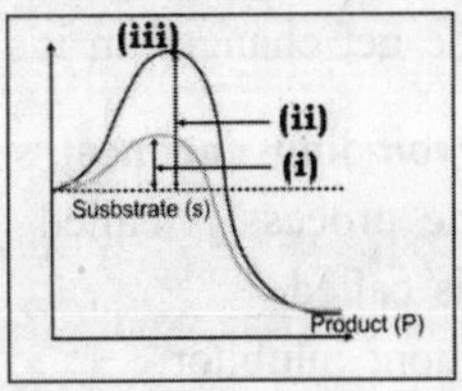

	(i)	(ii)	(iii)
A)	Activation energy without enzyme	Transition state	Activation energy with enzyme
B)	Transition state	Activation energy without enzyme	Activation energy with enzyme
C)	Activation energy with enzyme	Activation energy without enzyme	Transition state
D)	Activation energy without enzyme	Activation with enzyme	Transition state

126. Choose correct response
i) Y – axis represent potential energy
ii) X – axis represent substrate
iii) Y – axis represent progress of reaction
iv) X – axis represent state through transition state
(a) i) & ii) B) iii) & iv)
(c) i) & iv) (c) ii) & iii

127. If 'P' (product) is at lower level than 's' (substrate), the reaction is ______
(a) Endothermic reaction
(b) Exothermic reaction
(c) Spontaneous reaction
(d) A & C both

Topic 12.3 Nature of Enzyme Action

128. Which is correct way to represent enzyme action
(a) $E + S \rightleftharpoons ES \rightleftharpoons EP \rightleftharpoons E + P$
(b) $E + S \rightleftharpoons E + P$
(c) $E + S \rightleftharpoons ES \rightleftharpoons EP \rightleftharpoons E + P$
(d) $E + S \rightleftharpoons ES \rightleftharpoons EP \rightleftharpoons E + P$

129. ES complex is ______ and dissociates into ______ and ______
(a) Long lived; product; changed enzyme
(b) Short lived; reactant, changed enzyme
(c) Long lived, reactant, unchanged enzyme
(d) Short lived, product, unchanged enzyme

130. Arrange in correct sequence of catalytic cycle of an enzyme action-
i) The active site of the enzyme, now in

close proximity of the substrate breaks the chemical bonds of the substrate and the new enzyme product complex is formed

ii) The substrate binds to the active site of enzyme, fitting into the active site

iii) The enzyme release the products of the reaction and the free enzyme is ready to bind to another molecule of the substrate

iv) The binding of the substrate induces the enzyme to alter its shape, fitting more tightly around the substrate.

(a) i → ii → iii→ iv
(b) i → iii → ii → iv
(c) ii → iv → iii → I
(d) ii → iv → i → iii

Topic 12.4 Factors Affecting Enzyme Activity

131. Which of the following can change enzyme activities?
(a) All such activities that can alter the tertiary structure of the protein
(b) Temperature pH
(c) Substrate conditions
(d) All of these

132. Enzyme activity decline-
(a) Above the optimum value
(b) Below the optimum value
(c) A & B both
(d) Enzyme activity never decline

133. Optimum pH refer to –
(a) pH at which enzyme activity is lowest
(b) pH at which enzyme activity is highest
(c) pH at which enzyme activity started immediately
(d) pH at which enzyme activity ended completely

134. Choose response with respect to enzyme activities
i) low temperature destroy enzyme
ii) high temperature preserve enzyme in a temporarily inactive state
iii) optimum temperature is temperature at which enzyme activity is highest
iv) As temperature increase, enzyme activity increase until optimum and thereafter increase in temperature lead to decline in enzyme activities
v) As temperature increase enzyme activities is zero until optimum temperature & thereafter increase in temperature lead to increase in enzyme activities

(a) i, iii, iv (b) ii, v
(c) i, iv, v (d) iii, iv

135. As pH increase, enzyme activity-
(a) Constantly increase
(b) Constantly decrease
(c) No effect
(d) Increase until optimum and decrease further pH

136. With increase in substrate concentration, the velocity of the enzymatic reaction –
(a) Constantly increase
(b) Rise at first until Vmax and further no rise
(c) No effect
(d) Decrease first until Vmax and increase further

137. After reaching Vmax, the enzymatic reaction does not exceed by any further rise in concentration of substrate because-
(a) Enzymes molecules are fewer than substrate molecules
(b) After saturation of those enzyme molecules these are no free enzyme molecules to bind with additional substrate molecules
(c) A & B
(d) After saturation of those enzyme molecules, enzyme get changed in it's form.

138. When the binding chemical shut off enzyme activity, the process is called _____ and the chemical is called_____
(a) Inhibition; inhibitor

(b) Inhibition; cofactors
(c) Exhibition, exhibitor
(d) None of these

139. What effect is observe on enzyme activity due to inhibitor
(a) It fasten enzyme kinetics
(b) It decline enzyme kinetics
(c) It shut off enzyme kinetics
(d) No effect on enzyme kinetics

140. Inhibition of succinic dehydrogenase by malonate is due to
(a) Malonate closely resembles with substrate succinate in structure
(b) Malonate is competitive inhibitor
(c) It binds with active site of succinic dehydrogenase in place of substrate
(d) All of these

141. Competitive inhibitors are often used in the control of –
(a) Viral pathogen
(b) Bacterial pathogen
(c) Both A & B
(d) None of these

Topic 12.5 Classification & Nomenclature of Enzyme

142. Enzyme are divided into how many classes-
(a) 2 (b) 4
(c) 6 (d) 8

143. Each classes of enzyme were further classification into______ subclass and named by ___ digit
(a) 13; 4 – 13 (b) 4 – 13; 13
(c) 4 – 13; 4 (d) 4; 4 – 13

144. S reduced + S' oxidised → S oxidised + S' reduced
(a) Oxidoreductase (b) Dehydrogenase
(c) Transferase (d) A & B both

145. Enzyme catalysing a transfer of a group i.e. hydrogen between pair of substrate S and S' is-
(a) Transferase (b) Oxidoreductase
(c) Lyases (d) Ligases

146. Transferase enzyme catalyse a transfer of G between pair substrate S & S'. G is other than –
(a) Oxygen (b) Amino
(c) Hydrogen (d) Carbon

147. Hydrolases catalyse –
i) Hydrolysis of ester, ether, peptide, glycosidic,
ii) C – C breakdown
iii) C – halide breakdown
iv) P – N breakdown
(a) (i) only
(b) (i) & (ii) only
(c) (iii) & (iv) only
(d) i, ii, iii & iv

148. Lysase catalyse _ _ _ _ _ of groups from substrates by mechanism other than hydrolysis leaving _ _ _
A. Addition ; double
B. Removal ; double
C. Addition ; single
D. Removal ; triple

149. Isomerases catalyse inter-conversion of:
A. Optical isomer
B. Geometrical isomer
C. Position isomer
D. All of these

150. Linking of two compound is achived by-
(a) Lyases (b) Transferase
(c) Ligases (d) Hydrolase

151. Ligase catalyse-
(a) Joining of C-O
(b) Oxidation – reduction of substrate
(c) Hydrolysis of C-C
(d) Conversion of optical isomer

Topic 12.6 Co-factors

152. Cofactors are:-
(a) Proteinous part of enzyme
(b) Non-proteinous part of enzyme
(c) Bound to substrate
(d) Bound to enzyme to make enzyme catalytically retard

153. How many kind of cofactors may be identified-
(a) 1 (b) 2
(c) 3 (d) Zero

154. Cofactors are _ _ _ _ _ _ and apoenzyme are _ _ _ _ part of enzyme.
(a) Protein; protein
(b) Non-protein; non-protein
(c) Protein; non-protein
(d) Non-protein; protein

155. Prosthetic group are _ _ _ _ _ _ and are distinguished from other cofactors in that they are _ _ _ _ _ bound to apoenzyme.
(a) Organic compound; tightly
(b) Organic compound; loosely
(c) Inorganic compound; loosely
(d) Inorganic compound; tightly

156. Which of following is/are correct?
(i) Haem is prosthetic group.
(ii) Haem is apoenzyme.
(iii) Haem is not part of active site of peroxidase.
(iv) Haem catalyse the formation of hydrogen peroxide from water & oxygen.
(v) Haem is part of active site of peroxidase.
(vi) Haem catalyse the breakdown of hydrogen peroxide into water & oxygen.
(a) i , iii , vi (b) ii , iv , v
(c) i , v , vi (d) ii , v , vi

157. NAD & NADP contain-
(a) Vitamin niacin (b) Vitamin C
(c) Vitamin D (d) Vitamin K

158. Full form of NAD is:-
(a) Nicotinamide adenine nucleotide
(b) Nicotinamide adenine dinucleoside
(c) Nicotinamide adenine dinucleotide
(d) Nicotinamide adenine nucleoside

159. Choose correct response from following with respect to carboxypeptidase.
(a) Zinc are found as apoenzyme
(b) It is proteolytic enzyme
(c) Cofactor from covalent bond with side chain at active site
(d) Between cofactor and substrate ionic bond is formed

160. How many coordination found in activity of carboxypeptidase?
(a) Only one ; between cofactor and side chain at active site
(b) Two between cofactor and side chain at active site and at to many ; same time form one or more bond with substrate.
(c) Zero
(d) Only one ; between cofactor & substrate

161. Find mismatch.

	Column-I		Column-II
(a)	Carboxypeptidase	(i)	Zinc
(b)	NADP	(ii)	Niacin
(c)	Haem	(iii)	Peroxidase
(d)	NAD	(iv)	Zinc

162. When cofactor is removed from enzyme; what effect is observed.
(a) Catalytic activity lost
(b) Catalytic activity enhance
(c) Catalytic activity fix at optimum
(d) None of these

ANSWER KEY

1. (c) 2. (a) 3. (a) 4. (b) 5. (b) 6. (d) 7. (c) 8. (c) 9. (a) 10. (d)
11. (a) 12. (c) 13. (b) 14. (c) 15. (a) 16. (b) 17. (c) 18. (a) 19. (d) 20. (c)
21. (c) 22. (d) 23. (d) 24. (d) 25. (c) 26. (a) 27. (d) 28. (a) 29. (a) 30. (b)
31. (a) 32. (b) 33. (b) 34. (b) 35. (d) 36. (d) 37. (d) 38. (d) 39. (a) 40. (b)
41. (a) 42. (b) 43. (b) 44. (c) 45. (d) 46. (c) 47. (b) 48. (b) 49. (a) 50. (a)
51. (a) 52. (d) 53. (b) 54. (d) 55. (c) 56. (c) 57. (d) 58. (d) 59. (a) 60. (c)
61. (b) 62. (d) 63. (a) 64. (b) 65. (c) 66. (a) 67. (b) 68. (b) 69. (c) 70. (d)
71. (c) 72. (c) 73. (b) 74. (c) 75. (d) 76. (b) 77. (c) 78. (c) 79. (d) 80. (d)
81. (c) 82. (a) 83. (d) 84. (b) 85. (b) 86. (a) 87. (b) 88. (c) 89. (d) 90. (b)
91. (a) 92. (b) 93. (c) 94. (d) 95. (a) 96. (b) 97. (c) 98. (c) 99. (c) 100. (a)
101. (b) 102. (b) 103. (b) 104. (a) 105. (d) 106. (d) 107. (c) 108. (b) 109. (a) 110. (d)
111. (b) 112. (d) 113. (d) 114. (c) 115. (a) 116. (d) 117. (d) 118. (a) 119. (a) 120. (c)
121. (d) 122. (c) 123. (c) 124. (d) 125. (c) 126. (c) 127. (b) 128. (c) 19. (d) 130. (d)
131. (d) 132. (c) 133. (b) 134. (d) 135. (d) 136. (b) 137. (c) 138. (a) 139. (c) 140. (d)
141. (b) 142. (c) 143. (c) 144. (a) 145. (a) 146. (c) 147. (d) 148. (b) 149. (d) 150. (c)
151. (a) 152. (b) 153. (b) 154. (d) 155. (a) 156. (c) 157. (a) 158. (c) 159. (b) 160. (b)
161. (d) 162. (a)

4 Chapter

Digestion and Absorption

MULTIPLE CHOICE QUESTIONS

Human Physiology

1. What increases the use of physico – chemical concepts and techniques:-
(a) Forward approach
(b) Reductionist approach
(c) Both
(d) None of these

2. Majority of physico – chemical studies employed by employing:-
(a) Tissue model
(b) Cell free system
(c) Both
(d) None of these

3. Now a days it is realized that ________ would reveal the truth about biological processes or living phenomenon :-
(a) Purely organismic level
(b) Purely reductionistic molecular approach
(c) Both
(d) None

4. All living phenomenon are emergent properties due to ________.
(a) Interaction among components of the system.
(b) Defoliation of organs.
(c) Exchange of gases only.
(d) All of these

5. How many of the following creates emergent properties of living organism:-
i) Regulatory network of molecules
ii) Supra – molecular assemblies of cells; tissue; organs
iii) Population
iv) Communities
(a) Only two (b) Only three
(c) Only one (d) all of them

Digestion and Absorption Introduction

6. Which of the following components of our food are taken in small quantities?
(a) Carbohydrate and proteins
(b) Proteins and minerals
(c) Proteins and lipids
(d) Minerals and vitamins

7. Which of the following molecules can be used by us as a source of energy?
(a) Carbohydrates only
(b) Fats only
(c) Carbohydrates or fats
(d) Carbohydrates, fats and vitamins

8. Digestion is –
(a) Absorption of diffusible food
(b) Absorption of water
(c) Throwing out of non-diffusible food substances
(d) Conversion of non-diffusible complex food substances into simple absorbable forms

Topic 1	Digestive System- (Alimentary Canal)

9. Dental formula of adult person is-
(a) 2122/2122 (b) 2114/2114
(c) 2123/2123 (d) 2123/2124

10. Our teeth are –
(a) Acrodont and homodont

(b) Homodont and polyphyodont
(c) Thecodont, diphyodont and heterodont
(d) Acrodont, homodont and polyphyodont

11. Frenulum is –
(a) Adenoid present on pharyngeal wall
(b) Tonsils located on lateral wall of soft palate
(c) Fold attaching tongue to the floor of oral cavity
(d) V-shaped sulcus for terminalis on tongue

12. The hard chewing surface of teeth helping in mastication of food is called –
(a) Dentine (b) Frenulum
(c) Root (d) Enamel

13. The upper surface of the tongue has small projections, some of which bear taste buds. These projections are called- 4 Digestion and absorption
(a) Papillae (b) Taste pore
(c) Frenulum (d) Sulcus terminalis

14. The common passage for food and air is–
(a) Gullet (b) Glottis
(c) Larynx (d) Pharynx

15. The oesophagus and trachea (wind pipe) open into-
(a) Gullet (b) Glottis
(c) Larynx (d) Pharynx

16. A thin long tube extending posteriorly and passing through neck, thorax and a diaphragm and leading to stomach is called-
(a) Pharynx (b) Trachea
(c) Oesophagus (d) Larynx

17. Our stomach is -
(a) U-shaped (b) J-shaped
(c) C-shaped (d) Rod-shaped

18. A muscular sphincter regulating opening of oesophagus into the stomach is called
(a) Pyloric sphincter
(b) Cardiac Sphincter
(c) Sphincter of Oddi
(d) Boyden sphincter

19. Cardiac sphincter is –
(a) Gastro-oesophageal sphincter
(b) Pyloric sphincter
(c) Gastro-duodenal sphincter
(d) None

20. The stomach is located in the upper ____ portion of the ____ cavity-
(a) Right, thoracic
(b) Left abdominal
(c) Right, abdominal
(d) Left, thoracic

21. The narrow distal part of stomach leading to the intestine is called –
(a) Cardiac (b) Pyloric
(c) Fundus (d) None

22. The proximal part of stomach in which oesophagus opens is called -
(a) Cardiac (b) Pyloric
(c) Fundus (d) None

23. Which of the following is not the part of stomach?
(a) Caecum (b) Pyloric
(c) Fundus (d) Cardiac

24. Small intestine is distinguishable into 3 parts, a 'C' shaped ____, a long coiled middle portion ____ and a highly coiled ____.
(a) Ileum, jejunum, duodenum
(b) jejunum, Duodenum, ileum
(c) Duodenum, jejunum, ileum
(d) Caecum, duodenum, ileum

25. The opening of stomach into duodenum is guarded by-
(a) Cardiac sphincter
(b) Sphincter of Boyden
(c) Sphincter of Oddi
(d) Pyloric sphincter

26. Ileum is –
(a) First part of small intestine
(b) Last part of small intestine
(c) Middle part of small intestine
(d) First part of large intestine

27. Which of the following parts of small intestine opens into large intestine?
(a) Duodenum (b) Ileum
(c) Jejunum (d) Colon

28. All of the following is the part of large intestine except -
(a) Ileum (b) Caecum
(c) Colon (d) Rectum

29. Caecum is small blind sac which hosts some symbiotic micro-organisms. From it a small finger like vestigial organ arises. This organ is called -
(a) Parotid gland
(b) Vermis
(c) Vermiform appendix
(d) Lacteals

30. Caecum opens into -
(a) Rectum (b) Duodenum
(c) Colon (d) Jejunum

31. Which of the following organs has 3 parts (ascending, transverse and descending parts) -
(a) Colon (b) Caecum
(c) Small intestine (d) Large intestine

32. Which of the following sequence is correct?
(a) Descending part of colon → Rectum→Anus
(b) Stomach→ Jejunum→ Duodenum
(c) Ileum→ Colon→ Caecum
(d) Colon→ Anus→ Rectum

33.

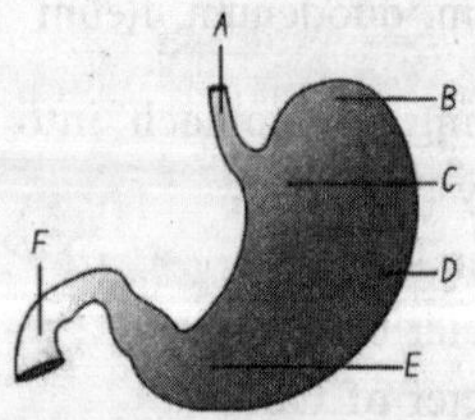

Anatomical regions of stomach are-
(a) B- Fundus, C- Cardiac, D- Body, EPyloric
(b) B- Cardiac, C- Fundus, D- Body, EPyloric
(c) B- Fundus, C- Cardiac, D- Pyloric, EBody
(d) B- Fundus, C- Body, D- Cardiac, EPyloric

34. The wall of alimentary canal from oesophagus to rectum posses four layers. The sequence of these layers is –
(a) Serosa-Mucosa-Submucosa- Muscularis
(b) Muscularis-Serosa-Mucosa- Submucosa
(c) Serosa-Muscularis-Mucosa- Submucosa
(d) Serosa-Muscularis-Submucosa- Mucosa

35. The below diagram represents the TS of Gut. Identify A, G, D and E –

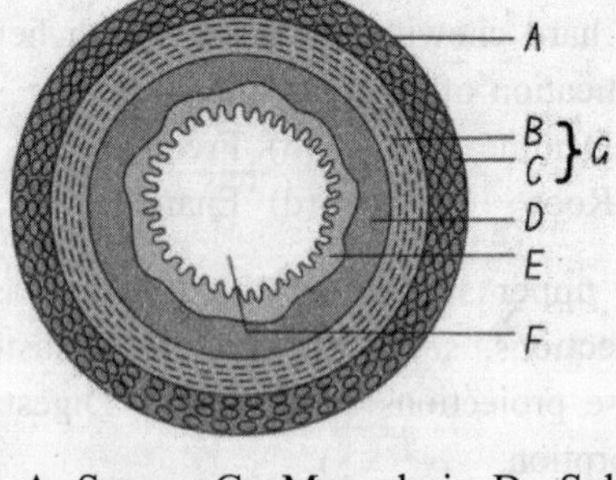

(a) A- Serosa; G - Muscularis; D - Submucosa; E – Mucosa
(b) A- Muscularis; G - Serosa; D - Submucosa; E – Mucosa
(c) A- Serosa; G - Muscularis; D - Mucosa; E – Submucosa
(d) A- Serosa; G - Submucosa; D - Muscularis; E – Mucosa

36. Epiglottis is a cartilaginous flap which prevents the entry of food into –
(a) Glottis (b) Gullet
(c) Oesophagus (d) None of the above

37. Duodenal glands/Brunner's glands are present in -
(a) Submucosa (b) Mucosa
(c) Muscularis (d) Serosa

38. Mucosa forms irregular folds (rugae) in the-
(a) Ileum (b) Stomach
(c) Jejunum (d) Colon

39. Mucosa forms many small finger like villi in the -
(a) Stomach (b) Colon
(c) Caecum (d) Small intestine

40. The many projections on the wall of small intestine function to –
(a) Secrete digestion enzymes
(b) Increase the surface area
(c) Hold products of digestion so they do not enter the large intestine
(d) Hold mucus, so ulcers do not form

41. Which layer of the gut is responsible for peristalsis?
(a) Smooth muscles (b) Mucosa
(c) Submucosa (d) Serosa

42. Which of the following statement is false?
(a) Mucosal epithelium has goblet cells which secrete mucus for lubrication
(b) Mucosa forms gastric glands in the stomach and crypts in between the bases of villi in intestine
(c) Cells lining the villi have brush border or microvilli
(d) All the four basic layer in the wall of gut never show modification in different parts of the alimentary canal

43. Lacteals, lymph capillaries are found in-
(a) Spleen (b) Intestinal villi
(c) Salivary gland (d) Mammary gland

44. Intestinal villi are supplied with –
(a) Only blood capillaries
(b) Only lacteals
(c) Lacteals and valves
(d) Blood capillaries and lacteals

45. The below diagram represents a section of small intestinal mucosa showing villi. Identify A, B, C and D –

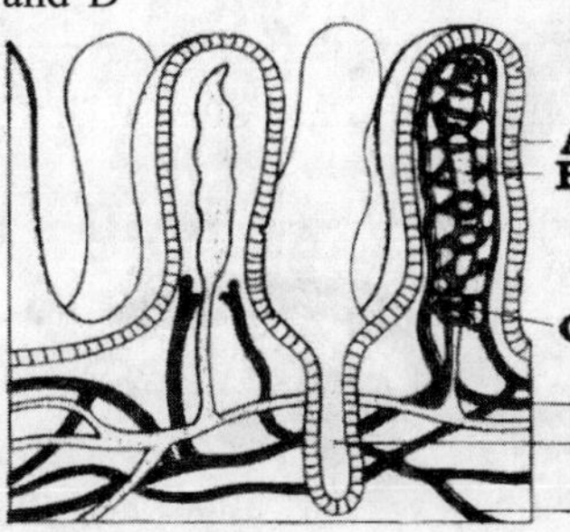

(a) A- Villi, B - Lacteal, C - Capillaries, D – Crypts
(b) A- Lacteal, B - Villi, C - Capillaries, D - Crypts
(c) A- Villi, B - Lacteal, C - Crypts, D – Capillaries
(d) A- Crypts, B - Lacteal, C – Capillaries, D – Villi

Topic 2 Digestive System- (Digestive Glands)

46. Number of salivary glands present in human being is –
(a) 5 pairs (b) 3 pairs
(c) 4 pairs (d) 2 pairs

47. Parotid glands are located below –
(a) Eye (b) Tongue
(c) Floor of mouth (d) In cheek near ear

48. Which of the following salivary gland is absent in human beings?
(a) Zygomatic
(b) Parotids
(c) The sub-maxillary/sub-mandibular (lower jaw)
(d) The sub-linguals (below the tongue)

49. Saliva is secreted by -
(a) Liver
(b) Gastric gland
(c) Duodenal gland
(d) None

50. Which one is the largest gland?
(a) Liver (b) Pancreas
(c) Salivary gland (d) Gastric gland

51. Liver secretes?
(a) No digestive enzymes
(b) Many digestive enzymes
(c) Hormones
(d) Succus entericus

52. Liver of man is-
(a) Bilobed (b) 3-lobed
(c) 4-lobed (d) 5-lobed

53. Digestive juice lacking enzyme but aiding digestion is -
(a) Chyle (b) Chyme
(c) Bile (d) Succus entericus

54. In adult human liver weighs -
(a) 2 kg (b) 2-3 kg
(c) 500 g (d) 1.2 to 1.5 kg

55. Liver is situated in -
(a) Thoracic cavity
(b) Above the thoracic cavity
(c) In abdominal cavity below diaphragm
(d) In abdominal cavity above diaphragm

56. Which of the following is the structural and functional unit of liver?
(a) Hepatic cells (b) Hepatic cord
(c) Hepatic lobule (d) Hepatic lobe

57. Find out the correct match –

	Column I		Column II
A	Hepatic lobule	I	Base of Villi
B	Crypts of leiberkuhn	II	Glisson's capsule
C	Sphincter of Oddi	III	Gall bladder
D	Cystic duct	IV	Hepato- pancreatic duct

(a) A-II, B-I, C- IV, D-III
(b) A-I, B-II, C- IV, D-III
(c) A-I, B-II, C-III, D- IV
(d) A - IV, B-III, C-II, D-I

58. Hepatocytes secrete -
(a) Lipase
(b) Bile, no digestive enzymes
(c) Bile with digestive enzymes
(d) Amylopsin

59. Bile is produced by -
(a) Gall bladder (b) Liver
(c) Hepatic duct (d) Blood

60. Cystic duct arises from -
(a) Liver (b) Kidney
(c) Pancreas (d) Gall bladder

61. Function of gall bladder is –
(a) Storage of bile
(b) Secretion of bile
(c) Formation of digestive enzyme
(d) Formation of bile salts

62. Common bile duct is formed when
(a) Right and left hepatic ducts are fused
(b) Bile duct is fused with pancreatic duct
(c) Cystic duct is fused with right hepatic duct.
(d) Cystic duct (duct of gall bladder) is fused with a common hepatic duct

63. In human beings which of the following opens into the duodenum -
(a) Hepatic duct and pancreatic duct separately
(b) Hepato-pancreatic duct
(c) 1st hepatic duct, then pancreatic duct
(d) 1st pancreatic duct then hepatic duct

64. Which of the following is incorrect about pancreas?
(a) It is compound gland as it has both exocrine and endocrine part
(b) Exocrine part secretes alkaline pancreatic juice having enzymes
(c) Endocrine part secretes hormones like insulin and glucagon
(d) It is surrounded by Glisson's capsule

65. The below diagram is a duct system of liver, gall bladder and pancreas. Write the names of ducts from A to D –

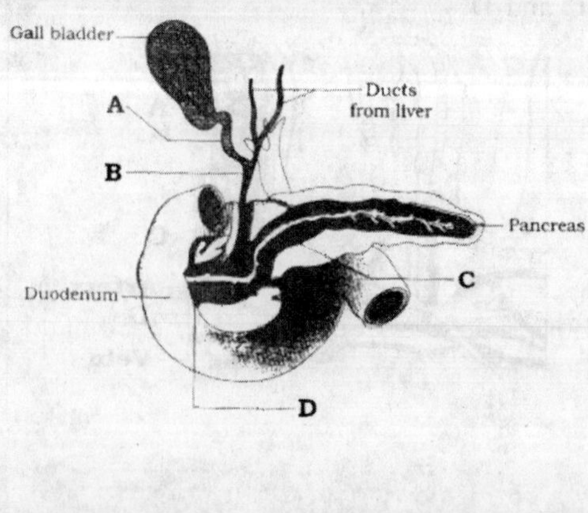

(a) Cystic duct, B - Bile duct, C - Pancreatic duct, D - Hepatopancreatic duct
(b) A- Bile duct, B - Cystic duct, C - Pancreatic duct, D - Hepatopancreatic duct
(c) A- Cystic duct, B - Bile duct, C - Hepato-pancreatic duct, D - Pancreatic duct.
(d) A- Cystic duct, B - Pancreatic duct, C - Bile duct, D - Hepato-pancreatic duct

Topic 3 Digestion of Food

66. The process of digestion is accomplished by:-
(a) Mechanical process
(b) Chemical process
(c) Both
(d) Chemical & Electrical

67. Mastication of food & facilitation of swallowing is the two major function of:-
(a) Teeth (b) Buccal Cavity
(c) Mouth (d) Trachea

68. _____(i)_____ in saliva helps in _____(ii)_____ & _____(iii)_____ the masticated food:-
(a) (i) Mucus (ii) Lubricating (iii) Adhering
(b) (i) Adhering (ii) Cohesion (iii) Surface tension
(c) (i) Surface tension (ii) Cohesion (iii) Adhesion
(d) (i) Lubrication (ii) Mucus (iii) Adhering

69. The bolus is conveyed into the pharynx and then into the ____(i)____ by ____(ii)
(a) (i) Swallowing (ii) Deglutition
(b) (i) Deglutition (ii) Swallowing
(c) (i) Oesophagus (ii) Deglutition
(d) (i) Oesophagus (ii) Mastication

70. The muscular contraction in oesophagus is known as:-
(a) Swallowing (b) Peristalsis
(c) Churning (d) Both (b) & (c)

71. What controls the passage of food into the stomach:-
(a) Gastro – oseophageal sphincter
(b) Pyloric sphincter
(c) Mucus in saliva
(d) All of the above

72. The salvia secreted into the oral cavity contains:-
→ Water; Amylase; Ptyalin; Lysozymes; Na+; K+; Cl-; HCO-3
How many of the above are composition of saliva:-
(a) Only 6 (b) Only 7
(c) Only 8 (d) Only 5

73. Which enzyme is responsible for initiation of digestion in the oral cavity:-
(a) Water splitting complex
(b) Mucus splitting enzyme
(c) Carbohydrate splitting enzyme
(d) Protein splitting enzyme

74. What percentage of starch is hydrolysed in oral cavity:-
(a) 20% (b) 30%
(c) 40% (d) 50%

75. In oral cavity starch is pyrolysed into ____ (i)____ ____(ii)____
(a) A monosaccharide; Maltose
(b) A disaccharide; Maltose
(c) A disaccharide; Galactose
(d) None of the above

76. Optimum pH required for the activation of carbohydrate – splitting enzyme is :-
(a) 5.8 (b) 6.8
(c) 7.8 (d) 4.8

77. Antibacterial agent present in saliva; that protects from bacterial infection is:-
(a) Ptyalin (b) Amlylase
(b) Lysozymes (d) Both

78. What major types of cells does the gastric gland contains:-
(a) Mucus neck cells
(b) Peptic of chief cells
(c) Parietal oxyntic cells
(d) All of the above

79. Factor essential for digestion of vitamin B_{12} is secreted by ____(i)____ & the factor is ____(ii)____
(a) (i) peptic cell (ii) Lysozyme
(b) (i) Intrinsic (ii) Peptic cell
(c) (i) Oxyntic (ii) Intrinsic
(d) (i) Parietal cell (ii) HCl

80. How many of the following statements are correct:-
(i) Proenzyme (Pepsinogen) is secreted by chief cells
(ii) Stomach stores food for 4 – 5 hours
(iii) Food thoroughly mixed up with acidic gastric juice is known as chyme.
(a) Only one (b) Only two
(c) All of them (d) None of them

81. Pepsinogen on the exposure of ____(i)____ converted into the active enzyme ____(ii)____
(a) (i) Proenzyme (ii) Lysozyme
(b) (i) HCl (ii) Pepsin
(c) (i) Lysozyme (ii) HCl
(d) (i) Churning (ii) Pepsin

82. Pepsin converts ____(i)____ into ____(ii)____ & ____(iii)____
(a) Proteins; Proteoses; Peptones
(b) Proteoses; Peptones; Proteins
(c) Peptones; Proteins; Proteoses
(d) Peptones; Proteoses; Proteins

83. What prevent the gastric epithelium from excoriation:-
(a) Mucus & Bicarbonates
(b) Bicarbonates only
(c) Mucus only
(d) HCl

84. The acidity in stomach for activation of pepsinogen required is
(a) 1.8 (b) 3.8
(c) 6.8 (d) 7.8

85. The proteolytic enzyme found in the milk for infants are:-
(a) Pepsin (b) Lectin
(c) Rennin (d) None of these

86. Lipases are also secreted by gastric glands in:-
(a) Small amount (b) Moderate amount
(c) Large amount (d) None of the above

87. ________ of movements are generated by the muscularis layer of the small intestine.
(a) A certain type
(b) Various type
(c) Churning type
(d) None of the above

88. How many of the following is released into the small instestine:-
(i) Bile juice
(ii) Gastric juice
(iii) Pancreatic juice
(iv) Intestinal juice
(a) Only One (b) Only Two
(c) Only Three (d) Only Four

89. Which of the following guards the release of Pancreatic and Bile juice into duodenum:-
(a) Plyloric sphincter
(b) Hepato – Pancreatic duct
(c) Sphincter of oddi
(d) Dust of Santorini

90. The contents of Pancreatic juice are:-
→ Trypsinogen; Chymotrypsinogen; Pepsinogen; Pro – carboxypeptidase; amylases; Lipases; nucleases
(a) All seven of the above
(b) Only five of the above
(c) Only six of the above
(d) Only four of the above

91. What are inactive enzymes of Pancreatic juice:-
Trypsinogen; Chymotrypsinogen; Pepsinogen; Pro – carboxypeptidase; amylases; Lipases; nucleases
(a) All seven of the above
(b) Only five of the above
(c) Only three of the above
(d) Only four of the above

92. What activates the enzymes of pancreas:-
(a) Enterokinase & Pepsin
(b) Enterokinase & Trypsin
(c) Enterokinase & HCl
(d) Chymotrypsin & Enterokinase

93. Bile released into duodenum contains the :-
(a) Bile salt (b) Bile pigment
(c) Both (d) Goblet cells

94. Intestinal mucosa secretes enzyme:-
(a) Lysozyme (b) Enterokinase
(c) Mucus (d) Both (b) & (c)

95. The composition of bile salt is:-
(a) Bilirubin & Biliverdin
(b) Bicarbonates & Cholesterol
(c) Phospholipids
(d) None of the them

96. The breaking down of fats into very small micelles is known as:-
(a) Digestion (b) Pyrolysis
(c) Emulsification (d) Absorption

97. The goblet cells of intestinal mucosal epithelium secretes:-
(a) Enterokinase (b) Mucus
(c) Lipase (d) All of the above

98. Succus entericus contains:-
(i) Disaccharide & Lipase
(ii) Dipeptidase & Nucleosidase
(iii) Mucus
(a) Only (i) & (ii) (b) All
(c) Only (ii) & (iii) (d) None

99. What is the pH of intestinal juice:-
(a) 7.0 (b) 7.8
(c) 6.0 (d) 6.8

100. Succus entericus is the combination of secretion of :-
(a) Mucus cells
(b) Brush bordered cells
(c) Both
(d) None

101. Which provides alkaline medium for enzymatic activities:-
(a) Mucus (b) Bicarbonates
(c) Both (d) None

102. Brunner's glands helps in :-
(a) Secretion of HCl
(b) Providing an Neutral medium
(c) Providing an alkaline medium
(d) Secretion of proteoses

103. Which of the following is partially hydrolysed protein:-
(a) Proteoses (b) Peptones
(c) Chyme (d) All

104. Which of the following statement is incorrect:-
(a) Carbohydrates in chyme is hydrolysed by salivary amylase.
(b) Fats are broken down by lipases.
(c) Bile helps in the break down process of fats
(d) None of the above

105. Which of the following is correct:-
(a) Final steps of digestion occur very close to the mucosal epithelium of the intestine
(b) Nucleic acids in pancreatic juice acts on nucleases to form nucleotides & Nucleosides
(c) Succus entericus acts on the start products of chyme
(d) All

106. Which of the following reaction in duodenal region:-
(a) Lactose $\xrightarrow{\text{Lactase}}$ Glucose + Galactose
(b) Nucleic acids $\xrightarrow{\text{Nucleases}}$ Nucleotides
(c) Starch $\xrightarrow{\text{amylase}}$ Disaccharids
(d) All

107. The undigested and unabsorbed substances are passed on to:-
(a) Jejunum (b) Ileum
(c) Caecum (d) Duodenum

108. Which of the following statements is incorrect:-
(a) No digestion occurs in the large intestine

(b) In large intestine absorption of water; minerals & certain drugs occurs.
(c) Mucus helps in adhesion of undigested particles
(d) None of the above

109. The undigested, unabsorbed substances are called:-
(a) Chyme (b) Faeces
(c) Bolus (d) Gullet

110. The entry of food into the caecum from Ileum is prevented by:-
(a) Pyloric sphincter
(b) Sphincter of oddi
(c) Ileo – caecal valve
(d) None

111. Which is the temporary storage region for faeces:-
(a) Ileum (b) Caecum
(c) Colon (d) Rectum

112. Which of the following statements is incorrect:-
(a) The activity of GIT are under neural and hormonal control for proper coordination of different part.
(b) The sight, smell and presence of food in oral cavity can stimulate secretion of saliva.
(c) Gastric and intestinal secretions are stimulated by neural signals.
(d) None of the above

113. Muscular activity of different part of the alimentary canal can be moderated by:-
(a) Local mechanism (b)CNS
(c) Both A and B (d) PNS

114. Hormonal control of secretion of digestive juices is carried out by local hormones produced by
(a) Gastric mucosa
(b) Intestinal mucosa
(c) Intestinal submucosa
(d) Both (a) & (b)

115. Which of the following is correct match :-

	Substrate	GCV	PCV
A)	Carbohydrate	4.0	4.21
B)	Protein	4.0	5.65
C)	Fat	9.45	9.0
D)	All		

⇒ All the values are in k cal / gm
⇒ PCV → Physiological calorific value
⇒ GCV → Gross calorific value

Topic 4 Absorption Digested Products

116. Absorption occurs through:-
(a) Passive transport
(b) Active transport
(c) Facilitated method
(d) All

117. Absorption of glucose, amino acids & some electrolytes like chloride ions occurs through simple diffusion in ______
(a) Small amount
(b) Moderate amount
(c) Large amount
(d) None

118. Which of the following is true:-
(a) Passage of substances into blood stream depends upon the concentration gradient.
(b) Glucose and amino acids are absorbed facilitatively
(c) Transport of water depends upon osmotic gradient
(d) All of the above

119. Which of the following is true:-
(a) Fat droplets → Micelle → Chylomicron
(b) Fat droplet (Lumen) → Micelle (Mucosa) → Chylomicron (Lacteal)
(c) Fat droplet (Mucosa) → Micelle (Sub – Mucosa) → Chylomicron (Lacteal)
(d) None of them

120. Principle organ for absorption of nutrients is:-
(a) Mouth (b) Stomach
(c) Small Intestine (d) Large Intestine

121. The absorbed substances finally reach the tissues which utilised them for their activities its known as:-
(a) Assimilation (b) Absorption
(c) Deglutition (d) Defecation

Topic 5 Disorders of Digestive system

122. Which is the ejection of intestinal content through the mouth:-
(a) Jaundice (b) Diarrhoea
(c) vomiting (d) None

123. Irregular bowel movement causes:-
(a) Jaundice (b) Vomiting
(c) Constipation (d) Indigestion

124. Dietary deficiency of proteins and total food calories are wide spread in
(a) North & North – east Asia
(b) South America & Central Africa
(c) East & south – east Asia
(d) North America & central Africa

125. Marasmus occurs in:-
(a) Children more than a year in age
(b) Infant more than a year in age
(c) Foetus
(d) Infant less than a year in age

126. Kwashiorkar occurs in
(a) Children more than a year in age
(b) Infant more than a year in age
(c) Foetus
(d) Infant less than a year in age

ANSWER KEY

1. (b)	2. (b)	3. (b)	4. (a)	5. (d)	6. (d)	7. (a)	8. (d)	9. (c)	10. (c)
11. (c)	12. (d)	13. (a)	14. (d)	15. (c)	16. (c)	17. (b)	18. (b)	19. (a)	20. (b)
21. (b)	22. (a)	23. (a)	24. (c)	25. (d)	26. (b)	27. (b)	28. (a)	29. (c)	30. (c)
31. (a)	32. (a)	33. (a)	34. (d)	35. (a)	36. (d)	37. (a)	38. (b)	39. (d)	40. (b)
41. (a)	42. (c)	43. (b)	44. (d)	45. (a)	46. (b)	47. (d)	48. (a)	49. (d)	50. (a)
51. (a)	52. (a)	53. (c)	54. (d)	55. (c)	56. (a)	57. (a)	58. (b)	59. (b)	60. (d)
61. (a)	62. (d)	63. (b)	64. (d)	65. (a)	66. (c)	67. (b)	68. (a)	69. (c)	70. (b)
71. (a)	72. (b)	73. (c)	74. (b)	75. (b)	76. (b)	77. (b)	78. (d)	79. (c)	80. (c)
81. (b)	82. (a)	83. (a)	84. (a)	85. (c)	86. (a)	87. (b)	88. (c)	89. (b)	90. (a)
91. (a)	92. (b)	93. (c)	94. (c)	95. (d)	96. (c)	97. (b)	98. (b)	99. (b)	100. (c)
101. (b)	102. (c)	103. (d)	104. (a)	105. (a)	106. (d)	107. (c)	108. (d)	109. (b)	110. (c)
111. (d)	112. (d)	113. (c)	114. (d)	115. (c)	116. (d)	117. (a)	118. (d)	119. (a)	120. (c)
121. (a)	122. (c)	123. (c)	124. (b)	125. (d)	126. (a)				

Chapter 5

Breathing and Exchange of Gases

MULTIPLE CHOICE QUESTIONS

Topic 1 Introduction and Respiratory Organs

1. In which of the following gaseous exchange between O_2 and CO_2 occurs through general body surface?

(a) Sponges (b) Coelenterates
(c) Flatworms (d) All

2. Match the followings correctly.

	Animals		Respiratory Organs
A.	Earthworms	1.	Lungs
B.	Most aquatic arthropods	2.	Trachea
C.	Fishes	3.	Gills
D.	Birds/Reptiles	4.	Moist cuticle
E.	Insects		

(a) A-IV, B and C-III, D-I, E-II
(b) A- IV, B - III C and D -I, E - III
(c) A-II, B and C - III, D - I, E - IV
(d) A-III, B and C-I, D-II, E-IV

3. Amphibians e.g. frogs respire –

(a) Through moist skin
(b) Lungs
(c) Both a and b
(d) Trachea

Topic 2 Human Respiratory System

4. Which is the correct sequence of air passages in man?

(a) External Nostril→ Nasal passage→ Internal nostril→ Pharynx →Larynx→ Trachea → Bronchi → Bronchioles → Alveoli
(b) Nose→ Larynx→ Pharynx→ Bronchioles→ Bronchi → Alveoli
(c) Nose→ Pharynx →Trachea→ Larynx→ Bronchi → Bronchioles→ Alveoli
(d) Nose → Larynx→ Bronchi→ Pharynx→ Trachea→ Bronchioles →Alveoli

5. Which is correct about nasopharynx?

(a) Internal nostrils open into nasopharynx
(b) It is the common passage for both air and food
(c) It opens through gullet of the larynx region into the trachea
(d) All

6. Which of the following options is wrong about the larynx (sound box)?

(a) It is a bony box
(b) Glottis is the opening into the larynx
(c) During swallowing of food glottis is covered by epiglottis to prevent food entry into the larynx
(d) All

7. Trachea divides into right and left primary bronchi at ______ thoracic vertebra.

(a) 4 (b) 5
(c) 6 (d) 9

8. Incomplete cartilaginous rings support all of the following except-

(a) Trachea
(b) Primary, secondary and tertiary bronchi

(c) Respiratory bronchioles
(d) Initial bronchioles

9. Which of the following has the smallest diameter?
(a) Trachea
(b) Terminal bronchiole
(c) Tertiary bronchus
(d) Secondary bronchus

10. Lungs are comprised by –
(a) Only alveoli
(c) Pleura
(c) Different types of bronchi
(d) Network of bronchi, bronchioles and alveoli

11. Sites of gaseous exchange in lungs are –
(a) Alveoli (b) Tracheoles
(c) Bronchioles (d) Pleura

12.
I. It is double layered and covers the lungs
II. Fluid between the layers reduces friction on lung-surface
III. Outer layer is in contact with thoracic wall
IV. Inner layer is in contact with lungs
The above features refer to –
(a) Pericardium (b) Peritoneum
(c) Pleura (d) None

13. The part starting with the external nostrils upto the terminal bronchioles constitute the -
(a) Respiratory or exchange part of respiratory system
(b) Inspiratory part
(c) Conducting part
(d) Expiratory part

14. Respiratory or exchange part of the respiratory system consists of-
(a) The parts starting with external nostrils upto terminal bronchioles
(b) Alveoli and their ducts
(c) All bronchi and terminal bronchioles
(d) All bronchioles

15. The conducting part of the respiratory system has functions.
(a) Filter, warm and moisten the air
(b) Gaseous exchange
(c) Filtering the air only
(d) Warm the air

16. The chamber formed dorsally by the vertebral column, ventrally by sternum, laterally by ribs and on the lower side by dome-shaped diaphragm is -
(a) Abdominal cavity
(b) Thoracic cavity
(c) Pelvic cavity
(d) Cranial cavity

17. Respiration involves following steps –
(I) Diffusion of gases O_2 and CO_2 across alveolar membrane
(II) Transport of gases by blood
(III) Utilization of O_2 by cell for catabolic reactions and resultant release of CO_2
(IV) Pulmonary ventilation by which atmospheric air is drawn in and CO_2 rich alveolar air is released out
(V) Diffusion of O_2 and CO_2 between blood and tissues.
The correct sequence of steps is –
(a) I)→ II)→III)→ IV)→V)
(b) V)→ IV)→III)→ II)→I)
(c) IV)→ I)→II)→ V)→III)
(d) III)→II)→ V)→I)→ IV)

18. Study the given diagram and identify A, B, C and D.

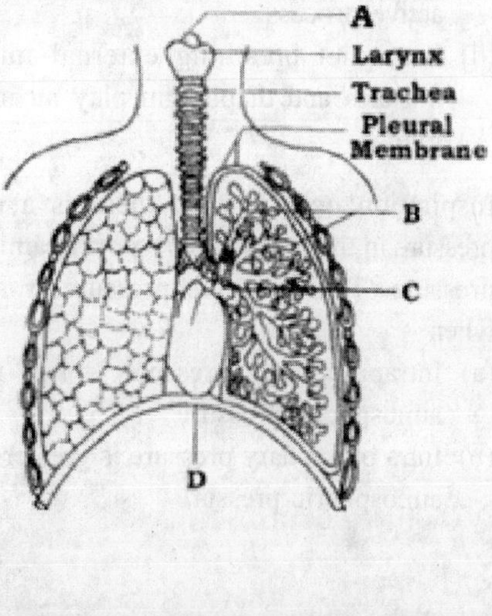

(a) A- Epiglottis, B -Alveoli, C- Bronchus, D - Diaphragm
(b) A- Epiglottis, B -Alveoli, C - Bronchioles, D - Diaphragm
(c) A- Sound box, B -Alveoli, C - Bronchus, D - Diaphragm
(d) A-Sound box, B-Alveoli, C- Bronchioles, D - Diaphragm

Topic 3 Mechanism of Breathing

19. The lungs expand in inspiration/inhalation because –
(a) Diaphragm contracts upward
(b) The volume of thoracic cavity increases
(c) External intercostal muscles relax
(d) Diaphragm relaxes

20. The process of exhalation / expiration is begun mainly due to –
(a) The contraction of intercostal muscles
(b) The contraction of the diaphragm
(c) The relaxation of muscles
(d) Low pressure in thoracic cavity

21. Which of the following statements about the mechanism of ventilation/breathing is false?
(a) As the diaphragm relaxes, air is expelled from the respiratory system
(b) During inspiration the lungs act as suction pump
(c) Inspiration is a passive and expiration is an active process.
(d) For quiet breathing external intercostal muscles and diaphragm play an important role.

22. Inspiration occurs when there is a negative pressure in the lungs with respect to atmospheric pressure. This negative pressure is achieved when –
(a) Intrapulmonary pressure is less than the atmospheric pressure
(b) Intra pulmonary pressure is greater than the atmospheric pressure
(c) Intrapulmonary pressure is equal to the atmospheric pressure
(d) Intrapleural pressure becomes more than the intra-alveolar pressure

23. Expiration takes place when the intrapulmonary pressure is –
(a) Greater than the atmospheric pressure
(b) Lesser than the atmospheric pressure
(c) Equal to atmospheric pressure
(d) Equal to intrapleural pressure

24. Which of the following sequences is correct to initiate inspiration?
I. The contraction of external intercostal muscles raises the ribs and sternum
II. Volume of thorax increases in the dorso-ventral axis
III. Intrapulmonary pressure decreases
IV. Diaphragm contraction
V. Air rushes into lungs
VI. Volume of thorax increases in the anterio-posterior axis
(a) I, II, IV, V, III, VI
(b) I, II, III, IV, V
(c) I, II, IV, VI, III, V
(d) VI, I, II, III, V

25. Which of the following sequences is correct to initiate expiration?
I. Relaxation of external intercostal muscles and return of diaphragm and sternum to their normal position
II. Air expelled from lungs
III. Volume of thorax decreases
IV. Intrapulmonary pressure increases
(a) I, III IV, II (b) II, IV, III, I
(c) IV, III, II, I (d) I, II, III, IV

26. On an average, a healthy human breathes ______ times /minute-
(a) 20 - 40 (b) 1-6
(c) 12-16 (d) 16-25

27. Additional muscles for forceful breathing are –
(a) Diaphragm and external intercostal muscles

(b) Abdominal muscles and internal intercostal .muscles
(c) Diaphragm and abdominal muscles
(d) External and internal intercostal muscles

28. Following illustration depicts the mechanism of breathing. In which of the following option all the parts A. B, C and D are correctly labelled?
(a) A-Air entering into lungs; B- Ribs and sternum raised; C- Diaphragm contracted; D-Volume of thorax raised
(b) A - Air expelled from lungs; B - Ribs and sternum return to original position; C - Diaphragm relaxed; D - Volume of thorax decreased
(c) A-Air expelled from lungs; B- Ribs and sternum raised; C - Diaphragm relaxed; D -Volume of thorax decreased
(d) A-Air expelled from lungs; B- Ribs and sternum raised; C- Diaphragm contracted; D-Volume of thorax decreased

Topic 4 Respiratory Volumes and Capacities

29. Match the following –

	Column A		Column B
1.	Tidal Volume	A.	Tidal volume and inspiratory reserve volume and expiratory reserve volume
2.	Residual Volume	B.	Additional amount of air inhaled beyond tidal volume when taking a very deep breath
3.	Expiratory reserve volume	C.	Amount of air remaining in lungs after expiratory reserve volume is expelled
4.	Inspiratory reserve volume	D.	Tidal volume and inspiratory reserve volume
5.	Inspiratory Capacity	E.	Volume of air in one breath
6.	Vital Capacity	F.	Amount of air exhaled in forced exhalation

(a) 1 - C, 2 - E, 3 - B, 4 - F, 5 - D, 6 -A
(b) 1 - E, 2 - F, 3 - C, 4 - B, 5 -A, 6 - D
(c) 1 - E, 2 - C, 3 - F, 4 - B, 5 - D, 6 -A
(d) 1 - E, 2 - C, 3 - B, 4 - F, 5 -A, 6 – D

30. Match the following -

	Column A		Column B
1.	Tidal Volume	A.	2500-3000 mL of air
2.	Inspiratory reserve volume	B.	1000 mL of air
3.	Expiratory reserve volume	C.	500 mL of air
4.	Residual volume	D.	3400-4800 mL of air
5.	Vital Capacity	E.	1200 mL of air

(a) 1 - C, 2 - D, 3 - B, 4 - A, 5 - E
(b) 1 - C, 2 -A, 3 - B, 4 - E, 5 - D
(c) 1 - C, 2 -A, 3 - D, 4 - E, 5 - B
(d) 1 - E, 2 -A, 3 - B, 4 - E, 5 – D

31. Arrange the following in order of increasing volume –
1. Tidal volume
2. Residual volume
3. Expiratory reserve volume
4. Vital capacity
(a) 1 < 2 < 3 < 4 (b) 1 < 4 < 3 < 2
(c) 1 < 3 < 2 < 4 (d) 1 < 4 < 2 < 3

32. Different respiratory volumes are given below-
I. Tidal Volume= 500 ml
II. Residual Volume= 1000 ml
III. Inspiratory Reserve Volume = 2500 ml
IV. Expiratory Reserve Volume = 1000 ml
The functional residual capacity (FRC) is-
(a) 3500 ml (b) 2000 ml
(c) 600 ml (d) 3000 ml

33. Expiratory capacity is equal to –
(a) TV+ ERV (b) ERV+ IRV
(c) ERV+ RV (d) ERV+ RV

34. A spirometer cannot be used to measure –
(a) IC (b) RV
(c) ERV (d) IRV

35. The maximum volume of air you can forcefully exhale after taking the deepest possible breath is called

(a) Tidal volume
(b) Total respiratory volume
(c) Residual volume
(d) Vital capacity

Topic 5 Exchange of Gases

36. Exchange of gases –
(a) Occurs between the alveoli and pulmonary blood capillary
(b) Occurs between blood and tissues
(c) By diffusion
(d) All

37. Which of the following factors affect the diffusion of gases?
(a) Partial pressure of diffusing gases
(c) Solubility of gases
(c) The thickness of diffusion membrane
(d) All

38. Which of the following statements about the partial pressure of CO_2 is true?
(a) It is higher in alveoli than in pulmonary artery
(b) It is higher in the systemic arteries than in tissues
(c) It is higher in systemic veins than in systemic arteries
(d) It is higher in the pulmonary veins than in pulmonary arteries

39. The partial pressure of CO_2 in the venous blood is –
(a) Greater than in the tissue spaces
(b) Lesser than in the tissue spaces
(c) Lesser than in the arterial blood
(d) Less than in alveoli

40. A section of an alveolus with a pulmonary capillary indicates the presence of major layers constituting diffusion membrane-
(a) 3 (b) 2
(c) 6 (d) 10

41. Partial pressures (in mmHg) of O_2 in atmospheric air, alveoli deoxygenated blood, oxygenated blood and tissues are-
(a) 40, 95, 40, 104, 159
(b) 104, 40, 40, 95, 159
(c) 159, 104, 40, 95, 40
(d) 195, 104, 95, 40, 40

42. Partial pressure (in mm Hg) of CO_2 in atmospheric air, alveoli, deoxygenated blood, oxygenated blood and tissues are-
(a) 0.3, 40, 45, 40, 45
(b) 40, 45, 40, 45, 0.3
(c) 40, 40, 45, 45, 0.3
(d) 0.3, 45, 45, 40, 40

43. Name the blood vessels A to D-

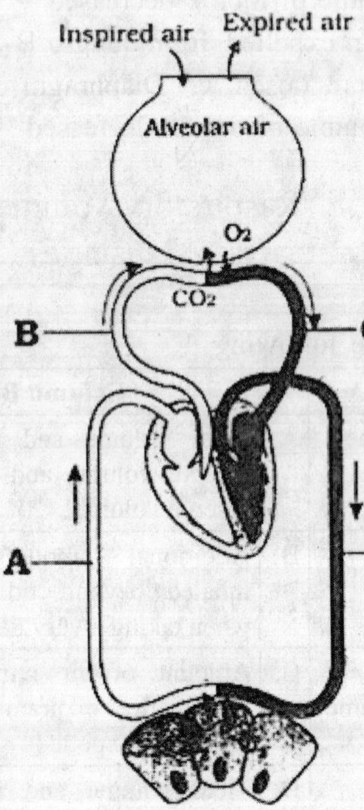

	A	B	C	D
A	Systemic vein	Pulmonary artery	Pulmonary vein	Systemic artery
B	Systemic artery	Pulmonary artery	Pulmonary vein	Systemic vein
C	Pulmonary artery	Systemic vein	Pulmonary vein	Systemic artery
D	Systemic vein	Pulmonary vein	Pulmonary artery	Systemic artery

44. In comparison to solubility of O_2 in blood the solubility of CO_2 is –
(a) 20 - 25 times lesser
(b) Slightly higher
(c) Slightly greater
(d) 20 - 25 times higher

45. Study the given figure and identify A to C.

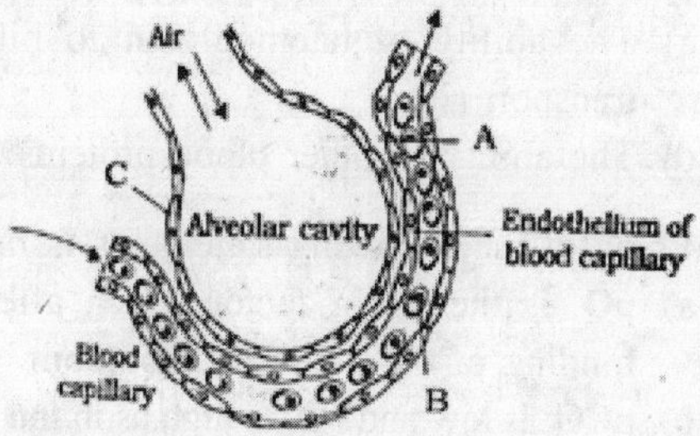

	A	B	C
(A)	Basement membrane	RBC	Alveolar wall
(B)	O_2	CO_2	Alveolar O_2
(C)	Pleura	RBC	Pericardium
(D)	Pleura	WBC	Pulmonary vein

46. The barrier between the air in alveolus and blood in pulmonary capillary consists of 3 layers and its total thickness is-
(a) 1 mm
(b) more than 1 mm
(c) much less than 1 mm
(d) 2 mm

Topic 6 Transport of Gases (Transport of Oxygen and Transport of Carbon dioxide)

47. Total percentage of O_2 transported by haemoglobin or RBC is –
(a) 3% (b) 97%
(c) 49% (d) 100%

48. Besides RBC blood plasma also carries O_2 in solution. The percentage is –
(a) 3% (b) 97%
(c) 49% (d) 25%

49. CO_2 is transported –
(a) By RBC
(b) As bicarbonates
(c) In a dissolved state through plasma
(d) All

50. The majority of CO_2 is transported as –
(a) Carbonates
(b) Bicarbonates
(c) Carbaminohaemoglobin
(d) Dissolved state in blood

51. Blood carries the CO_2 in 3 forms. The correct percentages of CO_2 in these forms are-

	As carbamino-haemoglobin in RBC	As bicarbonates	Dissolved form in plasma
A	20 -25%	70%	7%
B	70%	20 -25%	7%
C	20 -25%	7%	70%
D	7%	20 -25%	70%

52. Each molecule of haemoglobin when fully saturated carries how many molecules of O_2 –
(a) 1 (b) 2
(c) 4 (d) 20

53. Dissociation curve is associated with –
(a) Carbonic anhydrase
(b) CO
(c) $CHCl_3$
(d) Oxyhaemoglobin

54. Binding of O_2 with haemoglobin is primarily related to –
(a) pO_2 (b) pCO_2
(c) H^+ conc. (d) None

55. Besides pO_2 the other factors affecting the binding of O_2 with haemoglobin is/are–
(a) pCO_2 (b) H+conc.
(c) Temperature (d) All

56. Oxygen dissociation curve is –
(a) J-shaped (b) S-shaped
(c) L-shaped (d) Zig-zag

57. Which of the following statements is wrong?
(a) O_2 binds with haemoglobin in a reversible manner to form oxyhaemoglobin.
(b) Every 100 mL of oxygenated blood can deliver around 5 ml of O_2 to the tissue
(c) Occupational respiratory disorder are characterised by fibrosis (proliferation of fibrous tissues)
(d) None

58. What will be the pO_2 and pCO_2 in the atmospheric air compared to those in the alveolar?
(a) pO_2 lesser, pCO_2 higher
(b) pO_2 higher, pCO_2 lesser
(c) pO_2 higher, pCO_2 higher
(d) pO_2 lesser, pCO_2 lesser

59. Which of the following would have the same O_2 content?
(a) Blood entering the lungs - blood leaving the lungs
(b) Blood entering the right side of the heart - blood leaving the right side of the heart
(c) Blood entering the right side of the heart- blood leaving the left side of the heart
(d) Blood entering the tissue capillaries - blood leaving the tissue capillaries

60. CO_2 dissociates from carbaminohaemoglobin when –
(a) $pO_2\downarrow$, $pCO_2\downarrow$ (b) $pO_2\uparrow$, $pCO_2\uparrow$
(c) $pO_2\downarrow$, $pCO_2\uparrow$ (d) $pO_2\uparrow$, $pCO_2\downarrow$

61. Which of the following situations would result in the greatest degree of O_2 saturation for haemoglobin, assuming pO_2 remains constant –
(a) Increased CO_2 levels, decreased temperature
(b) Increased CO_2 levels, increased temperature
(c) Decreased CO_2 levels, decreased temperature
(d) Decreased CO2 levels, increased temperature

62. Which of the following factors favour the formation of oxyhaemoglobin in lungs?
(a) pO2 $\downarrow$, $pCO_2\uparrow$, H+$\uparrow$, Temperature$\uparrow$
(b) pO2 $\uparrow$, $pCO_2\uparrow$, H+$\downarrow$, Temperature$\uparrow$
(c) $pO_2\uparrow$, $pCO_2\downarrow$, H+$\downarrow$, Temperature$\downarrow$
(d) $pO_2\downarrow$, $pCO_2\uparrow$, pH$\uparrow$, Temperature$\downarrow$

63. All of the following favour the dissociation of oxyhaemoglobin to deliver O_2 to tissues except-
(a) $pO_2\uparrow$ (b) $pCO_2\uparrow$ OR H+$\uparrow$
(c) Temperature$\uparrow$ (d) $pO_2\downarrow$

64. The transport of CO_2 by the blood is primarily dependent on –
(a) The solubility of CO_2 in blood
(b) The presence of carbonic anhydrase in RBCs
(c) The ability of haemoglobin to bind and transport CO_2
(d) The ability of other blood proteins

65. Which of the following statements is false?
(a) pO_2 is the major factor which affects the binding of CO_2 with haemoglobin
(b) pCO_2 is low and pO_2 is high as in the tissues, more binding of CO_2 with Hb occurs
(c) RBC contains a very high conc. of carbonic anhydrase and minute quantities of the same in the plasma
(d) Every 100 mL of deoxygenated blood delivers approximately 4 mL of CO_2 to the alveoli.

66. Which of the following equation is correct?
(a) $CO_2 \rightarrow H_2SO_3 \rightarrow HCO_3^- + H^+$
(b) $CO_2 \rightarrow H_2O \overset{\text{Carbonic anhydrase}}{\rightleftharpoons} H_2CO_3$
$\overset{\text{Carbonic anhydrase}}{\rightleftharpoons} H^+ + HCO_3^-$
(c) $CO_2 \rightarrow H_2O \rightarrow CH_4 + 2O_2$
(d) $CO_2 \rightarrow H_2O \rightleftharpoons CO + H_2O_2$

67. Which of the following is incorrect about the given graph.

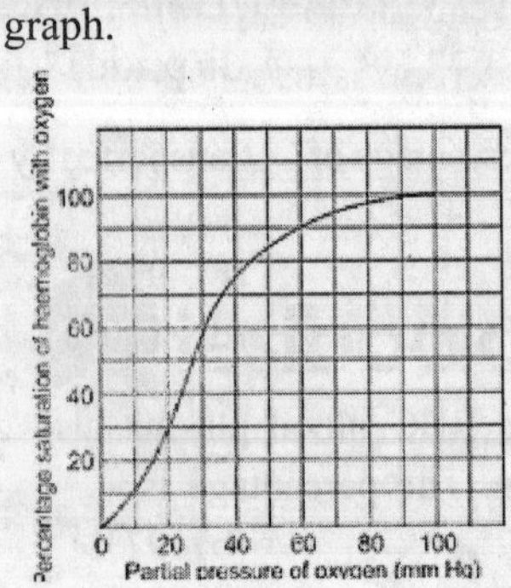

(a) Increase in partial pressure of CO_2 shift the curve to right.
(b) At low temperature the curve shifts to left.
(c) At high pH the curve shifts to right.
(d) Decrease in partial pressure of oxygen shifts the curve to right.

68. Assertion- The maximum pO_2 in alveoli is considerably less than in the atmosphere.
Reason- Lungs in mammals do not completely empty with each breath and inhalation occurs through the same airways as exhalation, so each inhalation mixes fresh air with oxygen depleted residual air.
(a) Both assertion and reason are true and reason is correct explanation of assertion.
(b) Both assertion and reason are true and reason is not correct explanation of assertion.
(c) Assertion is true but reason is false.
(d) Both assertion and reason are false.

69. How does an increase in the CO_2 concentration in the blood affect the pH of CSF?
(a) pH↑
(b) pH↓
(c) pH remains same.
(d) pH may increase or decrease.

70. Assertion- A drop in the blood pH causes an increase in heart rate.
Reason- Increased Heart Rate increases the rate at which CO_2 is delivered to the lungs, where CO_2 is removed.
(a) Both assertion and reason are true and reason is correct explanation of assertion.
(b) Both assertion and reason are true and reason is not correct explanation of assertion.
(c) Assertion is true but reason is false.
(d) Both assertion and reason are false.

Topic 7 Regulation of Respiration

71. Respiratory process is regulated by certain specialized centres in the brain. One of the following listed centres can reduce the inspiratory duration upon stimulation –
(a) Medullary inspiratory centre
(b) Pneumotaxic centre
(c) Chemosynthetic centre
(d) Apneustic centre

72. Pneumotaxic centre is present in –
(a) Pons (b) Medulla oblongata
(c) Cerebrum (d) Cerebellum

73. The breathing centre initiates ventilation in response to –
(a) A decrease in air pressure
(b) A decrease in O_2
(c) An increase in CO_2
(d) The rate of gas exchange in the alveoli

74. All of the following factors play role in the regulation of respiratory rhythm except –
(a) CO_2
(b) H^+ conc.
(c) O_2
(d) None of the above is correct

75. Receptors associated with aortic arch and carotid artery can recognise changes in _____ and ____ conc. and send necessary signal to _______ for remedial action.
(a) O_2, CO_2, Pneumothorax
(b) CO_2, H^+, rhythm centre
(c) CO_2, H^+, apneustic centre
(d) O_2, H^+, Pneumothorax

76. Respiratory centre of brain is stimulated by-
(a) CO_2 content in venous blood
(b) CO_2 content in arterial blood
(c) O_2 content in arterial blood
(d) O_2 content in venous blood

Topic 8 Disorders of Respiratory System-

77. Asthma is caused by –
(a) Infections of lungs
(b) Infection of trachea

(c) Spasm in bronchial muscles
(d) Infection in nose

78. One reason for emphysema is –
(a) Cigarette smoking
(b) Drug addiction
(c) Wine consumption
(d) Heavy exercise

79. Emphysema is characterised by –
(a) Permanent enlargement and destruction of alveolar area leading to reduction in respiratory surface
(b) Inhibition of respiratory centre
(c) Accumulation of fluid in lungs
(d) Spasm of muscles of trachea

80. Why do human beings have difficulty breathing at high elevations?
(a) O_2 makes up lower percentage of air there
(b) The temperature is lower there
(c) The barometric pressure is higher there
(d) pO_2 is lower there

81. Which of the following diseases are occupational respiratory disorder?
(a) Silicosis, Fibrosis and asbestosis
(b) Emphysema and mountain sickness
(c) Asthma and Emphysema
(d) Asthma and Hepatitis

82. If an injury tore a small hole in the membrane surrounding lungs, what effect on lung function would you expect?
(a) Pneumothorax with lung collapse
(b) Pneumothorax without lung collapse
(c) Silicosis with lung collapse
(d) Silicosis without lung collapse

ANSWER KEY

1. (d)	2. (a)	3. (a)	4. (a)	5. (a)	6. (c)	7. (b)	8. (c)	9. (b)	10. (d)
11. (a)	12. (c)	13. (a)	14. (a)	15. (a)	16. (b)	17. (c)	18. (a)	19. (b)	20. (c)
21. (c)	22. (a)	23. (a)	24. (a)	25. (a)	26. (c)	27. (b)	28. (a)	29. (c)	30. (b)
31. (c)	32. (a)	33. (a)	34. (b)	35. (d)	36. (d)	37. (d)	38. (d)	39. (c)	40. (a)
41. (c)	42. (c)	43. (a)	44. (d)	45. (a)	46. (a)	47. (b)	48. (a)	49. (d)	50. (b)
51. (a)	52. (c)	53. (d)	54. (a)	55. (d)	56. (b)	57. (d)	58. (b)	59. (b)	60. (c)
61. (d)	62. (c)	63. (a)	64. (a)	65. (b)	66. (b)	67. (d)	68. (b)	69. (b)	70. (a)
71. (b)	72. (a)	73. (c)	74. (c)	75. (b)	76. (a)	77. (c)	78. (a)	79. (a)	80. (d)
81. (a)	82. (a)								

6 Chapter

Body Fluids And Circulation

MULTIPLE CHOICE QUESTIONS

1. Which of the following use water from their environment as circulating fluid –
(a) Sponges (b) Coelenterates
(c) a and b (d) Fishes

Topic 1 Blood-Plasma and Formed Elements

2. Match the following –

	Column I		Column II
I.	Basophils	A.	Phagocytes
II.	Neutrophils	B.	Secrete histamine, serotonin, heparin and involved in inflammatory response
III.	Monocytes	C.	Resist infections and are also involved in allergic reaction
IV.	Eosinophils	D.	Immunity
V.	Lympho cytes	E.	

(a) I - B, II and III -A, IV - C, V - D
(b) I - B, II and III - C, IV -A, V - D
(c) I - C, II and III -A, IV - B, V - D
(d) I - D, II and III -C, IV -A, V – B

3. Megakaryocytes produce-
(a) Leucocytes
(b) Lymphocytes
(c) Bone cells
(d) Blood platelets (thrombocytes)

4. Which of the following is cell fragments?
(a) Leucocytes (b) RBCs
(c) Blood platelets (d) None

5. 1 mm^3 blood has how many blood platelets?
(a) 150000 - 350000
(b) 1.5 million to 3.5 million
(c) 1500-3000
(d) 10 to 15 lacs

6. During blood clotting, platelets release –
(a) Thrombin
(b) Fibrinogen
(c) Prothrombin
(d) Thrombokinase and other blood clotting factor

7. Find the correct descending order of percentage proportion of leucocytes in human blood.
(a) Neutrophils → Basophils → L → Lymphocytes → Acidophils → (Eosinophils) Monocytes
(b) Neutrophils → Monocytes → Lymphocytes → Acidophils → Basophils
(c) Neutrophils →Lymphocytes→ Monocytes → Acidophils → Basophils
(d) Neutrophils → Acidophils → Basophils → Lymphocytes → Monocytes

8. Assertion – A physician might order a white cell count for a patient with symptoms of an infection.
Reason - An increase in the number of white blood cells (leukocytes) may indicate that the person is combating an infection.
(a) Both assertion and reason are true and reason is correct explanation of assertion.
(b) Both assertion and reason are true and reason

is not correct explanation of assertion.
(c) Assertion is true but reason is false.
(d) Both assertion and reason are false.

9. Blood, a special type of connective tissue
(a) Consists of a fluid matrix (Plasma)
(b) Has formed elements
(c) Is the most commonly used body fluid by most of the higher organisms
(d) All

10. Plasma is a straw coloured, viscous fluid constituting nearly ____ % of blood –
(a) 55 (b) 45
(c) 90 (d) 10

11. The amount of water present in blood plasma is –
(a) 99% (b) 90-92%
(c) 10% (d) 55%

12. I. Proteins contribute 6 - 8% of the blood plasma
II. Plasma contains very high amount of minerals
III. Plasma without the clotting factors is called serum
IV. Glucose, amino acids, lipids, etc., are also present in the plasma as they are always in transit in the body. Of the above statements –
(a) All are correct
(b) Only II is false
(c) Only I, III, IV is correct
(d) All are false

13. Match List I with List II and select the correct option.

List I (Plasma protein)		List II (Functions)	
I.	Fibrinogen	A.	Defense mechanism
II.	Globulins	B.	Osmotic balance
III.	Albumins	C.	Coagulation of blood

(a) I-C, II -A, III- B
(b) I-A, II - C, III- B
(c) I-C, II - B, III -A
(d) I-B, II-A, III-C

14. Formed elements of blood include –
(a) RBC, WBC and blood platelets
(b) All solutes present in blood
(c) Proteins present in blood
(d) All minerals (elements)

15. Which of the following statements is false?
(a) Erythrocytes/RBC are the least abundant of all the cells in blood.
(b) The number of RBCs in adult man per mm^3 of blood is 5 million to 5.5 million.
(c) RSC are formed in the red bone marrow in the adults.
(d) RBCs are enucleate in most of the mammals.

16. Life span of human RBC is –
(a) 120 hours (b) 120 month
(c) 120 days (d) 102 days

17. What is the amount of haemoglobin present in 100 ml blood of human blood?
(a) 45g (b) 18-20g
(c) 12-16g (d) 10 -12g

18. Mammalian RBCs are in shape-
(a) Oval (b) Biconvex
(c) biconcave (d) Sickle like

19. All of the following statement are correct about WBCs except –
(a) They are nucleate and least constancy in shape
(b) They are lesser in number (6000 – 8000 per mm^3 blood)
(c) They are generally short lived
(d) They help in blood clotting

20. All of the following are granulocytes except-
(a) Neutrophils
(b) Eosinophils
(c) Basophils only
(d) Lymphocytes and monocytes

Topic 2 Body Fluids and Circulation

21. Match list I with list II correctly –

	List I (Types of leucocytes/ WBCs)		List II (Their% (of total WBC)
I.	Neutrophils	A.	20 - 25
II.	Basophils	B.	2 - 3
III.	Monocytes	C.	6 - 8
IV.	Eosinophils	D.	0.5 - 1
V.	Lymphocytes	E.	60 - 65

(a) I - E, II - D, III - C, IV -A, V - B
(b) I -A, II - B, III - C, IV - E, V - D
(c) I - E, II - D, III - C. IV - B, V –A
(d) I - B, II - D, III -A, IV - C, V –A

Topic 3 Blood Groups

22. ABO blood grouping is based on the presence or absence of surface antigens
(a) 2 (b) 3
(c) 6 (d) 12

23. Fill up gaps given below in the table-

Blood group	Antigens on RBCs	Antibody in Plasma	Donor groups
A	A	Anti- B	A,O
B	B	Anti-A	B,O
AB	AB	II	A,B,ABO
O	I_	III	IV

	I	II	III	IV
(a)	Nil	Nil	Nil	O
(b)	Nil	Nil	Anti-A,B	AB
(c)	Nil	Anti-A,B	Nil	O
(d)	Nil	Nil	Anti-A,B	O

24. Which of the following blood groups is universal donor and universal acceptors respectively?
(a) AB, O (b) O, AB
(c) AB, A (d) A, AB

25. Which of the following representations is correct about blood groups and donor compatibility?

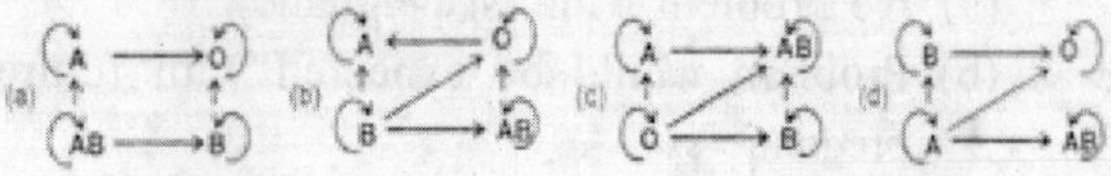

(a) 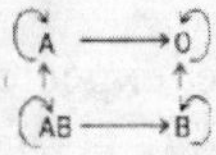
(b) 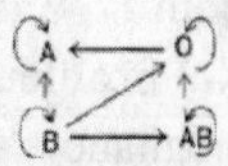
(c) 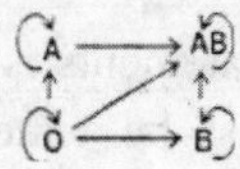
(d) 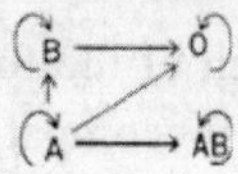

26. Rh factor is concerned with blood grouping. It derives its name from-
(a) Man (b) Chimpanzee
(c) Monkey (d) Rat

27. Rh factor is responsible for-
(a) Sickle cell anemia
(b) Erythroblastosis foetalis
(c) AIDS
(d) Turner syndrome

28. In developing foetus, erythroblastosis foetalis is caused by-
(a) Haemolysis
(b) Clumping of RBCs
(c) Failure of blood clotting
(d) Phagocytosis by WBC.

29. In erythroblastosis foetalis, which of the following factors passes through placenta into foetus -
(a) Rh antigens (b) Rh antibodies
(c) Agglutinins (d) ABO antibodies

30. A doctor suggested to a couple not to have more than one child because of –
(a) Rh+ male and Rh- female
(b) Rh- male and Rh+ female
(c) Rh- male and Rh- female
(d) Rh+ male and Rh+ female

31. In case of emergency which blood group could be safely transfused?

(a) AB Rh- (b) AB Rh+
(c) O Rh- (d) O Rh+

32. Which of the following is expected if husband is Rh+ and wife is Rh-?

(a) No problem with 1st pregnancy
(b) Problem would be expected with future pregnancies
(c) Both
(d) No problem could be expected in any pregnancy

33. Which of the following statements is correct?

(a) Rh compatibility must be tested before pregnancy establishment and blood transfusion
(b) Rh antibodies can cross placenta
(c) At the time of 1st delivery some of Rh+ RB Cs from the baby (Rh+) mix the mother's blood (Rh-) due to tear in placenta mother's blood for Rhantibodies
(d) All

Topic 4 Coagulation of Blood

34. What is the correct order of these events?

1. Conversion of fibrinogen to fibrin
2. Clot retraction and leakage of serum
3. Thromboplastin formation
4. Conversion of prothrombin to thrombin

(a) 3,2,1, 4 (b) 3,4,1,2
(c) 3,4,2,1 (d) 4,1,3,2

35. Which of the following statement are correct?

I. Ca^{+2} is necessary for blood coagulation
II. Coagulation in blood vessel is prevented during normal condition by heparin
III. Clotting of blood involves changes of fibrinogen to fibrin by thrombin
IV. Blood clotting involves cascading process involving a number of factors present in the active form always

(a) I, III, IV (b) II, IV
(c) I, II, III (d) III, IV

36. Which of the following pathways is correct for blood clotting

(a) 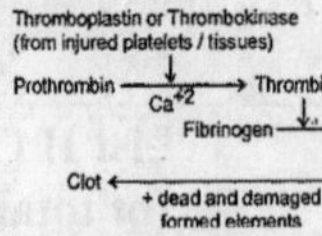

(b) 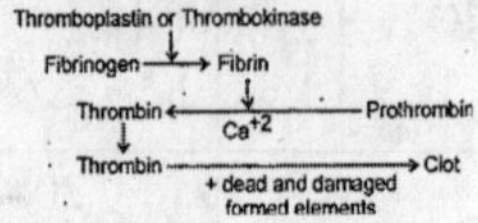

(c) 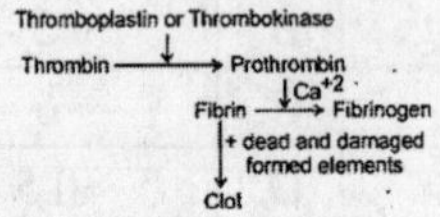

(d) 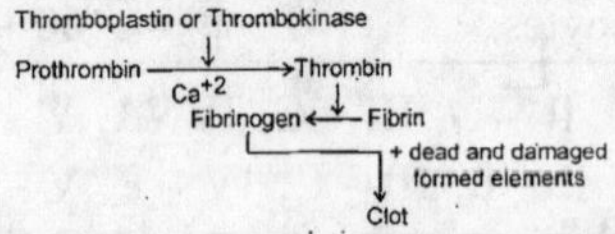

Topic 5 Lymph (Tissue fluid)

37. Which of following statements is wrong about lymph.

I. Lymph is colourful as it has haemoglobin but no RBC
II. The fluid present in lymphatic system is called lymph
III. It contains specialized lymphocytes which are responsible for immunity of the body
IV. Lymph is an important carrier for nutrients and hormones
V. Fats are absorbed through lymph in the lacteals present in the intestinalvilli

(a) Only I (b) III and IV
(c) II and III (d) Only IV

38. Which of the following statements is correct?

I. Lymphatic system collects tissue fluid/ interstitial fluid and drains it back to the major veins
II. Interstitial fluid (tissue fluid) and lymph

have almost similar composition

III. Lymph and interstitial fluid have no larger proteins and RBC

IV. Exchange of nutrients and gases, etc. between the blood and cells always occurs through tissue fluid

V. Interstitial fluid has the same mineral distribution as that in plasma

VI. Lymph can be defined as blood minus RBC but has specialized lymphocytes

(a) All (b) Only III and IV

(c) V and VI (d) I, III, V

Topic 6 Circulatory Pathways

39. Open circulatory system is found in –

(a) Arthropods and molluscs
(b) Annelids and Chordates
(c) Annelids and arthropods
(d) Fishes and molluscs

40. Closed circulatory system is found in –

(a) Arthropod and chordates
(b) Molluscs and chordates
(c) Amphibians and molluscs
(d) Annelids and chordates

41. In an open circulatory system –

(a) There is no heart
(b) There is no need of blood vessels
(c) There is no distinction between blood and tissue fluid
(d) There are no open spaces or sinuses in the body

42. Advantages of closed circulatory system over open circulatory system includes which of the following?

(a) Closed system can direct blood to specific tissues
(b) Exchange occurs more rapidly
(c) Close circulatory system can support higher levels of metabolic activity
(d) All

43. Which of the following statements is wrong about the closed circulatory system?

(a) Blood remains within blood vessels and never comes in direct contact with the body cells
(b) In it flow of fluid can be more precisely regulated
(c) There is no blood capillary
(d) Blood flow is more rapid due to higher pressure

44. Following are figures of hearts in different animals

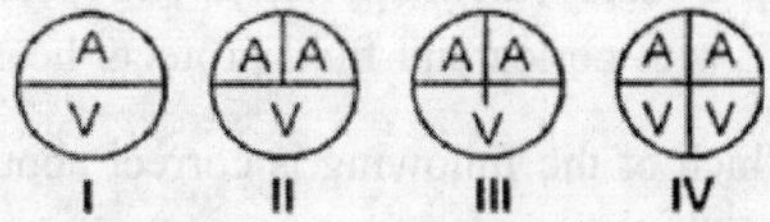

A = Auricle
V = Ventricle

Identify with their characteristic hearts –

	I	II	III	IV
(a)	Fishes	Reptiles	All reptiles	Birds, Mammals
(b)	Fishes	Birds	Reptiles, birds	Mammal
(c)	Fishes	Amphibians	Reptiles	Crocodiles, Birds, Mammals
(d)	Fishes	Crocodiles	Amphibians, Reptiles	Birds, Mammals

45. In fishes the blood circulation is represented as -

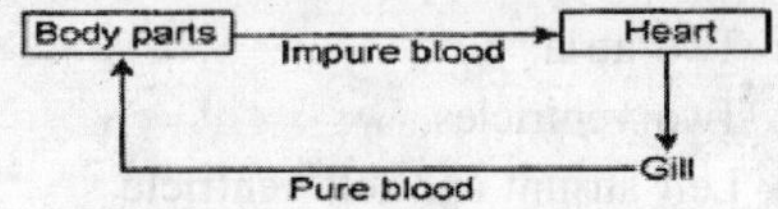

The above flow of blood indicates it is a

(a) Double circulation
(b) Single circulation
(c) Incomplete single circulation
(d) Incomplete double circulation

46. Incomplete double circulation is found in which of the following animals?

(a) Birds
(b) Mammals
(c) Birds and Mammals
(d) Amphibians and Reptiles

Topic 7 Human Circulatory System

47. Which of the following statements is not true?
(a) Heart is ectodermal in origin
(b) In human beings heart is situated in the thoracic cavity, in between the two lungs slightly lifted to the left
(c) Human heart has the size of a clenched fist.
(d) Double wall membranous bag (pericardium) with pericardial fluid protects heart

48. Which of the following is correct about human heart?
(a) The volume of both atria> the volume of both ventricles
(b) The volume of both ventricle> the volume of both atria
(c) The volume of both atria= the volume of both ventricles
(d) Ventricles are upper chambers and atria are lower chambers in our heart

49. Bicuspid valve / mitral valve is found between -
(a) Left atrium and left ventricle
(b) Right atrium and right ventricle
(c) Right atrium and left ventricle
(d) Left atrium and right ventricle

50. Tricuspid valve is present between the-
(a) Two atria
(b) Two ventricles
(c) Left atrium and left ventricle
(d) Right atrium and right ventricle

51. Chordae tendinae are found in –
(a) Joints
(b) Atria of heart
(c) Ventricles of heart
(d) Ventricles of brain

52. Ventricles are thick-walled as compared to atrium because -
(a) It is to receive blood from atria
(b) It is present on the posterior side
(c) It is to pump blood
(d) None

53. Which of the following has thickest wall?
(a) Left auricle (b) Left ventricle
(c) Right auricle (d) Right ventricle

54. Match the following.

	Column I		Column II
A.	Superior vena cava	p.	carries deoxygenated blood to lungs
B.	Inferior vena cava	q.	carries oxygenated blood from lungs
C.	Pulmonary artery	r.	brings deoxygenated blood from lower part of body to right atrium
D.	Pulmonary vein	s.	bring deoxygenated blood from upper part of body to right atrium

(a) A - q, B - s, C - r, D - p
(b) A - s, B - p, C - q, D - r
(c) A- s, B - r, C - p, D - q
(d) A - s, B - p, C - r, D – q

55. Origin of heart beat and its conduction is represented by -
(a) SA-node → Purkinje fibres → AV-node→ Bundle of His
(b) AV-node→ Bundle of His→ SAnode→ Purkinje fibres
(c) Purkinje fibres→ AV-node→ SAnode→ Bundle of His
(d) SA-node→ AV-node→ Bundle of His→ Purkinje fibres

56. 'Heart of heart' is -
(a) SA-node (b) AV-node
(c) Bundle of His (d) Purkinje fibres

57. SA node is located in –
(a) Upper lateral wall of left atrium
(b) Lower lateral wall of left atrium
(c) Lower lateral wall of right atrium
(d) Upper lateral wall of right atrium

58. SA node is called pace maker of the heart. Why?
(a) It can change contractile activity generated by AV node
(b) It delays the transmission of impulse between the atria and ventricles
(c) It gets stimulated when it receives neural signal
(d) It initiates and maintains the rhythmic contractile activity of heart

59. Sino-atrial node (SAN) can generate impulses -
(a) 70 - 75 min-1 (b) 50 - 55 min-1
(c) 35 - 40 min-1 (d) 100-150 min-1

60. The impulse of heart beat originate from –
(a) SAN (c) AVN
(c) Vagus nerve (d) Cardiac nerve

61. Rate of heart is determined by-
(a) SAN (b) AVN
(c) Purkinje fibres (d) Bundle of His

62. Bundle of His is a group of-
(a) Ganglia (b) Nerve fibres
(c) Muscular fibres (d) Connective tissue

63. Bundle of His / AV-bundle found in –
(a) Right auricle
(b) Left auricle
(c) Bone
(d) Interventricular septum

64. Atria-ventricular node (AVN) is situated in
(a) Lower left corner of left auricle, close to AV-septum
(b) Lower left corner of right auricle, close to AV-septum
(c) Upper left corner of right auricle, close to AV-septum
(d) Upper left corner of left auricle, close to AV-septum

65. Purkinje fibres are present in -
(a) Left auricle
(b) Right auricle
(c) Ventricular myocardium
(d) SAN

66. The chordae tendinae-
(a) Close the AV-valves
(b) Prevent the AV-valves flaps from everting
(c) Open semilunar valves
(d) Are present in auricles

67. Which of the following correctly traces the electrical impulses that trigger each heart beat
(a) Pacemaker → AV node→ Atria → Ventricles
(b) Pacemaker →Atria→ AV node → Ventricles
(c) AV node →Pacemaker→ Auricles →Ventricles
(d) Ventricle→ pacemaker→ AV node→ Auricle

68. An atrioventricular valve prevents the back flow or leakage of blood from –
(a) The right ventricle into the right atrium
(b) The left atrium into the left ventricle
(c) The aorta into the left ventricle
(d) The pulmonary vein into the right atrium

69. How many double circulations are normally completed by the human heart in one minute?
(a) 8 (b) 16
(c) 36 (d) 72

70. **Assertion-** If you trace the path of a molecule of carbon dioxide that starts in an arteriole in the right thumb and leaves the body in exhaled air, the minimum number of capillary beds the molecule encountered is 2.

Reason- The molecule of carbon dioxide would need to enter a capillary bed in the thumb before returning to the right atrium and ventricle, then travel to the lung and enter a capillary from which it would diffuse into an alveolus and be available to be exhaled.
(a) Both assertion and reason are true and reason is correct explanation of assertion.
(b) Both assertion and reason are true and reason is not correct explanation of assertion.
(c) Assertion is true but reason is false.
(d) Both assertion and reason are false.

Topic 8 Cardiac Cycle

71. The duration of cardiac cycle in a normal man is -
(a) 0.8 seconds (b) 80 seconds
(c) 60 seconds (d) 72 seconds

72. During systole of heart -
(a) Only atria contract
(b) only ventricles contract
(c) Auricles and ventricles contract separately
(d) Auricles and ventricles contract simultaneously

73. During ventricular systole –
(a) Oxygenated blood is pumped into the aorta and deoxygenated blood is pumped into the pulmonary artery
(b) Oxygenated blood is pumped into the pulmonary artery and deoxygenated blood is pumped into the artery
(c) Oxygenated blood is pumped into aorta and deoxygenated blood is pumped into pulmonary vein
(d) Oxygenated blood is pumped into pulmonary vein and deoxygenated blood is pumped into pulmonary artery

74. Contraction of right ventricle pumps blood into-
(a) Dorsal aorta
(b) Pulmonary vein
(c) Coronary artery
(d) Pulmonary artery

75. When ventricular systole occurs –
(a) Auricular diastole coincides
(b) Tricuspid and bicuspid valves close
(c) Semilunar valves guarding pulmonary artery and aorta are forced to open
(d) All

76. During cardiac cycle about % of ventricular filling occurs prior to atrial contraction. ____ % ventricular filling occurs due to atrial contraction–
(a) 50, 50 (b) 70, 30
(c) 30, 70 (d) 10, 90

77. Which of the following events do not occur during joint diastole?
I. All 4 chambers of heart are in relaxed state
II. Tricuspid and bicuspid valves open
III. Action potential is conducted from SAN to AVN
IV. Blood from the pulmonary veins and vena cava flows into the left and right ventricles respectively through the left and right atria
V. The Semilunar valves are closed
(a) Only V (b) Only III
(c) Only IV (d) Only I and II

78. The accompanying diagram shows three stages in the cardiac cycle- Which of the following sequence is correct?
(a) 2,3, 1 (b) 1,2, 3
(c) 2, 1, 3 (d) 3,1, 2

79. Cardiac output is determined by –
(a) Heart rate (b) Stroke volume
(c) Blood flow (d) Both a and b

80. The amount of blood to be pumped out by each ventricle/minute is-
(a) Stroke volume (b) Cardiac output
(c) Tidal volume (d) Residual volume

81. During cardiac cycle each ventricle pumps out about 70 ml of blood which is called -
(a) Stroke volume (b) Cardiac output
(c) Tidal volume (d) Residual volume

82. A red blood cell, entering the right side of the heart passes by or through the following structures –
1. Atrioventricular valves
2. Semi-lunar valves
3. Right atrium
4. Right ventricle
5. SAN
(a) 2→3→1→4→5 (b) 3→1→5→2→4
(c) 3→5→1→2→4 (d) 5→3→1→4→2

83. Cardiac output is –
(a) Stroke volume (SV) × Heart rate (HR) = 5L/ min
(b) SV × HR= 500 ml
(c) SV × HR= 72 ml/min
(d) SV × HR= 70 ml/min

84. Which of the following statement is not true?
(a) Cardiac output of an athlete is much higher than that of an ordinary man
(b) In each minute a single cardiac cycle is performed
(c) Cardiac sounds are of clinical diagnostic significances
(d) Cardiac cycle includes Auricular systole, ventricular systole and joint diastole/ complete diastole

85. First cardiac sound (lub) is associated with -
(a) Closure of tricuspid and bicuspid valves
(b) Opening of tricuspid and bicuspid valves
(c) Closure of semilunar valves
(d) Opening of semi lunar valves

86. Which of the following statement is wrong for second cardiac sound?
(a) It is heard as dup
(b) It is produced due to closure of semilunar valves
(c) It is clinically significant
(d) It is clinically non-significant

87. Assertion - The AV node delay the electrical impulse moving from the SA node and the atria to the ventricles.

Reason- The delay allows the atria to empty completely, filling ventricles fully before they contract.
(a) Both assertion and reason are true and reason is correct explanation of assertion.
(b) Both assertion and reason are true and reason is not correct explanation of assertion.
(c) Assertion is true but reason is false.
(d) Both assertion and reason are false.

88. Assertion - After exercising regularly for several months, our resting heart rate decreases, but our cardiac output at rest is unchanged.

Reason- The heart, like any other muscle, becomes stronger through regular exercise. The stronger heart would have a lesser stroke volume, which would allow for the decrease in heart rate.
(a) Both assertion and reason are true and reason is correct explanation of assertion.
(b) Both assertion and reason are true and reason is not correct explanation of assertion.
(c) Assertion is true but reason is false.
(d) Both assertion and reason are false.

Topic 9 Electrocardiograph(ECG)

89. Electrocardiogram is a measure of-
(a) Heart rate
(b) Ventricular contraction
(c) Volume of blood pumped
(d) Electrical activity of heart

90. Which of the following is a false statement?
(a) ECG is of a great clinical significance
(b) Electrocardiograph is the recording of electrical changes during the cardiac cycle
(c) To obtain a standard ECG, a patient is connected to the machine with 3 electrical electrodes (one to each wrist and to the left ankle)
(d) Normal activities of the heart are regulated intrinsically

91. P-wave represents -
(a) Depolarization of ventricles
(b) Repolarization of ventricle
(c) Repolarization of atria
(d) Depolarization of atria

92. QRS complex represents the -
(a) Depolarization of ventricles
(b) Repolarization of ventricles

(c) Repolarization of atria
(d) Depolarization of atria

93. T wave on an ECG represents –
(a) Depolarization of ventricles
(b) Repolarization of ventricle
(c) Repolarization of atria
(d) Depolarization of atria

94. The below figure is the diagrammatic representation of standard ECG.

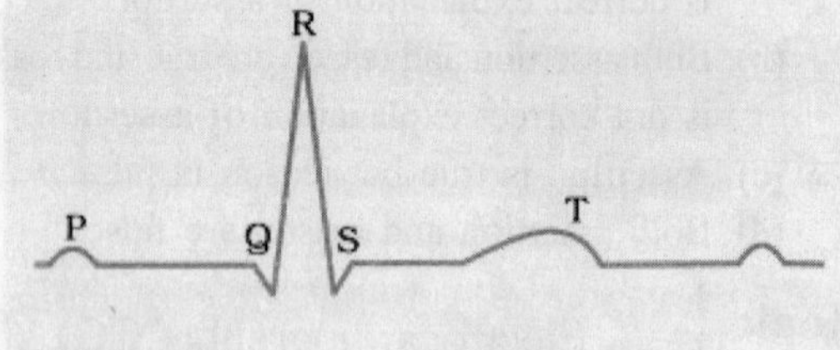

	Column I		Column II
A.	P-wave	I.	Ventricular depolarization followed by ventricular contraction
B.	QRS Complex	II.	Atrial depolarization followed by systole of both atria
C.	T- wave	III.	Ventricular repolarization followed by ventricular relaxation

(a) A-I, B-II, C-III
(b) A-III, B-II, C-I
(c) A-II, B-I, C-III
(d) A-II, B-III, C-I

95. Match the Column I with Column II –

	Column I		Column II
A.	Counting the number of QRS complex in a given time period	I.	A detailed evaluation of the heart function
B.	Potential generated by the recovery of ventricles from the depolarization state	II.	Determination of heart beat
C.	Multiple leads are attached to the chest region	III.	T-wave

Topic 10 Double Circulation

96. Which of the following options represents the pulmonary circulation in human being –
(a) Left Auricle —Oxygenated blood→ Lungs —Deoxygenated blood→ Right ventricle
(b) Left Auricle —Deoxygenated blood→ Lungs —Oxygenated blood→ Right Ventricle
(c) Right Ventricle —Deoxygenated blood→ Lungs —Oxygenated blood→ Left Auricle
(d) Right Ventricle —Oxygenated blood→ Lungs —Deoxygenated blood→ Left Auricle

97. Which of the following options represent correct systemic circulation in human being-
(a) Left Ventricle —Deoxygenated blood→ Tissues —Oxygenated blood→ Right Ventricle
(b) Right Ventricle —Oxygenated blood→ Tissues —Deoxygenated blood→ Right Auricle
(c) Left Ventricle —Deoxygenated blood→ Tissues —Oxygenated blood→ Right Auricle
(d) Left Ventricle —Oxygenated blood→ Tissues —Deoxygenated blood→ Right Auricle

98. Note the following blood vessels –
A. Arteriole B. Capillary
C. Aorta D. Muscular artery
E. Vein F. Venule

Choose the correct path that lists the blood vessels in. order, blood passes through them as it leaves the heart, travels to tissue and returns to heart -
(a) C, D, B, A, E, F (b) C, D, A, B, F, E
(c) D, C, A, B, E, F (d) D, C, B, A, E, F

99. Identify X, Y and Z?

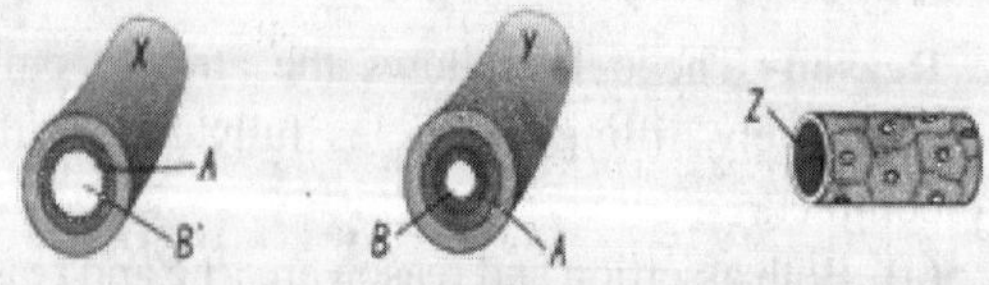

	X	Y	Z
(a)	Vein	Artery	Capillary
(b)	Capillary	Artery	Vein
(c)	Artery	Capillary	Vein
(d)	Vein	Capillary	Artery

100. Systemic circulation -
(a) Provides nutrient, O_2 and other essential substances to the tissues
(b) Takes CO_2 and other harmful substances away for elimination
(c) Both a and b
(d) Carries blood from heart to lungs

101. The blood circulation which starts and ends into capillaries is -
(a) Portal circulation
(b) Renal circulation
(c) Hepatic circulation
(d) Lymphatic circulation

102. Vascular connection between the digestive tracts and liver is called –
(a) Hepatic circulation
(b) Hepatic-portal system
(c) Both a and b
(d) Hepatic sinusoid

103. The hepatic-portal vein carries blood from to the ____ before it is delivered to the systemic circulation-
(a) Liver, intestine
(b) Pancreas, intestine
(c) Intestine, liver
(d) Hepatic artery, hepatic vein

104. A special coronary system of blood vessels present in our body exclusively for the circulation of blood to and from the-
(a) Corneocytes
(b) Cornea
(c) Cori cycle
(d) Heart/Cardiac musculature

105. **Assertion-** The heart of a normally developing human fetus has a hole between the left and right atria. In some cases, this hole does not close completely before birth. If the hole weren't surgically corrected, the O_2 content would be abnormally low.

Reason- In this case, some oxygen depleted blood returned to the right atrium from the systemic circuit would mix with the oxygen rich blood in the left atrium.
(a) Both assertion and reason are true and reason is correct explanation of assertion.
(b) Both assertion and reason are true and reason is not correct explanation of assertion.
(c) Assertion is true but reason is false.
(d) Both assertion and reason are false.

106. **Assertion** – There is low velocity of blood flow in the capillaries.

Reason – There is large total crosssectional area of the capillaries.
(a) Both assertion and reason are true and reason is correct explanation of assertion.
(b) Both assertion and reason are true and reason is not correct explanation of assertion.
(c) Assertion is true but reason is false.
(d) Both assertion and reason are false.

Topic 11 Regulation of Cardiac Activity

107. Cardiac centre lies in -
(a) Medulla oblongata
(b) Pons
(c) Cerebrum
(d) Epithalamus

108. Cardiac centre can moderate. the cardiac functions through -
(a) Somatic neural system
(b) Parasympathetic nervous system only
(c) Autonomic nervous system (ANS)
(d) Sympathetic nervous system only

109. Neural signal through the sympathetic nerve (part of ANS) increases cardiac output because of-
(a) Increasing the rate of heart beat
(b) Increasing the strength of ventricular contraction
(c) Both a and b
(d) Increasing the stimulation of vagus nerve

110. Parasympathetic neural signal decreases cardiac output by -
(a) Decreasing the rate of heart beat
(b) Decreasing the speed conduction of action potential
(c) Both
(d) Increasing adrenal medulla hormones secretion

111. Heart beat increases -
(a) On stimulation of sympathetic nerves
(b) On stimulation of vagus nerve (para sympathetic nerve)
(c) By adrenaline secreted by adrenal medulla
(d) Both a and c

Topic 12 Disorders of Circulatory System

112. In adult, normal blood pressure is –
(a) 80/120 mmHg (b) 100/80 mmHg
(c) 120/80 mmHg (d) 100/ 120 mmHg

113. Normal BP= 120 / 80 mmHg in an adult. In this measurement 120 mmHg is the ____ pressure and 80 mmHg is ______ pressure-
(a) Diastolic, systolic
(b) Systolic, diastolic
(c) Pulse, diastolic
(d) Pulse, systolic

114. Which one indicates B.P or hypertension?
(a) 120/ 80 mmHg
(b) 80/120 mmHg
(c) 140/90 mmHg or higher
(d) 40/60 mm Hg

115. Match the Column I with Column II –

	Column I		Column II
A.	Heart failure	I.	Heart muscle is suddenly damaged by an inadequate blood supply
B.	Cardiac arrest	II.	Chest pain due to inadequate 0 2 reaching the heart muscles
C.	Heart Attack	III.	Atherosclerosis
D.	Coronary Artery disease (CAD)	IV	Heart not pumping blood effectively enough to meet the needs of the body
E.	Angina pectoris	V.	Heart stops beating

	A	B	C	D	E
(a)	IV	V	III	I	II
(b)	V	IV	I	II	II
(c)	IV	V	III	I	III
(d)	V	IV	I	III	I

116. It is often referred as atherosclerosis, affects the blood vessels that supply blood to the heart muscles. It is caused by deposition of Ca, fat, cholesterol and fibrous tissues making the lumen of arteries narrow –

The above facts are related to-
(a) CAD (b) SCIO
(c) Blue baby (d) Heart arrest

117. Assertion – Nitroglycerin relieve chest pain caused by narrowing of the cardiac arteries.

Reason – The chest pain results from inadequate blood flow in coronary arteries. Vasodilation promoted by nitric oxide from nitroglycerin increases blood flow, providing the heart muscle with additional oxygen and thus relieving the pain.
(a) Both assertion and reason are true and reason is correct explanation of assertion.
(b) Both assertion and reason are true and reason is not correct explanation of assertion.
(c) Assertion is true but reason is false.
(d) Both assertion and reason are false.

ANSWER KEY

1. (c)	2. (a)	3. (d)	4. (d)	5. (b)	6. (d)	7. (c)	8. (a)	9. (d)	10. (a)
11. (b)	12. (c)	13. (a)	14. (a)	15. (a)	16. (c)	17. (c)	18. (c)	19. (d)	20. (d)
21. (c)	22. (a)	23. (d)	24. (b)	25. (c)	26. (c)	27. (b)	28. (a)	29. (b)	30. (a)
31. (c)	32. (c)	33. (d)	34. (b)	35. (c)	36. (a)	37. (a)	38. (a)	39. (a)	40. (b)
41. (b)	42. (d)	43. (c)	44. (c)	45. (d)	46. (d)	47. (a)	48. (b)	49. (a)	50. (d)
51. (c)	52. (c)	53. (d)	54. (c)	55. (d)	56. (a)	57. (d)	58. (d)	59. (a)	60. (a)
61. (a)	62. (c)	63. (d)	64. (b)	65. (c)	66. (b)	67. (b)	68. (a)	69. (d)	70. (a)
71. (a)	72. (c)	73. (a)	74. (d)	75. (d)	76. (c)	77. (b)	78. (c)	79. (d)	80. (a)
81. (a)	82. (d)	83. (a)	84. (b)	85. (a)	86. (d)	87. (a)	88. (c)	89. (d)	90. (d)
91. (d)	92. (a)	93. (b)	94. (c)	95. (a)	96. (c)	97. (b)	98. (a)	99. (a)	100. (c)
101. (c)	102. (b)	103. (c)	104. (d)	105. (a)	106. (a)	107. (a)	108. (c)	109. (b)	110. (b)
111. (d)	112. (c)	113. (b)	114. (c)	115. (b)	116. (a)	117. (a)			

7 Chapter

Excretory Products and their Elimination

MULTIPLE CHOICE QUESTIONS

Topic 1 Introduction

1. Least toxic nitrogenous waste is –
(a) NH_3 (b) Urea
(c) Uric acid (d) NH_3 and urea

2. Which of following in small amount is retained in kidney matrix of some animals to maintain a desired osmolarity?
(a) NH_3 (b) Urea
(c) Uric acid (d) NH_3 and uric acid

3. Terrestrial organisms must conserve water. The least amount of water is lost with the excretion of which nitrogenous waste product?
(a) NH_3 (b) Uric acid
(c) Urea (d) CO_2

4. The less amount of water is lost with the excretion of which nitrogenous product?
(a) NH_3 and urea (b) NH_3 and uric acid
(c) NH_3 (d) Urea and uric acid

5. Which of the following is correct about protonephridia/flame cells?
(a) Protonephridia are the excretory structures in Platyhelminthes (e.g. Planaria), rotifers and some annelids
(b) Protonephridia are the excretory structures in the cephalochordates e.g. Amphioxus
(c) Protonephridia are primarily concerned with ionic and fluid volume regulation i.e. osmoregulation
(d) All Excretory products and their elimination

6. Ammonia and urea are waste products derived from the metabolic breakdown of-
(a) Lipids
(b) Carbohydrates
(c) Proteins
(d) Sugars

7. Which of the following molecules is the most toxic to the cells?
(a) NaCl (b) Urea
(c) Uric acid (d) Ammonia

8. The terms "ammonotelic", "Ureotelic", and "Uricotelic" are used to describe-
(a) Modes of excretory system development
(b) The actions of hormones on the excretory systems
(c) The types of nitrogenous waste produced by various classes of vertebrates
(d) Modification of kidney tubules to enhance excretion

9. Which of the following statements is correct?
(a) Many bony fishes, aquatic amphibia and aquatic insects are ammoniotelic
(b) Ammonia is readily soluble
(c) NH_3 is generally excreted by the body surface or through gills (in fishes) as NH_4^+
(d) All

10. Which of the following statements is wrong?
(a) Kidney does not play any significant role in

the removal of ammonia
(b) Ureotelic animals excrete most of the nitrogenous waste as urea
(c) Ammonia and urea are the waste products derived from the metabolic breakdown of proteins
(d) None of the above is wrong

11. Urea and uric acid are –
(a) More toxic than NH_3
(b) Less toxic than NH_3
(c) Equally toxic to NH_3
(d) Non-toxic

12. Which of the following group of animals is ureotelic?
(a) many terrestrial amphibians
(b) Mammals
(c) Marine fishes
(d) All

13. NH_3 is converted into urea in –
(a) Kidney (b) Liver
(c) Spleen (d) Intestine

14. Which of the following groups of animals is uricotelic?
(a) Reptiles
(b) Insects
(c) Birds and land snail
(d) All

15. Excretion of nitrogenous products in semisolid forms by -
(a) Uricotelic animals
(b) Ureotelic animals
(c) Ammoniotelic animals
(d) Amniotes

16. Match the column I with column II.

	Column I		Column II
A.	Nephridia	I.	Crustaceans(Prawn)
B.	Malphigian tubules	II.	Annelids (Earthworm)
C.	Anteenal Gland or Green Glands	III.	Insects (Cockroach)

(a) A-I, B-II, C-III
(b) A-III, B-II, C-I
(c) A-II, B-III, C-I
(d) A-II, B-I, C-III

Topic 2 Human Excretory System

17. Which of the following statements is wrong about the human excretory system?
(a) Excretory system consists of one pair of bean shaped kidneys, one pair of ureter, a urinary bladder and a urethra.
(b) Kidneys are situated between the 12th thoracic and 3rd lumbar vertebrae close to the dorsal wall in abdominal cavity.
(c) Right kidney is a little higher level than the left one.
(d) All

18. Each kidney of adult human measures-

	Length	Width	Thickness	Weight
A)	10 - 12 cm	5 - 7cm	2 - 3cm	120-170 g
B)	10 – 20 cm	10 - 12 cm	6 - 12 cm	40-50 g
C)	2 – 6 cm	10 - 12 cm	6 - 12 cm	40-50 g
D)	10 - 12 mm	5 - 7 mm	2 - 3 mm	120-170 g

19. The part of kidney, gateway for ureter, nerves and blood vessels is-
(a) Hilum (b) Renal pore
(c) Minor calyx (d) Major calyx

20. Inner to the hilum of kidney is a broad funnel shaped space called-
(a) Cortex (b) Medulla
(c) Pelvis (d) Calyx

21. Which of the following statements is false?
I. Outer cortex and inner medulla are the two zones in kidney
II. Medulla is divided into about 8 to 18 renal pyramids
III. Pyramid projects into calyx

IV. Inwards extension of cortex between the pyramids is called renal column of Bertini

(a) I and IV (b) II and IV
(c) IV (d) None

22. Observe the following figure.

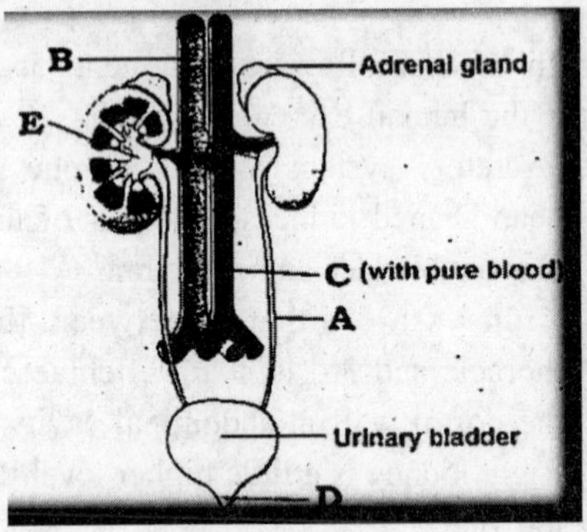

Identify A to E structure.

	A	B	C	D	E
A	Superior vena cava	Inferior vena cava	Dorsal Aorta	Urethra	Pelvis
B	Inferior vena cava	Superior vena cava	Dorsal Aorta	Urethra	Pelvis
C	Ureter	Inferior vena cava	Dorsal Aorta	Urethra	Pelvis
D	Dorsal Aorta	Inferior vena cava	Urethra	Cortex	Pelvis

23. Which one of the following is the structural and functional unit of kidney?

(a) Urethra (b) Urinary bladder
(c) Renal column (d) Nephron

24. Renal corpuscle or Malpighian body is-

(a) Glomerulus only
(b) Glomerulus along with Bowman's capsule
(c) Bowman's capsule
(d) Glomerulus with afferent arteriole

25. Which one of the following is a tube that carries urine from kidney to the urinary bladder?

(a) Loop of Henle (b) Ureter
(c) Urethra (d) Uvula

26. Go through the following figure-

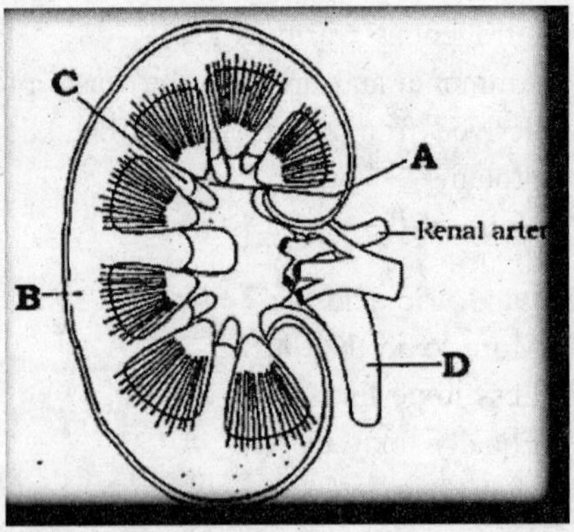

Identify A to DA

	A	B	C	D
A	Cortex	Calyx	Renal Column	Ureter
B	Calyx	Cortex	Renal Column	Ureter
C	Medulla	Cortex	Renal Column	Urethra
D	Calyx	Cortex	Renal Column	Urethra

27. Each kidney has how many nephrons?

(a) About 2 million (b) About 1 million
(c) About 5000 (d) About 50000

28. The bed of capillaries in the vertebrate kidney where water, urea and salts are filtered out of the blood is the –

(a) Bowman's capsule
(b) Collecting duct
(c) Glomerulus
(d) Loop of Henle

29. All of the following structures are situated in the renal cortex except –

(a) Loop of Henle
(b) Malpighian corpuscle
(c) PCT
(d) DCT

30. The DCTs of many nephrons open into a straight tube called –

(a) PCT
(b) Loop of Henle
(c) Collecting duct
(d) Bowman's capsule

31. Which of the following statements is false?
(a) Renal tubule starts with a double walled cup like structure called Bowman's capsule
(b) In majority of nephrons, the loop of Henle is too short and such nephrons are cortical nephrons
(c) Juxta medullary nephron has long loop of Henle
(d) None

32. Which is the correct pathway for passage of urine in humans?
(a) Collecting tubule→ ureter→ bladder→ urethra
(b) Renal vein→ renal ureter→ bladder→ urethra
(c) Pelvis→ Medulla → bladder→ urethra
(d) Cortex→ Medulla→ bladder→ ureter

33. Match the column I with column II.

	Column I		Column II
A.	Delivers blood to glomerulus	I.	Ascending and descending limb
B.	Carries urine to pelvis, also acts in water reabsorption	II.	Renal artery
C.	Collects filtrate from Bowman's capsule	III.	Collecting duct
D.	Loop of Henle	IV.	PCT

(a) A - II, B - III, C - IV, D - I
(b) A-I, B - III, C-II, D- IV
(c) A - II, B - IV, C - I, D - III
(d) A- IV, B - III, C - II, D – I

34. Which of the following is correct about Juxta medullary nephrons?
(a) Vasa recta is prominent
(b) Loop of Henle is long
(c) NaCl is returned to the interstitium by ascending limb of vasa recta
(d) All

35. Which of the following places the region of nephron in their correct sequence with respect to flow of tubular fluid?
(a) PCT→ Descending limb of Henle (DLH) → Ascending limb of Henle (ALH) → DCT→ Collecting duct(CD)
(b) PCT→ALH →DLH →OCT→CD
(c) ALH →DLH →PCT→OCT→CD
(d) OCT→ALH →DLH →PCT→CD

36. Vasa recta is –
(a) shaped (b) S-shaped
(c) U-shaped (d) J-shaped

37. In glomerulus, afferent arteriole –
(a) Is wider than efferent arteriole
(b) And efferent arteriole has similar diameter
(c) Is narrower than efferent arteriole
(d) Is narrow than efferent capillaries

38. Which of the following is incorrect?
(a) Blood vessel leading to glomerulus is called efferent arteriole
(b) Vasa recta, peritubular capillaries, Glomerulus all have blood
(c) Cortical nephron has no or highly reduced vasa recta
(d) Vasa recta runs parallel to the Henle's loop in juxtamedullary nephrons

Topic 3 Urine Formation

39. Urine formation involves-
(a) Ultra filtration and reabsorption occurring in different parts of nephron
(b) Ultrafiltration and reabsorption occurring in same part of nephron
(c) Ultrafiltration, reabsorption and secretion occurring in different parts of nephron
(d) Ultrafiltration, reabsorption and secretion occurring in same part of nephron

40. Match the column I with column II.

	Column I		Column II
a	PCT	I.	Concentrated urine formation
b	DCT	II.	Filtration of blood

c	Loop of Henle	III.	Reabsorption of 70 - 80% electrolytes
d	Counter- current mechanism	IV.	Ionic balance
e	Renal corpuscle	V.	Maintenance of conc. gradient in medulla

	a	b	c	d	e
(a)	III	IV	I	V	II
(b)	III	V	IV	II	I
(c)	I	III	II	V	IV
(d)	III	I	IV	V	II

41. Which of the following statements is correct?
I. Renal vein take blood away from kidney
II. Loop of Henle conserves water
III. Podocytes occur in inner wall of Bowman's capsule
IV. Ultrafiltrate I nephric filtrate is plasma minus proteins.
(a) I and II (b) I and Ill
(c) III and IV (d) I, II, III, IV

42. The glomerular capillaries cause filtration of blood through ____ layers –
(a) 1 (b) 2
(c) 3 (d) 6

43. The layers between the blood in glomerular blood Bowman's space are –
(a) Tunica media + Cuboidal epithelium+ Basement's membrane
(b) Endothelium + Epithelium of Bowman's capsule + Basement membrane between the 2 layers
(c) Endothelium of glomerular blood vessel + Endothelium of Bowman's capsule + Parietal layer of Bowman's capsule
(d) Tunica media + Epithelium of Bowman's capsule + Endothelium of Bowman's capsule

44. On average, _______ mL of blood is filtrated by the kidney per minute which constitute roughly __ of the blood pumped out by each ventricle of heart in a minute.
(a) 125 ml, 1/6th
(b) 100 -125 ml, 1/6th
(c) 1100 -1200 ml, 1/5th
(d) 5 L, 1/10th

45. The amount of the filtrate formed by the kidney / minute is called GFR (Glomerular Filtration Rate). The GFR of a healthy adult is-
(a) 80 mL/min (b) 125 mL/min
(c) 300 mL/min (d) 20 mL/min

46. The GFR/day in a healthy adult is –
(a) 5 L (b) 180 L
(c) 200 L (d) 20 L

47. Juxtaglomerular apparatus, a special sensitive cellular region is formed in –
(a) PCT and DCT
(b) PCT and DCT at the location of their contact
(c) PCT and loop of Henle at the location their contact
(d) DCT and afferent arteriole at the location of their contact

48. Of the filtrate, nearly how many of it is reabsorbed by the renal tubules?
(a) 5% (b) 99%
(c) 50% (d) 25%

Topic 4 Function of the Tubules

49. Which of following statements is false?
(a) The kidney has built in mechanism for regulation of GFR
(b) Tubular secretion does not play any significant role in urine formation
(c) The amount of urine output per day in normal adult is about 1.5 L
(d) During urine formation tubular cells secrete H+, K^+ and NH_3 in the filtrate

50. Which of the following statements about proximal convoluted tubule (PCT) is false?
(a) It is lined by simple cuboidal brush border epithelium which increases the surface area

(b) Nearly all the essential nutrients, 70 - 80% electrolytes, 70% H_2O are reabsorbed by PCT
(c) PCT is not the site of selective secretion
(d) PCT helps to maintain the pH and ionic balance of body fluids

51. PCT helps to maintain the pH and ionic balance of body fluids by -
(a) Selective secretion of H^+, NH_3 and K^+ ions in filtrate
(b) Reabsorption of HCO_3^- from filtrate
(c) Both a and b
(d) Secreting regulatory hormone like renin and angiotensinogen

52. If Loop of Henle were absent from mammalian nephrons, which of the following is to be expected?
(a) The urine will be more dilute
(b) There will be no urine formation
(c) The urine will be more concentrated
(d) There will be hardly any change in quality and quantity of urine formed

53. Use following diagram to complete the statements about the human nephron –

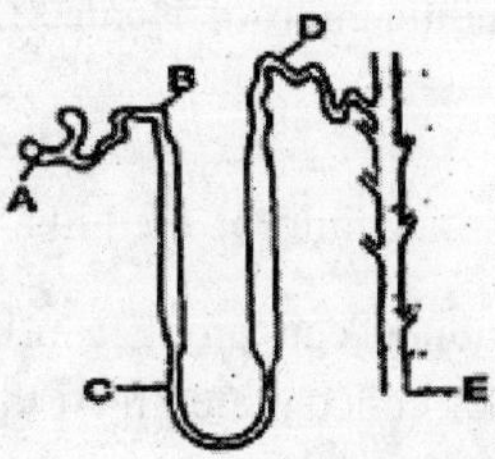

I. The composition of the filtrate would be most like plasma in the tubule next to the letter.
II. The urine would be most concentrated in the collecting duct next to letter
III. Most of the glomerular filtrate is reabsorbed into peritubular capillary next to the letter
IV. Conducting of urine to pelvis of the kidney from the structure next to the letter
V. Most water is reabsorbed by the structure next to the litter

(a) A C B E D (b) A E B C D
(c) A B E C D (d) A E B E B

54.
I. Reabsorption in this region is minimum.
II. This region plays a significant role in the maintenance of high osmolarity of intestinal fluid
III. Its descending limb is permeable to water but almost impermeable to electrolytes
IV. Its ascending limb is impermeable to water but allows transport of electrolyte actively or passively
V. In descending limb filtrate is hypertonic while in ascending limb filtrate is hypotonic

The above characteristics are associated with -
(a) PCT
(b) Loop of Henle
(c) DCT
(d) Bowman's capsule

55. Which of the following statements is correct?
I. Reabsorption of water occurs passively in the initial segment of nephron
II. Nitrogenous waste are absorbed by passive transport
III. Conditional reabsorption of Na^+ and water takes place in DCT
IV. DCT reabsorbs HCO_3^-
V. DCT is capable of selective secretion of H+, K^+ and NH_3 to maintain pH and Na^+ - K^+ balance in blood
VI. Substances like glucose, amino acids, Na^+, etc in the filtrate are reabsorbed actively

(a) I and II (b) II and III
(c) IV and V (d) All

56. Tubular secretion helps to maintain a proper acid-base balance by removing one of the following from blood –
(a) H+ and NH_3 (b) Uric acid
(c) H+ and urea (d) NH_3 and creatinine

57. Which of the following statements is false regarding the collecting duct?

I. Collecting duct is a straight duct
II. It extends from the cortex to medulla
III. Large amount of water could be reabsorbed from it to produce concentrated urine·
IV. Small amount of urea diffuses out from it into the medulla to keep up the osmolarity
V. It plays a role to maintain pH and ionic balance of blood by the selective secretion of H^+ and K^+ ions

(a) Only I (b) Only III
(c) IV and V (d) None

Topic 5 Mechanism of Concentration of the Filtrate

58. Mammals have the ability to produce ______ urine-
(a) Hypotonic (b) Hypertonic
(c) Isotonic (d) Alkaline

59. Which one plays an important role in counter current mechanism?
(a) Vasa recta (b) PCT
(c) Loop of Henle (d) A and C

60. In which of the following counter current operates-
(a) In ascending limb of loop of Henle
(b) In descending limb of loop of Henle
(c) In ascending limb or descending limb of vasa recta
(d) Between the 2 limb of Henle's loop and those of vasa recta

61. Medullary gradient is developed by all the following except -
(a) Reabsorption of Na^+ from ascending limb of Henle's loop into medullary interstitium
(b) Reabsorption of Na^+ from descending limb of Henle's loop
(c) Diffusion of small amount of urea from collecting duct into medullary interstitium
(d) Proximity between Henle's loop and vasa recta as well as the counter current in them

62. The medullary gradient is mainly caused by -
(a) Urea & K+ (b) H+ and K+
(c) NaCl and Urea (d) Urea and H+

63. The counter current mechanism helps to maintain a concentration gradient. This gradient helps in -
(a) Easy passage of water from medulla to collecting tubule and thereby concentrating urine
(b) Easy passage of water from collecting tubule and thereby concentrating urine
(c) Easy passage of water from medullary interstitial fluid to collecting tubule and thereby diluting urine
(d) Inhibition of passage of water between the collecting tubule and medulla and so isotonic urine is formed

64. NaCl is transported by the ascending limb of Henle's loop which is exchanged with -
(a) DCT
(b) PCT
(c) Ascending limb of vasa recta
(d) Descending limb of vasa recta

65. NaCl is returned to the by the ascending limb of vasa recta -
(a) Ascending limb of Henle's loop
(b) DCT
(c) PCT
(d) Interstitial fluid of medulla

66. Human kidney can produce urine nearly how many times concentrated than the initial filtrate formed?
(a) 4 (b) 2
(c) 10 (d) 100

67. The high osmolarity of the renal medulla is maintained by all of the following except -
I. Diffusion of salt from the ascending limb of the loop of Henle
II. Active transport of salt from the upper region of the ascending limb
III. The spatial arrangement of juxtamedullary nephrons
IV. Diffusion of urea from the collecting duct

V. Diffusion of salt from the descending limb of the loop of Henle

(a) Only I (b) Only V
(c) III and IV (d) I and V

Topic 6 Regulation of Kidney Function

68. Which one of the following is produced in the kidneys?

(a) Rennin (b) Renin
(c) Uricase (d) Arginase

69. Reabsorption of Na+ is controlled by –

(a) Vasopressin or ADH
(b) Aldosterone
(c) Renin
(d) Rennin

70. The reabsorption of water in the kidneys is under the control of a hormone –

(a) STH (b) ACTH
(c) LH (d) ADH/Vasopressin

71. Antidiuretic hormone secretion increases when the hypothalamus is stimulated by–

(a) Angiotensin receptors
(b) Glucose receptors
(c) Osmoreceptors
(d) Renin receptors

72. The kidneys help regulate acid-base balance by controlling the level of ____ in the blood-

(a) CO_2 (b) H+
(c) HCO_3^- (d) B and C

73. The functioning of the kidneys is efficiently monitored and regulated by hormonal feedback mechanisms involving -

(a) Hypothalamus only
(b) JGA only
(c) The heart only
(d) Hypothalamus, JGA and heart (to certain extent)

74. Osmoreceptors in the body are activated by changes in -

(a) Blood volume but not body fluid volume
(b) Body fluid volume but not blood volume
(c) Blood volume and body fluid volume
(d) Blood volume, body fluid volume and ionic concentration

75. Which of the following sequences is correct for regulation of kidney function?

(a) An excess loss of water from body→ Stimulates hypothalamus→ Osmoreceptors→ Neurohypophysis→ ADH→ Increases water permeability of DCT and CT→ Prevention of diuresis
(b) An excess loss of fluid from body→ Osmoreceptors→ Hypothalamus→ Neurohypophysis→ ADH→ Increases water permeability of DCT and CT→ Prevention of diuresis.
(c) An excess loss of fluid from body→ Osmoreceptors→ Hypothalamus→ Neurohypophysis→ Aldosterone→ Water permeability of DCT and CT increases→ Prevention of diuresis
(d) An excess loss of fluid from body→ osmoreceptor→ Hypothalamus→ Adenohypophysis→ ADH→ Increases water permeability of DCT and CT→ Prevention of diuresis

76. Osmoregulation is the function of-

(a) Oxytocin
(b) Prolactin
(c) Vasopressin (ADH)
(d) None of the above

77. ADH is synthesised by, ___________ released by ______ and acts on ________.

(a) Hypothalamus, Neurohypophysis, DCT and CT
(b) Hypothalamus, Neurohypophysis, Loop of Henle
(c) Hypothalamus, Adenohypophysis, DCT and CT
(d) Hypothalamus, Adenohypophysis, Loop of Henle

78. Which of the following sequence is correct?
(a) An increase in body fluid volume → switch off the Osmoreceptors → suppresses the ADH release
(b) ADH → Constricting effect on blood vessel→ B. P. high~ Glomerular blood flow more → GFR more
(c) Angiotensinogen→ Angiotensin I→ Angiotensin II→ Adrenal cortex → Aldosterone
(d) All

79. Which of the following factors can active the JG cells to release renin?
(a) A fall in glomerular blood pressure (GBP)
(b) A fall in glomerular blood flow (GBF)
(c) A fall in GFR
(d) A fall in GFR I GBP I GBF

80. Which of the following statements is false?
(a) Angiotensin II, being a powerful vasoconstrictor, increases glomerular pressure and thereby GFR
(b) Angiotensin II activates the adrenal cortex to release aldosterone
(c) Aldosterone promotes reabsorption of Na^+ and water from the DCT and CT leading to an increase in B.P. and GFR
(d) ANF causes vasoconstriction

81. RAAS (Renin -Angiotensinogen - Aldosterone System)-
(a) Is triggered when the juxtaglomerular cells of JGA releases renin in response to various stimuli
(b) Is responsible for regulation of kidney function
(c) Are stimulated when ANF is more in blood
(d) A and B are correct

82. Which of the following is true about Atrial Natriuretic factor (ANF)?
(a) An increase in blood volume and B. P. stimulates cardiac atria to release ANF
(b) ANF promotes vasoconstriction and thereby decrease B.P.
(c) ANF acts as a check on RAAS
(d) A and C

83. Renin-angiotensin pathway controls –
(a) Ultrafiltration
(b) Blood pressure
(c) Glucose reabsorption
(d) Cardia output

84. RAAS secretes which of the following hormones?
(a) Glucocorticoids
(b) Renin
(c) Mineralocorticoids
(d) All

Topic 7 Micturition

85. The expulsion of urine from the urinary bladder is called -
(a) Uricolysis (b) Micturition
(c) Uremia (d) Anuria

86. In micturition -
(a) Urethra relaxes (b) Ureter relaxes
(c) Ureter contracts (d) Urethra contracts

87. The outline of principal event of urination is given below in unordered manner-
I. Stretch receptors on the wall of urinary bladder send signal to the CNS
II. The bladder fills with urine and becomes distended
III. Micturition
IV. CNS passes on motor messages to initiate the contraction of smooth muscles of bladder and simultaneous relaxation of urethral Sphincter

The correct order of steps for urination is-
(a) I→ II→ III→ IV
(b) IV→ III→ II→I
(c) II→I→ IV→ III
(d) III→ II →I → IV

88. The neural mechanisms causing urination is called -
(a) Scarth reflex
(b) Withdrawal reflex
(c) Micturition reflex
(d) None

89. Average pH of human urine is –
(a) 6 (b) 9
(c) 3 (d) 7

90. Match the column I with column II.

	Column I		Column II
A.	Uremia	I.	Henle's loop
B.	Ketonuria	II.	Ketone bodies in urine
C.	Glycosuria	III.	Artificial kidney
D.	Blood dialyser	IV.	Glucose in urine
E.	Concentration of urine	V.	Accumulation of urea in blood

(a) A - V, B - II, C - IV, D – III, E - I
(b) A- Ill, B - II, C - IV, D - I, E - V
(c) A-I, B - II, C - IV, D - Ill, E - V
(d) A- I, B - II, C - IV, D - V, E – III

91. Diabetes mellitus is characterized by-
(a) Oilgonuria
(b) Ketonuria and glycosuria
(c) Anuria
(d) Haematuria

92. How much urea is excreted per day by a normal adult?
(a) 0 gm (b) 25 – 30 gm
(c) 50 gm (d) 1 – 2 gm

Topic 8 Role of other Organs in Excretion

93. Other than kidneys, which of the following also helps in the elimination of excretory wastes?
(a) Skin (b) Liver
(c) Lungs (d) All

94. How much CO_2 is removed per minute by our lungs-
(a) 18 Ml (b) 200 Ml
(c) 1L (d) 8 L

95. Which of the following statements is false?
(a) Micturition is carried out by a reflex
(b) Cholesterol is excreted in the bile and waxes are excreted in the sebum
(c) 8 L urine is excreted per day
(d) The primary function of sweat is excretion

96. Liver (largest gland) is both secretory and excretory organ. It secretes bile. Which of the following are major excretory products of bile?
(a) Degraded and steroid hormones
(b) Vitamins and drugs
(c) Bilirubin and Biliverdin
(d) Cholesterol

97. Most of excretory products of bile ultimately pass out along with-
(a) Urine (b) Digestive wastes
(c) Urea (d) Sweat

98.
I. The human skin possesses sweat and sebaceous glands which eliminate some wastes in their secretion.
II. Sweat is waxy protective secretion having sterols, hydrocarbons and fatty acid
III. Sebum is an aqueous fluid having NaCl, lactic acid, urea, amino acids, glucose

Which one of the above statement is correct?
(a) Only I (b) II and III
(c) Only II (d) I and II

Topic 9 Disorder of the Excretory System

99. In uremia, artificial kidney is used for removing accumulated waste products like urea by the process called-
(a) Micturition (b) Haemolysis
(c) Ureotelism (d) Hemodialysis

100. In artificial kidney dialysing fluid contains all the constituents as in plasma expect-

(a) Na^+
(b) Water
(c) Glucose
(d) Nitrogenous wastes

101. Kidney stone is produced by-
(a) Deposition of sand particles
(b) Crystallization of Ca- oxalate
(c) Precipitation of protein
(d) KCI or NaCl

102. Bright's disease/Glomerulonephritis is-
(a) Glycosuria
(b) Cystitis
(c) Inflammation of glomeruli
(d) Ketonuria

103. Following are the steps of dialysis-
A. Blood is passed into a vein.
B. Blood is mixed with heparin.
C. Blood is mixed with anti-heparin.
D. Blood is drained from convenient artery.
E. Blood is passed through a coiled and porous cellophane tube bathing in dialysis fluid.
F. Removal of nitrogenous wastes from blood.
The correct sequence of steps is-
(a) A → B → C → D → E → F
(b) F → C → E → B → A → D
(c) D → B → E → F → C → A
(d) D → C → E → F → B → A

ANSWER KEY

1. (c)	2. (b)	3. (b)	4. (d)	5. (d)	6. (c)	7. (d)	8. (c)	9. (d)	10. (d)
11. (b)	12. (d)	13. (b)	14. (d)	15. (c)	16. (c)	17. (d)	18. (a)	19. (a)	20. (c)
21. (d)	22. (c)	23. (d)	24. (b)	25. (b)	26. (b)	27. (b)	28. (c)	29. (b)	30. (c)
31. (d)	32. (a)	33. (a)	34. (d)	35. (a)	36. (c)	37. (a)	38. (a)	39. (c)	40. (a)
41. (d)	42. (c)	43. (b)	44. (c)	45. (b)	46. (b)	47. (d)	48. (b)	49. (b)	50. (c)
51. (c)	52. (a)	53. (d)	54. (b)	55. (d)	56. (a)	57. (d)	58. (b)	59. (d)	60. (d)
61. (b)	62. (c)	63. (b)	64. (d)	65. (d)	66. (a)	67. (b)	68. (b)	69. (b)	70. (d)
71. (c)	72. (d)	73. (d)	74. (d)	75. (b)	76. (c)	77. (a)	78. (d)	79. (d)	80. (d)
81. (d)	82. (d)	83. (b)	84. (c)	85. (b)	86. (a)	87. (c)	88. (c)	89. (a)	90. (a)
91. (b)	92. (b)	93. (d)	94. (b)	95. (c)	96. (c)	97. (b)	98. (a)	99. (d)	100. (d)
101. (b)	102. (c)	103. (c)							

8 Chapter

Locomotion and Movement

MULTIPLE CHOICE QUESTIONS

Topic 1 — Introduction and Type of Movement

1. Streaming of cytoplasm/cyclosis is seen in –
(a) Amoeba (b) Macrophages
(c) Leukocytes (d) All

2. Which of the following statements is correct?
(a) Cytoplasmic streaming helps in pseudopodia formation or amoeboid movement
(b) Cytoplasmic streaming is caused by contracting microfilament
(c) Both a and b
(d) Locomotion is not a voluntary movement

3. I. Paramecium employs cilia for pushing food in cytopharynx and in locomotion
II. Hydra takes help of tentacles for both food capturing and locomotion
III. All locomotion's are movements and vice-versa
IV. Methods of locomotion vary with habitats and the demands of situation
V. Ciliated epithelium is found in respiratory tract, renal tubules and reproductive tracts
Which of the above statements is false?
(a) I and III (b) III
(c) III and V (d) IV and V

Topic 2 — Muscle

4. The region between the ends of the Abands of 2-adjoining sarcomeres is called –
(a) The Z-band (b) The H-zone
(c) The T-tubule (d) The I-band

5. Choose the letter from the figure that most appropriately corresponds to the structure –

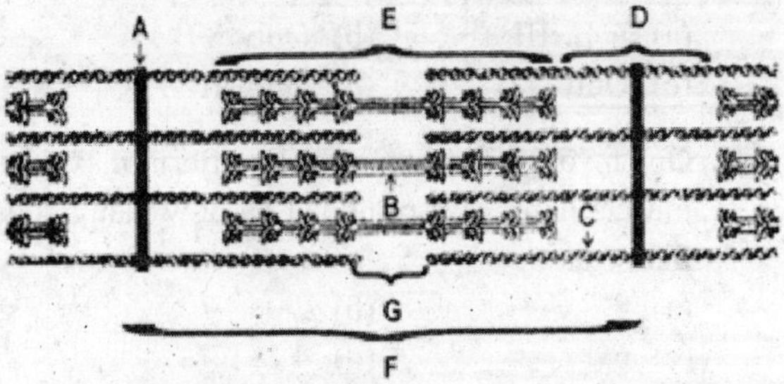

I. A-band
II. I-band
III. Sarcomere
IV. H-zone
V. Myosin
VI. Actin, Troponin, Tropomyosin
VII. Z- line
(a) I - E, II - D, III - F, IV - G, V - B, VI - C, VII -A
(b) I - E, II - D, III - C, IV - G, V - B, VI - A, VII - F,
(c) I - E, II - D, III - F, IV - G, V - C, VI - A, VII - B
(d) I - E, II - D, III - F, IV -A, V - B, VI - C, VII – G

6. An individual sarcomere consist of-
(a) A stack of actin fibres
(b) A stack of myosin units

(c) Overlapping actin and myosin
(d) Overlapping myosin arid membrane

7. Which of the following statements about the molecular arrangement of actin and myosin in myofibrils is false?
I. Each actin (thin filament) is made of 2F (filamentous) actins.
II. F-actin is the polymer of G (globular) actin.
III. 2F- actins are twisted into a helix
IV. Two strands of tropomyosin (protein) lie in the grooves of F-actin
V. Troponin molecules (complex proteins) are distributed at regular intervals on the tropomyosin
VI. Troponin forms the head of the myosin molecule
VII. The myosin is a polymerised protein
(a) I, II, III (b) Only VII
(c) Only VI (d) Only III

8. One myosin filament in the myofibril of skeletal muscle fibres is surrounded by how many actin filaments -
(a) 8 (b) 2
(c) 6 (d) 4

9. The cross bridges of the sarcomere in skeletal muscle are made up of –
(a) Actin (b) Myosin
(c) Troponin (d) Myelin

10. The functions of tropomyosin in skeletal muscle include -
(a) Sliding on actin to produce shortening.
(b) Release Ca+2 after initiation of contraction
(c) Acting as "relaxing protein" at rest by covering up the sites where myosin binds to actin
(d) Generates ATP

11. Tropomyosin is moved by which of following proteins -
(a) Calmodulin (b) Actin
(c) Troponin (D)Acetylcholine

12. Ca+2 bind ________ in the skeletal muscles and leads to exposure of the binding site for _______ on the filament ___________.
(a) Troponin, myosin, actin
(b) Troponin, actin, relaxin
(c) Actin, myosin, troponin
(d) Tropomyosin, myosin, actin

13. Following is the figure of actin (thin) filaments. Identify A, Band C.

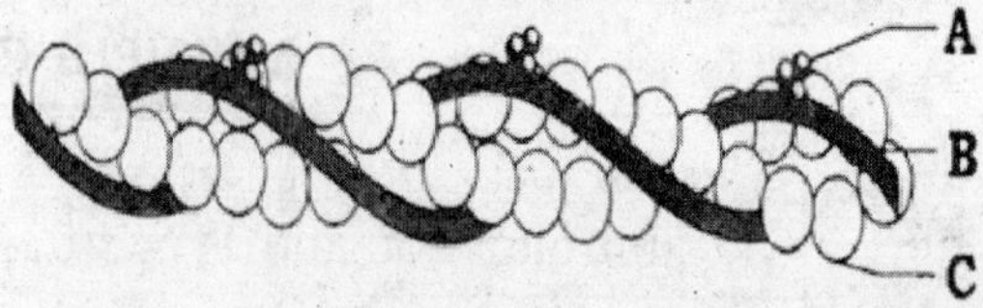

(a) A- Tropomyosin, B - Troponin, C - Factin
(b) A- Troponin, B - Tropomyosin, C - Myosin
(c) A- Troponin, B - Myosin, C - Tropomyosin
(d) A- Troponin, B - Tropomyosin, C - Factin

14. The above figure is related with myosin monomer (meromyosin). Identify A to C -

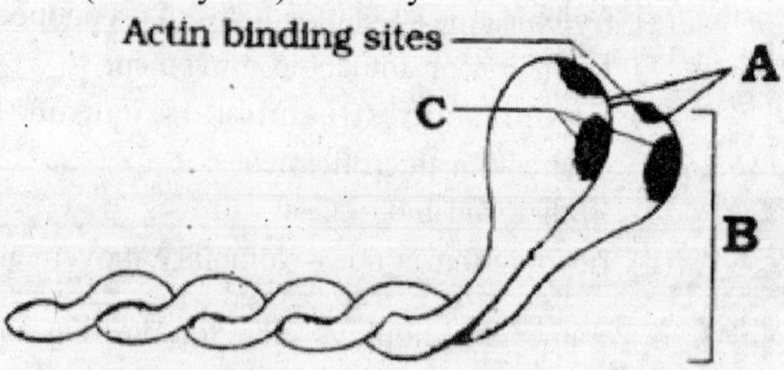

(a) A- head, B - cross arm, C - GTP binding sites
(b) A- head, B - cross arm, C - Ca+2 binding sites
(c) A- head, B - cross arm, C -ATP binding sites
(d) A- cross arm, B - head, C -ATP binding sites

15 Which of the following statements is false?
(a) Each myosin is a polymerised protein
(b) Many meromyosin constitute one thick filament (myosin)
(c) Each meromyosin's tail is called heavy meromyosin (HMM) and head is called light meromyosin (LMM)

(d) The globular head is an active ATPase enzyme and has binding sites for ATP and active sites for actin

16. Which of the following statements is false?
(a) Locomotion and many other movements required coordinated muscular activities
(b) Muscle is a specialised tissues of endodermal in origin
(c) There are about 639 muscles which contribute about 40 - 50 % of adult body weight
(d) Muscles show contractibility, excitability and flexibility

17. Which of the following statements about the skeletal muscles is correct?
(a) They are striated muscles
(b) They are voluntary muscles
(c) They are primarily involved in locomotory actions
(d) All

18. Which of the following statements about visceral muscles is correct?
(a) They are non-striated muscles (smooth muscles)
(b) They are involuntary muscles
(c) They have various functions
(d) All

19. Cardiac/heart muscles are -
(a) Striated and involuntary
(b) Not fatigued
(c) Branched
(d) All

20. Which of the following statements is false?
(a) Smooth muscles are found in urinary bladder, alimentary canal and genital tract
(b) A striated muscle is syncytium (multinucleate)
(c) The cytoplasm of striated muscle is called endoplasm
(d) The plasma membrane and ER of striated muscles are called sarcolemma and sarcoplasmic reticulum respectively

21. The source of Ca+2 for the muscle is –
(a) T-tubule (b) Sarcosome
(c) Sarcolemma (d) Sarcoplasmic reticulum

22. The fascia surrounding a muscle is made up of -
(a) Cartilage
(b) Collagenous connective tissues
(c) Adipose tissue
(d) Blood vessels

23. Contractile fibrils of muscles are called –
(a) Neurofibrils (b) Collagen fibres
(c) Myofibrils (d) Yellow fibres

24. Myofibrils show alternate dark and light bands in -
(a) Cardiac muscles (b) Smooth muscles
(c) Striped muscles (d) a and c

25. Select the true statement(s) -
(a) A-band is present in the middle of sarcomere
(b) H-zone is present in the middle of Aband
(c) M-line is present in the middle of Hzone
(d) All of the above

26. Which is the smallest one?
(a) Muscle fibre (b) Myofibril
(c) Actin (d) Sarcomere

27. Match Column I with Column II –

	Column I		Column II
A.	Structural and functional unit of a myofibril	I.	H-zone
B.	Protein of thin filament	II.	Myosin
C.	Protein of thick filament	III.	Sarcomere
D.	The central part of thick filament not overlapped by thin filament	IV.	Actin

(a) A-I, B-II, C-III, D-IV
(b) A-I, B-III, C-II, D-IV
(c) A- I, B - IV, C - III, D - II
(d) A- III, B - IV, C - II, D – I

28. Z-line divides the myofibrils into –
(a) Sarcomere (b) Sarcolemma

(c) Sarcosome (d) Microtubules

29. Sarcomere is the area between –
(a) 2 H-zones (b) 2 Z-lines
(c) 2 M-lines (d) 2A-bands

30. Light bands (thin filaments) contain actin and are called
(a) A-bands or Isotropic band
(b) A-bands or Anisotropic bands
(c) I-bands or Isotropic bands
(d) I-bands or Anisotropic bands

31. Dark bands (thick filaments) contain myosin and are called -
(a) A-bands or Isotropic band
(b) A-bands or Anisotropic bands
(c) I-bands or Isotropic bands
(d) I-bands or Anisotropic bands

32. Which of the following statements about the striated muscles is false?
I. In the centre of each I-band is an elastic fibre (Z-line) which bisects it
II. Thin filaments are firmly attached to the Z-line
III. M-line is a fibrous membrane in the middle of A-bands
IV. A sarcomere comprises one full Abands and 2 half I-bands
(a) All (b) IV
(c) I and II (d) None

Topic 3 Mechanism of Muscle Contraction

33. The action potential that triggers a muscle contraction travels deep within the muscle cell by means of ______.
(a) Sarcoplasmic reticulum
(b) Transverse tubules
(c) Synapse
(d) Motor end plates

34. ATP provides energy for muscle contraction by allowing for-
(a) An action potential formation in the muscle cell
(b) Cross-bridge detachment of myosin from actin
(c) Cross-bridge attachment of myosin to actin
(d) Release of ca+2 from sarcoplasmic reticulum

35. A motor unit is best described as –
(a) All the nerve fibres and muscle fibres in a single muscle bundle
(b) One muscle fibre and its single nerve fibre
(c) A single motor neuron and all the muscle fibres that it innervates
(d) It is the neuron which carries the message from muscle to CNS

36. Motor end plate is a - [Pg-307,E]
(a) Neuromuscular junction
(b) Dendron of motor neuron
(c) Plate of motor neuron
(d) Gradient of proton motive force

37. Electrical excitation in a muscle fibre most directly causes -
(a) Movement of tropomyosin
(b) Attachment of the cross bridges to actin
(c) Release of Ca+2 from sarcoplasmic reticulum
(d) Splitting of ATP

38. The energy for muscle contraction is most directly obtained from -
(a) Phosphocreatine
(b) ATP
(c) Anaerobic respiration
(d) Aerobic respiration

39. According to the sliding filament theory –
(a) Actin (thin filament) moves past myosin (thick filament)
(b) Myosin moves past actin
(c) Both myosin and actin move past each other
(d) None of these is correct

40. Put the following phrases in proper order to

describe what occurs at the neuromuscular junction to trigger muscle contraction.

I. Receptor sites on sarcolemma.

II. Nerve impulse.

III. Release of Ca+2 from sarcoplasmic reticulum

IV. The neurotransmitter acetylcholine is released

V. Sarcomere shorten

VI. Synaptic cleft

VII. Spread of impulses over sarcolemma on T-tubules

(a) II, IV, I, VI, VII, III, V
(b) II, IV, VI, I, VII, III, V
(c) I, II, III, IV, V, VI, VII
(d) VII, VI, V, IV, III, II, I

41. Go through the following diagram describing muscle contraction.

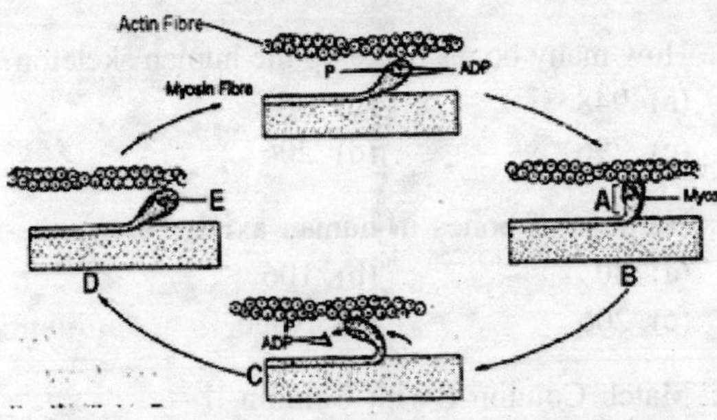

Now identify A to E.

(a) A- Cross bridge, B - Cross bridge formation, C-Breaking of cross bridge, D –Sliding (rotation), E -ATP
(b) A- Cross bridge, B - Cross bridge formation, C – Sliding/rotation, D – Breaking of cross bridge, E -ATP
(c) A- Cross bridge, B - Breaking of Cross bridge, C – Sliding/rotation, D - Cross bridge formation, E -AMP
(d) A- Cross bridge, B - Cross bridge formation, C – Sliding/rotation, D - ADP, E - Breaking of cross bridge

42. How does the troponin-tropomyosin complex affect cross-bridge cycling?

(a) When $[Ca^{2+}]$ is low, the troponintropomyosin complex blocks actin's binding site for myosin. When $[Ca^{2+}]$ is high, the complex rolls out of the way, allowing myosin to bind to actin and initiate the cross-bridge cycle.
(b) The troponin-tropomyosin complex regenerates ATP for the myosin ATPase.
(c) The troponin-tropomyosin complex regulates calcium release from the terminal cisternae.
(d) The troponin-tropomyosin complex binds to the myosin head, facilitating contact with the actin filaments

43. Relaxation of muscle is due to –

(a) Pumping of Ca+2 into sarcoplasmic cisternae
(b) Presence of ATP
(c) Conformational change in troponin and masking of actin filaments
(d) A and C

44. The diagrams given above show 3 different condition of sarcomeres. Identify these conditions-

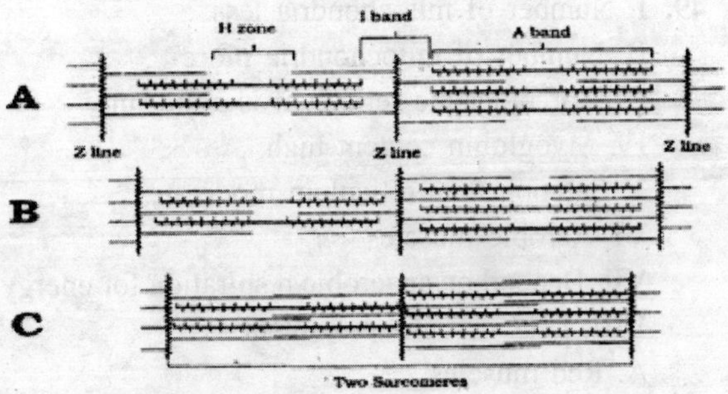

(a) A– contracting, B - relaxed, C – maximally contracted
(b) A – relaxed, B - contracting, C – maximally contracted
(c) A- maximally contracted, B – contracting, C - relaxed
(d) A- relaxed, B - maximally contracted, C– contracting

45. When a skeletal muscle shortens during contraction which of these statements is false?

(a) The I-band shortens
(b) The A-band shortens
(c) The H-zone becomes narrow

(d) The sarcomeres shorten

46. The muscle band that remains unchanged during muscle contraction and relaxation of the skeletal muscle is –

(a) I (b) A
(c) H (d) Z line

47. Which of the following statements is correct?

(a) During muscle contraction chemical energy changes into mechanical energy
(b) Muscle fatigue is due to lactic acid formation due to anaerobic respiration
(c) The reaction time of the fibres can vary in different muscles
(d) All

48. The compound or pigment acting as an oxygen store in skeletal muscles is –

(a) Myoglobin
(b) Haemoglobin
(c) Myokinase or ATP
(d) Cytochrome

49. I. Number of mitochondria less.
II. Number of mitochondria more
III. Sarcoplasmic reticulum is abundant
IV. Myoglobin content high
V. Sarcoplasmic reticulum moderate
VI. Aerobic muscles
VII. Depend on anaerobic respiration for energy
VIII. Less myoglobin content
A. Red muscles
B. White muscles

Identify above (I to VIII) traits as characteristic of A and B types of muscles- [Pg-307,308,M]

(a) A- I, III, VII, VIII; B -II, IV, V, VI
(b) A-II, IV, V, VI; B-I, III, VII, VIII
(c) A-I, III, IV, VII; B – II, V, VI, VIII
(d) A- II, V, VI, VIII; B - I, III, IV, VII

Topic 4 Skeletal System

50. Skeletal system consists of -

(a) Only bones
(b) Only cartilage
(c) A framework of bones and a few cartilage
(d) A framework of cartilage. and a few bones

51. Bone has a very hard matrix due to presence of-

(a) NaCl (b) Ca-salts
(c) K-salts (d) Fe-salts

52. Cartilage has slightly pliable matrix due to -

(a) Chondroitin salts
(b) Osteoblast
(c) Chondroblasts
(d) Osteoclast

53. How many bones make up the human skeleton?

(a) 948 (b) 96
(c) 796 (d) 206

54. Number of bones in human axial skeleton is -

(a) 80 (b) 106
(c) 206 (d) None

55. Match Column I with Column II –

	Column I		Column II (Number of bones)
A.	Cranium/ Brainbox	I.	29
B.	Skull (Cranial and facial bones)	II.	8
C.	Face	III.	14
D.	Hind limb	IV.	12 pairs
E.	Ribs	V.	30

(a) A-I, B-II, C-III, D-V, E-IV
(b) A- II, B - I, C - III, D - V, E - IV
(c) A - I, B - II, C- III, D - IV, E - V
(d) A- V, B - IV, C - III, D - II, A- I

56. Hyoid/Tongue bone is -
(a) T-shaped (b) J-shaped
(c) U-shaped (d) L-shaped

57. A normal human being has how many ear ossicle?
(a) 3 (b) 6
(c) 9 (d) None

58. Which one of the following is not included under ear ossicles -
(a) Malleus (b) Ileum
(c) Incus (d) Stapes

59. Human Cranium has small protuberance(s) at the posterior end called __________ and ____ in number. that articulates with first vertebra (atlas vertebra)-
(a) occipital condyle, 6
(b) occipital condyle, 2
(c) occipital condyle, 4
(d) occipital condyle, 3

60. Human skull is -
(a) Dicondylic (b) Monocondylic
(c) Procoelous (d) Hetercoelous

61. Which of the following statements about human vertebral column is false?
(a) Vertebral column consists of 26 vertebrae
(b) It is ventrally placed
(c) It extends from the base of skull and constitutes the main framework of the trunk
(d) Neural canal in vertebra is the passage for spinal cord

62. Human adult vertebral formula is-
(a) $C_4\ T_8\ L_4\ S_8\ C_8$ (b) $C_7\ T_8\ L_5\ S_6\ C_7$
(c) $C_7\ T_{12}\ L_2\ S_1\ C_2$ (d) $C_7\ T_{12}\ L_5\ S_1\ C_1$

63. Which of the following vertebra in adult human are fused ones?
(a) Thoracic and lumber
(b) Thoracic and cervical
(c) Sacral and coccygeal
(d) Cervical and coccygeal

64. Which of the following is not the function of vertebral column?
(a) Protects spinal cord and supports the head
(b) Serves as the point of attachment for ribs and musculature of the back
(c) Both
(d) Supports Tarsals and Metacarpals

65. Which of the following is not correct about sternum?
(a) It is commonly called breast bone
(b) It is flat bone
(c) It is 2 in number
(d) It is located on the ventral mid line of thorax

66. Each typical rib is a thin flat bone connected ____ to the vertebral column and __________ to the sternum-
(a) Dorsally, ventrally
(b) Ventrally, dorsally
(c) Dorsally, dorsally
(d) Ventrally, Ventrally

67. Typical ribs are -
(a) Monocephalic (b) Dicephalic
(c) Tricephalic (d) Tetracephalic

68. Match Column I with Column II –

	Column I		Column II
A.	True ribs	I.	3 pairs
B.	False ribs	II.	2 pairs
C.	Floating ribs	III.	7 pairs

(a) A-I, B-II, C-III
(b) A-III, B-I, C-II
(c) A- III, B - II, C - I
(d) A-ll, B-l, C-III

69. Match Column I with Column II –

	Column I		Column II
A.	False ribs	I.	1st to 7th pair
B.	True ribs	II.	11th and 12th pair
C.	Floating ribs	III.	8th to 10th pair
D.	Sternum	IV.	One

(a) A-I, B-II, C-III, D- IV
(b) A-IV, B-III, C-II, D- I
(c) A- I, B - III, C - II, D- IV
(d) A-III, B-I, C-II, D- IV

70. Identify the ribs -

a. Ribs are attached to the sternum ventrally and to the vertebrae dorsally.
b. Ribs are attached to sternum through costal cartilage (hyaline) of 7th rib
c. Ribs are not attached to sternum

I. True ribs
II. False ribs
III. Floating ribs

(a) a-I, b-II, c-III (b) a-I, b-III, c-II
(c) a-II, b-I, c-III (d) a-III, b -II, c-I

71. Which of the following ribs are called vertebrochondral ribs?

(a) True ribs (b) False ribs
(c) GorIIIa ribs (d) Floating ribs

72. Rib cage is formed by all except –

(a) Thoracic vertebrae
(b) Lumbar vertebrae
(c) Ribs
(d) Sternum

73. Each limb (upper or lower) consists of how many bones –

(a) 30 (b) 60
(c) 101 (d) 8

74. The accompanied figure is rib cage. Identify A, Band C respectively-

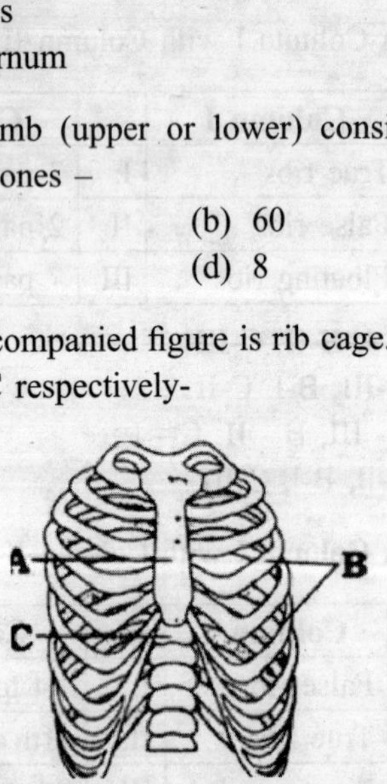

(a) Coccyx, ribs, vertebral column
(b) Sternum, ribs, vertebral column
(c) Scapula, ribs, vertebral column
(d) Tarsal, ribs, vertebral column

75. Number of bones in human appendicular skeleton is -

(a) 80 (b) 120
(c) 126 (d) 206

76. Number of bone in each upper limb is –

(a) 1, 1, 1
(b) 8, 5, 14
(c) 2, 2, 2, 16, 10, 28
(d) 1, 1, 1, 8, 5, 14

77. Phalangeal/digital formula for human hand/foot is-

(a) 0, 2, 2, 3 (b) 0, 2, 3, 3 ,3
(c) 2, 2, 3, 3, 3 (d) 2, 3, 3, 3, 3

78. The hand contains ____ carpals (wrist bones), ____ metacarpals (palm bones), and ___ phalanges.

(a) 14, 5, 8 (b) 5, 8, 14
(c) 8, 5, 14 (d) 1, 5, 5

79. The accompanied diagram shows right pectoral girdle and upper arm (frontal view). Identify A to G

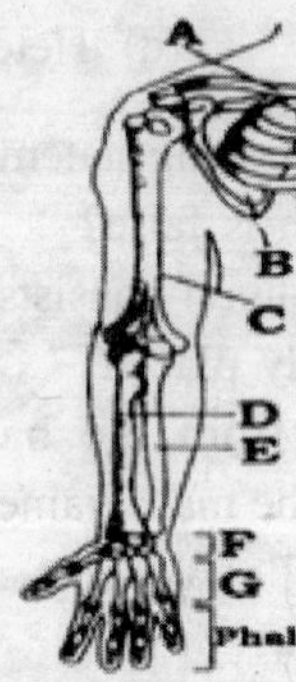

(a) A- 1st Vertebra, B - Scapula, C - Humerus, D - Radius, E - Ulna, F - Carpals, G - Metacarpals
(b) A- Scapula, B - Clavicle, C - Humerus, D – Radius, E - Ulna, F - Carpals, G - Metacarpal
(c) A- Ilium, B - Scapula, C - Humerus, D - Radius, E - Ulna, F - Carpals, G – Metacarpals
(d) A- Clavicle, B - Scapula, C - Humerus, D - Radius, E - Ulna, F - Carpals, G –

Metacarpals

80. An acromion process is characteristically found in -
(a) Pelvic girdle of mammals
(b) Pectoral girdle of mammals
(c) Skull bone
(d) Vertebrae of mammals

81. The shoulder blade is large triangular bone situated in the dorsal part of the thorax between the 2nd and the 7th ribs. It is called -
(a) Clavicle (b) Ilium
(c) Scapula (d) Carpals

82. For articulation of head of humerus a depression found in scapula is called –
(a) Acetabulum
(b) Manubrium
(c) Occipital condyle
(d) Glenoid cavity

83. Which of the following statement is correct?
(a) Pectoral and pelvic girdle bones help in the articulation of the upper and lower limbs respectively with the axial skeleton
(b) Each girdle is formed of 2 halves
(c) Each half of pectoral girdle consists of a clavicle (collar bone) and Scapula
(d) All

84. Which of the following statements is false?
(a) Scapula has the spine which projects as acromion process
(b) Below acromion process is a glenoid cavity
(c) Each clavicle (collar bone) articulates with acromion
(d) Clavicle is long S-shaped bone with 4 curvatures

85. Which one of the following is the longest bone in human?
(a) Radius
(b) Tibia
(c) Femur (Thigh bone)
(d) Clavicle (Collar bone)

86. Human foot consists of 26 bones. What are the number of tarsals (ankle bones), metatarsals and phalanges?
(a) 7, 5, 14 (b) 5, 7, 14
(c) 1, 1, 5 (d) 5, 5, 5

87. A cup shaped bone covering knee ventrally is called -
(a) Cuneiform (b) Tarsal
(c) Patella (d) Carpal

88. Study the accompanying figure. Identify A, B, C and D –

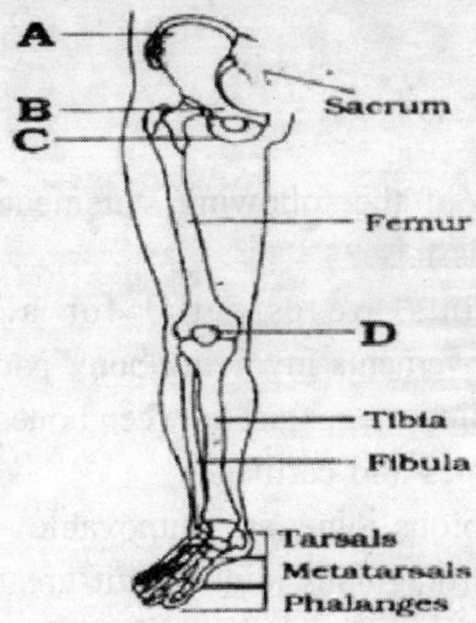

(a) A- Pubis, B - ilium, C - Ischium, D - Patella
(b) A- Ischium, B - Pubis, C - ileum, D– Patella
(c) A- ileum, B - Pubis, C - Ischium, D– Patella
(d) A- ilium, B - Pubis, C - Ischium, D– Patella

89. Acetabulum occurs in -
(a) Cranium (b) Pectoral girdle
(c) Pelvic girdle (d) Vertebrae

90. Pelvic girdle (hip girdle) is composed ____ coxal (hip) bones-
(a) 3 (b) 2
(c) 4 (d) 5

91. Pelvic girdle consists of-
(a) Ileum, ischium and pubis
(b) Ilium, ischium and pubis
(c) Ilium, ischium and clavicle
(d) Coracoid, ischium and pubis

92. Two halves of pelvic girdle articulate ventrally

at a fibrocartilaginous joint called -
(a) Pubic symphysis
(b) Synchodroses
(c) Gomphoses
(d) Sutures

93. Each coxal bone is formed by the fusion of 3 bones named as -
(a) Ileum, ischium and pubis
(b) Ilium, ischium and pubis
(c) Ilium, ischium and clavicle
(d) Coracoid, ischium and pubis

Topic 5 Joints

94. Which of the following statements about the joints is false?
(a) Joints are essential for all types of movements involving bony parts
(b) Joints are contact between bones or between bones and cartilages
(c) Fibrous joints are immovable
(d) Cartilaginous joint permit great movement

95. Match Column I with Column II –

	Column I		Column II
A.	Hinge joint	I.	Between humerus and pectoral girdle
B.	Pivot joint	II.	Between carpals and Metacarpals of thumb
C.	Gliding joint	III.	Between the carpals
D.	Saddle joint	IV.	Between atlas and axis
E.	Saddle joint	V.	Knee joint

(a) A- V, B - IV, C - III, D - II, E - I
(b) A- I, B - II, C - II, D - V, E - IV
(c) A- I, B - III, C - II, D - V, E - IV
(d) A-V, B-III, C-II, D-I, E-II

96. Which of the following statements is correct?
(a) Synovial joints are characterised by synovial cavity with fluid between the articulating surface of two· bones
(b) Synovial joints are freely movable
(c) Ball and socket, hinge joint, gliding joints, pivot joints and saddle joints are the types of synovial joints
(d) All

97. Joint between bones in the form of sutures of human skull is
(a) Hinge joint
(b) Synovial joint
(c) Cartilaginous joint
(d) Fibrous joint

98. Which of the following statements is correct?
(a) Movable skull bone is mandible
(b) We move our hands while walking for balancing
(c) Cartilaginous joints have little mobility due to fibrocartilage disc between its articular ends e.g. intervertebral disc between centre of vertebrae
(d) All

Topic 6 Disorders of Muscular and Skeletal System

99. A disease associated with joint is –
(a) Glaucoma (b) Arthritis
(c) Paget's disease (d) Homer's syndrome

100. Gout is the inflammation of joints due to accumulation of -
(a) Urea crystal
(b) NH_3
(c) Uric acid crystal
(d) $CaCO_3$ crystals

101. I. Age-related disorder characterised by decreased bone mass and increased chances of fracture
II. Causative factor deficiency of estrogen is common.
The above characters are associated with–
(a) Gout (b) Osteoporosis
(c) Arthritis (d) Polio

102. Myasthenia is an ____ disorder affecting neuromuscular junction leading to fatigue, weakening and paralysis of skeletal muscles -
(a) Arthritis (b) Autoimmune
(c) Agnosic (d) Amnesic

103. Tetany is the rapid spasm in muscles due to –
(a) High Ca^{+2} in body fluid
(b) Low Ca^{+2} in body fluid
(c) High uric acid in body fluid
(d) High urea in blood

104. Progressive degeneration of skeletal muscles due to genetic disorder is called –
(a) Myasthenia gravis
(b) Tetany
(c) Muscular dystrophy
(d) Myopia

105. Arthritis is -
(a) Inflammation of muscles
(b) Inflammation of bone
(c) Inflammation of joints
(d) Inflammation of tongue

ANSWER KEY

1. A	2. C	3. D	4. D	5. A	6. C	7. A	8. C	9. B	10. C
11. C	12. A	13. D	14. C	15. C	16. B	17. D	18. D	19. D	20. C
21. D	22. B	23. C	24. D	25. D	26. D	27. D	28. A	29. B	30. C
31. B	32. D	33. B	34. B	35. C	36. A	37. C	38. B	39. C	40. B
41. B	42. A	43. A	44. B	45. D	46. B	47. D	48. A	49. B	50. C
51. B	52. A	53. D	54. A	55. B	56. C	57. B	58. B	59. B	60. A
61. B	62. D	63. C	64. C	65. C	66. A	67. B	68. B	69. D	70. C
71. B	72. B	73. A	74. B	75. C	76. B	77. D	78. C	79. D	80. B
81. C	82. D	83. D	84. D	85. C	86. A	87. C	88. D	89. C	90. B
91. B	92 A	93. B	94. D	95. D	96. D	97. D	98. D	99. B	100. C
101. B	102. B	103. B	104. C	105. C					

9 Chapter

Neural Control and Coordination

MULTIPLE CHOICE QUESTIONS

Topic 1 Introduction

1. The ganglia found in insects are
(a) masses of fat bodies
(b) aggregated neurons which gives off nerves
(c) point where numerous neurons meet
(d) degenerated neuron masses

2. Assertion: Nervous system and endocrine system jointly coordinate and integrate activities of organs.
Reason: Endocrine system regulate all the activities of nervous system.
(a) Both Assertion and Reason are true and Reason is correct explanation of Assertion.
(b) Both Assertion and Reason are true, but Reason is not the correct explanation of Assertion.
(c) Assertion is true, but Reason is false.
(d) Assertion is false, but Reason is true.

3. Assertion: Neural organisation become complex in vertebrates as compared to invertebrates.
Reason: The ganglion in insects acts as a brain.
(a) Both Assertion and Reason are true and Reason is correct explanation of Assertion.
(b) Both Assertion and Reason are true, but Reason is not the correct explanation of Assertion.
(c) Assertion is true, but Reason is false.
(d) Assertion is false, but Reason is true.

4. Coordination is considered as an important process in an animal body because
(a) it helps to maintain homeostasis.
(b) it enables different organs to interact and function efficiently.
(c) it ensures the normal functioning of vital organs.
(d) all of these

5. The neural system within human body provide
(a) chemical coordination through hormones
(b) point to point connections
(c) electrical coordination
(d) both (b) and (c)

6. Neurons are the specialised cells of nervous system in
(a) humans only
(b) all vertebrates only
(c) mostly all animals including vertebrates and invertebrates
(d) both (a) and (B)

7. The neurons in all animals are capable of
i. detecting various stimuli in environment
ii. receiving and transmitting stimuli to CNS
iii. processing various internal and external stimuli
Select the most appropriate option.
(a) I, II and III are correct
(b) Only II is correct
(c) II and III are correct
(d) I and II are correct

8. The nervous system of Hydra is composed of
(a) brain and peripheral nerves
(b) network of neurons
(c) ganglia and plexuses
(d) brain and nerve nets

9. Why is nervous system of insects considered better organised as compared to Hydra and Planaria?
(a) Insects possess brain, ganglia and neural tissues.
(b) Hydra and planaria do not possess nervous system at all.
(c) Hydra possess the least developed nerve cord while insects contain highly developed ventral nerve cord.
(d) Planaria do not possess brain while a rudimentary brain is found in insects.

Topic 2 Human Neural System

10. The two major divisions of human neural system are
(a) CNS and brain
(b) ANS and PNS
(c) CNS and PNS
(d) Brain and spinal cord

11. Consider the following statements.
(a) Central nervous system is the major site of information processing.
(b) Central nervous system is composed of brain and cranial nerves.
Select the correct option.
(a) A is true, B is false.
(b) A is false, B is true.
(c) Both A and B are true.
(d) Both A and B are false.

12. The major structural component of peripheral nervous system is
(a) spinal cord (b) nerves
(c) visceral organs (d) all of these

13. Refer to the given diagram and choose the correct option accordingly.

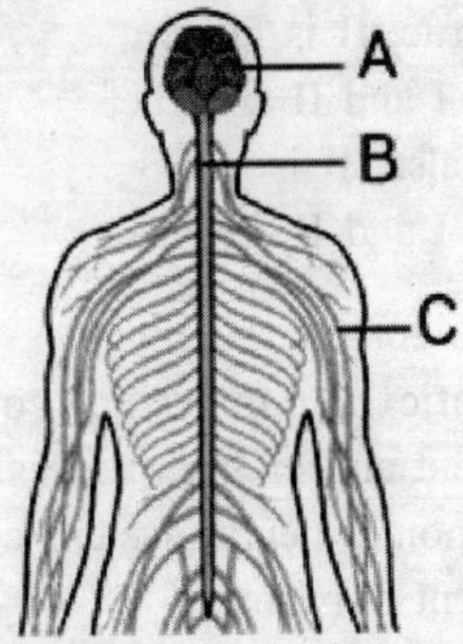

(a) C is the major site of information processing.
(b) A, B and C constitute central nervous system.
(c) C represents neurons.
(d) B is a component of CNS while 'C' constitutes PNS.

14. The nerve fibres of PNS are
(a) afferent
(b) efferent
(c) both afferent and efferent
(d) only motor

15. Refer to the diagram representing the transmission of impulse through PNS.

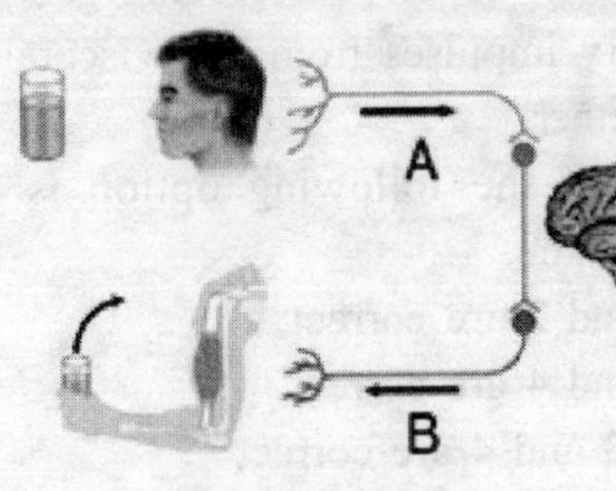

Which of the following option is most pertinent?
(a) A and B are afferent nerves.
(b) A is afferent nerve while B is efferent nerve
(c) A is efferent nerve while A is afferent nerve.
(d) Both A and B are efferent nerves.

16. Consider the following statements.
I. Two major division of CNS are somatic neural system and autonomic neural system.
II. Both somatic and autonomic neural system are antagonistic in their functions.

Select the correct option.
(a) I is true, II is false.
(b) Both I and II are true.
(c) I is false, II is true.
(d) Both I and II are false.

17. Select the correct statement.
(a) Somatic neural system consists of both afferent and efferent nerves.
(b) Autonomic neural system consists of only afferent fibres.
(c) Only efferent nerves are found in somatic and autonomic neural system.
(d) Both afferent and efferent fibres are found in somatic and autonomic neural system.

18. The two divisions of autonomic nervous system are
(a) antagonistic to each other
(b) complementary to each other
(c) highly reduced and non-functional in humans
(d) functional as a single system in humans

19. Visceral nervous system within human body
(1) is the division of peripheral nervous system.
(2) is the division of central nervous system.
(3) consists of nerve fibres and ganglia.
(4) carry impulses from one visceral organ to another.
Which of the following option is the most appropriate?
(a) 1 and 2 are correct
(b) 2 and 4 are correct
(c) 1, 3 and 4 are correct
(d) 2, 3, 4 are correct

20. Which component of neural system would control the functioning of heart and stomach?
(a) Somatic neural system.
(b) Only sympathetic nervous system.
(c) Only parasympathetic nervous system.
(d) Both sympathetic and parasympathetic nervous system.

21. Match the following columns:

	Column-I		Column-II
(a)	Afferent fibres	(1)	Involuntarily controlled muscles
(b)	Somatic neural system	(2)	Carry impulse away from the CNS
(c)	Autonomic neural system	(3)	Voluntarily controlled muscles
(d)	Efferent fibres	(4)	Carry impulse towards the CNS

Which of the following is the correct option?

	a	b	c	d
(a)	2	1	3	4
(b)	2	3	1	4
(c)	4	3	1	2
(d)	4	1	3	2

22. Identify A-C. [Pg-316,M]

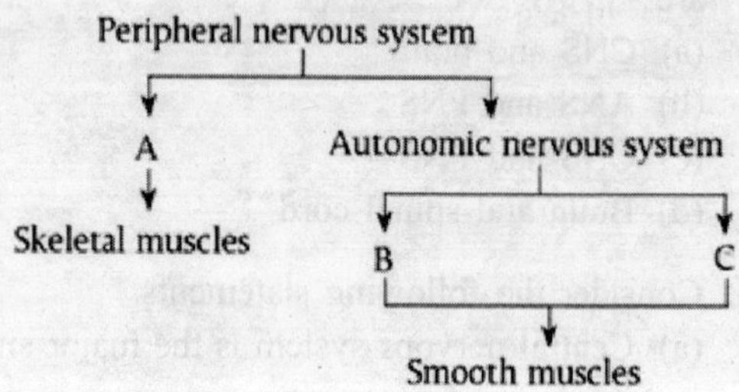

Select the correct option.

	A	B	C
(a)	Spinal cord	Afferent nerves	Efferent nerves
(b)	Somatic nervous system	Parasympathetic nervous system	Afferent and efferent nerves
(c)	Sympathetic nervous system	Efferent nerves	Afferent nerves

(d)	Somatic neural system	Sympathetic nervous system	Parasympathetic nervous system

23. Assertion: A person would not be able to perceive hot or cold sensation if afferent fibres of PNS are degenerated in him.
Reason: All the afferent nerve fibres are motor nerves. [Pg-316,H]
(a) Both Assertion and Reason are true and Reason is correct explanation of Assertion.
(b) Both Assertion and Reason are true, but Reason is not the correct explanation of Assertion.
(c) Assertion is true, but Reason is false.
(d) Assertion is false, but Reason is true.

24. Assertion: Autonomic nervous system is controlled voluntarily by humans.
Reason: All the muscular activities are inhibited by sympathetic nervous system.
(a) Both Assertion and Reason are true and Reason is correct explanation of Assertion.
(b) Both Assertion and Reason are true, but Reason is not the correct explanation of Assertion.
(c) Assertion is true, but Reason is false.
(d) Assertion is false, but Reason is true.

Topic 3 Functional Unit Of Neural System

25. The basic structural components of a neuron are
(a) cell body and axon
(b) cell body and dendrites
(c) axon and dendrites
(d) cell body, axon and dendrites

26. Identify the incorrectly matched pair of cell organelle found in neuron and its function.
(a) Nucleus – Contain DNA
(b) Mitochondria – Energy production
(c) Golgi apparatus – Protein synthesis
(d) Lysosomes – Scavengers of neurons

27. Neurons differ from a typical cell because of the presence of
(a) nucleolus
(b) Nissl's granules
(c) protein synthesizing machinery
(d) microfilaments

28. Nissl bodies are mainly composed of
(a) proteins and lipids
(b) DNA and RNA
(c) nucleic acids and SER
(d) free ribosomes and RER

29. The dendrites of a neuron are
(a) long, unbranched processes associated with cell body and axon
(b) short, highly branched processes of cell body
(c) long and branched processes of cell body
(d) short, unbranched processes of cell body and axon

30. Which of the following characteristic is correct about dendrites?
(a) non-functional, degenerating axons
(b) involved in carrying impulse away from the cell body
(c) involved in transmitting impulse toward the cell body
(d) involved in transmitting impulse toward and away from the cell body

31. The only similarity between the cell body and dendrites is the presence of
(a) golgi bodies
(b) Nissl's granules
(c) nucleus
(d) mitochondria

32. While studying the structural details of a neuron, how would you distinguish an axon from a dendrite?
(a) Axon is a long process whose distal end is branched.
(b) Dendrites are branched irregularly while the axon gives off long branches alternatively.

(c) Axon contains Nissl's granules while dendrites do not.
(d) Axons are numerous while each neuron contain only two dendrites.

33. Synaptic knob [Pg-317,M]
(1) is terminal bulb-like structure of dendrites and axons.
(2) contains neurotransmitter – filled vesicles.
(3) is a protoplasmic extension of cell body.
Which of the following option is most appropriate?
(a) 1 and 2 are correct
(b) Only 3 is correct
(c) Only 2 is correct
(d) 1 and 3 are correct

34. Consider the following statements.
(a) Axons carry impulse away from the cell body.
(b) The synaptic knob of axons are found in contact with neuro-muscular junctions.
Select the correct option.
(a) A is true, B is false.
(b) Both A and B are false.
(c) Both A and B are true.
(d) A is false, B is true.

35. Refer to the given diagram of the structure of a neuron and identify A, B and C.

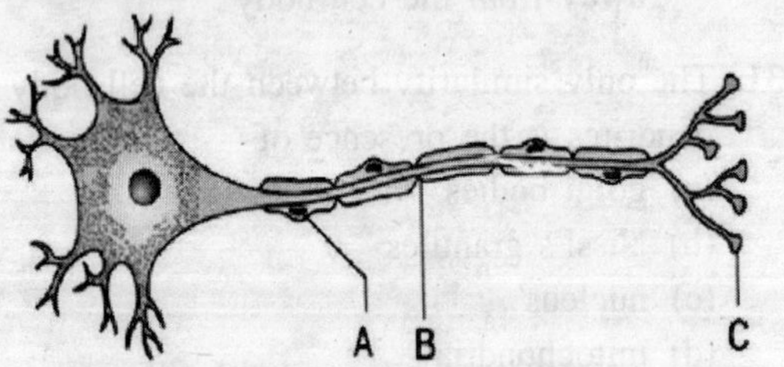

Select the correct option.

	A	B	C
(a)	Nissl's granule	Axon	Schwann Cell
(b)	Schwann cell	Nodes of Ranvier	Synaptic Knob
(c)	Synaptic Knob	Dendrite	Synaptic Knob
(d)	Nucleus	Myelin sheath	Nissl's granule

36. Match the following columns.

	Column-I		Column-II
(a)	Neurotransmitters	(1)	Ribosomal granules
(b)	Nissl's granules	(2)	Short and branched
(c)	Dendrites	(3)	Contained in synaptic knob
(d)	Axon	(4)	Carry impulse away from cell body

Select the correct option.

	a	b	c	d
(a)	1	3	2	4
(b)	3	1	4	2
(c)	3	1	2	4
(d)	1	3	4	2

37. On what basis, neurons are classified as unipolar, bipolar or multipolar?
(a) Transmission of impulse
(b) Number of axons and dendrites
(c) Sensory or motor nature
(d) Number of nucleus within cell body

38. A neuron with one axon and one dendrite is known
(a) unipolar (b) bipolar
(c) nonpolar (d) multipolar

39. A multipolar neuron contains multiple
(a) dendrites
(b) axons
(c) axons and dendrites
(d) synaptic bulbs

40. Select the correct statement regarding unipolar neuron.
(a) It contains cell body only.
(b) It contains one dendrite and one axon.
(c) It contains one dendrite and one cell body only.
(d) It contains one axon only.

41. Match the following columns.

	Column-I (Types of Neurons)		Column-II (Location)
(a)	Multipolar neuron	(1)	Embryonic stages
(b)	Bipolar neuron	(2)	Retina of eyes
(c)	Unipolar neuron	(3)	Cerebral cortex

Select the correct option.

	a	b	c
(a)	3	2	1
(b)	2	1	3
(c)	1	3	2
(d)	2	3	1

42. Consider the following statements.
(a) Myelinated and non-myelinated neurons are differentiated on the basis of type of neurons.
(b) Humans contain only myelinated neurons.

Select the correct option.
(a) A is true, B is false.
(b) A is false, B is true.
(c) Both A and B are true.
(d) Both A and B are false.

43. The myelin sheath around the axons is formed by
(a) osteocytes and astrocytes
(b) astrocytes and Schwann cells
(c) Schwann cells and oligodendrocytes
(d) oligodendrocytes and osteoclasts

44. Nodes of Ranvier are
(a) granulated bodies in cytoplasm
(b) gaps between adjacent myelin sheath on axons
(c) modulated bodies at the ends of dendrites
(d) vesicles at the terminal ends of axons

45. The myelinated neurons are found in
(a) cranial nerves
(b) spinal nerves
(c) nerves of ANS
(d) cranial and spinal nerves

46. Consider the following statements.
(a) Unmyelinated nerve fibres are commonly found in the cranial nerves.
(b) Unmyelinated nerve fibres transmit impulse at slower rate.

Select the correct option.
(a) Both A and B are true.
(b) A is true, B is false.
(c) Both A and B are false.
(d) A is false, B is true.

47. Assertion: The axons of neurons can receive signals from other neurons.
Reason: A multipolar neuron contains numerous axons.
(a) Both Assertion and Reason are true and Reason is correct explanation of Assertion.
(b) Both Assertion and Reason are true, but Reason is not the correct explanation of Assertion.
(c) Assertion is true, but Reason is false.
(d) Assertion is false, but Reason is true.

48. Assertion: The speed of nerve impulse along axon would slowdown in the absence of Schwann cells.
Reason: Schwann cells are abundantly found in cells body of neurons.
(a) Both Assertion and Reason are true and Reason is correct explanation of Assertion.
(b) Both Assertion and Reason are true, but Reason is not the correct explanation of Assertion.
(c) Assertion is true, but Reason is false.
(d) Assertion is false, but Reason is true.

Topic 4 Generation, Conduction and Transmission of Nerve Impulses

49. Select the incorrect statement.
(a) Neurons possess the excitability due to their polarised membranes.
(b) Neural membrane contains different types of ion channels.

(c) A resting neuron is not permeable to any ion.
(d) During polarised state, conduction of nerve impulse does not occur along axonal membrane.

50. The potential difference across an axonal membrane during rest is maintained by
(a) Na+ and Cl- ions
(b) Na+ and K+ ions
(c) K+ and Cl- ions
(d) Na+ and HCO3- ions

51. The resting axonal membrane is
(1) permeable to K+ ions.
(2) permeable to Na+ ions.
(3) impermeable to negatively charged proteins of axoplasm.
Select the correct option.
(a) 1 and 2 are correct.
(b) 1, 2 and 3 are correct.
(c) 2 and 3 are correct.
(d) 1 and 3 are correct.

52. Consider the following statements.
(a) At rest, the axoplasm inside the axon contain low concentration of K+ ions.
(b) The concentration gradient across axonal membrane is generated due to the different concentration of Na+ and K+ ions across it.
Select the correct option.
(a) A is true, B is false.
(b) Both A and B are false.
(c) A is false, B is true.
(d) Both A and B are true.

53. When a neuron is not conducting any impulse i.e. resting, the axonal membrane is –
(a) Comparatively more permeable to K^+ and impermeable (nearly impermeable) to Na^+
(b) Impermeable to negatively charged proteins present in the axoplasm
(c) Both
(d) More permeable to Na^+ ions than K^+ ion.

54. In a resting axonal membrane
(a) both outside and inside fluid is positively charged.
(b) both outside and inside fluid is negatively charged.
(c) outside fluid is positively charged while inside is negatively charged.
(d) inside fluid is positively charged while outside is negatively charged.

55. Match the following columns.

	Column-I		Column-II
(a)	Na+ ions during resting stage	(1)	More inside, less inside
(b)	K+ ions during resting stage	(2)	More outside, less inside
(c)	Action potential	(3)	Depolarised Membrane
(d)	Resting potential	(d)	Polarised membrane

Select the correct option.

	a	b	c	d
(a)	2	1	4	3
(b)	2	1	3	4
(c)	1	2	3	4
(d)	1	2	4	3

56. What change would you find in a polarised membrane after a stimulus is applied at a particular point (P)?
(a) The whole membrane becomes impermeable to Na^+ ions.
(b) At point P, the membrane allows movement of Na^+ and K^+ equally.
(c) At point P, the membrane becomes permeable to Na^+ ions.
(d) At point P, the membrane becomes impermeable to both Na^+ and K^+ ions.

57. A depolarised axonal membrane contains
(a) equal amount of Na^+ and K^+ ions at outside and inside.
(b) positive charge inside and negative charge outside.

(c) positive charge outside and negative charge inside.
(d) negative charge at both outside and inside.

58. $Na^+ - K^+$ pump [Pg-317,E]
I. Needs energy (ATP) to work
II. Expels 3 Na^+ for every $2K^+$ ions imported
III. Works against a concentration gradient
IV. Maintains resting potential
(a) All are correct
(b) Only II and III are correct
(c) Only I and III are correct
(d) None is correct

59. Consider the following statements.
(a) The stimulus-induced increased permeability of Na^+ ions helps in the conduction of action potential.
(b) Increased permeability of K^+ ions helps to restore the resting potential of the membrane.
Select the correct option.
(a) A is true, B is false.
(b) Both A and B are true.
(c) A is false, B is true.
(d) Both A and B are false.

60. Match the following columns.

	Column-I		Column-II
(a)	Synaptic vesicles	(1)	Bind to neurotransmitters
(b)	Receptors of postsynaptic membrane	(2)	Contain neurotransmitters
(c)	Electrical synapse	(3)	Rare in human system
(d)	Chemical synapse	(d)	Slow conduction of nerve impulse

Select the correct option.

	a	b	c	d
(a)	2	1	3	4
(b)	2	1	4	3
(c)	1	2	3	4
(d)	1	2	4	3

61. Which of the following set of structures constitutes a synapse?
(a) Axon terminal, synaptic vesicles, receptors.
(b) Presynaptic and pest synaptic membranes.
(c) Synaptic cleft, receptors, post synaptic membrane.
(d) Presynaptic membrane, synaptic cleft, post synaptic membrane.

62. During an action potential
(1) impulse is conducted along the axons
(2) Na+ ions move outwards
(3) permeability of membrane to K+ ions decreases
Select the most appropriate option.
(a) 1, 2, 3 are correct.
(b) 1 and 2 are correct.
(c) 1 and 3 are correct.
(d) Only 1 is correct.

Topic 5 Transmission of impulses

63. Synapse is a junction between
(a) two neurons
(b) CNS and PNS
(c) spinal cord and nerves
(d) cell body and axon

64. Select the incorrect statement
(a) Synaptic cleft is not necessarily found between all the neurons.
(b) At synapse, the impulse travels along the single direction, i.e., from presynaptic to post synaptic membrane.
(c) Electrical synapses are rare in human body.
(d) Chemical synapses are faster than the electrical synapses.

65. The chemicals found in the synaptic cleft are known as
(a) prohormones
(b) hormones
(c) neurotransmitters
(d) proenzymes

66. Consider the following statements.
(A) The neurotransmitters help in the transmission of impulses at chemical synapses.
(B) Neurotransmitters have no role in electrical synapses. Select the correct statement.
(a) A is true, B is false.
(b) A is false, B is true.
(c) Both A and B are false.
(d) Both A and B are true.

67. The neurotransmitter-filled synaptic vesicles are found in
(a) post synaptic membranes
(b) receptor sites of post-synaptic vesicles
(c) axon-terminal
(d) tips of dendrites

68. From the following diagram of axon terminal and synapse, identify at least two correctly labelled structures.

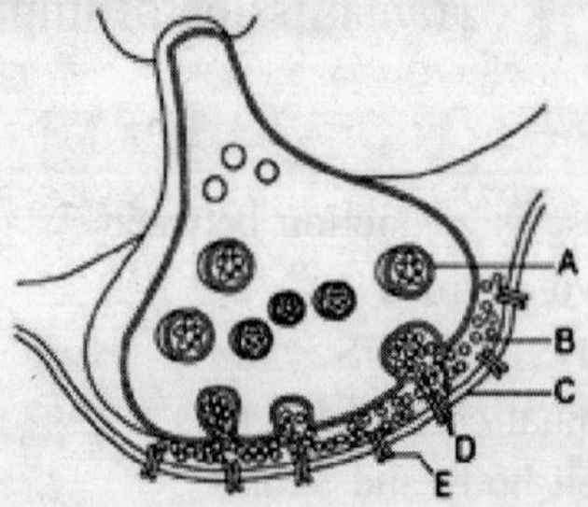

(a) B – Receptor, C – Neurotransmitter
(b) A – Synaptic Vesicles, E – Receptor
(c) C – Post synaptic membrane, D – K^+ ions
(d) D – Na^+ ions, A – Neurotransmitters

69. To release the neurotransmitters, synaptic vesicles
(1) get burst open
(2) require stimulation through action potential
(3) get digested by the lysosomes at axon terminals
Which of the following option is most appropriate?
(a) 1 and 2 are correct.
(b) 2 is correct.
(c) 2 and 3 are correct.
(d) 1 is correct.

70. Receptor sites for neurotransmitters are present on
(a) membranes of synaptic vesicles
(b) pre-synaptic membrane
(c) tips of axons
(d) post-synaptic membranes

71. The opening of ion-channels on postsynaptic membrane generates
(a) excitatory potential
(b) inhibitory potential
(c) either (a) or (b)
(d) no action potential

72. Assertion: Neurons are excitable cells.
Reason: The membrane of neurons remain in polarised state.
(a) Both Assertion and Reason are true and Reason is correct explanation of Assertion.
(b) Both Assertion and Reason are true, but Reason is not the correct explanation of Assertion.
(c) Assertion is true, but Reason is false.
(d) Assertion is false, but Reason is true.

73. Assertion: The resting axonal membrane possess positive charge outside.
Reason: The concentration of K+ ions is higher outside the axonal membrane at rest.
(a) Both Assertion and Reason are true and Reason is correct explanation of Assertion.
(b) Both Assertion and Reason are true, but Reason is not the correct explanation of Assertion.
(c) Assertion is true, but Reason is false.
(d) Assertion is false, but Reason is true.

74. Assertion: Synaptic cleft is the point of fusion of pre-synaptic and post synaptic membrane at synapse.
Reason: Impulse transmission across chemical synapse is faster than that across an electrical synapse.
(a) Both Assertion and Reason are true and Reason is correct explanation of Assertion.

(b) Both Assertion and Reason are true, but Reason is not the correct explanation of Assertion.
(c) Assertion is true, but Reason is false.
(d) Assertion is false, but Reason is true

Topic 6 Central Nervous System

75. Consider the following statements.
(A) Brain is the central control system of human body.
(B) Brain can control both voluntary movements and functioning of vital involuntary organs.
Select the correct option.
(a) A is true, B is false.
(b) Both A and B are false.
(c) A is false, B is true.
(d) Both A and B are true.

76. Select the incorrect statement.
(a) Brain is protected by the skull.
(b) Human brain can regulate thermoregulation and circadian rhythm of body.
(c) Inside the skull, humans possess two cranial meninges.
(d) Processing of vision and speech occur in human brain.

77. The cranial meninges from outer to inner region of brain are
(a) dura mater, arachnoid, pia mater
(b) pia mater, dura mater, arachnoid
(c) arachnoid, Pia mater, dura mater
(d) arachnoid, dura mater, pia mater

78. Which cranial meninges is in contact with brain tissue?
(a) Dura mater and arachnoid
(b) Arachnoid and pia mater
(c) Pia mater
(d) Dura mater

79. All the given structures are included in forebrain except
(a) cerebrum (b) hypothalamus
(c) pons (d) thalamus

80. Match the following columns.

	Column-I		Column-II
(a)	Cerebrum	(1)	Grey matter of brain
(b)	Cerebral cortex	(2)	Major part of human brain
(c)	Corpus callosum	(3)	Tract of nerve fibres
(d)	Association areas	(d)	Neither sensory nor motor in function

Which of the following is the correct option?

	a	b	c	d
(a)	4	3	2	1
(b)	2	1	3	4
(c)	1	4	1	2
(d)	3	2	4	1

81. The two longitudinal cerebral hemispheres are connected by
(a) cerebral cortex (b) association area
(c) corpus callosum (d) corpus albicans

82. Cerebral cortex is
(a) outer folded layer of cerebrum
(b) non-functional area of cerebrum
(c) inner white layer of cerebrum and cerebellum
(d) only functional area of cerebrum

83. Select the correct statement regarding cerebral cortex.
(a) It is white in appearance due to the presence of axons in it.
(b) It is white in appearance due to the presence of cell bodies of neurons.
(c) It is grey in appearance due to the presence of cell bodies of neurons.
(d) It is grey in appearance due to the presence of axons in it.

84. The cerebral cortex contains
(a) motor areas
(b) sensory areas
(c) motor and sensory areas
(d) motor, sensory and association areas

85. The inner region of cerebral hemisphere
(a) is grey in appearance
(b) contain axonal fibres covered by myelin sheath
(c) contain cell bodies of the neurons
(d) both (b) and (c)

86. Refer to the diagram showing sagittal section of human brain.

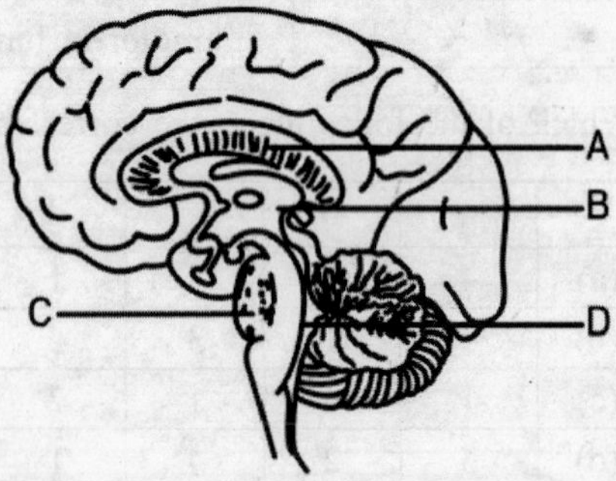

Identify the parts labelled as A–D.

	A	B	C	D
(a)	Cerebrum	Medulla	Cerebellum	Medulla
(b)	Hypothalamus	Cerebellum	Medulla	Pons
(c)	Corpus callosum	Thalamus	Pons	Cerebra 1 aqueduct
(d)	Thalamus	Corpus callosum	Medulla	Cerebellum

87. Thalamus in human brain [Pg-321,M]
(1) is surrounded by cerebrum.
(2) acts as a major coordinating centre for sensory and motor signalling.
(3) is under the direct control of hypothalamus.

Which of the following option is most appropriate?
(a) 1 and 2 are correct.
(b) 2 and 3 are correct.
(c) only 3 is correct.
(d) 1, 2 and 3 are correct.

88. The structure found at the base of thalamus
(a) is vestigial organ.
(b) controls urge of eating and drinking.
(c) involved in thermoregulation.
(d) both (b) and (c).

89. Hypothalamic hormones are secreted by
(a) glial cells (b) Schwann cells
(c) oligodendrocytes (d) neurosecretory cells

90. Which part of the brain is responsible for thermoregulation?
(a) Hypothalamus
(b) Corpus callosum
(c) Medulla oblongata
(d) Cerebrum

91. Which of the following structure or region is incorrectly paired with its function?
(a) Medulla oblongata: Controls respiration and cardiovascular reflexes.
(b) Limbic system: Consists of fibre tracts that interconnect different regions of brain; controls movement.
(c) Hypothalamus: Production of releasing hormones and regulation of temperature, hunger and thirst.
(d) Corpus callosum: Band of fibres connecting left and right cerebral hemispheres.

92. Limbic system within human brain is found
(a) at the base of brain stem
(b) inner portion of cerebral hemispheres
(c) adjacent to cerebellum
(d) above the cerebral cortex

93. All the listed structures are the parts of limbic system except
(a) Hippocampus (b) amygdala
(c) medulla (d) hypothalamus

94. Consider the following statements.
(a) Limbic system regulates endocrine activities along with hypothalamus.
(b) Limbic system helps to regulate excitement, pleasure, rage and fear.

Select the correct option.
(a) A is true, B is false.
(b) Both A and B are true.
(c) A is false, B is true.
(d) Both A and B are false.

95. Match the following columns.

	Column-I		Column-II
(a)	Hypothalamus	(1)	Inner part of cerebral cortex
(b)	White matter	(2)	Neurosecretory cells
(c)	Amygdala	(3)	Between thalamus and pons
(d)	Midbrain	(4)	Part of limbic system

Select the correct option.

	A	B	C	D
(a)	3	2	1	4
(b)	2	1	4	3
(c)	1	3	2	4
(d)	4	2	3	1

96. Cerebral aqueduct is found in
(a) forebrain
(b) midbrain
(c) hindbrain
(d) in between forebrain and midbrain

97. Corpora quadrigemina are
(a) four rounded swellings at dorsal portion of midbrain.
(b) two plate-like structures which separate forebrain and midbrain.
(c) circular hollow brain ventricles containing cerebrospinal fluid.
(d) elongated, cylindrical canals which connect midbrain to brainstem. Hindbrain

98. Which of the following structures is not found in hindbrain?
(a) Pons (b) Cerebellum
(c) Medulla (d) Hippocampus

99. What is the function of pons?
(a) It conveys information from midbrain to target organs.
(b) It controls involuntary smooth muscles and voluntary skeletal muscles.
(c) It interconnects different regions of the brain.
(d) All of these.

100. How can cerebellum be differentiated from the pons of hindbrain?
(a) Cerebellum has white matter outside while pons contain grey matter outside.
(b) Cerebellum has highly convoluted surface while pons contain fibre tracts.
(c) Cerebellum is smaller in size as compared to pons.
(d) All of these.

101. Consider the following statements.
(a) Cerebellum is a part of hindbrain.
(b) Cerebellum helps to maintain body posture and equilibrium.

Select the correct option.
(a) A is true, B is false.
(b) Both A and B are false.
(c) Both A and B are true.
(d) A is false, B is true.

102. The medulla oblongata
(1) is a part of hindbrain.
(2) control autonomic functions like breathing, heart rate, etc
(3) relay motor and sensory signals between spinal cord and higher brain regions.

Select the most appropriate option.
(a) 1, 2 and 3 are correct.
(b) 1 and 2 are correct.
(c) Only 1 is correct.
(d) Only 2 is correct.

103. The brain stem is formed by
(a) forebrain, midbrain, hindbrain
(b) midbrain, hindbrain
(c) cerebrum, cerebellum, spinal cord
(d) midbrain, pons, medulla oblongata

104. Match the following columns.

	Column I		Column II
(a)	Brain stem	(1)	Emotions
(b)	Cerebellum	(2)	Brain ventricle
(c)	Limbic system	(3)	Breathing and consciousness
(d)	Cerebral aqueduct	(4)	Balance and coordination

Select the correct option.

	A	B	C	D
(a)	4	3	2	1
(b)	2	1	4	3
(c)	3	4	1	2
(d)	1	2	3	4

105. Assertion: Cerebral cortex appear grey in colour.

Reason: It contains the cell bodies of the neurons.

(a) Both Assertion and Reason are true and Reason is correct explanation of Assertion.
(b) Both Assertion and Reason are true, but Reason is not the correct explanation of Assertion.
(c) Assertion is true, but Reason is false.
(d) Assertion is false, but Reason is true.

106. Assertion: Association areas can carry out complex functions like communication and memory.

Reason: These areas are completely motor in nature.

(a) Both Assertion and Reason are true and Reason is correct explanation of Assertion.
(b) Both Assertion and Reason are true, but Reason is not the correct explanation of Assertion.
(c) Assertion is true, but Reason is false.
(d) Assertion is false, but Reason is true.

107. Assertion: Damage of limbic system would affect the emotional behaviour of a person.

Reason: The amygdale is the emotion centre of the brain.

(a) Both Assertion and Reason are true and Reason is correct explanation of Assertion.
(b) Both Assertion and Reason are true, but Reason is not the correct explanation of Assertion.
(c) Assertion is true, but Reason is false.
(d) Assertion is false, but Reason is true.

Topic 7 Reflex Actions and Reflex arc

108. Reflex action is

(a) voluntarily controlled response of CNS.
(b) involuntary response to peripheral nervous stimulation.
(c) involuntary response to environmental stimulus which does not involve CNS.
(d) Both (b) and (c)

109. The reflex actions are controlled by

(a) CNS (b) PNS
(c) ANS (d) Both (b) and (c)

110. The reflex pathway is composed of

(a) afferent neurons only
(b) efferent neurons only
(c) motor neurons only
(d) Both (a) and (b)

111. Consider the following statements.

(a) Afferent neurons are found close to the sensory organs.
(b) The efferent neurons carry signals from CNS to the effectors.

Select the correct option.

(a) (a) is true, (b) is false.
(b) Both (a) and (b) are false.
(c) (a) is false, (b) is true.
(d) Both (a) and (b) are true.

112. Which of the following option represents the correct sequence of nerve impulse transmission in a reflex arc?

(a) Interneuron Dorsal root ganglion Sensory organ

(b) CNS Efferent neuron Interneuron
(c) Sensory organ Afferent neuron Dorsal root ganglion
(d) Efferent neuron CNS Afferent neuron

113. In a reflex arc, what is the role of an interneuron?
(a) It relays impulse to effector at motor end plate.
(b) It transmits impulse from the white matter to grey matter of spinal cord.
(c) It receives impulse from afferent neuron and transmits it to motor neuron.
(d) It connects two dorsal root ganglions.

114. Among the following listed structures, knee-jerk doesn't involve
(a) motor neuron (b) spinal cord
(c) interneuron (d) brain

115. In the given diagram of reflex action, identify A, B and C.

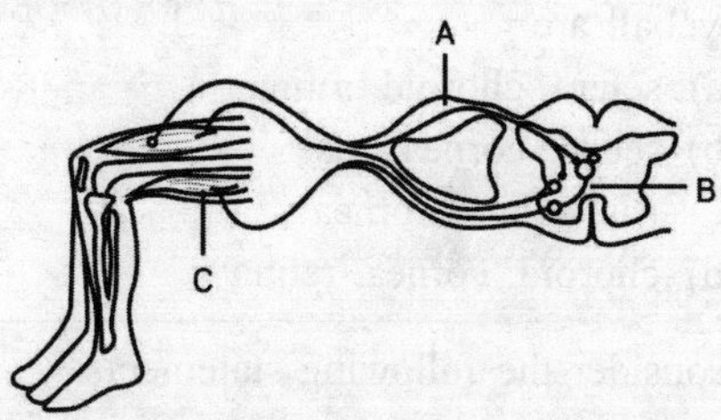

Select the correct option.

	A	B	C
(a)	White matter	Afferent nerve	Efferent nerve
(b)	Dorsal root ganglion	Interneuron	Motor endplate
(c)	Grey Matter	White matter	Interneuron
(d)	Motor neuron	Grey matter	receptor

116. Assertion: All autonomic actions of body are reflex action.

Reason: Reflex actions do not require CNS.
(a) Both Assertion and Reason are true and Reason is correct explanation of Assertion.
(b) Both Assertion and Reason are true, but Reason is not the correct explanation of Assertion.
(c) Assertion is true, but Reason is false.
(d) Assertion is false, but Reason is true.

117. Assertion: Sneezing in response to an allergen is reflex action

Reason: It is an involuntary action.
(a) Both Assertion and Reason are true and Reason is correct explanation of Assertion.
(b) Both Assertion and Reason are true, but Reason is not the correct explanation of Assertion.
(c) Assertion is true, but Reason is false.
(d) Assertion is false, but Reason is true.

118. Assertion: In a reflex arc, efferent neuron receives signal from sensory organ.

Reason: The afferent neuron relays impulses to PNS in a reflex arc.
(a) Both Assertion and Reason are true and Reason is correct explanation of Assertion.
(b) Both Assertion and Reason are true, but Reason is not the correct explanation of Assertion.
(c) Assertion is true, but Reason is false.
(d) Assertion is false, but Reason is true.

Topic 8 Sense Organs: Nose and Tongue

119. Consider the following statements.
(a) The olfactory receptors help us to receive the sense of smell.
(b) The olfactory receptors are coated by mucus.

Select the correct option.
(a) A is true, B is false.
(b) Both A and B are false.
(c) A is false, B is true.
(d) Both (a) and (b) are true.

120. The olfactory epithelium is made up of
(a) sensory cells
(b) basal cells
(c) sensory, sustentacular and basal cells

(d) sustentacular and sensory cells

121. The olfactory epithelium neurons transmit the signals from
(a) environment to olfactory bulb
(b) one olfactory bulb to another
(c) olfactory bulb to PNS
(d) environment to CNS

122. To which region of brain, olfactory bulb is connected?
(a) Cerebrum
(b) Limbic system
(c) Cerebral aqueduct
(d) Corpus callosum

123. Gustation and olfaction are achieved through
(a) proprioceptors (b) mechanoreceptors
(c) baroreceptors (d) chemoreceptors

124. Consider the following statements.
(a) The gustatory receptors are found at the base of tongue.
(b) The tongue detects taste through taste buds.

Select the correct option.
(a) (a) is true, (b) is false.
(b) Both (a) and (b) are false.
(c) (a) is false, (b) is true.
(d) Both (a) and (b) are true.

125. Assertion: The sense of olfaction is concerned with limbic system.

Reason: Olfactory bulb is a part of cerebrum.
(a) Both Assertion and Reason are true and Reason is correct explanation of Assertion.
(b) Both Assertion and Reason are true, but Reason is not the correct explanation of Assertion.
(c) Assertion is true, but Reason is false.
(d) Assertion is false, but Reason is true.

126. Assertion: Gustatory cells gets directly activated after being exposed to dissolved chemicals.

Reason: Three cranial nerves carry taste stimulus to brain.
(a) Both Assertion and Reason are true and Reason is correct explanation of Assertion.
(b) Both Assertion and Reason are true, but Reason is not the correct explanation of Assertion.
(c) Assertion is true, but Reason is false.
(d) Assertion is false, but Reason is true.

Topic 9 Eye

127. Select the incorrect statement.
(a) The sockets of skull containing eyes are orbits.
(b) The wall of eyes is made up of two layers.
(c) Retina is the innermost layer of eye ball.
(d) None of these.

128. The outermost, middle and innermost layers of eyeball are
(a) sclera, choroid, retina
(b) retina, cornea, iris
(c) iris, sclera, cornea
(d) choroid, cornea, retina

129. Consider the following statements.
(a) Cornea is the anterior portion of sclera.
(b) Cornea helps to refract the light entering the eyes.

Select the correct option.
(a) (a) is true, (b) is false.
(b) Both (a) and (b) are false.
(c) (a) is false, (b) is true.
(d) Both (a) and (b) are true.

130. Which of the following statement is correct?
(a) Cornea consists of dense connective tissue of elastin and can repair itself.
(b) Cornea is convex, transparent layer which is highly vascularised.
(c) Cornea consists of dense matrix of collagen and is the most sensitive portion of the eye.
(d) Cornea is an external, transparent and protective proteinaceous covering of the eyeball.

131. Match the following columns.

	Column I		Column II
(a)	Retina	(1)	Contains blood vessels
(b)	Sclera	(2)	Neural cells
(c)	Choroid	(3)	Dense connective tissue

Select the correct option.

	A	B	C
(a)	1	2	3
(b)	3	1	2
(c)	1	3	2
(d)	2	3	1

132. Which of the following statement is correct regarding choroid?

(a) It contains numerous blood vessels.
(b) It has no role in vision.
(c) It is the major component of the lens of eyes.
(d) It secretes aqueous humor.

133. The ciliary body is the anterior part of

(a) sclera (b) retina
(c) cornea (d) choroid

134. What is the correct description about iris?

(a) Nonvascular, opaque portion of eye
(b) Pigmented, opaque and vascular structure of eye.
(c) Nonvascular, visible coloured portion of eye.
(d) Vascular, non-pigmented posterior most portion of eye.

135. The transparent lens in the human eye is held in its place by

(a) ligament attached to ciliary body
(b) ligaments attached to the iris
(c) smooth muscles attached to the iris
(d) smooth muscles attached to the ciliary body

136. An aperture in front of lens is called

(a) fovea (b) blind spot
(c) pupil (d) iris

137. The diameter of pupil is regulated by

(a) aqueous humor (b) retina
(c) rods (d) iris

138. How many neural cells are found in retina?

(a) Two (b) Five
(c) Three (d) Four

139. The arrangement of neural cells in retina from inside to outside is

(a) ganglion cells, bipolar cells, photoreceptors cells.
(b) photoreceptor cells, ganglion cells, bipolar cells.
(c) bipolar cells, ganglion cells, photoreceptor cells.
(d) bipolar cells, photoreceptor cells, ganglion cells.

140. Within retina, the rods and cones are the type of

(a) bipolar and photoreceptor cells, respectively
(b) ganglion cells
(c) photoreceptor cells
(d) photoreceptor and ganglion cells

141. Consider the following statements.

(a) Photopigments are light sensitive proteins in rods and cones.
(b) Rods contain rhodopsin while cones contain three different type of photopigments.

Select the correct option.

(a) (a) is true, (b) is false.
(b) Both A and (b) are true.
(c) (a) is false, (b) is true.
(d) Both (a) and (b) are false.

142. Match the following columns.

	Column I		Column II
(a)	Iris	(1)	Secrete aqueous humor
(b)	Ciliary body	(2)	Photosensitive layer of eye
(c)	Retina	(3)	Devoid of receptor cells
(d)	Blind spot	(4)	Regulate the amount of light entering into eyes

Select the correct option.

	A	B	C	D
(a)	2	3	4	1
(b)	4	1	2	3
(c)	1	2	3	4
(d)	3	4	1	2

143. The cone cells of retina provide
(a) photopic vision
(b) colour vision
(c) photopic and colour vision
(d) scotopic vision

144. Which of the following statements is not correct?
(a) In the knee-jerk reflex, stimulus is the stretching of muscle and response is its contraction.
(b) An action potential in an axon does not move backward because the segment behind is in a refractory phase.
(c) Depolarisation of hair cells of cochlea results in the opening of the mechanically gated potassium-ion channels.
(d) Rods are very sensitive and contribute to daylight vision.

145. Rhodopsin in the rod cells of retina
(1) is purplish-red protein.
(2) contains opsin protein.
(3) contains retinene which is an aldehyde of vitamin D.

Select the most appropriate option.
(a) 1 and 2 are correct.
(b) 1 and 3 are correct.
(c) Only 2 is correct.
(d) 1, 2 and 3 are correct.

146. Good vision depends on adequate intake of carotene rich food.

Select the best option from the following statements.
(a) Vitamin A derivatives are formed from carotene.
(b) The photopigments are embedded in the membrane discs of the inner segment.
(c) Retinal is derivative of Vitamin A.
(d) Retinal is a light absorbing part of all the visual photopigments.

Options:
(a) B, C, A (b) A, B
(c) A, C, D (d) A, C

147. Which of the following options is wrong?
(a) Eye muscles are attached with sclera
(b) Visual purple 1s concerned with dim light, while visual violet is concerned with bright light
(c) The colour differentiation is done by cones
(d) None

148. What would happen if the three photopigments of conc cells are stimulated equally?
(a) No colour vision and daylight vision would be produced.
(b) Sensation of black light would be produced
(c) Sensation of white light would be produced.
(d) Mosaic vision of different colours would be produced.

149. Photosensitive compound in human eye is made up of
(a) opsin and retinal
(b) opsin and retinol
(c) transducin and retinene
(d) guanosine and retinol

150. Consider the following statements.
(a) The optic nerves and blood vessels enter the eyes through blind spot.
(b) Blind spot contains abundant rod cells but no cone cells. Select the correct option.
(a) (a) is true, (b) is false.
(b) Both (a) and (b) are false.
(c) (a) is false, (b) is true.
(d) Both (a) and (b) are true.

151. Select the correct statement.
(a) Macula lutea is a red coloured spot at the entrance of optic nerves.
(b) Fovea is the central portion of macula lutea, containing abundant cone cells.

(c) Macula lutea is the only structure of retina which contain rod and cone cells.
(d) No true image is formed at fovea due to the overlapping of photoreceptor cells.

152. The point of greatest visual acuity in human eye is
(a) fovea (b) blind spot
(c) iris (d) pupil

153. Consider the following statements.
(a) Aqueous chamber containing aqueous humor is found in the space between lens and retina.
(b) Vitreous chamber containing vitreous humor is found in the space between lens and cornea.

Select the correct option.
(a) (a) is true, (b) is false.
(b) Both (a) and (b) are false.
(c) (a) is false, (b) is true.
(d) Both (a) and (b) are true.

154. In the given structure of human eye, identify the location of fovea, cornea choroid and sclera.

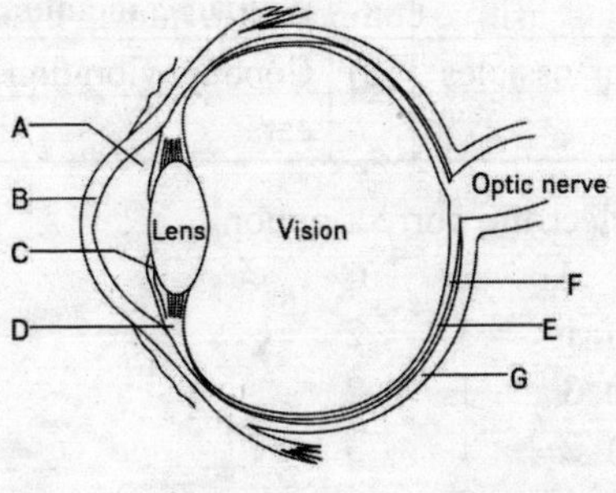

	Fovea	Cornea	Choroid	Sclea
(a)	B	C	A	D
(b)	G	B	D	A
(c)	F	B	E	G
(d)	D	A	B	C

155. The light rays of visible wavelength are focused on
(a) retina through cornea
(b) lens through pupil
(c) pupil through iris
(d) retina through iris

156. How does the potential differences generated in photoreceptor cells when light of suitable wavelength enters the human eye?
(a) Light cause chemical modification of rhodopsin and iodopsin so that they form a new compound.
(b) Light causes the conversion of opsin to retinal.
(c) Light causes dissociation of opsin and retinal so as to cause structural changes of opsin.
(d) Light causes destruction of opsin and retinal so that iodopsin can change membrane potential.

157. The route of transmission of action potential by optic nerves to visual cortex of brain is
(a) Photoreceptor cells » Bipolar cells » Ganglion cells.
(b) Ganglion cells » Bipolar cells » Photoreceptor cells
(c) Bipolar cells » Photoreceptor cells » Ganglion cells
(d) Bipolar cells » Ganglion cells » Photoreceptor cells

158. Assertion: Sclera and cornea form the fibrous coat of human eye.

Reason: Sclera and cornea contain abundant blood vessels.
(a) Both Assertion and Reason are true and Reason is correct explanation of Assertion.
(b) Both Assertion and Reason are true, but Reason is not the correct explanation of Assertion.
(c) Assertion is true, but Reason is false.
(d) Assertion is false, but Reason is true.

159. Assertion: The posterior region of the eyeball possesses the points of no vision and maximum visual resolution.

Reason: Fovea contain abundant rod cells but lack cone cells.
(a) Both Assertion and Reason are true and Reason is correct explanation of Assertion.

(b) Both Assertion and Reason are true, but Reason is not the correct explanation of Assertion.
(c) Assertion is true, but Reason is false.
(d) Assertion is false, but Reason is true.

160. Assertion: Neural impulses are analysed by the visual cortex of the brain.

Reason: Image formed on retina is recognised based on earlier memory and experience.
(a) Both Assertion and Reason are true and Reason is correct explanation of Assertion.
(b) Both Assertion and Reason are true, but Reason is not the correct explanation of Assertion.
(c) Assertion is true, but Reason is false.
(d) Assertion is false, but Reason is true. Ear

161. The structural component of external ear is
(a) pinna
(b) pinna and auditory meatus
(c) pinna, auditory meatus and eardrum
(d) auditory meatus and tympanic membrane

162. Consider the following statements.
(a) Fine hairs and wax secreting glands are found in the skin of pinna and auditory meatus.
(b) Pinna and auditory meatus are vestigial organs of human ear.

Select the correct option.
(a) (a) is true, (b) is false.
(b) Both (a) and (b) are false.
(c) (a) is false, (b) is true.
(d) Both (a) and (b) are true.

163. The wax-secreting glands in auditory meatus is
(a) Weber's glands
(b) Ebner's glands
(c) Ceruminous glands
(d) Krause's glands

164. The membrane found between the outer and middle ear is
(a) basilar membrane
(b) Reissner's membrane
(c) tympanic membrane
(d) tectorial membrane

165. The outer and inner surface of tympanic membrane is composed of
(a) connective tissues
(b) mucus membrane
(c) mucus membrane outside and connective tissue inside
(d) connective tissue outside and mucus membrane inside

166. The middle ear consists of
(a) ear ossicles
(b) cochlea and labyrinth
(c) auditory meatus and tympanic membrane
(d) ear ossicles and cochlea

167. Match the following columns.

	Column I		Column II
(a)	Auditory meatus	(1)	Receive sound vibrations from environment
(b)	Pinna	(2)	Conduct vibrations to tympanic membrane
(c)	Ear ossicles	(3)	Conduct vibrations to inner ear

Select the correct option.

	A	B	C
(a)	1	3	2
(b)	3	1	2
(c)	2	1	3
(d)	1	2	3

168. The arrangement of ear ossicles from outer to inner ear is
(a) stapes, malleus, incus
(b) incus, stapes, malleus
(c) malleus, stapes, incus
(d) malleus, incus, stapes

169. Select the incorrect statement.
(a) The stapes is found attached to the oval window of cochlea.
(b) Ear ossicles increase the efficiency of sound wave transmission to inner ear.

(c) Eustachian tube connects inner ear to the pharynx.
(d) Eustachian tube helps in equalising the pressure on either side of tympanic membrane.

170. Consider the following statements.
(a) The fluid-filled inner ear has two parts - bony labyrinth and membranous labyrinth.
(b) The membranous labyrinth surrounds the bony labyrinth in inner ear.

Select the correct option. [Pg-325,M]
(a) (a) is true, (b) is false.
(b) (a) is false, (b) is true.
(c) Both (a) and (b) are false.
(d) Both (a) and (b) are true.

171. Match the following columns.

	Column I		Column II
(a)	Labyrinth	(1)	Ear drum
(b)	Tympanic membrane	(2)	Filled with perilymph
(c)	Bony labyrinth	(3)	Inner ear
(d)	Membranous labyrinth	(4)	Filled with endolymph

Choose the most appropriate match.

	A	B	C	D
(a)	3	1	4	2
(b)	3	1	2	4
(c)	1	3	2	4
(d)	1	3	4	2

172. Identify the incorrectly matched pair.
(a) Membranous labyrinth Surrounded by perilymph
(b) Scala vestibuli Coiled portion of labyrinth
(c) Cochlea Responsible for hearing
(d) Scala media Filled with endolymph

173. Within the bony labyrinth, the three chambers (upper to lower) are
(a) scala vestibuli, scala media, scala tympani
(b) scala tympani, scala corti, scala media
(c) scala utricle, scala media, scala vestibule
(d) scala corti, scala media, scala vestibuli

174. Match the following columns.

	Column I		Column II
(a)	Reissner's membrane	(1)	In between outer and middle ear
(b)	Basilar membrane	(2)	A component of organ of corti
(c)	Tympanic membrane	(3)	In between scala vestibuli and scala media
(d)	Tectorial membrane	(4)	In between scala media and scala tympani

Select the most appropriate option.

	A	B	C	D
(a)	4	2	1	3
(b)	2	3	4	1
(c)	4	3	1	2
(d)	3	4	1	2

175. The scala vestibuli and scala tympani terminates at
(a) oval window of cochlea.
(b) round window of middle ear.
(c) oval window and round window, respectively.
(d) round window and oval window, respectively.

176. In the given diagram of cochlea, identify A, B and C.

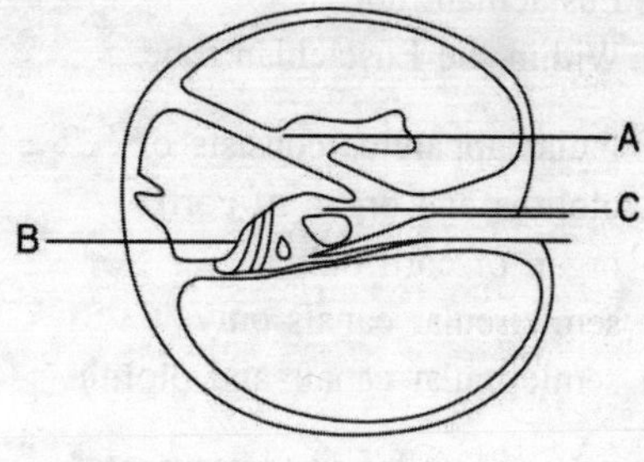

Select the correct option.

	A	B	C
(a)	Reissner's membrane	Organ of corti	Tectorial membrane
(b)	Tectorial membrane	Utricule	Basilar membrane

(c)	Basilar membrane	Tectorial membrane	Reissner's membrane
(d)	Basilar membrane	Macula	Organ of Corti

177. The organ of corti within cochlea can be located on
(a) tectorial membrane
(b) basilar membrane
(c) Reissner's membrane
(d) tympanic membrane

178. Select the incorrect statement regarding the structure of organ of corti.
(a) It contains hair cells that acts as auditory receptors.
(b) The hair cells are found on the inner side, arranged in rows.
(c) The basal end of hair cells is closely associated with the afferent nerve fibre.
(d) In between ends of hair cells and afferent nerves, tectorial membrane is present.

179. What is the location of vestibular apparatus in human ear?
(a) Adjacent to malleus of middle ear.
(b) Above the cochlea of inner ear.
(c) At the junction of round window and Eustachian tube.
(d) Within the Eustachian tube.

180. Vestibular apparatus consists of
(a) otoliths and organ of corti
(b) organ of corti only
(c) semicircular canals only
(d) semicircular canals and otolith

181. Consider the following statements.
(a) The three semicircular canals of inner ear are found in different planes at right angle to each other.
(b) The two semicircular canals play major role in hearing while the third canal is vestigial.

Select the correct option.
(a) (a) is true, (b) is false.
(b) (a) is false, (b) is true.
(c) Both (a) and (b) are false.
(d) Both (a) and (b) are true.

182. The sensory part of otoliths is
(a) utricle (b) saccule
(c) macula (d) Both (a) and (b)

183. Crista ampullaris containing hair cells is found
(a) within otoliths.
(b) at the base of semicircular canals.
(c) within the semicircular canals.
(d) at the tip of semicircular canals and otoliths.

184. Which of the following receptors are specifically responsible for maintenance of balance of body and posture?
(a) Crista ampullaris and macula
(b) Basilar membrane and otoliths
(c) Hair cells and organ of corti
(d) Tectorial membrane and macula

185. Refer to the given diagram and identify the correct function of the labeled structures.

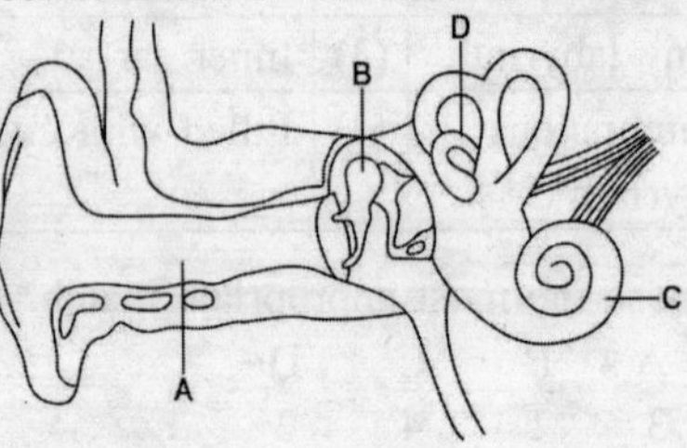

(a) B – Stapes – Vibration of oval window
(b) A – Auditory meatus – Transmit neural signals to auditory cortex
(c) D – Semicircular canals – Amplification of auditory signals
(d) C – Cochlea – Main hearing organ

186. Refer to the following events which occur during hearing.
(I) Rippling in basilar membrane
(II) Waves in the lymph of cochlea
(III) Vibration of ear drum
(IV) Pressing of hair cells against the tectorial membrane

Select the correct order in which the following events occur.
(a) II, III, I, IV (b) I, IV, II, III
(c) III, II, I, IV (d) II, IV, I, III

187. How does the nerve impulses in ear are generated?
(a) Due to the bending of hair cells towards tectorial membrane.
(b) Due to the vibrations in utricle and otoliths.
(c) Due to the streaming of perilymph in semicircular canals.
(d) Due to the contact between crista ampularis and macula.

188. Assertion: Stapes is found attached to the tympanic membrane.

Reason: Stapes is the largest ear bone.
(a) Both Assertion and Reason are true and Reason is correct explanation of Assertion.
(b) Both Assertion and Reason are true, but Reason is not the correct explanation of Assertion.
(c) Assertion is true, but Reason is false.
(d) Assertion is false, but Reason is true.

189. Assertion: The hair cells of organ of corti acts as auditory receptors.

Reason: The base of hair cells is in close contact with afferent fibres of auditory nerves.
(a) Both Assertion and Reason are true and Reason is correct explanation of Assertion.
(b) Both Assertion and Reason are true, but Reason is not the correct explanation of Assertion.
(c) Assertion is true, but Reason is false.
(d) Assertion is false, but Reason is true.

190. Assertion: Vestibular apparatus has no role in hearing.

Reason: Crista and macula helps in maintaining body posture and balance.
(a) Both Assertion and Reason are true and Reason is correct explanation of Assertion.
(b) Both Assertion and Reason are true, but Reason is not the correct explanation of Assertion.
(c) Assertion is true, but Reason is false.
(d) Assertion is false, but Reason is true.

ANSWER KEY

1. (b)	2. (c)	3. (b)	4. (d)	5. (d)	6. (c)	7. (a)	8. (b)	9. (a)	10. (c)
11. (a)	12. (b)	13. (d)	14. (c)	15. (b)	16. (c)	17. (d)	18. (a)	19. (a)	20. (d)
21. (d)	22. (d)	23. (c)	24. (c)	25. (d)	26. (c)	27. (b)	28. (d)	29. (b)	30. (c)
31. (c)	32. (a)	33. (c)	34. (c)	35. (b)	36. (c)	37. (b)	38. (b)	39. (a)	40. (d)
41. (a)	42. (a)	43. (c)	44. (b)	45. (d)	46. (d)	47. (d)	48. (c)	49. (c)	50. (b)
51. (d)	52. (c)	53. (c)	54. (b)	55. (b)	56. (c)	57. (b)	58. (a)	59. (b)	60. (a)
61. (d)	62. (c)	63. (a)	64. (d)	65. (c)	66. (d)	67. (c)	68. (b)	69. (b)	70. (d)
71. (c)	72. (a)	73. (c)	74. (d)	75. (d)	76. (c)	77. (a)	78. (c)	79. (c)	80. (b)
81. (c)	82. (a)	83. (d)	84. (b)	85. (b)	86. (c)	87. (a)	88. (d)	89. (d)	90. (a)
91. (b)	92. (b)	93. (c)	94. (b)	95. (b)	96. (b)	97. (a)	98. (d)	99. (c)	100. (b)
101. (c)	102. (a)	103. (d)	104. (c)	105. (a)	106. (c)	107. (a)	108. (b)	109. (a)	110. (d)
111. (d)	112. (c)	113. (c)	114. (d)	115. (b)	116. (d)	117. (b)	118. (d)	119. (d)	120. (c)
121. (a)	122. (b)	123. (d)	124. (c)	125. (c)	126. (b)	127. (b)	128. (a)	129. (d)	130. (c)
131. (d)	132. (a)	133. (d)	134. (b)	135. (a)	136. (c)	137. (d)	138. (c)	139. (a)	140. (c)
141. (b)	142. (b)	143. (c)	144. (d)	145. (a)	146. (d)	147. (d)	148. (c)	149. (a)	150. (a)
151. (b)	152. (a)	153. (d)	154. (c)	155. (a)	156. (c)	157. (a)	158. (c)	159. (c)	160. (b)
161. (b)	162. (a)	163. (c)	164. (c)	165. (d)	166. (a)	167. (c)	168. (d)	169. (c)	170. (a)
171. (b)	172. (b)	173. (a)	174. (d)	175. (c)	176. (a)	177. (b)	178. (d)	179. (b)	180. (d)
181. (a)	182. (c)	183. (b)	184. (a)	185. (c)	186. (c)	187. (a)	188. (d)	189. (a)	190. (b)

10 Chapter

Chemical Control and Coordination

MULTIPLE CHOICE QUESTIONS

Topic 1	Human Endocrine System Introduction

1. Both neurotransmitters and hormones
   (a) produce a slower response.
   (b) bind to receptors on or in their target cells.
   (c) are transported by interstitial fluid.
   (d) produce a rapid response.

2. Consider the following statements:
   (a) Exocrine glands are the ductless glands.
   (b) Sweat glands and gastric glands release their secretions directly into the blood.
   Select the correct option.
   (a) Both (a) and (b) are true.
   (b) (a) is true but (b) is false.
   (c) Both (a) and (b) are false.
   (d) (a) is false but (b) is true.

3. Which of the following gland types is correctly matched with its examples?
   (a) Adrenal gland: exocrine gland
   (b) Placenta: exocrine gland
   (c) Pancreas: endocrine gland only
   (d) Thyroid : endocrine gland

4. Which of the following statements about hormones is/are correct?
   I. Hormones are non-nutrient chemicals
   II. Hormones act as intercellular messengers
   III. Hormones are produced in trace amount
   IV. Hormones may be proteins, steroids, glycoproteins and bigenic mines
   (a) All (b) I, II, III
   (c) IV (d) I, III

5. The organ system that helps regulate the metabolic activities through secretion of hormones is
   (a) cardiovascular system
   (b) endocrine system
   (c) nervous system
   (d) digestive system

6. The nervous and endocrine systems interact to coordinate the functions of all the body systems and help to maintain
   (a) homeostasis
   (b) reflex actions
   (c) fight or flight response
   (d) stress responses

7. Responses of the endocrine system are ______ and ______ than the responses of the nervous system.
   (a) rapid, long-lasting
   (b) rapid, briefer
   (c) slower, long-lasting
   (d) slower, briefer

8. Which of the following statements correctly differentiate the endocrine and nervous systems?
   (a) The endocrine system regulates all types of body cells.
   (b) Nerve fibers innervate all the body cell types.
   (c) Endocrine system acts on specific muscle cells only.
   (d) The influence of the nervous system is much broader.

Topic 2 Gland : Endocrine Gland

9. Assertion: The nervous system produces a delayed response that lasts for a longer period.

Reason: Neurotransmitters are released from postsynaptic neurons.

(a) Both Assertion and Reason are true and Reason is correct explanation of Assertion.
(b) Both Assertion and Reason are true, but Reason is not the correct explanation of Assertion.
(c) Assertion is true, but Reason is false.
(d) Assertion is false, but Reason is true.

10. Which of the following statements is false?

(a) Hormones provide chemical coordination, integration and regulation in the human body
(b) Hormones regulate metabolism, growth and development of our organs
(c) Besides hypothalamus, pituitary, pineal, thyroid, adrenal, parathyroid, thymus, etc., GIT, heart, kidney, etc also produce hormones.
(d) Hormone can be used again and again like biocatalyst

Topic 3 Chemical Control and Coordination

11. The small part of the brain that is present below the thalamus and serves as the main link between the nervous and endocrine system is

(a) pons
(b) hypothalamus
(c) brain stem
(d) medulla oblongata

12. Consider the following events:

(a) Production of regulatory hormones from neurosecretory cells.
(b) Hormones are secreted into the portal system.
(c) Hormones move down the axons to axon endings.

Arrange them in sequential order and select the correct option.

(a) A, B, C (b) A, C, B
(c) B, C, A (d) C, A, B

13. The hormones produced by hypothalamic nucle

(a) regulate the functions of the anterior pituitary.
(b) regulate the functions of the posterior pituitary.
(c) regulate the functions of both anterior and posterior pituitary.
(d) inhibit the secretion of posterior pituitary hormones.

14. Which of the following statements about hypothalamic hormones is incorrect?

(a) Hypothalamic releasing hormones stimulate the secretion of anterior pituitary hormones.
(b) Hypothalamic releasing hormones stimulate the secretion of posterior pituitary hormones.
(c) Hypothalamic inhibiting hormones inhibit the secretion of anterior pituitary hormones.
(d) Somatostatin is a hypothalamic inhibitory hormone.

15. GnRH, a hypothalamic hormone, needed in reproduction, acts on

(a) the posterior pituitary gland and stimulates secretion of LH and relaxin.
(b) the anterior pituitary gland and stimulates secretion of LH and oxytocin.
(c) the anterior pituitary gland and stimulates secretion of LH and FSH.
(d) the posterior pituitary gland and stimulates secretion of oxytocin and FSH.

16. Which of the following hypothalamic hormones is incorrectly matched with its function?

(a) TRH: Stimulates secretion of thyrotropin
(b) PIH: suppresses the secretion of prolactin
(c) GH-RH: stimulates secretion of growth hormone
(d) CRH: stimulates the release of prolactin

17. How many of the following are correct?
(i) Somatostatin is released from hypothalamus.
(ii) Hypothalamic hormones reach the pituitary gland through hypophyseal-hypoportal system.
(iii) Anterior pituitary is under direct neural regulation of hypothalamus.
(a) 1 (b) 2
(c) 3 (d) None of these

18. Portal blood vessels connect the _____ to the _____
(a) Hypothalamus, brain
(b) Hypothalamus, posterior pituitary
(c) hypothalamus, anterior pituitary
(d) Anterior pituitary, posterior pituitary

Topic 4 Pituitary Gland

19. The pituitary gland is a pea-shaped gland that lies in the hypophyseal fossa of
(a) sella turcica of the glenoid bone
(b) sella turcica of the sphenoid bone
(c) sella turcica of the parietal bone
(d) sella turcica of the frontal bone .

20. The pituitary gland is connected to the hypothalamus by
(a) infundibulum (b) bony cavity
(c) hyaline cartilage (d) elastic cartilage

21. Following is the diagrammatic representation of the pituitary gland and its connection with the hypothalamus.

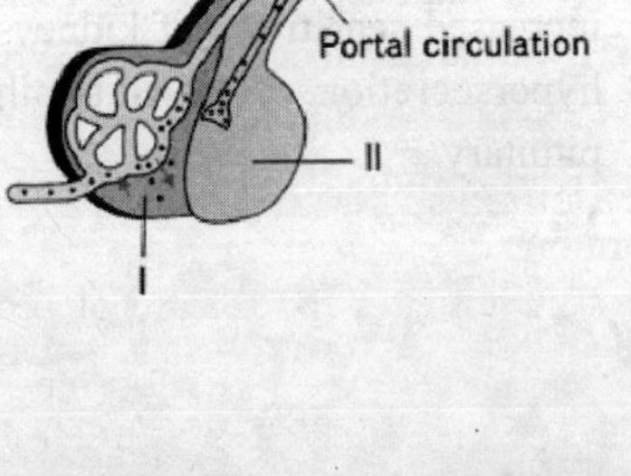

Select the correct option regarding the same.
(a) I: Anterior pituitary that is composed of neural tissues.
(b) I: Anterior pituitary that is composed of epithelial tissues.
(c) II: Posterior pituitary that is composed of epithelial tissues.
(d) II: Posterior pituitary that is composed of connective tissues.

22. In adults, adenohypophysis accounts for 75% of the total weight of the gland and consists of
(a) pars distalis
(b) pars tuberalis and pars intermedia
(c) pars nervosa
(d) pars intermedia

23. Which of the following hormone is secreted by pars intermedia?
(a) Prolactin
(b) Thyroid-stimulating hormone
(c) Adrenocorticotrophic hormone
(d) Melanocyte stimulating hormone

24. Which of the following statement is incorrect about the posterior pituitary?
(a) The posterior pituitary lobe is not a true endocrine gland.
(b) It serves as a hormone storage region.
(c) The posterior pituitary lobe and infundibulum together make neurohypophysis.
(d) Oxytocin and vasopressin hormones are synthesized in the posterior pituitary lobe.

25. Malfunctioning of endocrine gland deviates the body from homeostasis and causes several disorders. Which of the following disorder is correctly matched with its respective endocrine gland?
(a) Gigantism: Hyposecretion of growth hormone
(b) Dwarfism: Hypersecretion of growth hormone
(c) Acromegaly: Hypersecretion of growth hormone
(d) Gigantism: Hypersecretion of somatostatin

26. Hypersecretion of growth hormone in adults does not cause a further increase in height, because
(a) muscle fibers do not grow in size after birth.
(b) growth hormone becomes inactive in adults.
(c) epiphyseal plates close after adolescence.
(d) bones lose their sensitivity to growth hormone in adults.

27. A person is diagnosed with hypersecretion of growth hormone due to a pituitary tumor. Select the incorrect statement about his medical condition.
(a) Hypersecretion of growth hormone would cause hyperglycemia.
(b) There will be a reduced secretion of insulin.
(c) Insulin secretion would remain unaffected.
(d) Hypersecretion of growth hormone results in a diabetogenic effect.

28. Hypersecretion of prolactin hormone in females causes inappropriate lactation and absence of menstrual cycle. Based on the given information, select the pair of correct statements.
(a) Prolactin is required for milk ejection from mammary glands.
(b) The blood level of prolactin is increased just before menstruation.
(c) During pregnancy, prolactin inhibiting hormone suppresses the release of prolactin.
(d) Sucking action of newborn inhibits the release of PIH.
(a) A and B (b) B and C
(c) A and C (d) B and D

29. Select the option that correctly matches the hormone with its source endocrine gland and respective function/target organs.

	Hormone	Endocrine gland	Function/target organ
(A)	TSH	Anterior pituitary	All body cells
(B)	ACTH	Posterior pituitary	Stimulates secretion of glucocorticoids from the adrenal cortex
(C)	LH	Anterior pituitary	Triggers secretion of androgens in males
(D)	FSH	Posterior pituitary	Stimulates the growth of ovaria follicles

30. Which of the following set of hormones are called gonadotropins?
(a) GH and LH (b) LH and FSH
(c) LH and ACTH (d) FSH and TSH

31. The target cells/structures of FSH and LH in human males are
(a) interstitial cells and sertoli cells respectively
(b) sertoli cells and interstitial cells respectively
(c) interstitial cells only.
(d) sertoli cells only.

32. LH is required for fertility in females because
(a) it maintains corpus luteum.
(b) it induces ovulation.
(c) it stimulates the release of FSH from corpus luteum.
(d) both (a) and (b).

33. Which of the following hormone is incorrectly matched with its target organ/cells?
(a) MSH: melanocytes
(b) Oxytocin: Smooth muscles
(c) Vasopressin: distal tubules of kidneys
(d) ADH: Glomerulus

34. Diuresis refers to [Pg- 334,E]
(a) loss of water through urine
(b) reduced loss of water through urine
(c) process of urine formation in nephrons
(d) contraction of muscles of the urinary bladder

35. Diabetes insipidus is caused by
(a) hyposecretion of insulin by the pancreas
(b) hyposecretion of ADH by the posterior pituitary
(c) increased sensitivity of kidneys for ADH
(d) hypersecretion of ADH by the posterior pituitary

36. Assertion: TSH stimulates the thyroid gland to secrete thyroid hormones.

Reason: The hormones of anterior pituitary that regulate the secretions of other endocrine glands are called tropic hormones.
(a) Both Assertion and Reason are true and Reason is correct explanation of Assertion.
(b) Both Assertion and Reason are true, but Reason is not the correct explanation of Assertion.
(c) Assertion is true, but Reason is false.
(d) Assertion is false, but Reason is true.

37. Assertion: Hyposecretion of growth hormone causes stunted growth in infants.

Reason: Epiphyseal plates are closed after the childbirth.
(a) Both Assertion and Reason are true and Reason is correct explanation of Assertion.
(b) Both Assertion and Reason are true, but Reason is not the correct explanation of Assertion.
(c) Assertion is true, but Reason is false.
(d) Assertion is false, but Reason is true.

38. Assertion: Hypersecretion of GH during childhood causes acromegaly.

Reason: The growth hormone regulates the fluid-electrolyte balance of the body.
(a) Both Assertion and Reason are true and Reason is correct explanation of Assertion.
(b) Both Assertion and Reason are true, but Reason is not the correct explanation of Assertion.
(c) Assertion is true, but Reason is false.
(d) Assertion is false, but Reason is true.

39. Assertion: ADH secretion is stimulated under the conditions of dehydration.

Reason: ADH prevents water loss from the body by decreasing the urine volume.
(a) Both Assertion and Reason are true and Reason is correct explanation of Assertion.
(b) Both Assertion and Reason are true, but Reason is not the correct explanation of Assertion.
(c) Assertion is true, but Reason is false.
(d) Assertion is false, but Reason is true.

Topic 5 — Pineal Gland

40. Which of the following set of functions is not regulated by the hormone of the pineal gland?
(a) Diurnal rhythm and body temperature
(b) Metabolism and pigmentation
(c) Growth of bones and defense capability
(d) Diurnal rhythm and defense capability

41. Which of the following hormone exhibits an anti-gonadotropic effect in humans?
(a) ADH (b) Thyroxin
(c) Melatonin (d) ACTH

Topic 6 — Thyroid Gland

42. Following is the diagrammatic view of the position of endocrine glands.

Select the option that correctly labels the glands and their respective hormones.
(a) I - Thyroid gland - Thyroxine and TSH
(b) II - Parathyroid gland - PTH
(c) II - Thyroid gland - Thyroxine and calcitonin
(d) I - Parathyroid gland - PTH and calcitonin

43. The thyroid gland is composed of
(a) stromal tissues only
(b) follicles only
(c) stromal tissues and follicles
(d) isthmus

44. Most of the T_4 is converted into T_3 in the target tissues because
(a) T_4 is more active.
(b) T_3 is more active.
(c) T_4 has a shorter half-life.
(d) follicular cells cannot synthesize T_3.

45. Thyroid hormones are required for normal growth and development of humans because
(a) thyroid hormones increase the basal metabolic rate in most body tissues.
(b) thyroid hormones are regulated by negative feedback systems.
(c) thyroid hormones contain iodine atoms.
(d) thyroid hormones stimulate glycogen formation.

46. A group of tadpoles with no thyroxin hormone production exhibited disrupted metamorphosis. Which of the following statement correctly describes the role of thyroid hormones in metamorphosis.
(a) Thyroid hormones lower the blood calcium levels.
(b) Thyroid hormones increase blood calcium levels.
(c) Thyroid hormones lower the blood phosphate levels.
(d) Thyroxin stimulates protein synthesis.

47. Which of the following disorders of the endocrine system is incorrectly matched with its description?
(a) Hypothyroidism: Iodine deficiency
(b) Goitre: Enlarged thyroid gland
(c) Hyperthyroidism: Cretinism
(d) Exopthalmic goitre: Hyperthyroidism

48. Which of the given statement correctly differentiates Myxedema from Graves' disease?
(a) Hypothyroidism in adult ages causes Graves' disease.
(b) Graves' disease is more common among males than females.
(c) Myxedema causes swelling in facial tissues due to the accumulation of interstitial fluid.
(d) Myxedema is an auto-immune disorder.

49. Graves' disease is caused due to
(a) hyposecretion of the thyroid gland
(b) hypersecretion of the thyroid gland
(c) hyposecretion of the adrenal gland
(d) hypersecretion of the adrenal gland

50. Assertion: Thyroid hormones regulate oxygen consumption and basal metabolic rate of the body cells.

Reason: Thyroid hormones reduce the number of active mitochondria in body cells.
(a) Both Assertion and Reason are true and Reason is correct explanation of Assertion.
(b) Both Assertion and Reason are true, but Reason is not the correct explanation of Assertion.
(c) Assertion is true, but Reason is false.
(d) Assertion is false, but Reason is true.

Topic 7 Parathyroid Gland

51. The physiological role of parathyroid gland does not include
(a) increased activity of osteoclasts
(b) bone resorption
(c) reduced loss of Ca2+ and Mg^{2+} ions into urine
(d) reduced loss of $HPO_4 2^-$ into the urine

52. Which of the following hormones can play significant role in osteoporosis?
(a) Aldosterone and Prolactin
(b) Progesterone and Aldosterone
(c) Estrogen and Parathyroid hormone
(d) Parathyroid hormone and Prolactin

53. Which of the following pairs of hormones have antagonistic effects?
(a) T_3 and T_4
(b) ACTH and glucocorticoids
(c) PTH and TCT (thyrocalcitonin)
(d) T_3 and TSH

54. Assertion: Calcitonin and PTH exhibit an antagonistic effect on blood levels of calcium ions.

Reason: Insulin and glucagon exhibit antagonistic effects on the blood levels of glucose.
(a) Both Assertion and Reason are true and Reason is correct explanation of Assertion

(b) Both Assertion and Reason are true, but Reason is not the correct explanation of Assertion.
(c) Assertion is true, but Reason is false.
(d) Assertion is false, but Reason is true.

Topic 8 Thymus Gland

55. The endocrine gland that functions as a component of the lymphatic system is
(a) thyroid gland
(b) thymus gland
(c) parathyroid gland
(d) pineal gland

56. Which of the following endocrine gland is responsible for reduced immune responses in old ages?
(a) Thyroid gland (b) Pineal gland
(c) Thymus gland (d) Pituitary gland

Topic 9 Adrenal Gland

57. Adrenal cortex and adrenal medulla differ from each other in terms of
(a) histology (b) function
(c) origin (d) all of these

58. The endocrine gland present at the top of kidneys and involved in the fluidelectrolyte balance of the body is
(a) adrenal gland
(b) pineal gland
(c) parathyroid gland
(d) pancreas

59. Which of the following categories of hormones is correctly matched with its examples?
(a) Catecholamines: adrenaline and noradrenaline
(b) Emergency hormones: adrenaline and insulin
(c) Glucocorticoids: aldosterone
(d) Mineralocorticoids: cortisol

60. Consider the following statements:
(a) Some chemicals act as both neurotransmitters and hormones.
(b) Norepinephrine is released as a neurotransmitter by sympathetic division.
(c) Norepinephrine is released as a hormone by the thyroid gland.
(d) Norepinephrine is released as a hormone by the parathyroid gland.
Which of the two statements are correct?
(a) a and b (b) a and c
(c) b and c (d) c and d

61. Secretion of hormones from adrenal medulla is controlled by
(a) parasympathetic nervous system
(b) pituitary gland
(c) sympathetic nervous system
(d) peripheral nervous system

62. The fight or flight response generated by the release of hormones by adrenal medulla includes
(a) dilation of pupils and increased heart rate and blood pressure.
(b) glycogenolysis and lipid synthesis.
(c) increased muscular movement of the gastrointestinal tract.
(d) reduced blood glucose levels.

63. Which of the following hormones of the adrenal cortex are correctly matched with their source?
(a) Mineralocorticoids - Zona fasciculata cells
(b) Glucocorticoids - Zona glomerulosa cells
(c) Androgens - Zona glomerulosa cells
(d) Glucocorticoids - Zona fasciculata cells

64. Glucocorticoids are involved in
(a) fluid electrolyte balance
(b) carbohydrate metabolism
(c) water reabsorption from kidneys
(d) regulation of blood glucose levels

65. Which of the following sets of physiological functions correctly describes the role of cortisol in the human body?
(a) Anti-inflammatory response and suppression of the immune response

(b) Breakdown of RBCs in spleen
(c) Upregulation of uptake of amino acids
(d) Reabsorption of Na+ from kidneys

66. Target organ/structure of aldosterone is
(a) renal calyces
(b) renal tubules
(c) glomerulus
(d) Bowman's capsule

67. **Assertion:** Adrenal medullary hormones regulate the fluid-electrolyte balance of the body.

Reason: Aldosterone is a glucocorticoid that raises blood sodium levels.
(a) Both Assertion and Reason are true and Reason is correct explanation of Assertion.
(b) Both Assertion and Reason are true, but Reason is not the correct explanation of Assertion.
(c) Assertion is true, but Reason is false.
(d) Assertion is false, but Reason is true.

Topic 10 Pancreas

68. Which of the following pairs of endocrine glands are composite glands?
(a) Pancreas and thymus gland
(b) Adrenal glands and thymus gland
(c) Pancreas and adrenal gland
(d) Adrenal gland and pineal gland

69. The cells that make about 70% of the cells of the pancreatic islets serve to secrete
(a) insulin
(b) glucagon
(c) somatostatin
(d) pancreatic polypeptide

70. Alpha and delta cells of pancreatic islands are the source of
(a) glucagon and insulin respectively
(b) somatostatin and glucagon respectively
(c) glucagon and somatostatin respectively
(d) insulin and glucagon respectively

71. Which of the following hormones is correctly matched with its description/effects?
(a) Hyperglycemic hormone: Glucagon
(b) Hypoglycemia: Glucagon
(c) Hypercalcemic hormone: Calcitonin
(d) Addison's disease: Catecholamines.

72. Which of the given statements correctly differentiates glycogenolysis from glycogenesis?
(a) Glycogenolysis is the formation of glycogen from glucose and is triggered by glucagon.
(b) Glycogenolysis is a breakdown of glycogen into glucose and is triggered by insulin.
(c) Glycogenesis is the conversion of glucose into glycogen and is stimulated by insulin.
(d) Glycogenesis is the formation of glucose from non-carbohydrate substrates.

73. Diabetes mellitus is a disorder of carbohydrate metabolism and is characterised by
(a) polyuria (b) polydipsia
(c) polyphagia (d) all of these

74. The water-soluble fuels that are normally exported by the liver but overproduced during fasting or diabetes mellitus when not treated are known as
(a) fatty acids (b) glucose
(c) ketone bodies (d) amino acids

75. Match the following hormones with the respective disease.

	Column-I		Column-II
(a)	Insulin	(1)	Addison's disease
(b)	Thyroxin	(2)	Diabetes Insipidus
(c)	Corticoids	(3)	Acromegaly
(d)	Growth hormone	(4)	Goitre
		(5)	Diabetes mellitus

Select the correct option.

	(a)	(b)	(c)	(d)
(A)	5	1	2	3
(B)	2	4	3	1
(C)	5	4	1	3
(D)	2	4	1	3

Topic 11 Testes

76. The ________ are the oval glands present in the scrotum and serve to secrete the hormones.
(a) ovaries and androgen
(b) testes and testosterone
(c) pineal gland and melatonin
(d) adrenal gland and cortisol

77. Which of the following is not a function of testes?
(a) Spermatogenesis
(b) Sperm production
(c) Testosterone secretion
(d) All of these

78. Which of the following structures/cells of testes is incorrectly matched with its functions?
(a) Seminiferous tubules - Spermatogenesis
(b) Sustentacular cells - Secretion of FSH
(c) Interstitial cells - Secretion of testosterone
(d) Sertoli cells - Nourishment of developing spermatogenic cells

79. Testes serve as an endocrine gland. Which of the given statement about testicular hormones is correct?
(a) LH from the posterior pituitary stimulates interstitial cells to secrete testosterone.
(b) Testosterone suppresses the secretion of LH and GnRH by the positive feedback mechanism.
(c) FSH and testosterone stimulate the sustentacular cells to stimulate the secretion of the androgen-binding protein.
(d) Testosterone stimulates the process of spermiogenesis.

80. The function of inhibin hormone is to
(a) inhibit the FSH secretion
(b) inhibit the testosterone secretion
(c) inhibit spermiogenesis
(d) stimulate spermatogenesis

81. Which of the given sets of organs represent the male accessory sex organs?
(a) Epididymis and testes
(b) Vas deferens and testes
(c) Seminal vesicles and prostate gland
(d) Urethra and oviduct

82. The hormone responsible for the descent of testes into the scrotum is
(a) FSH (b) LH
(c) testosterone (d) inhibin

83. The function of testosterone in human males is/are
(a) development of male secondary sexual characters
(b) male sexual behaviour
(c) stimulation of protein breakdown
(d) both (a) and (b)

84. Assertion: In human males, LH is required for sperm production.

Reason: FSH and testosterone stimulate interstitial cells to secrete androgen binding protein.
(a) Both Assertion and Reason are true and Reason is correct explanation of Assertion.
(b) Both Assertion and Reason are true, but Reason is not the correct explanation of Assertion.
(c) Assertion is true, but Reason is false.
(d) Assertion is false, but Reason is true.

Topic 12 Ovary

85. Androgens are responsible for libido in both human males and females. The source of androgens in human females is
(a) ovaries (b) oviducts
(c) corpus luteum (d) adrenal cortex

86. Which of the following set of hormones is known as female sex hormones?

(a) FSH and LH
(b) Estrogen and progesterone
(c) FSH and estrogen
(d) LH and estrogen

87. Match the following structures in Column-I with the correct description in Column-II.

	Column-I		Column-II
(a)	Ovarian follicles	(1)	progesterone
(b)	Corpus luteum	(2)	LH
(c)	Graafian follicle	(3)	prolactin
(d)	Mammary glands	(4)	estrogen

Select the correct option.

	(a)	(b)	(c)	(d)
(a)	4	3	2	1
(b)	4	1	2	3
(c)	3	4	2	1
(d)	3	1	4	2

88. A temporary endocrine gland in the human body is
(a) pineal gland
(b) corpus cardiacum
(c) corpus luteum
(d) corpus allatum

89. Which of the following statement is incorrect about the effects and regulation of female sex hormones?
(a) Ovarian and uterine cycles are under the regulation of Gonadotropinreleasing hormone.
(b) In females, FSH triggers the development of follicles.
(c) Ovarian follicles secrete estrogen under influence of FSH and LH.
(d) Estrogen from ovarian follicles is converted into progesterone.

90. Which of the following hormones regulates the formation of corpus luteum and stimulates it to release hormones?
(a) FSH (b) LH
(c) Estrogen (d) Androgens

91. Which of the following hormones is secreted by corpus luteum?
(a) Estrogens (b) Progesterone
(c) Relaxin (d) LH

92. Which of the following statements is incorrect?
(a) Testis and ovary function as a primary sex organ as well as endocrine gland
(b) Ovaries are located in thoracic cavity of females
(c) Ovary produces ovum, 2 groups of steroid hormones (estrogen and progesterone)
(d) Ovary is composed of ovarian follicles and stromal tissue

93. Which of the following hormone is synergistic to human growth hormone?
(a) Estrogen
(b) Progesterone
(c) Inhibin
(d) Androgen binding protein

94. After ovulation, the ruptured follicle is converted into-
(a) Graafian follicle
(b) Corpus callosum
(c) FSH
(d) LH

95. The hormone responsible to prepare and maintain endometrium for implantation of a fertilized ovum is:
(a) FSH (b) LH
(c) Progesterone (d) Inhibin

96. Assertion: Corpus luteum is a temporary endocrine gland.

Reason: Corpus luteum secretes female sex hormones.
(a) Both Assertion and Reason are true and Reason is correct explanation of Assertion.
(b) Both Assertion and Reason are true, but Reason is not the correct explanation of Assertion.

(c) Assertion is true, but Reason is false.
(d) Assertion is false, but Reason is true.

Topic 13 Hormones of Heart, Kidney and Gastrointestinal Tract

97. Atrial natriuretic hormone I factor (ANF) secreted by atrial wall of our heart has exactly the opposite function of this hormone secreted by zona glomerulosa –
(a) ADH (b) Aldosterone
(c) Androgen (d) Calcitonin

98. Match the hormones in Column-I with their functions in Column-II.

	Column-I		Column-II
(a)	Progesterone	(1)	Inhibits uterine contraction
(b)	Atrial natriuretic factor	(2)	Formation of RBCs
(c)	Erythropoietin	(3)	Formation of
(d)	Relaxin	(4)	Lowers blood pressure

Select the correct option.

	(a)	(b)	(c)	(d)
(a)	4	3	2	1
(b)	4	1	2	3
(c)	3	4	2	1
(d)	2	1	4	3

99. ANF-
(a) Decrease B.P.
(b) Causes vasodilation
(c) Is secreted when B.P. increases
(d) All

100. Gastrin, secretin, cholecystokinin (CCK) and gastric inhibitory peptide (GIP) are 4 major peptide hormone secreted by-
(a) Only stomach
(b) Only small intestine
(c) Gastro-intestinal tract
(d) Only pancreas

101. Which of the following hormones of the gastrointestinal tract is wrongly matched with its function?

	Column-I		Column-II
(a)	Gastrin	(1)	Inhibition of gastric secretions
(b)	Gastric inhibitory peptide (GIP)	(2)	Stimulates secretion of pancreatic juice and bile juice
(c)	Secretin	(3)	Secretion of gastric juice
(d)	Cholecystokinin	(4)	Stimulates secretion of pancreatic juice

Select the correct option.

	(a)	(b)	(c)	(d)
(a)	3	1	4	2
(b)	2	1	4	3
(c)	3	4	2	1
(d)	2	4	4	1

102. Identify the hormone with its correct matching of source and function.
(a) Oxytocin - posterior pituitary, growth, and maintenance of mammary glands.
(b) Melatonin - pineal gland, regulates the normal rhythm of the sleep-wake cycle.
(c) Progesterone - corpus luteum, stimulation of growth and activities of female secondary sex organs.
(d) Atrial natriuretic factor - ventricular wall, increases the blood pressure.

103. Assertion: Hormones are also secreted by tissues that are not the endocrine glands.
Reason: Kidneys secrete the hormone atrial natriuretic factor.
(a) Both Assertion and Reason are true and Reason is correct explanation of Assertion.
(b) Both Assertion and Reason are true, but Reason is not the correct explanation of Assertion.
(c) Assertion is true, but Reason is false.
(d) Assertion is false, but Reason is true.

104. Assertion: Duodenum serves endocrine function and secretes secretin hormone.

Reason: Gastrin hormone from the mucosa of the stomach stimulates the secretion of gastric glands.

(a) Both Assertion and Reason are true and Reason is correct explanation of Assertion.
(b) Both Assertion and Reason are true, but Reason is not the correct explanation of Assertion.
(c) Assertion is true, but Reason is false.
(d) Assertion is false, but Reason is true.

Topic 14 Mechanism of Hormone Action

105. Match the Column I with Column II.

	Column I		Column II
A.	Peptide, polypeptide protein	I.	Epinephrine, nor-epinephrine
B.	Steroid	II.	T_3 and T_4 (thyroid hormones)
C.	Iodothyronines	III.	Cortisol, testosterone, estradiol, progesterone
D.	Amino acid derivatives	IV.	Pituitary hormones, pancreatic hormones, hypothalamic hormones

(a) A-I , B - II, C - III , D - IV
(b) A- IV, B - III, C - II , D - I
(c) A- IV, B – III , C - I, D - II
(d) A- I, B - II, C - IV, D - III

106. Steroid hormones initiate the production of target cell substances in which manner?
(a) They initiate second messenger activity
(b) They bind with membrane protein
(c) They initiate DNA transcription
(d) They activate enzyme pathways

107. Why do some hormones (first messenger) need to trigger a "second messenger" to activate a target cell?
(a) The first messenger needs activation of ATP
(b) The first messenger cannot cross a plasma membrane
(c) There are no specific cell surface receptors for first messenger
(d) The first messenger is not a watersoluble molecule

108. Which of the following category of hormones is incorrectly matched with its examples?
(a) Protein hormone: Insulin
(b) Steroids: Cortisol
(c) Iodothyronines: Thyroid hormones
(d) Amino-acid derivatives: Oxytocin

109. Which of the following categories of the hormones is water-insoluble?
(a) Eicosanoids (b) Peptide hormones
(c) Amines (d) Steroid hormones

110. Which of the given hormones is/are second messengers?
(a) cAMP (b) IP_3
(c) Ca^{+2} (d) All

111. Which of the following hormones does not act by a second messenger system?
(a) Glucagon (b) Epinephrine
(c) FSH (d) Testosterone

112. Which of the given sets of endocrine gland secrete/ release only water-soluble hormones?
(a) Pancreas and thyroid gland
(b) Parathyroid gland and pineal gland
(c) Adrenal gland and thyroid gland
(d) Parathyroid gland and pancreas

113. Epinephrine is ________ derivative.
(a) amino acid (b) carbohydrate
(c) steroid (d) nucleic acid

114. Which of the given hormones is incorrectly matched with its source amino acid?
(a) Serotonin: Tryptophan
(b) Histamine: Histidine

(c) Epinephrine: Tyrosine
(d) Melatonin: Alanine

115. The amino acid tryptophan is the precursor for the synthesis of
(a) estrogen and progesterone
(b) cortisol and cortisone
(c) melatonin and serotonin
(d) thyroxin and triiodothyronine

116. Which of the following statement correctly differentiate the transport of water-soluble and lipid soluble hormones in blood?
(a) Most of the lipid-soluble hormones are bound to transport proteins.
(b) Most of the water-soluble hormones are bound to transport proteins.
(c) Transport proteins enhance the rate of hormone loss in urine
(d) Transport proteins are synthesized in muscles.

117. Receptors for protein hormones arc mostly present at/in
(a) nucleus (b) nuclear envelop
(c) cell surface (d) cytoplasm

118. Which of the given statement is correct about the mechanism of water soluble hormones?
(a) Hormones bind to their cytoplasmic receptors present in the target cells.
(b) The hormone-receptor complex alters the gene expression.
(c) Hormones serve as the first messenger and cause the production of a second messenger.
(d) The newly formed proteins produce a physiological response.

119. Which of the following molecules serve as the second messenger in the mechanism of action of a protein hormone?
(a) T_3 (b) cAMP
(c) T_4 (d) protein kinases

120. Following is the diagrammatic representation of the mechanism of action of a protein hormone. Which of the given options correctly describes the labeled events?
(a) I-hormone receptor
(b) II-hormone
(c) III-second messenger
(d) IV-physiological response

121. Following is the diagrammatic representation of the mechanism of action of a steroid hormone. Which of the given options correctly describes the labeled events?
(a) I-Hormone, II-second messenger
(b) I-Hormone, III-nucleus
(c) II-Hormone-receptor complex, IVDNA
(d) II-Second messenger, IV-mRNA

122. Assertion: Aldosterone can diffuse freely through the lipid bilayer.
Reason: Steroid hormones are lipidsoluble.
(a) Both Assertion and Reason are true and Reason is correct explanation of Assertion.
(b) Both Assertion and Reason are true, but Reason is not the correct explanation of Assertion.
(c) Assertion is true, but Reason is false.
(d) Assertion is false, but Reason is true.

123. Assertion: Receptors for steroid hormones are present at the cell surface.
Reason: Receptors for protein hormones are present in the nucleus.
(a) Both Assertion and Reason are true and Reason is correct explanation of Assertion.
(b) Both Assertion and Reason are true, but Reason is not the correct explanation of Assertion.
(c) Assertion is true, but Reason is false.
(d) Assertion is false, but Reason is true.

124. Assertion: cAMP serves as the second messenger for protein hormones.
Reason: Insulin is a protein hormone.
(a) Both Assertion and Reason are true and Reason is correct explanation of Assertion.
(b) Both Assertion and Reason are true, but Reason is not the correct explanation of Assertion.
(c) Assertion is true, but Reason is false.
(d) Assertion is false, but Reason is true.

ANSWER KEY

1. (b)	2. (c)	3. (d)	4. (a)	5. (b)	6. (a)	7. (c)	8. (a)	9. (d)	10. (d)
11. (b)	12. (b)	13. (a)	14. (b)	15. (c)	16. (d)	17. (b)	18. (c)	19. (b)	20. (a)
21. (a)	22. (a)	23. (d)	24. (d)	25. (c)	26. (c)	27. (c)	28. (d)	29. (c)	30. (b)
31. (b)	32. (d)	33. (d)	34. (a)	35. (b)	36. (a)	37. (c)	38. (d)	39. (a)	40. (c)
41. (c)	42. (b)	43. (c)	44. (b)	45. (a)	46. (d)	47. (c)	48. (c)	49. (b)	50. (c)
51. (d)	52. (c)	53. (c)	54. (b)	55. (b)	56. (c)	57. (d)	58. (a)	59. (a)	60. (a)
61. (c)	62. (a)	63. (b)	64. (b)	65. (a)	66. (b)	67. (d)	68. (c)	69. (a)	70. (c)
71. (a)	72. (c)	73. (d)	74. (c)	75. (c)	76. (b)	77. (d)	78. (b)	79. (c)	80. (a)
81. (c)	82. (c)	83. (d)	84. (c)	85. (d)	86. (b)	87. (b)	88. (c)	89. (d)	90. (b)
91. (b)	92. (b)	93. (a)	94. (c)	95. (c)	96. (a)	97. (b)	98. (c)	99. (d)	100. (c)
101. (a)	102. (b)	103. (c)	104. (d)	105. (b)	106. (c)	107. (b)	108. (d)	109. (d)	110. (d)
111. (d)	112. (b)	113. (a)	114. (d)	115. (c)	116. (a)	117. (c)	118. (c)	119. (b)	120. (c)
121. (b)	122. (a)	123. (d)	124. (b)						

11 Chapter

Reproduction in Organisms

MULTIPLE CHOICE QUESTIONS

1. Which one of the following is correct statement with respect to life span of organisms?
   (a) Mango tree has a much longer lifespan as compared to peepal tree
   (b) Single celled organism doesn't show natural death
   (c) Lifespan is correlated with complexity and habit of plant
   (d) Parrot has a shorter lifespan as compared to crow

2. Asexual reproduction-
   (a) Involvement of gamete formation
   (b) Fusion of male & female gamete
   (c) Without involvement of gamete formation
   (d) Biparental (opposite sex)

3. Life span represents-
   (a) Period from birth to natural death of an organism
   (b) Period from birth to reproductive maturity of an organism
   (c) Period from reproductive maturity to death of an organism
   (d) Period from adolescence to senescent phase of an organism

4. Life span depends on-
   (a) Size of organism (b) Shape of organism
   (c) (a) and (b) both (d) None of these

5. Arrange the following organism on basis of their ascending life span.
   (a) Parrot, Crocodile, Crow, Butterfly
   (b) Tortoise, Crocodile, Crow, Butterfly
   (c) Butterfly, Crocodile, Parrot
   (d) (a) and (b) both

Topic 1 Asexual Reproduction

6. Choose most appropriate statement about Water hyacinth.
   (i) An aquatic plant that can propagate vegetatively at a phenomenal rate
   (ii) Introduced in India because of beautiful flowers and shape of fruit
   (iii) It reproduce by offset
   (iv) It drains oxygen from water, which leads to death of fishes
   (v) It spread all over the water body is a short period of time
   (vi) It is very difficult to get rid off them
   (vii) It is invasive weeds found growing wherever there is standing water
   (a) i, ii, iii, iv, v, vii, vii
   (b) i, iii, iv, v, vi, vii
   (c) ii, iv, v, vi, vii
   (d) None of these

7. Term 'vegetative reproduction' is frequently used in-
   (a) Plants (b) Animals
   (c) Monerans (d) (a) and (c) both

8. Banana is reproduced vegetatively by-
   (a) Stolon (b) Runner
   (c) Rhizome (d) Offset

9. 'Eyes' of Potato arise from-
   (a) Node (b) Internode
   (c) Both (a) and (b) (d) Root

10. Bryophyllum arise vegetatively by
    (a) adventitious bud arise from notches present

at leaves margin.

(b) adventitious bud arise from notches present at node.

(c) adventitious bud arise from notches present at internode.

(d) None of these

11. Arrange in sequence-

(i) Gamete transfer

(ii) Formation of gamete

(iii) Post-fertilization

(iv) Fertilization

(a) ii-i-iv-iii (b) i-iii-ii-iv

(c) iii-iv-ii-I (d) iv-i-iii-ii

12. Choose correct statement with respect to sexual reproduction-

(a) It involves formation of male & female gametes only by different individuals of the opposite sex.

(b) It is an elaborate, simple and slow process as compared to asexual reproduction.

(c) Zygotes are formed by fusion of gametes which develop to form the new organisms.

(d) All of these

13. Clone are

(a) Produce generally by sexual reproduction

(b) Individual genetically similar to parent

(c) Individual morphologically similar to parent

(d) (b) and (c) both

14. Given below is figure represent-

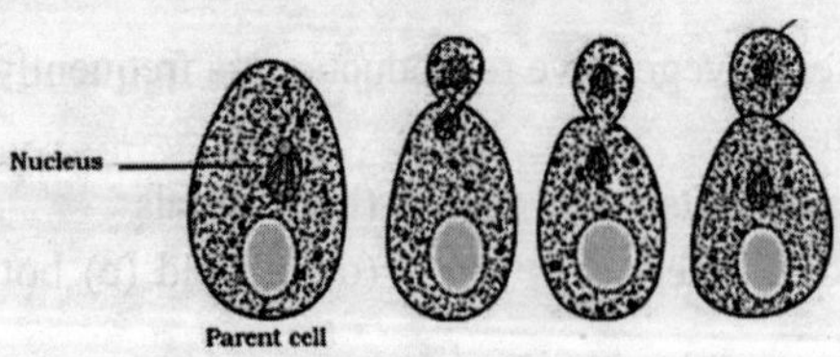

(a) Binary fission in Amoeba

(b) Budding in Amoeba

(c) Budding in Yeast

(d) Binary fission in Yeast

15. Cell division is itself a mode of reproduction in-

(a) Protists (b) Moneran

(c) Fungi (d) (a) and (b) both

16. Zoospores is mode of asexual reproduction in-

(a) Chlamydomonas (b) Penicillium

(c) Hydra (d) Sponges

17. Choose the correct statement-

(a) Conidia of Penicillium is endogenous borne

(b) Gemmules of sponges is example of external budding

(c) In Yeast, the division is unequal and small buds are produced

(d) Zoospore is non-motile structure

18. Match Column-I and Column-II and select correct option.

	Column I		Column II
(a)	Tuber	(i)	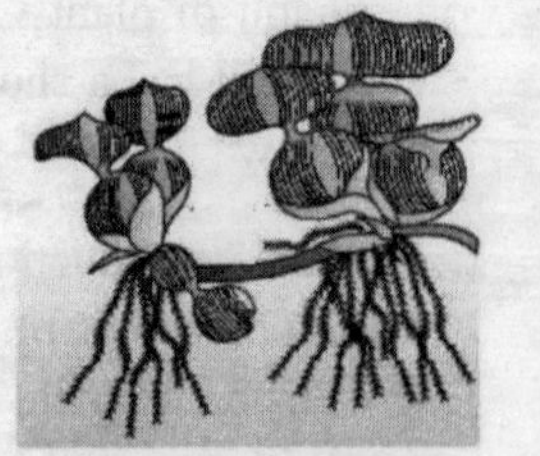
(b)	Rhizome	(ii)	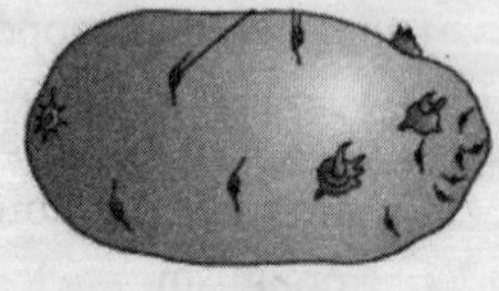
(c)	Bulbil	(iii)	
(d)	Leaf buds	(iv)	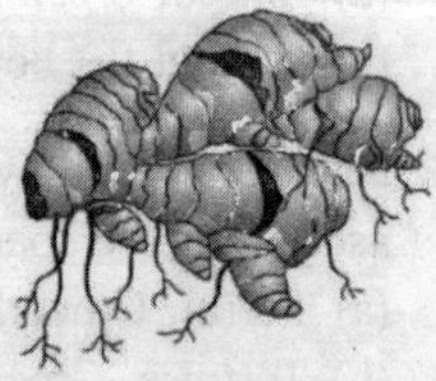

(e)	Offset	(v)	

(a) a-ii, b-iv, c-iii, d-iv, e-i
(b) a-ii, b-iv, c-v, d-iii, e-i
(c) a-ii, b-iv, c-i, d-v, e-iii
(d) None of these

19. Which of the following is not meant for vegetative propagation in Angiospermic plants.
(a) Gemmules (b) Runner
(c) Bulb (d) Sucker

20. Vegetable propagation involve-
(a) Uniparent
(b) Biparental
(c) Zoospores
(d) Reproduction by flower

21. Label A and B

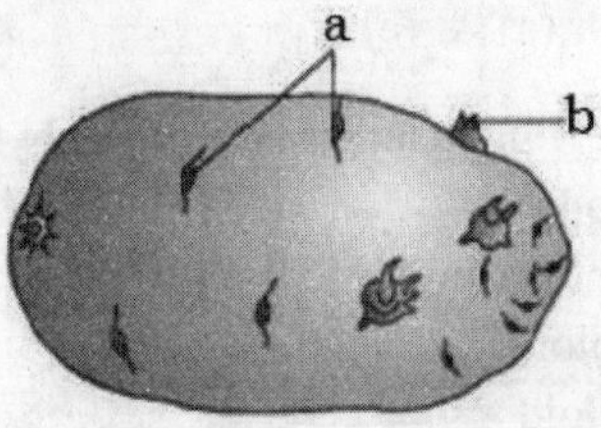

	a	b
(a)	Eyes	Germinating eye bud
(b)	Germinating eye	Eyes bud
(c)	Buds	Adventitious root
(d)	Adventitious root	Buds

Topic 2 Sexual Reproduction

22. The period of growth where an organism is not nature for sexual reproduction is-
(a) Juvenile phase (b) Vegetative phase
(c) Both (a) and (b) (d) None of these

23. Choose incorrect statement among following-
(a) In perennial species its show clear cut vegetative, reproductive and senescent phase.
(b) In annual & biennial species it is show cut vegetative, reproductive & senescent phase.
(c) In perennial species it is very difficult to define vegetative, reproductive & senescent phase.
(d) (a) and (b) both

24. Neelakuranji flower once bloom in-
(a) 12 years (b) 12 months
(c) 12 days (d) 36 years

25. Bamboo show flowering is
(a) 10-20 years (b) 50-100 years
(c) Once in every year (d) B and C both

26. Choose correct statement about Strobilanthus Kunthianan.
(a) This plant flowered during September-October 2006.
(b) Its mass flowering transformed large tracks of hilly areas in Kerala, Karnataka and Tamil Nadu into Pink stretches.
(c) (a) and (b) both
(d) Its flower during April-May 2006.

27. Oestrus cycle observed is-
(a) Primates (b) Non-primate
(c) Both (a) and (b) (d) Human beings

28. Choose odd ones among following on basis of hormonal changes is reproductive phase.
(a) Ape (b) Cow
(c) Deer (d) Tiger

29. Juvenile phase is related with all, except.
(a) It is pre-reproductive phase of an individual.
(b) It is of different durations in different organisms.
(c) It is period of vegetative growth.
(d) It involves appearance of flowers in higher plant.

Topic 2.1 Pre-fertilization Events Gametogenesis

30. Isogametes present not in-
(a) Cladophora a fungi (b) Fucus an alga
(c) Human beings (d) All of these

31. The given figure is of-

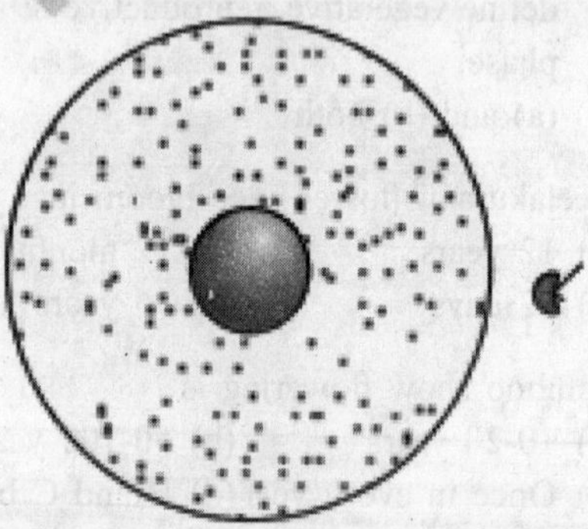

(a) Heterogametes of Human beings
(b) Heterogametes of Fucus, a fungi
(c) Homogamete of Fucus, an alga
(d) None of these

32. Choose the correct statement-
(a) The process of formation of two types of gametes is known as gametogenesis.
(b) In Cladophora the two gametes are so similar in appearance that it is not possible to categorise them into male & female gametes; such known as heterogametes.
(c) The male and female morphological distinct types of gamete is character of Fucus.
(d) (a) & (c) both

33. Plants may have both male female reproductive structures in same pant is–
(i) Bisexual (ii) Unisexual
(iii) Monoecious (iv) Dioecious
(a) i & iii (b) i & iv
(c) ii & iii (d) ii & iv

34. Staminate and pistillate flower bear on different plant, term as-
(i) Bisexual (ii) Unisexual
(iii) Monoecious (iv) Dioecious
(a) i & iii (b) i & iv
(c) iv & iii (d) iv & ii

35. Choose odd one with respect to sexuality.
(a) Cucurbit (b) Cockroach
(d) Chara (d) None of these

36. How many of following is monoecious/bisexual- Papaya, Cucurbit, Datepalm, Cockroach, Marchantia, Earthworm, Sweet potato, Chara, Sponge, Tapeworm, Leech.
(a) 4 (b) 8
(c) 7 (d) 12

37. Which of the following is / are hermaphrodites
(a) Cockroach (b) Capeworm
(c) Earthworm (d) Leech

38. Choose correct statement
(a) A diploid parent produces gametes by mitotic division
(b) A haploid parent produces gametes by mitotic division
(c) A haploid parent produces gametes by meiotic division
(d) Both (a) & (c)

39. Monera, Fungi, algae and bryophytes parental plant body is-
(a) Same as in pteridophytes
(b) Haploid
(c) Diploid
(d) Both (a) & (b)

40. Meiocytes of human beings-
(a) Undergo meiosis for gamete formation
(b) Undergo mitosis only
(c) Both ()a & (b)
(d) None of these

41. Meiocytes have-
(a) One set of chromosomes
(b) Two set of chromosomes
(c) Three set of chromosomes
(d) All of these

42. Choose correct response among following-

(a) The female sex organ of Chara positioned above antheridium
(b) Testis sac of earthworm positioned below its ovary
(c) Testis of cockroach positioned above ovary of same cockroach individual.
(d) (a) and (b) both

43. Given diagram is of-

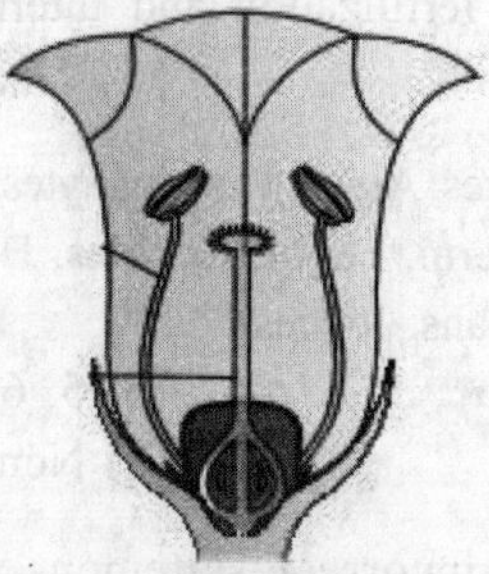

(a) Bisexual flower of potato
(b) Unisexual flower of sweet potato
(c) Unisexual flower of potato
(d) None of these

44. Among following maximum number of chromosome is meiocytes of-
(a) Rice (b) Maize
(c) Potato (d) Onion

45. Arrange is ascending order, number of chromosomes in gamete-
(a) Housefly, Fruitfly, Maize, Cat
(b) Cat, Rat, Butterfly, Apple
(c) Onion, Rice, Rat, Ophioglossum
(d) Human, Potato, Dog, Butterfly

46. Ophioglossum is-
(a) Algae (b) Bryophytes
(c) Pteridophyte (d) None of these

47. Find mismatched column.

Column-I (Organism)	Column-II (Chromosome number in meiocyte)	Column-III (Chromosome number in gamete)
(a) Maize	24	12
(b) Onion	16	8
(c) Dog	78	39
(d) Apple	34	17

48. In bryophyte, pteridophyte gamete transfer need-
(a) Air (b) Water
(c) Biotic agent (d) None of these

49. Choose the correct statement-
(a) In a majority of organisms, male gamete is stationary and female gamete is motile.
(b) A large number of male gametes liberated during gamete transfer to compensate the loss male gamete during transport.
(c) In a majority of organisms, male gametes and female gametes are stationary.
(d) In majority of organism including human male gametes and female gametes are motile.

50. Pollen grains carries-
(a) Male gametes (b) Female gametes
(c) Meiocytes (d) (a) and (c) both

51. Pollen grain produces in-
(a) Stigma (b) VBCarpel
(c) Anther (d) (b) and (c) both

52. Dioecious plants show-
(a) Self-pollination (b) Cross pollination
(c) A and B both (d) Autogamy

53. Peas-
(a) Self-fertilizing
(b) Bisexual
(c) (a) and (b) both
(d) Cross-pollination only

54. Pollination is-
(a) Transfer of pollen grain from anther to stigma before it can lead to fertilization
(b) Transfer of egg from female reproductive part to anther before it can lead to fertilization.
(c) Transfer of pollen grain from androecium to anther before it can lead to fertilization
(d) Essential step during sexual reproduction of mammals

55. Most critical event in sexual reproduction-
(a) Gametogenesis (b) Fertilization
(c) Gamete transfer (d) Embryogenesis

56.

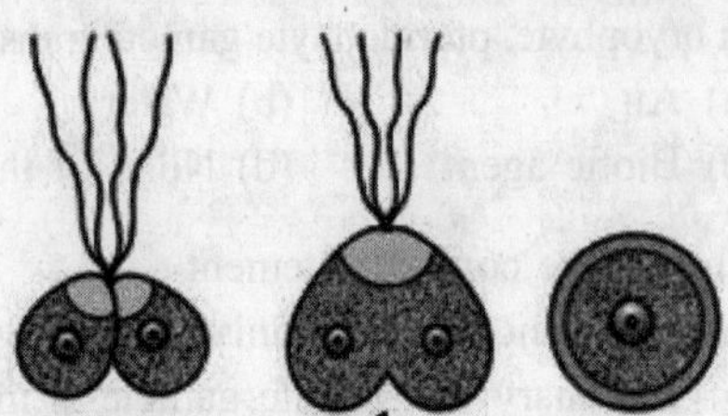

(a) a = Fusion of gametes
b = Zygote
c = New individuals
(b) a = Parent organisms
b = Fusion of gametes
c = Zygote
(c) a = Parent Organisms
b = Fusion of gametes
c = Formation of two new individuals
(d) a = Zygote
b = Fusion of gamete
c = New individual

Topic 2.2 Fertilization

57. Choose the correct statement-
(I) Fusion of gametes is syngamy
(II) Formation of diploid zygote is result of fertilization
(III) Fusion of gametes is fertilization
(a) I and II are correct, III is incorrect
(b) II and III are correct, I is incorrect
(c) I, II, and III are correct
(d) I, III are correct, II is incorrect

58. Which bird of following show parthenogenesis-
(a) Parrot (b) Turkey
(c) Ostrich (d) None of these

59. Parthenogenesis is-
(a) Development of new organism from male gamete only without fertilization
(b) Development of new organisms from both male and female gametes without fertilization
(c) Development of new organisms from unfertilized female gametes
(d) All of these

60. From enlist of organisms, how man shows external fertilization and internal fertilization respectively.

Bryophytes, Algae, Pteridophytes, Gymnosperms, Angiosperm, Fungi, Reptiles, Birds, Mammals, Amphibians, Fishes.
(a) 3, 8 (b) 5, 6
(c) 6, 5 (d) None of these

61. Choose incorrect statement about external fertilization.
(a) Release a large number of gametes into the surrounding media.
(b) The offspring resulted from external fertilization are extremely vulnerable to predators threatening their survival up to adulthood
(c) Amphibians of plant kingdom show external fertilization
(d) None of these

62. Choose the correct statement regarding gametes involve in internal fertilization-
(a) Egg is formed inside female body is motile structure.
(b) In seed plants, the non-motile male gametes are carried to female gametes.
(c) There is significant reduction in number of female gametes.
(d) Both male and female gametes produced in equal number and both are motile.

Topic 2.3 Post-fertilization Events

63. Post-fertilization events involves-

(a) Formation of zygote
(b) Embryogenesis
(c) Gamete transfer
(d) Both (a) and (b)

Topic 2.3.1 The Zygote

64. Which is universally occurring event during sexual reproduction-
(a) Pollination
(b) Endosperm formation
(c) Zygote formation
(d) All of these

65. Choose the correct statements-
(a) In some organisms of fungi and algae, Zygote develop thin wall that is resistant to desiccation and damage.
(b) In haplontic life cycle diploid spore is form after mitosis of zygote that lead to development of new individuals.
(c) Haploid spores are formed after meiosis in zygote that lead to development of new individuals during haplontic life cycle
(d) All of these

66. Connecting link between one generation to other generation is-
(a) Gametes (b) Spores
(c) Zygote (d) All of these

67. Life began as ____ in human beings.
(a) Zygote (b) Gametes
(c) Foetus (d) None of these

Topic 2.3.2 Embryogenesis

68. Embryogenesis is-
(a) Development of embryo from spore
(b) Development of embryo from zygote
(c) Development of embryo from gametes
(d) None of these

69. Embryogenesis involves-
(a) Meiosis and cell differentiation
(b) Mitosis and cell differentiation
(c) Cell division (Meiosis & Mitosis) and cell differentiation
(d) Cell division (Meiosis & Mitosis) only

70. Statement-I : Cell divisions increase the number of cells in developing embryo.

Statement-II : Cell differentiation helps groups of cells to undergo certain modification to form specialized tissues and organ to form an organisms.
(a) Statement-I and Statement-II are correct
(b) Statement-I is correct while Statement-II is incorrect
(c) Statement-I is incorrect while Statement-II is correct
(d) Statement-I and Statement-II are incorrect.

71. In oviparous animals-
(a) Unfertilized egg laid in safe place is environment
(b) Fertilized eggs are covered by hard calcareous shell
(c) Unfertilized eggs undergoes for incubation & hatching of new individual
(d) (a) and (b) both

72. Choose incorrect about viviparous.
(a) Zygote develop into young one inside body of female organism.
(b) Chance of survival than viviparous organism is lesser.
(c) Change of survival of young ones of human is more than young ones of reptilians.
(d) None of these

73. In flowering plant, zygote is formed inside-
(a) Ovule
(b) Stamen
(c) Stigma
(d) Embryo sac present in stamen

74. After fertilization, which of following floral part does not either and fall off in every plant-
(a) Calyx
(b) Corolla
(c) Androecium
(d) Gynoecium

75. Seed and fruit develop from-
(a) Ovary & Ovule respectively
(b) Ovule & Ovary respectively
(c) Ovary & Ovary respectively
(d) Ovule & Ovule respectively

76.

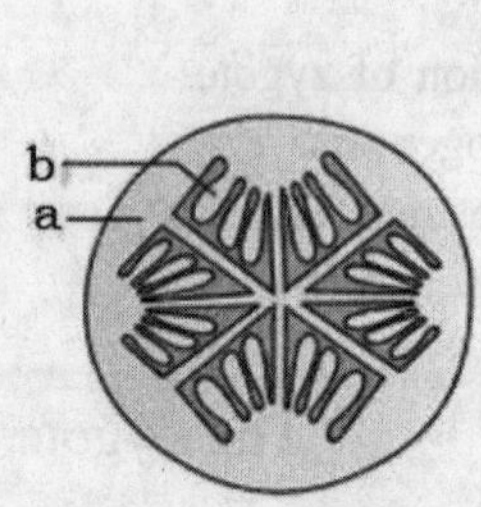

(a) a = Seed b = Pericarp
(b) a = Pericarp b = Seed
(c) a = Pericarp b = Pericarp
(d) a = Seed b = Seed

ANSWER KEY

1. (b)	2. (c)	3. (a)	4. (d)	5. (c)	6. (b)	7. (a)	8. (c)	9. (a)	10. (a)
11. (a)	12. (c)	13. (d)	14. (c)	15. (d)	16. (a)	17. (c)	18. (a)	19. (a)	20. (a)
21. (a)	22. (c)	23. (a)	24. (a)	25. (b)	26. (a)	27. (b)	28. (a)	29. (d)	30. (d)
31. (d)	32. (d)	33. (a)	34. (b)	35. (b)	36. (b)	37. (a)	38. (b)	39. (b)	40. (a)
41. (b)	42. (a)	43. (d)	44. (c)	45. (d)	46. (c)	47. (a)	48. (b)	49. (b)	50. (a)
51. (c)	52. (b)	53. (c)	54. (a)	55. (b)	56. (a)	57. (c)	58. (b)	59. (c)	60. (a)
61. (c)	62. (b)	63. (b)	64. (c)	65. (c)	66. (c)	67. (a)	68. (b)	69. (b)	70. (a)
71. (b)	72. (b)	73. (a)	74. (d)	75. (b)	76. (b)				

12 Chapter Human Reproduction

MULTIPLE CHOICE QUESTIONS

Topic 1 The Male Reproductive System

1. Among the following which one is not a male accessory gland?
(a) Seminal vesicle
(b) Ampulla
(c) Prostate
(d) Bulbourethral gland

2. Match the Column-I (parts) to Column-II (feature) and choose the correct option from the codes given below.

	Column I		Column II
(a)	Sertoli cells	(1)	Testicular hormones
(b)	Leydig cells	(2)	External opening of urethra
(c)	Epididymis	(3)	Nutrition to the germ cells
(d)	Urethral meatus	(4)	Male sex accessory duct

Codes

	a	b	c	d
(a)	4	3	2	1
(b)	3	1	4	2
(c)	1	2	3	4
(d)	2	4	1	3

3. Seminal plasma is contributed by:
(I) Seminal vesicle
(II) Prostate
(III) Urethra
(IV) Bulbourethral gland
(a) I and II (b) I, II and IV
(c) II, III and IV (d) I and IV

4. Read the following statements about seminal plasma and choose the correct statement(s) from the given options.
(I) Seminal plasma is secreted by seminal vesicles, prostate and bulbourethral glands.
(II) It is rich in sucrose and calcium.
(III) It contains certain enzymes also.
(a) I and II (b) II and III
(c) I and III (d) All of these

5. Read the following statements about male reproductive system and choose the incorrect statements from the given options.
(I) It is located in the pelvis region.
(II) The testes are situated outside the abdominal cavity within a pouch called scrotum.
(III) Each testis has about 350 testicular tubules.
(IV) Penis, the male external genitalia is made up of special tissues to facilitate insemination.
(a) I and III (b) III and IV
(c) I and IV (d) Only III

6. Assertion: The scrotum helps in maintaining the low temperature of the testes.

Reason: The low temperature of the testes is necessary for spermatogenesis.
(a) Both assertion and reason are true and reason is the correct explanation of assertion.
(b) Both assertion and reason are true, but reason is not the correct explanation of assertion.
(c) Assertion is true, but reason is false.
(d) Both assertion and reason are false.

7. Assertion: The enlarged part of penis is called glans penis.

Reason: The glans penis is covered by a loose fold of skin called foreskin.
(a) Both assertion and reason are true and reason is the correct explanation of assertion.
(b) Both assertion and reason are true, but reason is not the correct explanation of assertion.
(c) Assertion is true, but reason is false.
(d) Both assertion and reason are false.

8. The testes are situated outside the abdominal cavity within a pouch called
(a) urethra (b) scrotum
(c) penis (d) none of these

9. In humans, sperms are produced in
(a) epididymis
(b) rete testis
(c) seminiferous tubules
(d) vas deferens

10. Sertoli cells which line the seminiferous tubules from inside
(a) undergo meiotic division to produce sperms
(b) provide nutrition to the germ cells
(c) synthesise and secrete testicular hormones
(d) All of these

11. Refer to the given figure showing sectional view of seminiferous tubule. In the figure, some parts are labelled as A, B, C and D. Identify the part which provides nutrition to the developing sperms.

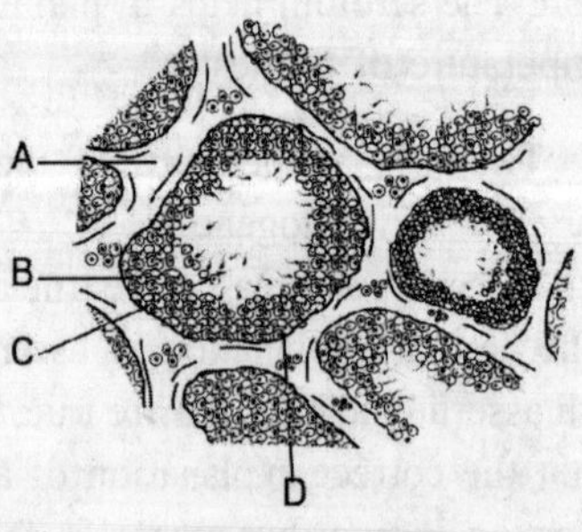

(a) A (b) C
(c) D (d) B

12. The regions outside the seminiferous tubules that contain Leydig cells are called
(a) interstitial spaces (b) antrum
(c) scrotum (d) none of these

13. Testicular hormones called androgens are secreted by
(a) interstitial cells (b) Leydig cells
(c) Sertoli cells (d) both (a) and (b)

14. Which one is odd from the following structures with reference to the male reproductive system.
(a) Rete testis (b) Epididymis
(c) Vasa efferentia (d) Isthmus

15. The vas deferens opens into urethra as [NCERT Exemplar]
(a) epididymis (b) ejaculatory duct
(c) efferent ductule (d) ureter

16. Which of the following depicts the correct pathway of transport of sperms?
(a) Rete testis → Efferent ductules → Epididymis → Vas deference
(b) Rete testis → Epididymis → Efferent ductules → Vas deference
(c) Rete testis → Vas deference → Efferent ductules → Epididymis
(d) Efferent ductules → Rete testis → Vas deference → Epididymis

17. Among the following which one is not an accessory duct of male reproductive system?
(a) Rete testis (b) Vasa efferentia
(c) Vas deferens (d) Urethra

18. The ejaculatory duct transports the sperms to the outside through
(a) urethra (b) rete testis
(c) vasa efferentia (d) none of these

19. Refer to the given figure and choose the correct option for the parts labelled as A, B, C and D.

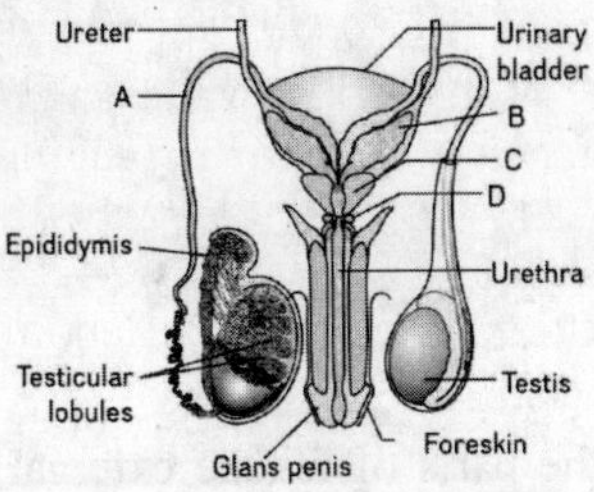

	a	b	c	d
A	Vas deferens	Seminal vesicle	Prostate gland	Bulbourethral gland
B	Vasa efferentia	Prosta te gland	Seminal vesicle	Bulbourethral gland
C	Prostate gland	Semin al vesicle	Bulbourethral gland	Vas deferens
D	Bulbourethral gland	Vas deferens	Prostate gland	Vasa efferentia

20. Match the parts given in Column-I to their characteristic features in Column-II and choose the correct option from the codes given below.

	Column I		Column II
(a)	Penis	(1)	Loose fold of skin
(b)	Glans penis	(2)	Male external genitalia
(c)	Foreskin	(3)	External opening urethra
(d)	Urethral meatus	(4)	Enlarged end of penis

Codes

	a	b	c	d
(a)	2	4	1	3
(b)	3	4	1	2
(c)	2	4	3	1
(d)	4	3	2	1

21. Urethral meatus is/are **[NCERT Exemplar]**
(a) the urinogenital duct
(b) opening of vas deferens into urethra
(c) external opening of the urinogenital duct
(d) muscles surrounding the urinogenital duct

Topic 2 The Female Reproductive System

22. The primary female sex organ is/are
(a) vagina (b) uterus
(c) ovaries (d) external genitalia

23. Among the following which one is not the part of female reproductive system?
(a) Cervix (b) Sertoli cells
(c) Mammary glands (d) Oviducts

24. The parts that constitute the female accessory ducts include
(a) Fallopian ducts (b) vagina
(c) ovaries (d) both (a) and (b)

25. The funnel-shaped part of Fallopian tube that remains closer to the ovary is
(a) infundibulum (b) fimbriae
(c) ampulla (d) isthmus

26. The part of the oviduct that joins the uterus is
(a) ampulla (b) isthmus
(c) fimbriae (d) infundibulum

27. The uterus is also called
(a) womb (b) cervix
(c) cervical canal (d) none of these

28. The inner glandular layer that lines the uterine cavity is
(a) perimetrium (b) myometrium
(c) endometrium (d) ectometrium

29. The uterine layer that undergoes cyclical changes during menstrual cycle is
(a) myometrium (b) endometrium
(c) perimetrium (d) both (a) and (b)

30. Which uterine layer exhibits strong contractions during the delivery of the baby?
(a) Endometrium (b) Perimetrium
(c) Myometrium (d) Both (a) and (c)

31. The female external genitalia include
(a) mons pubis (b) labia majora
(c) clitoris (d) all of these

32. The opening of the vagina is often covered partially by a membrane called
(a) hymen (b) clitoris
(c) labia minora (d) none of these

33. A tiny finger-like structure which lies at the upper junction of the two labia minora is
(a) hymen (b) mons pubis
(c) clitoris (d) none of these

34. How many mammary lobes are found in each breast?
(a) 20–25 (b) 15–20
(c) 10–15 (d) 25–30

35. The alveoli of mammary glands open into
(a) mammary tubules (b) mammary duct
(c) lactiferous duct (d) mammary lobes

36. The milk is sucked out through
(a) mammary duct (b) lactiferous duct
(c) alveoli (d) none of these

37. Match the parts of female reproductive system given in Column-I with their functions in Column-II and choose the correct option from the codes given below.

	Column I		Column II
(a)	Ovary	(1)	Delivery of baby
(b)	Fimbriae	(2)	Steroid hormone
(c)	Myometrium	(3)	Secretion of milk
(d)	Cells of alveoli	(4)	Collection of ovum

Codes

	a	b	c	d
(a)	2	4	1	3
(b)	4	3	2	1
(c)	3	4	1	2
(d)	1	4	3	2

38. Match the layers of uterus given in Column-I with their characteristic features given in Column-II and choose the correct option from the codes given below.

	Column I		Column II
(A)	Perimetrium	(1)	Thick layer of smooth muscles
(B)	Myometrium	(2)	Thick membranous layer
(C)	Endometrium	(3)	Glandular layer
		(4)	Thin membranous layer

Codes

	a	b	c
(a)	2	4	3
(b)	4	1	3
(c)	1	2	3
(d)	3	1	2

39. Match the parts of female external genitalia in Column-I with their characteristic features in Column-II. Choose the correct option from the codes given below.

Column I Column II

	Column I		Column II
(a)	Mons pubis	(1)	Fleshy folds of tissue
(b)	Labia majora	(2)	Cushion of fatty tissue
(c)	Hymen	(3)	Tiny fingerlike structure
(d)	Clitoris	(4)	Covers opening of vagina

Codes

	a	b	c	d
(a)	2	1	4	3
(b)	4	3	2	1
(c)	1	4	3	2
(d)	2	1	3	4

40. The edges of the infundibulum possess finger like projections that
(a) are the sight of fertilisation.
(b) help in the collection of ovum after fertilisation.
(c) are responsible for the release of egg.
(d) none of these

41. Read the following statements about uterus and choose the correct option from the codes given below.
(I) The shape of the uterus is like inverted pear.
(II) The uterus opens into vagina through a narrow cervix.
(III) The uterus along with cervix forms the birth canal.

Codes
(a) I and III (b) II and III
(c) I and II (d) All of these

42. Read the following statements about mammary glands and choose the incorrect statement.
(I) The mammary glands contain glandular tissue and fat.
(II) The mammary lobes of breasts contain alveoli which secrete milk.
(III) The milk secreted by alveoli is stored in lactiferous duct.
(a) Only II (b) Only III
(c) Only I (d) I and III

43. Assertion: Ovaries are the primary sex organs.
Reason: Ovaries produce the female gamete.
(a) Both assertion and reason are correct and reason is the correct explanation of assertion.
(b) Both assertion and reason are correct but reason is not the correct explanation of assertion.
(c) Assertion is correct, but reason is incorrect.
(d) Both assertion and reason are incorrect.

44. Assertion: Ovaries produce gamete as well as steroid hormones.

Reason: The oviducts, ovaries and cervix constitute the female accessory ducts.
(a) Both assertion and reason are correct and reason is the correct explanation of assertion.
(b) Both assertion and reason are correct but reason is not the correct explanation of assertion.
(c) Assertion is correct, but reason is incorrect.
(d) Both assertion and reason are incorrect.

45. Assertion: The uterus opens into vagina through a narrow cervix.

Reason: The cavity of cervix is called cervical canal.
(a) Both assertion and reason are correct and reason is the correct explanation of assertion.
(b) Both assertion and reason are correct but reason is not the correct explanation of assertion.
(c) Assertion is correct, but reason is incorrect.
(d) Both assertion and reason are incorrect.

46. Assertion: Mons pubis is a cushion of fatty tissue covered by skin and pubic hair.

Reason: The labia majora are paired folds of tissue under the labia minora.
(a) Both assertion and reason are correct and reason is the correct explanation of assertion.
(b) Both assertion and reason are correct but reason is not the correct explanation of assertion.
(c) Assertion is correct, but reason is incorrect.
(d) Both assertion and reason are incorrect.

47. Assertion: The alveoli of mammary lobes open into their lumen.

Reason: Several lactiferous ducts join to form a mammary duct through which milk is sucked out.
(a) Both assertion and reason are correct and reason is the correct explanation of assertion.
(b) Both assertion and reason are correct but reason is not the correct explanation of assertion.
(c) Assertion is correct, but reason is incorrect.
(d) Both assertion and reason are incorrect.

Topic 3 Gametogenesis

48. The process of producing gametes by primary sex organs is known as
(a) gametogenesis
(b) spermatogenesis
(c) oogenesis
(d) none of these

49. The immature, diploid male germ cells that produce sperms are
(a) spermatogonia
(b) secondary spermatocytes
(c) spermatids
(d) spermatozoa

50. Which of the following cells during gametogenesis

is normally diploid? **[AIPMT-2015]**

(a) Spermatid
(b) Spermatogonia
(c) Secondary polar body
(d) Primary polar body

51. Among the following, identify the cell(s) which undergo mitotic division during spermatogenesis?
(a) Primary spermatocytes
(b) Secondary spermatocytes
(c) Spermatids
(d) Spermatogonia

52. Spermatogenesis is the process in which immature male germ cells undergo division to produce sperms. Choose the correct one with reference to above.
(a) Spermatogonia have 46 chromosomes and always undergo meiotic cell division.
(b) Primary spermatocytes divide by mitotic cell division.
(c) Secondary spermatocytes have 23 chromosomes and undergo second meiotic division.
(d) Spermatozoa are transformed into spermatids.

53. After spermiogenesis, sperm heads become embedded in
(a) Leydig cells (b) antrum
(c) Sertoli cells (d) interstitial cells

54. During spermiation the sperms are released from
(a) seminiferous tubules
(b) vas deferens
(c) epididymis
(d) prostate gland

55. The difference between spermiogenesis and spermiation is **[NEET-2018]**
(a) in spermiogenesis spermatids are formed, while in spermiation spermatozoa are formed.
(b) in spermiogenesis spermatozoa are formed, while in spermiation spermatids are formed.
(c) in spermiogenesis spermatozoa from Sertoli cells are released into the cavity of seminiferous tubules, while in spermiation spermatozoa are formed.
(d) in spermiogenesis spermatozoa are formed, while in spermiation spermatozoa are released from Sertoli cells into the cavity of seminiferous tubules.

56. Spermatogenesis starts due to significant increase in the secretion of
(a) FSH (b) GnRH
(c) LH (d) oxytocin

57. Refer to the given figure showing diagrammatic sectional view of a seminiferous tubule. In the figure, some parts are labeled as A, B, C and D. Identify the part which gets activated by FSH.

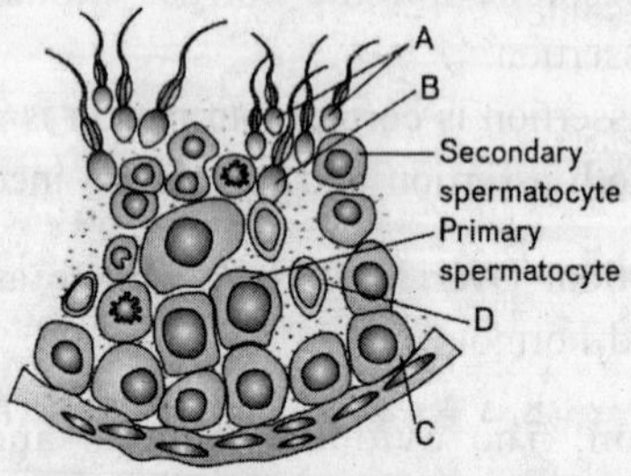

(a) A (b) B
(c) D (d) C

58. Refer to the given flowchart. In it, some spaces are mentioned as A, B, C and D. Identify the correct option for them from the codes given below.

Codes

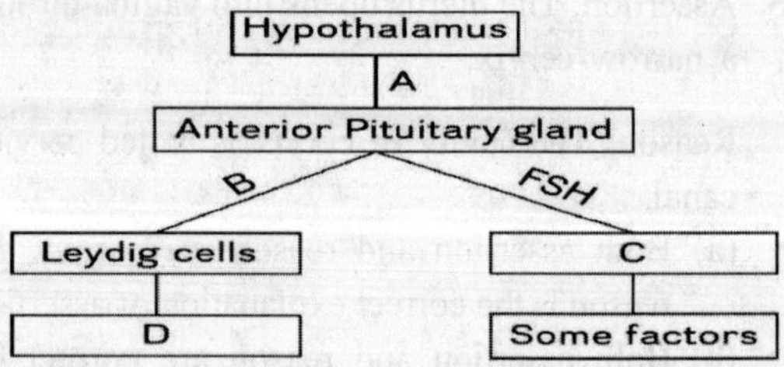

	A	B	C	D
A	GnRH	LH	Sertoli cells	Androgens
B	LH	GnRH	Androgens	Sertoli cell

C	Androgens	LH	Interstitial cells	FSH
D	FSH	Sertoli cells	GnRH	LH

59. The anterior portion of the sperm head is covered by a cap-like structure called

(a) middle piece (b) antrum
(c) acrosome (d) none of these

60. Match the Column-I representing parts of the sperm to Column-II showing their functions and choose the correct option.

	Column I		Column II
(a)	Head	(1)	Enzymes
(b)	Middle piece	(2)	Sperm motility
(c)	Acrosome	(3)	Energy
(d)	Tail	(4)	Genetic material

Codes

	A	B	C	D
(a)	2	4	1	3
(b)	4	3	1	2
(c)	4	1	2	3
(d)	2	1	3	4

61. The semen of human male contains

(a) seminal plasma (b) sperms
(c) enzymes (d) both (a) and (b)

62. Refer to the given figure showing structure of a sperm. The figure is followed by four (I–IV) statements. Choose the incorrect statement(s) about it.

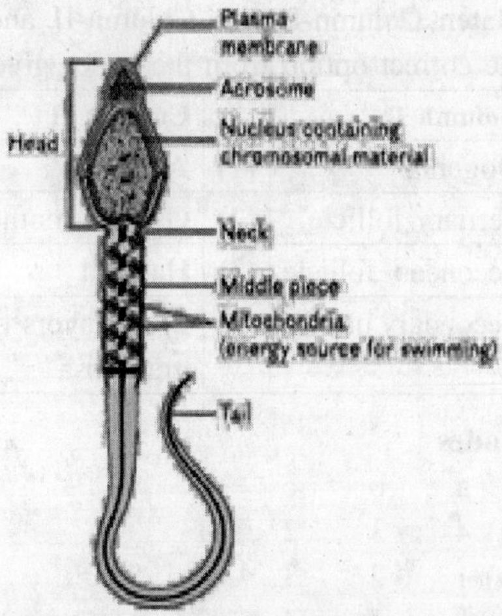

(I) The acrosome is filled with enzymes that help in fertilizing the ovum.
(II) The neck possesses numerous mitochondria.
(III) Tail is responsible for sperm motility.
(IV) The human male ejaculates 50–100 million sperms during a coitus.

(a) II and IV (b) I and III
(c) I and II (d) III and IV

63. Which among the following has 23 chromosomes?

(a) Spermatogonia (b) Zygote
(c) Secondary oocyte (d) Oogonia

64. The oogenesis is markedly different from spermatogenesis because

(a) it is initiated during embryonic development stage.
(b) it produces diploid gametes.
(c) it produces sperms also in special conditions.
(d) none of these

65. The tertiary follicle is characterized by a fluid filled cavity called

(a) antrum (B) corpus luteum
(c) matrix (d) none of these

66. In which stage primary oocyte completes its first meiotic division?

(a) Primary follicle
(b) Secondary follicle
(c) Tertiary follicle
(d) None of these

67. Match Column-I with Column-II and choose the correct option from the codes given below.

	Column I		Column II
(a)	Oogonia	(1)	Antrum
(b)	Tertiary follicle	(2)	Gamete mother cells
(c)	Secondary follicle	(3)	Haploid
(d)	Secondary oocyte	(4)	More layers of granulosa

Codes

	a	b	c	d
(a)	4	3	2	1
(b)	1	2	3	4
(c)	2	1	4	3
(d)	3	4	2	1

68. Extrusion of second polar body from egg nucleus occurs **[NEET 2019]**
(a) after entry of sperm, but before fertilisation
(b) after fertilisation
(c) before entry of sperm into ovum
(d) simultaneously with first cleavage

69. Refer to the given figure showing diagrammatic section view of ovary. The encircled part of figure is showing a process of oogenesis. Identify it as well as the follicle which is involved in this process.

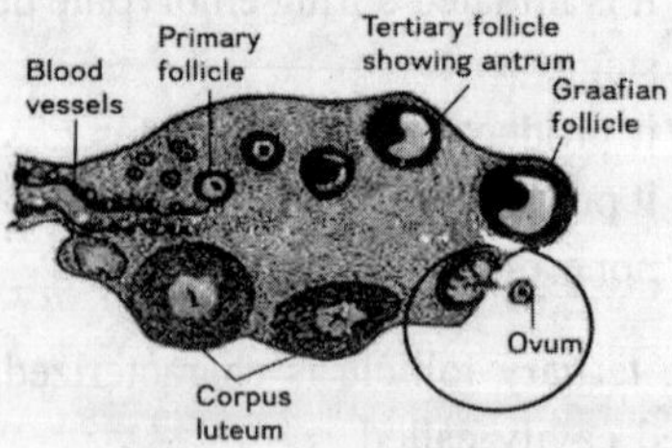

(a) Spermiation, Secondary follicle
(b) Menstruation, Primary follicle
(c) Ovulation, Graafian follicle
(d) Ovulation, Tertiary follicle

70. Assertion: Spermatogenesis starts at the age of puberty.

Reason: There is a significant increase in the secretion of gonadotropin releasing hormone at puberty.
(a) Both assertion and reason are true and reason is the correct explanation of assertion.
(b) Both assertion and reason are true but reason is not the correct explanation of assertion.
(c) Assertion is true, but reason is false.
(d) Both assertion and reason are false.

71. Assertion: Oogenesis is initiated at puberty.

Reason: Millions of oogonia are formed within each ovary every month.
(a) Both assertion and reason are true and reason is the correct explanation of assertion.
(b) Both assertion and reason true but reason is not the correct explanation of assertion.
(c) Assertion is true but reason is false.
(d) Both assertion and reason are false.

72. Assertion: A large haploid secondary oocyte is formed due to unequal division.

Reason: A tiny second polar body is formed during this division.
(a) Both assertion and reason are true and reason is the correct explanation of assertion.
(b) Both assertion and reason true but reason is not the correct explanation of assertion.
(c) Assertion is true but reason is false.
(d) Both assertion and reason are false.

Topic 4 Menstrual Cycle

73. The reproductive cycle in the female primates is called
(a) menstrual cycle
(b) oestrous cycle
(c) reproduction cycle
(d) none of these

74. Menarche that begins at puberty is
(a) cessation of menstrual cycle
(b) first menstruation
(c) period of pregnancy in which menstruation

ceases
(d) none of these

75. The cycle of events starting from one menstruation till the next one is called
(a) menopause (b) menarche
(c) menstrual cycle (d) oestrous cycle

76. The menstrual phase of menstrual cycle lasts for:
(a) 8–15 days (b) 1–2 days
(c) 1 day (d) 3–5 days

77. Menstrual flow results due to breakdown of
(a) endometrial lining
(b) blood vessels
(c) myometrial lining
(d) both (a) and (b)

78. Menstruation only occurs if
(a) implantation has occurred
(b) the released ovum is fertilized
(c) the released ovum is not fertilized
(d) both (A) and (D)

79. The lack of menstruation may be due to
(a) pregnancy (b) stress
(c) poor health (d) all of these

80. The menstrual phase is followed by
(a) follicular phase (b) luteal phase
(c) secretory phase (d) both (b) and (c)

81. What change(s) occur in ovary and/or uterus during follicular phase of menstrual cycle?
(a) Formation of Graafian follicle
(b) Formation of corpus luteum
(c) Regeneration of endometrium
(d) Both (a) and (c)

82. The changes in the ovary and uterus during proliferative phase are induced by changes in the levels of
(a) pituitary hormone (b) ovarian hormone
(c) pineal hormone (d) both (a) and (b)

83. In the ovary of a healthy human female mature Graafian follicle is generally present around
(a) 5–8 day of menstrual cycle
(b) 11–17 day of menstrual cycle
(c) 18–23 day of menstrual cycle
(d) 24–28 day of menstrual cycle

84. During proliferative phase, the growing follicles secrete
(a) LH (b) FSH
(c) gonadotropins (d) estrogens

85. Refer to the given flowchart, in which three parts are labeled as A, B and C. Identify them and choose the correct option from the codes given below.

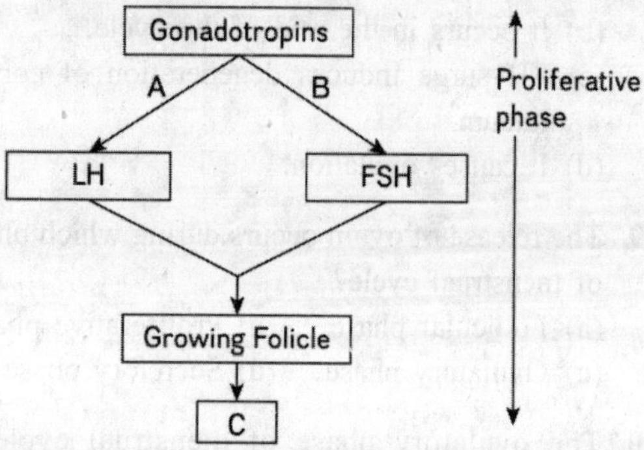

Codes

	A	B	C
(A)	Increase	Increase	Estrogens
(B)	Decrease	Increase	Estrogens
(C)	Decrease	Decrease	Testosterone
(D)	Increase	Decrease	TSH

86. Match the items given in Column-I with those in Column-II and select correct option from the codes given below. **[NEET–2018]**

	Column I		Column II
(a)	Proliferative phase	(1)	Breakdown of endometrial lining
(b)	Secretary phase	(2)	Follicular phase
(c)	Menstruation	(3)	Luteal phase

Codes

	a	b	c
(a)	3	2	1

(b) 1 3 2
(c) 2 3 1
(d) 3 1 2

87. Both LH and FSH attain a peak level on about 'A' day of cycle. Choose the correct option for 'A'.
(a) 10th day (b) 14th day
(c) 19th day (d) 5th day

88. Read the following statements about LH surge and choose the incorrect one.
(a) LH surge is rapid secretion of luteal hormone.
(b) It occurs in the mid of the cycle.
(c) LH surge induces degeneration of corpus luteum.
(d) It causes ovulation.

89. The release of ovum occurs during which phase of menstrual cycle?
(a) Follicular phase (b) Proliferative phase
(c) Ovulatory phase (d) Secretory phase

90. The ovulatory phase of menstrual cycle is followed by
(a) luteal phase (b) follicular phase
(c) proliferative phase (d) menstrual phase

91. No new follicles develop in the luteal phase of the menstrual cycle because **[NEET Odisha-2019]**
(a) Both FSH and LH levels are low in the luteal phase
(b) follicles do not remain in the ovary after ovulation
(c) FSH levels are high in the luteal phase.
(d) LH levels are high in the luteal phase.

92. Match the phases of menstrual cycle given in Column-I with the hormones secreted during that phase in Column-II.

Choose the correct option from the codes given below.

	Column I		Column II
(a)	Follicular phase	(1)	Progesterone
(b)	Ovulatory phase	(2)	Gonadotropins
(c)	Luteal Phase	(3)	LH surge
		(4)	Estrogens

Codes

	a	b	c
(a)	4	3, 1	2
(b)	2,4	3	1
(c)	2	4,1	3
(d)	2	1	4, 3

93. Among the following which change occurs during luteal phase?
(a) Corpus luteum → Graafian follicle
(b) Graafian follicle → Corpus luteum
(c) Primary follicle → Secondary follicle
(d) Secondary follicle → Graafian follicle

94. Read the following statements about corpus luteum and choose the correct ones from the following options.
(I) It is formed during ovulatory phase of menstrual cycle.
(II) It secretes large amounts of progesterone.
(III) In the absence of ovulation, the corpus luteum degenerates.
(IV) The degeneration of corpus luteum causes disintegration of endometrium.
(a) I and III (b) II and III
(c) II and IV (d) I and IV

95. In human beings, permanent cessation of menstrual cycle is called:
(a) Menopause (b) Menarche
(c) Ovulation (d) None of these

96-97. Refer to the given figure to answer the question no 96–97. The figure is showing diagrammatic presentation of various events during a menstrual cycle. In the figure, A, B and C, D show the levels of pituitary and ovarian hormones respectively.

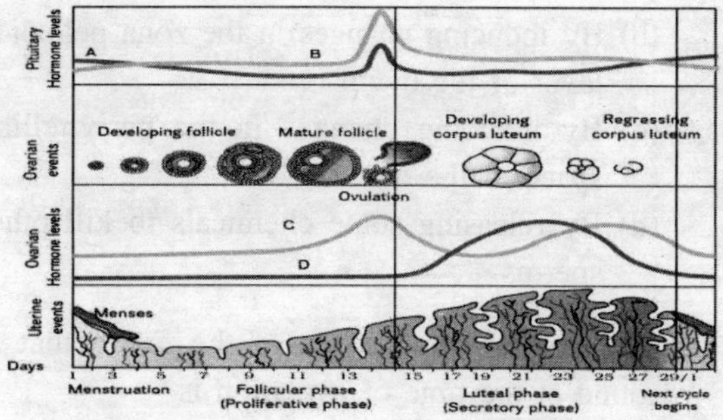

96. The gradual increase of which hormone stimulates the secretion of hormone C
(a) A (b) B
(c) D (d) Both (a) and (b)

97. The rapid increase of which hormone will induce rupture of Graafian follicle and thereby the release of ovum?
(a) A (b) C
(c) B (d) D

98. Choose the incorrect statement from the following.
(I) High levels of estrogen triggers the ovulatory surge.
(II) Oogonial cells start to proliferate and give rise to functional ova in regular cycles from puberty onwards.
(III) Sperms released from seminiferous tubules are poorly motile/non-motile.
(IV) Progesterone level is high during the post ovulatory phase of menstrual cycle.
(a) I and III (b) II and IV
(c) I and IV (d) I and II

99. Consider the following features.
(i) Transformation of Graafian follicle into corpus luteum.
(ii) Secretion of large amount of progesterone from corpus luteum.
(iii) Maintenance of endometrial lining of the uterus.

Select the correct phase of menstrual cycle that possesses all the above characteristics.
(a) Follicular phase (b) Secretory phase
(c) Proliferative phase (d) Ovulatory phase

100. Assertion: The lack of mensuration may be indicative of pregnancy.

Reason: Mensuration only occurs if the released ovum is not fertilised.
(a) Both assertion and reason are true and reason is the correct explanation of assertion.
(b) Both assertion and reason are true but reason is not the correct explanation of assertion.
(c) Assertion is true but reason is false.
(d) Both assertion and reason are false.

101. Assertion: During secretory phase the levels of LH and FSH gradually increase.

Reason: The increased levels of FSH and LH induce Graafian follicles to secrete progesterone.
(a) Both assertion and reason are true and reason is the correct explanation of assertion.
(b) Both assertion and reason are true but reason is not the correct explanation of assertion.
(c) Assertion is true but reason is false.
(d) Both assertion and reason are false.

102. Assertion: During pregnancy, all events of the menstrual cycle stop.

Reason: In the absence of fertilisation, the corpus luteum degenerates.
(a) Both assertion and reason are true and reason is the correct explanation of assertion.
(b) Both assertion and reason are true but reason is not the correct explanation of assertion.
(c) Assertion is true but reason is false.
(d) Both assertion and reason are false.

Topic 5 Fertilization and Implantation

103. The sperms released during copulation, finally reach to which part of the Fallopian tube?
(a) Infundibulum (b) Isthmus
(c) Ampulla (d) Ampullary-isthmic

104. All copulations not lead to the fertilisation and pregnancy. Choose the correct reason for the same from the following options.

(a) Fertilisation can only occur if sperms are transported to the ampullaryisthmic junction before the ovum.
(b) Fertilisation can only occur if the ovum and sperms are transported simultaneously to the ampullaryisthmic junction.
(c) Fertilisation can only occur if ovum is transported to the ampullary–isthmic junction before the sperms.
(d) None of these

105. Capacitation occurs in **[NEET–2017]**
(a) epididymis
(b) vas deferens
(c) female reproductive tract
(d) rete testis

106. Capacitation refers to changes in the **[AIPMT-2015]**
(a) ovum before fertilisation
(b) ovum after fertilisation
(c) sperm after fertilisation
(d) sperm before fertilisation

107. Refer to the given figure showing an ovum surrounded by few sperms. Sperm 'A' in the figure is trying to fertilise the ovum. How will this sperm 'A' ensure that no other sperm can fertilise the ovum?

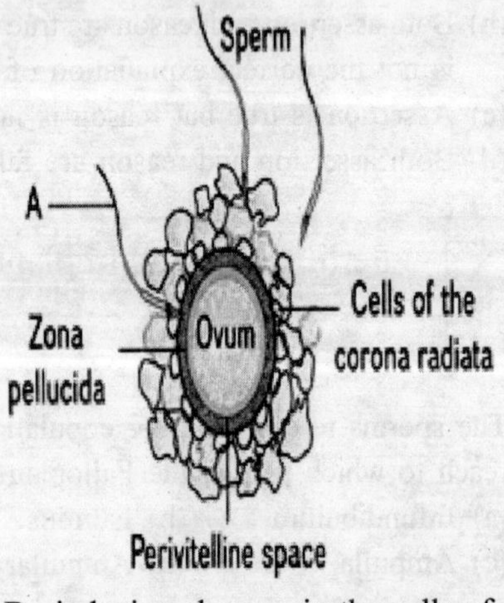

(a) By inducing changes in the cells of corona radiata.
(b) By inducing changes in the zona pellucida layer of the ovum.
(c) By inducing changes in the perivitelline space of the ovum.
(d) By releasing some chemicals to kill other sperms.

108. The membranous cover of the ovum that is found at the time of ovulation is
(a) corona radiata (b) zona radiata
(c) zona pellucida (d) chorion

109. During acrosomal reaction the sperm
(a) comes in contact with zona pellucida of the ova.
(b) undergoes reactions within the uterine environment of female.
(c) undergoes reactions within the epididymal environment of the male.
(d) produces androgens in the uterus.

110. The secretions of the acrosome help the sperm to enter into the cytoplasm of ovum through
(a) corona radiata (b) zona pellucida
(c) chorion (d) amnion

111. Match the events given in Column-I with their characteristic features in Column-II and choose the correct option from the codes given below.

	Column I		Column II
(a)	Fertilisation	(1)	Female reproductive tract
(b)	Capacitation	(2)	Contact of sperm with zona pellucida
(c)	Acrosomal reaction	(3)	Before fertilisation and after ovulation
(d)	Second polar body	(4)	Ampullary-isthmic junction

Codes

	a	b	c	d
(a)	1	4	2	3
(b)	4	1	2	3
(c)	3	2	4	1
(d)	4	1	3	2

112. The embryo having 8–16 blastomeres is called

(a) blastula (b) gastrula
(c) morula (d) trophoblast

113. Morula is known as a developmental stage
(a) between the zygote and blastocyst
(b) between the blastocyst and gastrula
(c) after the implantation
(d) between implantation and parturition

114. Refer to the given flowchart. It has some blank spaces mentioned as A, B, C and D. Choose the correct option for these A, B, C and D.

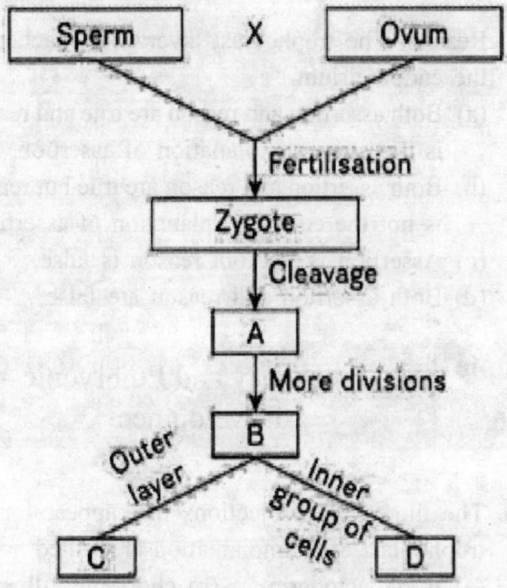

(a) Morula Blastocyst Trophoblast Inner cell mass
(b) Blastocyst Morula Trophoblast Inner cell mass
(c) Morula Blastocyst Inner cell mass Trophoblast
(d) Blastocyst Morula Inner cell mass Trophoblast

115. The blastomeres in the blastocyst are arranged into
(a) an outer trophoblast and inner cell mass
(b) an outer cell mass and inner trophoblast
(c) an outer cell mass and inner cell mass
(d) none of these

116. During embryogenesis which part of blastocyst gets differentiated into embryo?
(a) Trophoblast (b) Inner cell mass
(c) Morula (d) Both (a) and (b)

117. The embedding of blastocyst in the endometrium of uterus is called
(a) pregnancy (b) lactation
(c) embryogenesis (d) implantation

118. Extrusion of second polar body from egg nucleus occurs **[NEET–2015]**
(a) after entry of sperm but before fertilisation
(b) after fertilisation
(c) before entry of sperm into ovum
(d) simultaneously with first cleavage

119. All the haploid gametes produced by the female have
(a) X-chromosomes
(b) Y-chromosomes
(c) 50% gametes have X and 50% have Ychromosomes
(d) 25% gametes have X and 75% have Ychromosomes

120. Cleavage that occurs in the zygote as it moves through the isthmus of oviduct towards the uterus is
(a) meiotic division
(b) mitotic division
(c) reductional division
(d) none of these

121. Match the following and choose the correct option from the codes given below.

	Column I		Column II
(a)	Trophoblast	(1)	Embedding of Blastocyst in the endometrium
(b)	Cleavage	(2)	Group of cells that would differentiate as embryo
(c)	Inner cell mass	(3)	Outer layer of blastocyst attached to the endometrium
(d)	Implantation	(4)	Mitotic division of zygote

Codes

	a	b	c	d
(a)	2	1	3	4
(b)	3	4	2	1

(c) 3 1 2 4
(d) 2 4 3 1

122. Refer to the given figure showing transport of ovum, fertilisation and passage of growing embryo through Fallopian tube. The figure is followed by four statements. Choose the incorrect statement about it.

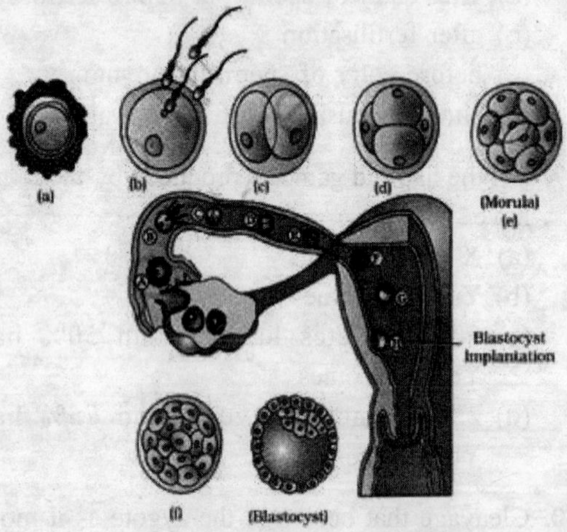

(a) The second meiotic division of secondary oocyte occurs after fertilisation.
(b) The embryo with 8–16 blastomeres is called morula.
(c) The trophoblast layer of blastocyst gets embedded in the endometrium.
(d) The cells of inner cell mass differentiate to form embryo.

123. Assertion: All copulations do not lead to fertilisation and pregnancy.

Reason: Fertilisation can only occur if the ovum and sperms are transported simultaneously to the ampullary–isthmic junction. [Pg-51,H]
(a) Both assertion and reason are true and reason is the correct explanation of assertion.
(b) Both assertion and reason are true but reason is not the correct explanation of assertion.
(c) Assertion is true but reason is false.
(d) Both assertion and reason are false.

124. Assertion: The secretions of the acrosome help the sperm enter into the cytoplasm of the ovum.

Reason: The entry of sperm into the cytoplasm of the ovum is responsible for the capacitation of sperms.
(a) Both assertion and reason are true and reason is the correct explanation of assertion.
(b) Both assertion and reason are true but reason is not the correct explanation of assertion.
(c) Assertion is true but reason is false.
(d) Both assertion and reason are false.

125. Assertion: The blastomeres in the blastocyst are arranged into trophoblast and inner cell mass.

Reason: The trophoblast layer gets attached to the endometrium.
(a) Both assertion and reason are true and reason is the correct explanation of assertion.
(b) Both assertion and reason are true but reason is not the correct explanation of assertion
(c) Assertion is true but reason is false.
(d) Both assertion and reason are false.

Topic 6 Pregnancy and Embryonic Development

126. The finger-like projections that appear on the trophoblast after implantation are called
(a) trophoectoderm (b) chorionic villi
(c) placenta (d) none of these

127. The structural and functional unit formed between the developing embryo (foetus) and maternal body is called
(a) placenta (b) trophoblast
(c) chorionic villi (d) trophoectoderm

128. The embryo remains connected to the placenta through
(a) chorionic villi (b) trophoblast
(c) umbilical cord (d) none of these

129. Choose the incorrect statement about placenta
(a) The placenta facilitates the supply of oxygen and nutrients to the embryo.
(b) It helps in the removal of CO_2 and excretory

waste materials produced by the embryo.

(c) The placenta is connected to the embryo through umbilical cord.

(d) Placenta acts like an endocrine tissue and produces several enzymes also.

130. Which of the following hormones is not secreted by human placenta?

(a) hCG (b) Estrogen

(c) Progesterone (d) LH

131. Several hormones like hCG, hPL, estrogen, progesterone are produced by **[NEET–2016]**

(a) Fallopian tube (b) pituitary

(c) ovary (d) placenta

132. A hormone that is secreted by ovary in the later phase of pregnancy is

(a) estrogen (b) FSH

(c) relaxin (d) hCG

133. The hormone(s) that is/are produced during pregnancy only

(a) hCG (b) hPL

(c) relaxin (d) all of these

134. The levels of estrogens, progestogens, cortisol, prolactin, etc., increase many fold in blood during pregnancy. It is necessary for

(a) supporting the fetal growth

(b) metabolic changes in the mother

(c) the maintenance of pregnancy

(d) all of these

135. Match Column-I with Column-II and choose the correct option from the codes given below.

	Column I		Column II
(a)	Chorionic villi	(1)	Secretes relaxin hormone
(b)	Placenta	(2)	Finger-like projections on the trophoblast
(c)	Umbilical cord	(3)	Structural and functional unit between foetus and mother
(d)	Ovary	(4)	Connects embryo to placenta

Codes

	a	b	c	d
(a)	2	3	4	1
(b)	3	2	4	1
(c)	2	3	1	4
(d)	4	1	3	2

136. Immediately after implantation, which part of blastocyst differentiates into ectoderm and endoderm?

(a) Trophoblast (b) Chorionic villi

(c) Inner cell mass (d) None of these

137. Refer to the given figure showing human foetus within the uterus. How will the removal of 'A' in the figure affect the growth of foetus?

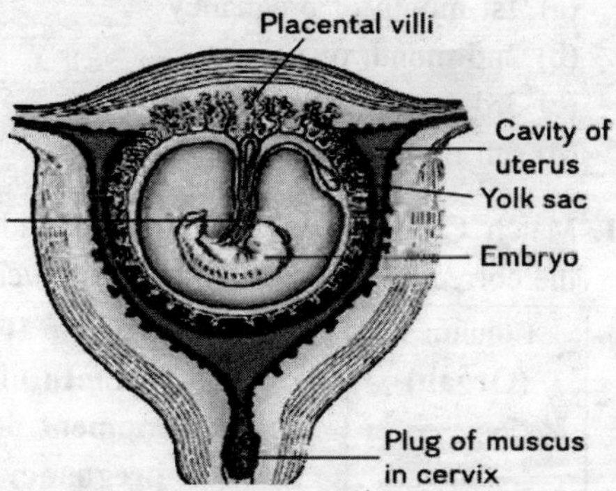

(a) Secretion of hCG hormone will stop

(b) Secretion of relaxin hormone will stop

(c) Transportation of substances to and from the embryo will stop

(d) All of these

138. The inner cell mass contains certain cells called stem cells which have the potency to give rise to

(a) specific tissues and organs

(b) all tissues and organs

(c) only cardiac cells and heart

(d) none of these

139. Match Column-I with Column-II and select the correct option using the codes given below. **[NEET–2016]**

	Column I		Column II
(a)	Mons pubis	(1)	Embryo formation
(b)	Antrum	(2)	Sperm
(c)	Trophectoderm	(3)	Female external genitalia
(d)	Nebenkern	(4)	Graafian follicle

Codes

	a	b	c	d
(a)	3	1	4	2
(b)	1	4	3	2
(c)	3	4	2	1
(d)	3	4	1	2

140. The foetus develops limbs and digits by the end of

(a) 1st month of pregnancy
(b) 2nd month of pregnancy
(c) 3rd month of pregnancy
(d) 5th month of pregnancy

141. Match Column-I with Column-II and choose the correct option from the codes given below.

	Column I (Organ)		Column II (Month of development during pregnancy)
(a)	Heart	(1)	Second month
(b)	Limbs and digits	(2)	First month
(c)	External genitalia	(3)	Fifth month
(d)	Appearance of hair on head	(4)	Third month

Codes

	a	b	c	d
(a)	1	2	4	3
(b)	2	1	3	4
(c)	2	1	4	3
(d)	3	2	4	1

142. Read the following statements about major features of embryonic development at various months of pregnancy. Choose the incorrect statement(s) about it.

(I) The embryo's heart is the first formed organ
(II) Most of the major organ systems are formed by the end of 2nd month of pregnancy.
(III) The first movement of foetus is observed during 7th month of pregnancy.
(IV) The eyelashes are formed by the end of second trimester.

(a) II and III (b) I and II
(c) III and IV (d) Only IV

143. Select the correct sequences of events. **[Odisha, NEET-2019]**

(a) Gametogenesis → Gamete transfer → Syngamy → Zygote cell differentiation → cell division (cleavage) → organogenesis
(b) Gametogenesis → Gamete transfer → Syngamy → Zygote cell division (cleavage) → cell differentiation → organogenesis
(c) Gametogenesis → Gamete transfer → Syngamy → Zygote cell division (cleavage) → organogenesis → cell differentiation
(d) Gametogenesis → Syngamy → Gamete transfer → Zygote cell division (cleavage) → cell differentiation → organogenesis

144. Assertion: The placenta is connected to the embryo through an umbilical cord.

Reason: The umbilical cord helps in the transport of substances to and from the embryo.

(a) Both assertion and reason are true and reason is the correct explanation of assertion.
(b) Both assertion and reason are true, but reason is not the correct explanation of assertion.
(c) Assertion is true, but reason is false.
(d) Both assertion and reason are false.

145. Assertion: Placenta also acts as an endocrine gland.

Reason: In the later phase of pregnancy, relax is secreted by placenta.

(a) Both assertion and reason are true and reason is the correct explanation of assertion.
(b) Both assertion and reason are true, but reason

is not the correct explanation of assertion.
(c) Assertion is true, but reason is false.
(d) Both assertion and reason are false.

146. Assertion: The first movements of foetus are observed during the third month of pregnancy.

Reason: By the end of first trimester, eyelids separate and eye-lashes are formed.
(a) Both assertion and reason are true and reason is the correct explanation of assertion.
(b) Both assertion and reason are true, but reason is not the correct explanation of assertion.
(c) Assertion is true, but reason is false.
(d) Both assertion and reason are false.

Topic 7 Parturition and Lactation

147. Which of the following hormones is responsible for both the milk ejection reflex and the foetal ejection reflex? [Odisha, NEET-2019]
(a) Relaxin (b) Estrogen
(c) Prolactin (d) Oxytocin

148. The milk produced during the initial few days of lactation is called
(a) colostrum (b) first milk
(c) milky water (d) none of these

149. Choose the incorrect statement from the following.
(a) Internal fertilisation takes place, in birds and mammals.
(b) Colostrum contains antibodies and nutrients.
(c) Polyspermy is prevented by chemical changes on the egg surface.
(d) In the human female implantation occurs almost seven days after fertilisation.

150. Colostrum contains
(a) antibodies (b) nutrients
(c) enzymes (d) both (a) and (b)

151. Assertion: Parturition is a complex neuroendocrine mechanism.

Reason: The signals of parturition originate from the fully developed fetus and placenta.
(a) Both assertion and reason are true and reason is the correct explanation of assertion.
(b) Both assertion and reason are true, but reason is not the correct explanation of assertion.
(c) Assertion is true, but reason is false.
(d) Both assertion and reason are false.

152. Assertion: The mammary glands of the female undergo differentiation after parturition.

Reason: Lactation always starts after childbirth.
(a) Both assertion and reason are true and reason is the correct explanation of assertion.
(b) Both assertion and reason are true, but reason is not the correct explanation of assertion.
(c) Assertion is true, but reason is false.
(d) Both assertion and reason are false.

153. Assertion: Breast-feeding during the initial period of infant growth is recommended.

Reason: During initial few days after delivery, colostrum is produced.
(a) Both assertion and reason are true and reason is the correct explanation of assertion.
(b) Both assertion and reason are true, but reason is not the correct explanation of assertion.
(c) Assertion is true, but reason is false.
(d) Both assertion and reason are false

154. The duration of pregnancy is called
(a) parturition (b) lactation
(c) gestation period (d) none of these

155. The process of delivery of the foetus is called
(a) lactation
(b) parturition
(c) foetal ejection reflex
(d) none of these

156. The signals for parturition originate from the
(a) fully developed foetus
(b) placenta
(c) umbilical cord
(d) both (a) and (b)

157. Match Column-I with Column-II and choose the correct option from the codes given below.

	Column I		Column II
(a)	Gestation period	(1)	Mild uterine contractions
(b)	Parturition	(2)	Duration of pregnancy
(c)	Foetal ejection reflex	(3)	Process of delivery of the foetus
(d)	After birth	(4)	Placental expulsion

Codes

	a	b	c	d
(a)	2	3	1	4
(b)	2	4	1	3
(c)	4	3	2	1
(d)	1	2	3	4

158. The process of milk production is called
(a) lactation (b) parturition
(c) after birth (d) colostrum

159. Choose the incorrect statement from the following:
(a) Parturition is induced by a complex neuroendocrine mechanism.
(b) The foetal ejection reflex triggers release of oxytocin from the hypothalamus.
(c) Oxytocin acts on the uterine muscles.
(d) The placenta is also expelled out of the uterus after the delivery of infant.

ANSWER KEY

1. (b)	2. (b)	3. (b)	4. (c)	5. (d)	6. (a)	7. (b)	8. (b)	9. (c)	10. (b)
11. (d)	12. (a)	13. (b)	14. (d)	15. (b)	16. (a)	17. (d)	18. (a)	19. (a)	20. (a)
21. (c)	22. (c)	23. (b)	24. (d)	25. (a)	26. (b)	27. (a)	28. (c)	29. (b)	30. (c)
31. (d)	32. (a)	33. (c)	34. (b)	35. (a)	36. (b)	37. (a)	38. (b)	39. (a)	40. (b)
41. (c)	42. (b)	43. (a)	44. (c)	45. (b)	46. (c)	47. (d)	48. (a)	49. (a)	50. (b)
51. (d)	52. (c)	53. (c)	54. (a)	55. (d)	56. (b)	57. (d)	58. (a)	59. (c)	60. (b)
61. (d)	62. (a)	63. (c)	64. (a)	65. (a)	66. (c)	67. (c)	68. (a)	69. (c)	70. (a)
71. (d)	72. (c)	73. (a)	74. (b)	75. (c)	76. (d)	77. (d)	78. (c)	79. (d)	80. (a)
81. (d)	82. (d)	83. (b)	84. (d)	85. (a)	86. (c)	87. (b)	88. (c)	89. (c)	90. (a)
91. (a)	92. (b)	93. (b)	94. (c)	95. (a)	96. (d)	97. (c)	98. (a)	99. (b)	100. (a)
101. (d)	102. (b)	103. (c)	104. (b)	105. (c)	106. (d)	107. (b)	108. (a)	109. (a)	110. (b)
111. (b)	112. (c)	113. (a)	114. (a)	115. (a)	116. (b)	117. (d)	118. (a)	119. (a)	120. (b)
121. (b)	122. (a)	123. (a)	124. (c)	125. (b)	126. (b)	127. (a)	128. (c)	129. (d)	130. (d)
131. (d)	132. (c)	133. (d)	134. (d)	135. (a)	136. (c)	137. (c)	138. (b)	139. (d)	140. (b)
141. (c)	142. (a)	143. (b)	144. (b)	145. (c)	146. (d)	147. (d)	148. (a)	149. (c)	150. (d)
151. (b)	152. (a)	153. (a)	154. (c)	155. (b)	156. (d)	157. (a)	158. (a)	159. (b)	

13 Chapter

Reproductive Health

MULTIPLE CHOICE QUESTIONS

Topic 1	Problems and Strategies of Reproductive Health

1. Amniocentesis can be used to detect all the listed diseases except

(a) Hemophilia (b) Malaria

(c) Sickle cell anemia (d) Down syndrome

2. In the context of amniocentesis, which of the following statement is incorrect? **[NEET-2016]**

(a) It is used for prenatal sex determination.

(b) It can be used for detection of Down syndrome.

(c) It can be used for detection of cleft palate.

(d) It is usually done when a woman is 14–16 weeks pregnant.

3. Assertion: Sex education is provided to adolescents at school only.

Reason: Sex education is introduced to teach adolescents about myths related of sexual practices only.

(a) Both assertion and reason are true and the reason is the correct explanation of assertion.

(b) Both assertion and reason are true but the reason is not the correct explanation of assertion.

(c) Assertion is true but reason is false.

(d) Both assertion and reason are false.

4. Assertion: Amniocentesis helps to detect foetal disorders during early pregnancy.

Reason: Jaundice can be prevented in foetus by amniocentesis.

(a) Both assertion and reason are true and the reason is the correct explanation of assertion.

(b) Both assertion and reason are true but the reason is not the correct explanation of assertion.

(c) Assertion is true but reason is false.

(d) Both assertion and reason are false.

5. According to World Health Organisation, a reproductively healthy individual possesses

(a) Functionally and morphologically normal reproductive organs

(b) Normal emotional and behavioural interactions with society

(c) Total well-being in all aspects of reproductions

(d) All of these

6. Which country commenced the action plans at national level to achieve total reproductive health as a social goal?

(a) China (b) Japan

(c) India (d) Argentina

7. Consider the following statements:

a. In India, family planning program was initiated in 1982.

b. Reproductive and Child Health Care (RCH) program is an improved version of family planning programs.

Select the correct option

(a) (a) is true, (b) is false

(b) Both (a) and (b) are false

(c) (a) is false, (b) is true

(d) Both (a) and (b) are true

8. In what way awareness can be created among

people about reproduction related aspects?

a. Through audio-visual and print media advertisements.
b. Creating fear among adolescents about sexual practices.
c. Introduction of sex education at school.

Select the most appropriate option
(a) (a), (b) and (c) are correct
(b) (b) and (c) are correct
(c) (a) and (b) are correct
(d) (a) and (c) are correct

9. While providing sex education to adolescents at school, which attribute would be least significant?
(a) Safe and hygienic sexual practices.
(b) Information about postnatal care of mother and child.
(c) Information about STDs and AIDS.
(d) Discouraging children from believing in myths related to sexual practices.

10. RCH programs helps to curb
(a) Population explosion
(b) Female foeticide
(c) Sex abuse
(d) All of these

11. Statutory ban on amniocentesis helps to check or prevent
(a) Female foeticides and massive child immunization
(b) Sexually transmitted diseases
(c) Genetic disorders in foetus
(d) All of these

12. Consider the following statements:

a. In amniocentesis, small sample is taken from the foetus blood.
b. Foetal sex can be determined through amniocentesis by studying the chromosomal pattern in amniotic fluid.

Select the correct option.
(a) (a) is true, (b) is false
(b) Both (a) and (b) are false
(c) (a) is false, (b) is true
(d) Both (a) and (b) are true

Topic 2 Population Stabilisation and Birth Control

13. An ideal contraceptive should have all the listed characteristics, except
(a) User friendly (b) Reversible
(c) Easily available (d) Decrease sex drive

14. Consider the following statements:
(a) Natural method of contraception is a traditional method to prevent pregnancy.
(b) The basic principle of natural method of contraception is to avoid the contact between ovum and sperms

Select the correct option.
(a) a is true, b is false
(b) a is false, b is true
(c) Both a and b are true
(d) Both a and b are false

15. In periodic abstinence method
(a) Female has to take contraceptive pills periodically
(b) Couple should avoid coitus periodically
(c) Couple should avoid coitus for few months or years
(d) female do not ovulate for few months.

16. During which period of a typical menstrual cycle of 28 days, coitus should be avoided to prevent conception naturally?
(a) 0–5 days, i.e., during menstrual flow
(b) 5–8 days after menstrual flow
(c) 11–16 days after menstrual flow
(d) 23–26 days after menstrual flow

17. How does coitus interruptus method is practiced by couples?
(a) By avoiding coitus at the time of ovulation
(b) By avoiding insemination just before

ejaculation
(c) By ejaculation semen in the condoms
(d) By avoiding the insertion of penis into the vagina

18. In lactational amenorrhea, chances of conception are nil during
(a) 40 days following parturition
(b) The time till mother breastfeeds her child
(c) Six months following parturition
(d) The gestational period of female

19. Match the following columns.

	Column I (Natural methods of contraception)		Column II (Characteristics)
(a)	Coitus interruptus	(1)	Avoid intercourse during ovulation
(b)	Lactational amenorrhea	(2)	Avoid insemination into vagina
(c)	Periodic abstinence	(3)	Ovum is not available for fertilisation

Select the correct option.

	a	b	c
(a)	3	1	2
(b)	1	3	2
(c)	3	2	1
(d)	2	3	1

20. Select the incorrect statement.
(a) Natural methods of contraception are riskier but reliable.
(b) The act of breastfeeding interrupt ovulation during menstrual cycle.
(c) The chances of conception are higher during the ovulation period.
(d) Natural methods of contraception do not interfere with the sexual drive of an individual.

21. The barrier method of contraception
(a) can be used by males only
(b) prevents physical meeting of sperm and ovum
(c) are used at the time of coitus

Select the most appropriate option.
(a) (a) and (c) are correct
(b) (a), (b), (c) are correct
(c) (b) and (c) are correct
(d) Only (b) is correct

22. Select the incorrect statement regarding condoms.
(a) These are made up of thin latex sheath.
(b) These are used to cover penis or cervix.
(c) They help the ejaculate to enter uterus but not into fallopian tube.
(d) Nirodh is a popular brand of male condoms.

23. Use of condoms can help to prevent
(a) AIDS and other STDs
(b) Spermatogenesis
(c) Coitus
(d) All of these

24. What is the similarity between vaults and diaphragms?
(a) They both prevent ovulation in females.
(b) They are used to cover penis in males.
(c) They are used to cover cervix during coitus.
(d) They release copper ions to suppress sperm motility.

25. The trend of world's population growth from the year 2000 to 2011 is
(a) Increasing (b) Decreasing
(c) Stable (d) Fluctuating

26. Match the following columns.

	Column I (Year)		Column II (World population)
(a)	2011	(1)	2000 million
(b)	1900	(2)	7.2 billion
(c)	2000	(3)	6 billion

Codes

	a	b	c
(a)	3	2	1
(b)	1	3	2
(c)	2	1	3
(d)	3	1	2

27. Consider the following statements.

a. The world population increased significantly in between the years 1900–2000 as compared to the years 2000–2011.

b. Increased health services had an explosive impact on world's population growth.

Select the correct option

(a) a is false, b is true (b) Both a and b are true

(c) a is true, b is false

(d) Both a and b are false

28. Which among the following graph represents accurate population growth pattern of India?

a)

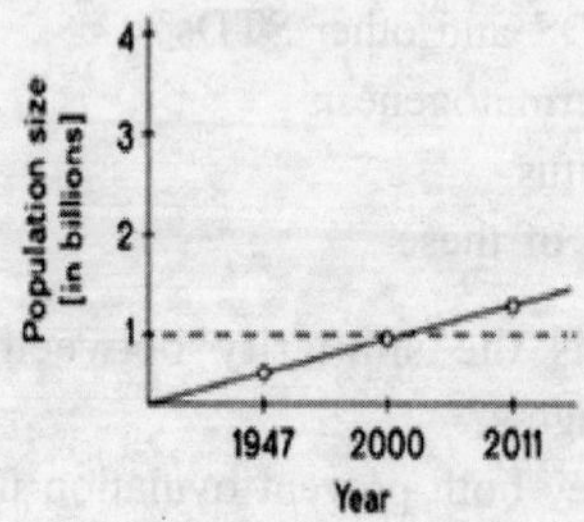

b)

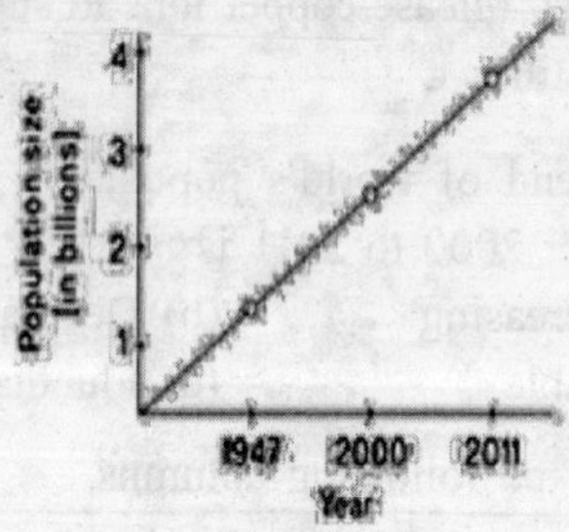

c)

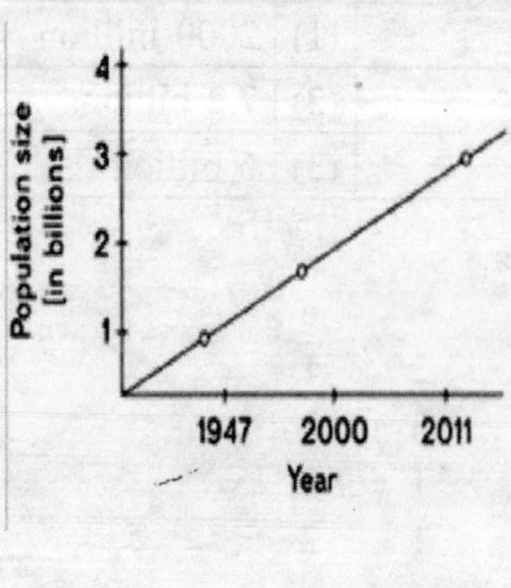

d)

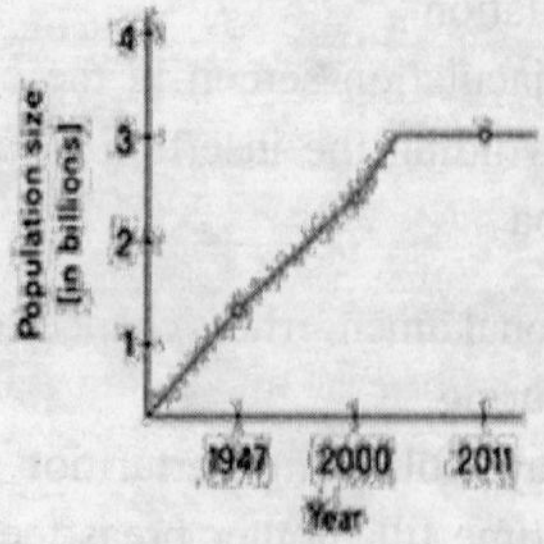

29. According to the 2011 census report, the population growth rate of India was

(a) More than five percent

(b) Less than one percent

(c) Less than two percent

(d) More than six percent

30. Under what circumstances, the growth of population declines?

a. Increased number of people in reproducible age.

b. Decreased infant mortality rate.

c. Increased maternal mortality rate.

Select the most appropriate option.

(a) (a), (b), (c) are correct

(b) (a) and (b) are correct

(c) Only (c) is correct

(d) Only (b) is correct

31. A population with declined infant mortality rate would show

(a) Zero growth

(b) Increased growth

(c) Decreased growth

(d) Fluctuating growth

32. Which among the following measure can be opted to check the population growth rate?

(a) Contraceptives (b) MTP

(c) Amniocentesis (d) ART

33. Consider the following statements.

a. 'One child norm' is a threat to population growth rate of India.

b. Statutory raising of marriageable age of males and females check the population growth rate.

Select the correct option.
(a) a is true, b is false
(b) Both a and b are true
(c) a is false, b is true
(d) Both a and b are false

34. Marriageable age of males and female in India is
(a) 18 years
(b) 21 years
(c) 21 years and 18 years, respectively
(d) 18 years and 21 years, respectively

35. Consider the following statements:
(a) Cervical caps and vaults are not cost effective as they cannot be reused.
(b) Contraceptive efficiency of vaults and cervical caps can be increased by using spermicidal creams along with these barriers.

Select the correct option
(a) a is true, b is false
(b) Both a and b are true
(c) a is false, b is true
(d) Both a and b are false

36. Select the correct statement regarding intra uterine devices (IUDs).
(a) They are used by females at the time of copulation.
(b) IUDs are inserted in the uterus by doctors.
(c) IUDs are not reliable contraceptives.
(d) All IUDs release hormones which suppress sperm motility.

37. Match the following columns.

	Column I		Column II
(a)	LNG 20	(1)	Hormone-releasing IUD
(b)	Lippes loop	(2)	Copper releasing IUD
(c)	Multiload 375	(3)	Non medicated IUD

Select the correct option.

	a	b	c
(a)	2	1	3
(b)	2	3	1
(c)	1	2	3
(d)	1	3	2

38. All the listed IUDs release copper ions except
(a) LNG 20 (b) Cu 7
(c) Cu T (d) Multiload 375

39. What is significance of copper ions in contraception?
(a) They delay ovulation in females.
(b) They suppress sperm motility and fertilising capacity of sperms.
(c) They increase the level of prolactin so as to suppress the activity of LH and FSH.
(d) They decrease the libido and sexual desire of female.

40. Select the hormone-releasing Intra-Uterine Devices. **[NEET-2019]**
(a) Multiload 375, Progestasert
(b) Progestasert, LNG-20
(c) Lippes Loop, Multiload 375
(d) Vaults, LNG-20

41. Which of the following is a correct statement? **[NEET-2019]**
(a) IUDs suppress gametogenesis
(b) IUDs once inserted need not be replaced
(c) IUDs are generally inserted by the user herself
(d) IUDs increase phagocytosis of sperms in the uterus

42. The hormones used for the purpose of contraception in pills and IUDs are
(a) Progesterone and estrogen
(b) Estrogen and testosterone
(c) LH and progesterone
(d) Prolactin and testosterone

43. For the effective results of contraception pills, they should be taken
(a) Once every month on the day of ovulation
(b) For 21 days daily, followed by a gap of seven days
(c) For seven days daily, during the time of ovulation
(d) Daily, without any gap

44. Consider the following statements.
(a) Oral contraceptives inhibit ovulation and implantation to induce contraception.
(b) Menstrual flow does not occur in females after taking oral contraceptives.

Select the correct option.
(a) a is true, b is false
(b) a is false, b is true
(c) Both a and b are false
(d) Both a and b are true

45. Select the incorrect statement regarding 'Saheli'.
(a) It was developed by CDRI, Lucknow, India.
(b) It contains high amount of progesterone and estrogen.
(c) It is taken once in a week.
(d) It has least side effects and high contraceptive value.

46. The contraceptive 'saheli' **[NEET-2018]**
(a) Blocks estrogen receptors in the uterus, preventing eggs from getting implanted.
(b) Increases the concentration of estrogen and prevents ovulation in females
(c) Is an IUD.
(d) Is a post-coital contraceptive.

47. Which of the following statement is incorrect regarding implants?
(a) They are used by both males and females.
(b) These are placed under the skin.
(c) They contain progestogens or combinations of pills.
(d) They are effective for longer period as compared to pills.

48. All the listed contraceptives are effective during emergency to avoid pregnancy, except
(a) Progestasert
(b) Progestogen pills
(c) Vaults
(d) Progesterone-estrogen pills

49. Which of the following contraceptive methods do involve a role of hormone? **[NEET-2019]**
(a) Barrier method, Lactational amenorrhea, Pills
(b) CuT, Pills, Emergency contraceptives
(c) Pills, Emergency contraceptives, Barrier methods
(d) Lactational amenorrhea, Pills, Emergency contraceptives

50. Consider the following statements.
(a) Emergency contraceptives are taken within 72 hours of coitus to avoid pregnancy.
(b) Sterilisation is the terminal method to prevent any pregnancy as it is irreversible.

Select the correct option.
(a) a is true, b is false
(b) Both a and b are true
(c) a is false, b is true
(d) Both a and b are false

51. Which of the following statement is incorrect?
(a) Vasectomy is the sterilisation procedure in males.
(b) Sterilisation method blocks the transport of gametes to prevent conception.
(c) In tubectomy, fallopian tubes are removed from female body completely.
(d) After tubectomy, menstrual cycle is not disrupted in females.

52. Which of the following approaches does not give the defined action of contraceptive? **[NEET-2016]**
(a) Intra uterine devices - Increase phagocytosis of sperms, suppress sperm motility and fertilizing capacity of sperm
(b) Hormonal contraceptives - Prevent/retard entry of sperm, prevent ovulation and fertilisation
(c) Vasectomy - Prevents spermatogenesis
(d) Barrier methods - Prevents fertilisation

53. The semen of male after vasectomy
(a) Does not contain sperms
(b) Is not produced in body
(c) Lacks fructose and prostaglandins
(d) Both (a) and (c)

54. Among the listed options, contraception is achieved through
(a) Tubectomy (b) Ovariectomy
(c) Hysterectomy (d) Castration

55. Which of the following diagram depict vasectomy accurately?

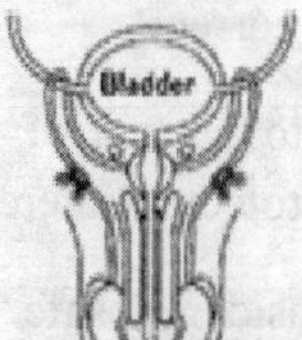

(a)

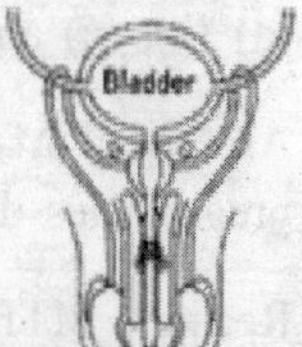

(b)

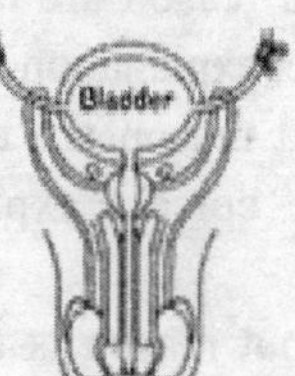

(c)

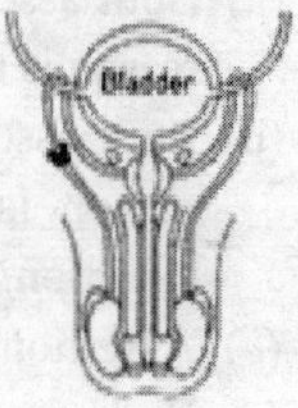

(d)

56. Assertion: Menstration does not occur during the intense period of lactation.

Reason: Chances of conception are higher after 1–2 months following parturition.
(a) Both assertion and reason are true and the reason is the correct explanation of assertion.
(b) Both assertion and reason are true but the reason is not the correct explanation of assertion.
(c) Assertion is true but reason is false.
(d) Both assertion and reason are false.

57. Assertion: Use of condoms can prevent the transmission of AIDS.

Reason: AIDS is a fluid-transmitted disease.
(a) Both assertion and reason are true and the reason is the correct explanation of assertion.
(b) Both assertion and reason are true but the reason is not the correct explanation of assertion.
(c) Assertion is true but reason is false.
(d) Both assertion and reason are false.

58. Assertion: Emergency contraceptives contain high level of steroidal preparation.

Reason: Saheli is an emergency contraceptive.
(a) Both assertion and reason are true and the reason is the correct explanation of assertion.
(b) Both assertion and reason are true but the reason is not the correct explanation of assertion.
(c) Assertion is true but reason is false.
(d) Both assertion and reason are false.

59. Assertion: After vasectomy, sperms are not found in the semen or ejaculate.

Reason: Vas deferens is cut and tied in vasectomy.
(a) Both assertion and reason are true and the reason is the correct explanation of assertion.
(b) Both assertion and reason are true but the reason is not the correct explanation of assertion.
(c) Assertion is true but reason is false.
(d) Both assertion and reason are false.

Topic 3 Medical Termination of Pregnancy

60. Medical termination of pregnancy is
(a) Carried out by tubectomy in females
(b) Induced abortion
(c) Inability to conceive due to certain problem
(d) Menopause stage in females

61. Consider the following statements.
(a) One-fifth of total number of conceived pregnancies in a year, all over the world are terminated through MTP.
(b) MTP restrictions help to check illegal female foeticide in India.

Select the correct option.
(a) Both a and b are false
(b) a is true, b is false

(c) a is false, b is true
(d) Both a and b are true

62. Government of India legalised MTP in the year
(a) 1988 (b) 1978
(c) 1985 (d) 1971

63. To reduce the incidences of illegal abortions, recent MTP (Amendment) Act in India was enacted in
(a) 2009 (b) 2011
(c) 2017 (d) 2019

64. According to MTP Act 2017, a pregnancy may be terminated
(a) Within first 12 weeks of pregnancy on the opinion of one medical practitioner
(b) Before first six weeks of pregnancy, without any opinion
(c) Within first 24 weeks of pregnancy on the opinion of one medical practitioner
(d) All of these

65. On what grounds, pregnancy of more than 12 weeks is terminated through MTP in India?
(a) If it imposes life threatening risks to pregnant woman.
(b) If the female already has three girl children.
(c) If the family history of parents has certain genetic disorder.
(d) If the female is widow or divorced.

66. MTP is considered safe upto
(a) 12 weeks (b) 24 weeks
(c) 26 weeks (d) 28 weeks

67. A pregnant woman was denied for undergoing MTP as she was not having opinion documents from two registered medical practitioners. The duration of her pregnancy might be
(a) 8 weeks (b) 12 weeks
(c) 21 weeks (d) 29 weeks

68. Select the correct statement regarding MTP.
(a) It is carried out up to third trimester of pregnancy.
(b) It is always surgical.
(c) It is widely used as a method of contraception.
(d) It requires the assistance of registered medical practitioner.

Select the most appropriate option
(a) (a) and (b) are correct
(b) Only (d) is correct
(c) (b) and (d) are correct
(d) (a), (c) and (d) are correct

69. Assertion: MTP helps to get rid of unwanted pregnancies due to unprotected intercourse.

Reason: MTPs are considered safe up to 28 weeks of pregnancy.
(a) Both assertion and reason are true and the reason is the correct explanation of assertion.
(b) Both assertion and reason are true but the reason is not the correct explanation of assertion.
(c) Assertion is true but reason is false.
(d) Both assertion and reason are false.

70. Assertion: Ban on amniocentesis helped to reduce the incidences of MTP in India.

Reason: Amniocentesis helps to determine the sex of unborn child.
(a) Both assertion and reason are true and the reason is the correct explanation of assertion
(b) Both assertion and reason are true but the reason is not the correct explanation of assertion.
(c) Assertion is true but reason is false.
(d) Both assertion and reason are false.

Topic 4 Sexually Transmitted Infections

71. Sexually transmitted diseases are alternatively known as
(a) Contagious diseases
(b) Venereal diseases
(c) Viral diseases
(d) Degenerative diseases

72. RTI stands for
(a) Rubella-Typhoid Infection
(b) Rotavirus-Trichomoniasis Infection
(c) Reproductive Tract Infection
(d) Rotavirus

73. Among the given options, which set represents sexually transmitted infections (STI) only?
(a) Gonorrhea, Chlamydiasis, Trichomoniasis
(b) Syphilis, Giardiasis, Genital warts
(c) Taeniasis Genital herpes, Leishmaniasis
(d) Hepatitis B, Trichinosis, Gonorrhea

74. Consider the following statements.
(a) AIDS and Hepatitis B are STIs which are caused by virus.
(b) Gonorrhoea and syphilis are caused by bacteria.

Select the correct option.
(a) a is true, b is false
(b) a is false, b is true
(c) Both a and b are false
(d) Both a and b are true

75. Which among the following STIs can be transmitted by blood transfusion and sharing of infected needles with infected person?
(a) AIDS
(b) Genital warts
(c) Gonorrhoea
(d) Both (a) and (c)

76. A pregnant lady was suffering from a STI and various tests confirmed that her baby has also contracted the disease. The pregnant lady must be suffering from
(a) Genital herpes (b) Trichomoniasis
(c) Chlamydiasis (d) Hepatitis B

77. Which of the following sexually transmitted diseases is not completely curable?

[NEET-2017]

(a) Genital warts (b) Genital herpes
(c) Chlamydiasis (d) Gonorrhoea

78. The incidences of STDs are more frequent
(a) In age group of 15–24 years
(b) Before the beginning of puberty age
(c) In a menopausing woman
(d) In elder generation

79. Consider the following statements.
(a) Sexually transmitted diseases can be cured completely if detected early and treated properly.
(b) Early symptoms of STIs include itching, fluid discharge and swelling in genital areas.

Select the correct option.
(a) a is true, a is false
(b) Both a and b are true
(c) a is false, b is true
(d) Both a and b are false

80. The consequences of delayed treatment of STIs include
(1) ectopic pregnancy
(2) still births
(3) petric inflammatory diseases
(4) uncontrolled growth of cells

Select the most appropriate option.
(a) (1) and (2) are correct
(b) (3) and (4) are correct
(c) (1), (2), (3) and (4) are correct
(d) (1), (2) and (3) are correct

81. Match the following sexually transmitted diseases in Column-I with their causative agent in Column II and select the correct option.

[NEET-2017]

Column I Column II

	Column I		Column II
(a)	Gonorrhea	(1)	HIV
(b)	Syphilis	(2)	Neisseria
(c)	Genital Warts	(3)	Treponema
(d)	AIDS	(4)	Human papilloma-virus

Select the correct option.

	a	b	c	d
(a)	4	3	2	1
(b)	2	3	4	1

(c) 3 4 1 2
(d) 4 2 3 1

82. W What measures can be taken to prevent STDs among population?
(a) Use condom during coitus
(b) Avoid sex with multiple partners
(c) Do not share needles and syringes
(d) All of these

83. Which of the following sexually transmitted diseases do not specially affect reproductive organs? **[NEET-2019]**
(a) Chlamydiasis and AIDS
(b) Genital warts and Hepatitis-B
(c) Syphilis and Genital herpes
(d) AIDS and Hepatitis B

84. Assertion: Gonorrhoea spreads through infected needles during blood transfusion.

Reason: Gonorrhoea cannot be cured completely.
(a) Both assertion and reason are true and the reason is the correct explanation of assertion.
(b) Both assertion and reason are true but the reason is not the correct explanation of assertion.
(c) Assertion is true but reason is false.
(d) Both assertion and reason are false.

85. W Assertion: AIDS can be transmitted through body fluids of affected individual.

Reason: Hepatitis B virus can be found in blood and semen of affected individual.
(a) Both assertion and reason are true and the reason is the correct explanation of assertion.
(b) Both assertion and reason are true but the reason is not the correct explanation of assertion.
(c) Assertion is true but reason is false.
(d) Both assertion and reason are false.

Topic 5 Infertility

86. Consider the following statements.
(a) An infertile couple is unable to produce children due to immunological disorders only.
(b) In human population, males are always fertile while females are either fertile or infertile.

Select the correct option
(a) a is true, b is false
(b) Both a and b are true
(c) a is false, b is true
(d) Both a and b are false

87. How does assisted reproductive techniques, (ART) help an infertile couple to have children?
(a) By easing the process and procedure of adoption.
(b) By treating the disease associated with infertility.
(c) By carrying out certain fertilisation processes through special techniques.
(d) Any of these

88. W IVF stands for
(a) In vitro fertilisation
(b) In vivo fertilisation
(c) Inter–vaginal fertilisation
(d) Intra vaginal fusion

89. Where does fertilisation occur in case of IVF?
(a) Within the fallopian tube
(b) In uterus
(c) Outside the body
(d) Inside vagina

90. In test tube baby program, embryo transfer
(a) Is not possible
(b) Can be done at eight blastomere stage
(c) Can be done after the eight blastomere stage
(d) Both (b) and (c)

91. Which among the listed ARTs are included under embryo transfer (ET) technique?
(a) ICSI and IUI (b) ZIFT and IUT
(c) GIFT and ZIFT (d) ICSI and GIFT

92. W A childless couple can be assisted to have a child through a technique called GIFT. The

full form of this technique is **[NEET-2015]**
(a) Germ cell internal fallopian transfer
(b) Gamete inseminated fallopian transfer
(c) Gamete intra fallopian transfer
(d) Gamete internal fertilisation and transfer

93. Select the correct statement.
(a) ZIFT involves the transfer of male's sperm into fallopian tube of female.
(b) In IUT, embryo with more than eight blastomeres is extracted from donor's body.
(c) ZIFT involves the transfer of embryo into fallopian tube at eight blastomere stage.
(d) In both ZIFT and IUT, embryo is formed inside the female's body.

94. Consider the following statements.
(a) Embryo formed by in-vivo fertilisation could be used in intra uterine transfer.
(b) Fully formed embryo is transferred into the fundus region of uterus in ZIFT. Select the correct option
(a) a is true, b is false
(b) a is false, b is true
(c) Both a and b are false
(d) Both a and b are true

95. Which ART a woman should adept if she does not ovulate or produce a viable egg?
(a) ICSI (b) AI
(c) GIFT (d) IUI

96. Match the following columns.

	Column I		Column II
(a)	ZIFT	(1)	Transfer of ovum into fallopian tube
(b)	GIFT	(2)	Transfer of embryo into the uterus Transfer of embryo into the uterus
(c)	IUT	(3)	Transfer of embryo into fallopian tube

Select the correct option.

	a	b	c
(a)	3	2	1
(b)	1	2	3
(c)	2	3	1
(d)	3	1	2

97. 'ICSI' stands for
(a) Infertility case study and inspection
(b) Intra cytoplasmic sperm injection
(c) Intra cervix sperm insemination
(d) Intra cytoplasmic sperm interaction

98. In which technique, sperm is injected directly into the ovum in the laboratory?
(a) GIFT (b) ICSI
(c) IUI (d) AI

99. Under what circumstances, artificial insemination could be employed?
(a) When females do not ovulate
(b) Male partner fails to inseminate
(c) Female has blocked fallopian tube
(d) Any of these

100. In case of a couple where the male is having a very low sperm count, which technique will be suitable for fertilisation? **[NEET-2017]**
(a) Intracytoplasmic sperm injection
(b) Intrauterine transfer
(c) Gamete intracytoplasmic fallopian transfer
(d) Artificial insemination

101. In artificial insemination, sperms are transferred into
(a) Vagina (b) Uterus
(c) Ovum (d) Vagina or uterus

102. How intra uterine insemination (IUI) differ from intra uterine transfer (IUT)?
(a) Embryo is transferred into uterus in IUI.
(b) IUI is a type of IVF technique.
(c) IUI involves the transfer of semen into female's uterus.
(d) Sperm is directly injected into ovum in IUI.

103. W Ectopic pregnancies are referred to as **[NEET-2015]**
(a) Pregnancies terminated due to hormonal imbalance
(b) Pregnancies with genetic abnormality
(c) Implantation of embryo at site other than uterus

(d) Implantation of detective embryo in the uterus

104. Assertion: In all ART, fertilisation occurs outside the female's body.

Reason: Infertile couples cannot produce gametes.

(a) Both assertion and reason are true and the reason is the correct explanation of assertion.

(b) Both assertion and reason are true but the reason is not the correct explanation of assertion

(c) Assertion is true but reason is false.

(d) Both assertion and reason are false.

105. Assertion: ICSI technique is different from AI.

Reason: ICSI is carried out in laboratory.

(a) Both assertion and reason are true and the reason is the correct explanation of assertion.

(b) Both assertion and reason are true but the reason is not the correct explanation of assertion.

(c) Assertion is true but reason is false.

(d) Both assertion and reason are false.

ANSWER KEY

1. B	2. C	3. D	4. C	5. D	6. C	7. C	8. D	9. B	10. D
11. A	12. C	13. D	14. C	15. B	16. C	17. B	18. C	19. D	20. A
21. C	22. C	23. A	24. C	25. A	26. C	27. B	28. A	29. C	30. C
31. B	32. A	33. C	34. C	35. C	36. B	37. D	38. A	39. B	40. B
41. D	42. A	43. B	44. A	45. B	46. A	47. A	48. C	49. D	50. B
51. C	52. C	53. A	54. A	55. A	56. C	57. A	58. C	59. A	60. B
61. C	62. D	63. C	64. A	65. A	66. A	67. C	68. B	69. C	70. B
71. B	72. C	73. A	74. D	75. A	76. D	77. B	78. A	79. B	80. D
81. B	82. D	83. D	84. D	85. B	86. D	87. C	88. A	89. C	90. D
91. B	92. C	93. C	94. A	95. C	96. D	97. B	99. B	99. B	100. D
101. D	102. C	103. C	104. D	105. A					

Chapter 14

Evolution

MULTIPLE CHOICE QUESTIONS

Topic 1	Origin of Life and Evolution of Life Forms

1. Conventional religious literature tells us about the theory of
(a) Religious Creation
(b) Abiogenetic creation
(c) Spontaneous creation
(d) Special creation

2. Special creation theory has three connotations. Select the correct connotation among the following.
(a) All living organisms (species or types) that we see today were created as such.
(b) The diversity was always the same since creation and will be the same in future also.
(c) Earth is about 40000 years old
(d) Both (a) and (b)

3. Select the correct statement among the following.
(a) All the ideas of Special creation theory were strongly challenged during the eighteenth century.
(b) Based on observations made during a sea voyage in a sail ship called H.M.S. Beagle round the world, Charles Darwin concluded that existing living forms share similarities to varying degrees not only among themselves but also with life forms that existed millions of years ago. Many such life forms do exist anymore.
(c) There had been extinctions of different life forms in the years gone by just as new forms of life arose at same periods of history of earth.
(d) Any population has built in variation in characteristics.

4. How many statements among the following are incorrect?
i. Those characteristics which enable some to survive better in natural conditions (climate, food, physical factors, etc.) would outbreed others that are less-endowed to survive under such natural conditions.
ii. The fitness, according to Darwin, refers ultimately and only to reproductive fitness.
iii. Those who are better fit in an environment leave more progeny than others. These, therefore, will survive more and hence are selected by nature. Darwin called it natural selection and implied it as a mechanism of evolution.
iv. Alfred Wallace, a physicist worked in Malay Archipelago.
v. The geological history of earth closely correlates with the physical history of earth.
(a) 3 (b) 4
(c) 2 (d) 1

5. The history of Earth can be studied in terms of
(a) Epochs (b) Periods
(c) Eras (d) All of these

6. Select the correct option about the given figure.
(a) The given figure represents diagrammatic representation of Miller's Experiment.
(b) X- Vacuum pump
(c) Y – Water containing inorganic compounds
(d) Z – Spark discharge

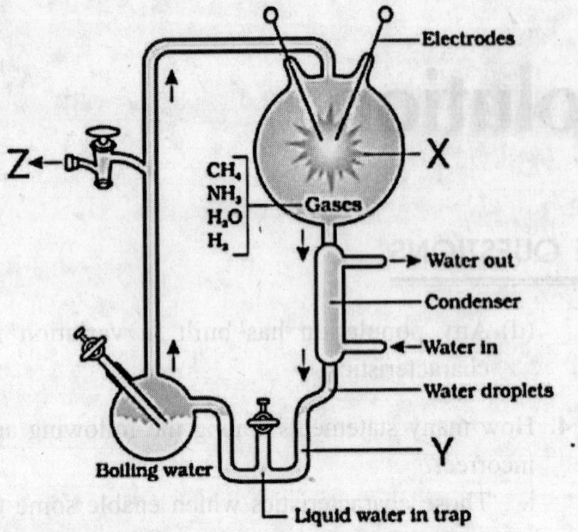

7. The study of the history of life forms of the earth is called
(a) Evolutionary Biology
(b) Ecology
(c) Environmental biology
(d) Comparative anatomy

8. Match Column-I with Column-II.

	Column I		Column II
(a)	Origin of the universe	(1)	4.5 billion years ago
(b)	Origin of earth	(2)	4 billion years ago
(c)	Origin of life	(3)	2.7 billion years ago
(d)	Origin of first eukaryotes	(4)	20 billion years ago

Select the correct option.

	a	b	c	d
(a)	1	4	2	3
(b)	4	1	2	3
(c)	4	2	1	3
(d)	4	1	3	2

9. Select the correct statement among the following.
(a) The universe is very old – almost 20000 million years old.
(b) Huge clusters of galaxies comprise the universe. Galaxies contain stars and clouds of gas and dust.
(c) The Big Bang theory attempts to explain to us the origin of universe.
(d) All of these.

10. Select the correct statement about the Big Bang theory.
(a) It talks of a many huge explosions unimaginable in physical terms.
(b) The universe expanded and hence, the temperature increased. Hydrogen and Helium formed sometime later.
(c) The gases condensed under gravitation and formed the galaxies of the present day universe.
(d) In the solar system of the Milky Way galaxy, earth was supposed to have been formed about 4.5 million years back.

11. Select the correct statement about early Earth 4.5 billion year ago.
(a) There was a thick atmosphere on early Earth.
(b) Water vapour, hydrogen, carbondioxide and ammonia released from molten mass covered the surface.
(c) The UV rays from the sun brokeup water into Hydrogen and Oxygen and the lighter O2 escaped. Oxygen combined with ammonia and methane to form water, CO2 and others.
(d) The ozone layer was formed. As it cooled, the water vapor fell as rain, to fill all the depressions and form oceans.

12. When did life appear on earth?
(a) 500 billion years after the formation of Earth
(b) Almost four billion years back.
(c) Both (a) and (b)
(d) Almost three billion years back

13. Select the correct statement among the following.
(a) Some scientists believe that life came from outside.
(b) Early Greek thinkers thought units of life called spores were transferred to different planets including earth.

(c) 'Panspermia' is still a favourite idea for some astronomers.
(d) All of these

14. For a long time it was also believed that life came out of decaying and rotting matter like straw, mud, etc. This was the theory of ________.
(a) Biogenesis
(b) Spontaneous generation
(c) Non-Spontaneous generation
(d) Both A and B

15. Select the incorrect statement among the following.
(a) Louis Pasteur by careful experimentation demonstrated that life comes only from pre-existing life.
(b) Louis Pasteur showed that in presterilised flasks, life did not come from killed yeast while in another flask open to air, new living organisms arose from 'killed yeast'.
(c) Spontaneous generation theory was dismissed by Louis Pasteur.
(d) Louis Pasteur answered how the first life form came on earth.

16. How many statements are incorrect among the following?
i. Oparin of England and Haldane of Russia proposed that the first form of life could have come from pre-existing non-living organic molecules (e.g. RNA, protein, etc.) and that formation of life was preceded by chemical evolution
ii. The conditions on earth were – high temperature, volcanic storms, reducing atmosphere containing CH_4, O_2, etc.
iii. In 1952, S.L. Miller, an American scientist created similar conditions in a laboratory scale
iv. S.L. Miller created electric discharge in a closed flask containing CH_4, H_2, NH_3 and water vapour at 800°C.
v. With limited evidence, the first part of the conjectured story, i.e., chemical evolution was more or less accepted.

(a) 2 (b) 3
(c) 4 (d) 5

17. In S.L. Miller's electric discharge experiment, he observed formation of ________.
(a) Amino sugar (b) Glucosamine
(c) Amino acid (d) Proteins

18. In experiments similar to S.L. Miller's discharge experiment others observed, formation of how many of the following substances.
Sugars, Nitrogen gas, Pigment, Fats, Nitrogen bases, Amino sugar
(a) 3 (b) 4
(c) 5 (d) 6

19. Select the incorrect statement among the following.
(a) We have no idea about how the first self replicating metabolic capsule of life arose.
(b) The first non-cellular forms of life could have originated 3 billion years back. They would have been giant molecules (RNA, Protein, Polysaccharides, etc.). These capsules reproduced their molecules perhaps.
(c) The first cellular form of life did not possibly originate till about 2 million years ago. These were probably single-cells.
(d) All life forms were in water environment only. This version of biogenesis, i.e., the first form of life arose slowly through evolutionary forces from non-living molecules is accepted by majority.

20. Chemical evolution refers to-
(a) Formation of diverse inorganic molecules from organic constituents
(b) Formation of diverse organic molecules from radioactive constituents
(c) Formation of diverse organic molecules from inorganic constituents
(d) Formation of diverse non-reducing molecules from inorganic constituents

Topic 2 The Evidences for Evolution

21. Fossils are ____

(a) The preserved remains of the past life forms.
(b) Mineralized form of hard body parts such as bones, teeth, etc.
(c) Found mainly in the layers of sedimentary rocks.
(d) all of these

22. Consider the following statements about fossils:

(a) Rock sediments of different ages contain fossils of life forms that died during the formation of that particular layer of sediment.
(b) Fossil records are based on the sequence of occurrence of fossils in various strata of sedimentary rocks. Select the correct option.

(a) Both (a) and (b) are true.
(b) (a) is true but (b) is false.
(c) Both (a) and (b) are false.
(d) (a) is false but (b) is true.

23. Which of the following represents the paleontological evidence of evolution?

(a) Preserved remains of Dinosaurs obtained from sedimentary rocks.
(b) Presence of homologous structures.
(c) Presence of analogous structures.
(d) Presence of vestigial parts.

24. Fossils serve as one of the evidence of evolution because

(a) Fossil records show that various groups of organisms dominated earth during the different course of evolution.
(b) Many organisms are extinct today.
(c) Certain groups of organisms are restricted to a certain geological time period.

Select the option with all correct statements.

(a) Only a (b) Only b
(c) Both b and c (d) a, b and c

25. Given below is the family tree of dinosaurs and their living present day counterpart organisms. Which of the given options correctly name A, B and C?

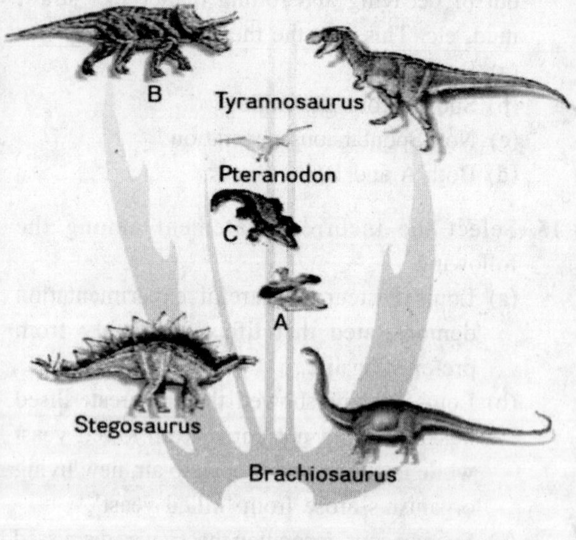

(a) A: Triceratops; B: Archaeopteryx; C: Crocodilian
(b) A: Triceratops; B: Crocodilian; C: Archaeopteryx
(c) A: Archaeopteryx; B: Triceratops; C: Crocodilian
(d) A: Archaeopteryx; B: Crocodilian; C: Triceratops

26. Embryological support for evolution was proposed by ______ based upon the observation of certain features during embryonic stage common to all vertebrates that are absent in adult.

(a) Ernst Mayr (b) JBS Haldane
(c) Ernst Haeckel (d) Charles Lyell

27. Sweet potato is a ___(I)___ modification whereas potato is a ___(II)___ modification. [Pg 131,E]

(a) (I)- Stem (b) (II)- Root
(c) Both A and B (d) None of these

28. Which of the given statement is incorrect about radiometric dating?

(a) A method to determine the absolute age of

the fossils.
(b) It is based on the process of decay of radioactive isotopes.
(c) A method to determine the relative age of the fossils.
(d) Organisms accumulate isotopes of substances during their lifetime.

29. Consider the following statements:
(a) Comparative anatomical and morphological details of different but related organisms demonstrate a basic similarity.
(b) The same structures in different organisms are derived from a common ancestor.

Choose the correct option.
(a) Both (a) and (b) are true.
(b) (a) is true but (b) is false.
(c) Both (a) and (b) are false.
(d) (a) is false but (b) is true.

30. The similar structures with different functions present in different species because of their common ancestry is known as _____
(a) Analogous structures
(b) Homologous structures
(c) Vestigial organs
(d) Homoplasy

31. Presence of homologous structures in different but related organisms is the result of _____
(a) Common ancestry
(b) Divergent evolution
(c) Convergent evolution
(d) Both (a) and (b)

32. The image shows the forelimbs of a whale, bat, cheetah and human (all mammals). Choose the correct option regarding the same.

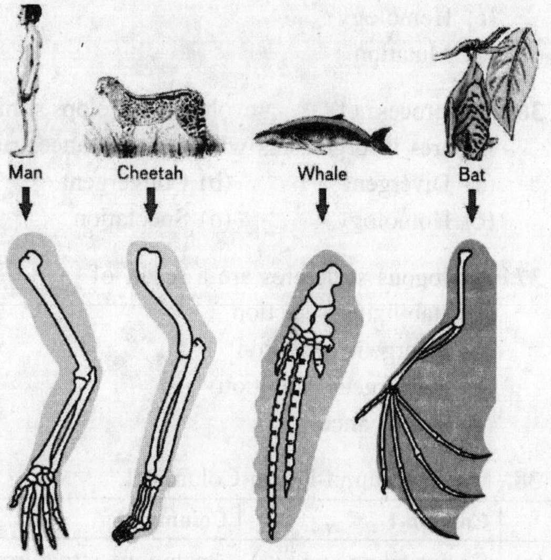

(a) The basic similarities in the arrangement of bones in forelimbs of whales, bats, cheetahs and humans reflect their common ancestry.
(b) The forelimbs of whales, bats, cheetahs and humans are analogous structures.
(c) The image represents structures that evolved from convergent evolution.
(d) The image represents vestigial structures.

33. Which of the following structures represent homology?
(a) Wings of butterflies and birds
(b) Eyes of octopus and mammals
(c) Thorns and tendrils of Bougainvillea and Cucurbita
(d) Flippers of penguins and dolphins

34. Which of the following structures is homologous to the wing of a bird?
(a) Wing of a moth
(b) Hind limb of the rabbit
(c) Flipper of the whale
(d) The dorsal fin of a shark

35. Independent evolution of structures with similar functions but different anatomy in distantly related organisms is known as _____
(a) Convergent evolution
(b) Divergent evolution

(c) Homology
(d) Mutation

36. The process of ______ evolution develops similar features in organisms with separate ancestries.
(a) Divergent (b) Convergent
(c) Homology (d) Speciation

37. Analogous structures are a result of ______
(a) stabilizing selection
(b) divergent evolution
(c) convergent evolution
(d) shared ancestry

38. Match Column-I with Column-II.

	Column I		Column II
(a)	Cactus spine and pea tendril	(1)	Analogous structures
(b)	Wings of insects and birds	(2)	Vestigial structures
(c)	Hind limb bones of whales	(3)	Natural selection
(d)	Darwin's finches	(4)	Homologous structures

Select the correct option.

	a	b	c	d
(a)	1	4	2	3
(b)	4	1	2	3
(c)	4	2	1	3
(d)	4	1	3	2

39. The wings of a bird and the wings of an insect are
(a) homologous structures and represent divergent evolution
(b) phylogenetic structures and represent divergent evolution
(c) analogous structures and represent convergent evolution
(d) homologous structures and represent convergent evolution

40. Which of the following examples does not represent analogous organs?
(a) Eyes of octopus and mammals.
(b) Sweet potato and potatoes.
(c) The spine of barberry and thorns of hawthorn.
(d) Eyes of cave-dwelling animals.

41. Match Column-I with Column-II.

	Column I		Column II
(a)	Common proteins and genes present in diverse organisms	(1)	Artificial selection
(b)	Vertebrates heart or brain	(2)	Anatomical evidence of evolution
(c)	Animal husbandry and plant breeding	(3)	Natural selection
(d)	Industrial melanism	(4)	Biochemical evidence of evolution

Select the correct option.

	a	b	c	d
(a)	1	4	2	3
(b)	4	1	2	3
(c)	4	2	1	3
(d)	4	1	3	2

42. Select the incorrect match from the following options.
(a) Tendrils and passionflower and thorns of pomegranate - homologous organs.
(b) Tendrils of Vitis and thorns of Carissa – homologous organs.
(c) Tendrils of balloon vine and bulbils of Agave – homologous organs.
(d) Scale leaves of onion and spines of Opuntia – analogous organs.

43. Industrial melanism is an example of
(a) Neo Darwinism (b) Natural selection
(c) Mutation (d) Neo Lamarckism

44. The given image shows white-winged and dark-winged moths.

I II

(a) Image I represents the survival advantage of white winged moths in unpolluted areas.
(b) Image II represents the survival advantage of dark winged moths in polluted areas.

Select the correct option.
(a) Both (a) and (b) are true.
(b) (a) is true but (b) is false.
(c) Both (a) and (b) are false.
(d) (a) is false but (b) is true.

45. Which of the following serves as an indicator of atmospheric pollution?
(a) Lichens (b) Rhizopus
(c) Penicillin (d) Lycopodium

46. Match Column-I with Column-II.

Column I	Column II
A) Herbicide resistant varieties	(1) A herbivore sauropod
B) Brachiosaurus	(2) Bony plates on the back
C) Stegosaurus	(3) A flying reptile
D) Pteranodon	(4) Evolution by anthropogenic action

Select the correct option.

	a	b	c	d
(a)	1	4	2	3
(b)	4	1	2	3
(c)	4	2	1	3
(d)	4	1	3	2

47. Assertion: Archaeopteryx is a dead connecting link between reptiles and birds.

Reason: Archaeopteryx is the earliest known bird.
(a) Both Assertion and Reason are true and Reason is correct explanation of Assertion.
(b) Both Assertion and Reason are true, but Reason is not the correct explanation of Assertion.
(c) Assertion is true, but Reason is false.
(d) Assertion is false, but Reason is true.

48. Assertion: Evolution is not a directed process in the sense of determinism.

Reason: Evolution is based on chance events that occur in nature.
(a) Both Assertion and Reason are true and Reason is correct explanation of Assertion.
(b) Both Assertion and Reason are true, but Reason is not the correct explanation of Assertion.
(c) Assertion is true, but Reason is false.
(d) Assertion is false, but Reason is true.

49. Select the correct order of extinction of following dinosaurs?
I. Stegosaurus
II. Brachiosaurus
III. Pteranodon
IV. Triceratops
V. Tyrannosaurus
(a) II, I, III, V, IV (b) II, I, V, IV, III
(c) III, V, IV, II, I (d) None of these

Topic 3 Adaptive Radiation

50. The mechanism of adaptive radiation was first explained by
(a) Darwin (b) Morgan
(c) Lamarck (d) Hugo de Vries

51. Consider the following statements.
(a) Darwin compared the animals and plant species of Galapagos Islands with those of mainland of South America.
(b) He found observable and distinct similarities and differences among them.

(c) Flora and fauna of the Galapagos Islands do not exhibit resemblances to that of the South American mainland.

Select the correct option.
(a) Both (a) and (b) are true.
(b) (a) is true but (b and c) are false.
(c) Both (a) and (b) are false.
(d) All are false.

52. Lamarck was a ________.
(a) French zoologist (b) German botanist
(c) French naturalist (d) British botanist

53. Diversification of single ancestral species into many species in a relatively short period of time is called
(a) Artificial selection (b) Adaptive radiation
(c) Homology (d) Natural selection

54. Adaptive radiation occurs when
(a) lineages of an ancestral species encounter an empty niche.
(b) populations of a species exhibit random mating.
(c) gene flow between the populations of a species continues.
(d) adaptive radiation does not occur in nature.

55. Darwin's finches are an example of
(a) adaptive radiation (b) microevolution
(c) genetic drift (d) gene flow

56. Evolution of different species in a given area starting from a point and spreading to other geographical areas is known as
(a) Adaptive radiation
(b) Natural selection option
(c) Migration option
(d) Divergent evolution

57. Consider the following statements:
(a) Adaptive radiation occurs when numerous unexploited ecological opportunities are available.
(b) Islands have a large number of empty ecological niche as compared to the mainland present in the same geographical location.

Choose the correct answer.
(a) Both (a) and (b) are true.
(b) (a) is true but (b and c) are false.
(c) Both (a) and (b) are false.
(d) All are false.

58. ________ and ________ are the two key concepts of Darwinian theory of Evolution.
(a) Branching descent
(b) Natural Selection
(c) Both A and B
(d) None of these

59. The evolution of multiple species by the process of adaptive radiation cannot occur
(a) on remote islands
(b) in Darwin's finches
(c) in Australian marsupials
(d) in a habitat with a large number of existing species

60. Select the incorrect statement among the following.
(a) Lamarck said that evolution of life forms had occurred but driven by use and disuse of organs.
(b) Lamarck gave the examples of Giraffes who in an attempt to forage leaves on tall trees had to adapt by elongation of their necks.
(c) Giraffes passed on this acquired character of elongated neck to succeeding generations, Giraffes, slowly, over the years, came to acquire long necks.
(d) Everybody believes Lamarck's conjecture today.

61. Which of the following statements does not reflect adaptive radiation in marsupials?
(a) Australia was geographically isolated from the rest of the world during the continental drift.
(b) The geographical isolation of Australia prevented the immigration of placental mammals to the continent.

(c) Marsupials were the dominant mammals on the land of the Australian continent.
(d) Geographical isolation of Australia followed the migration of placental mammals to the continent.

62. Select the correct option for the given image:

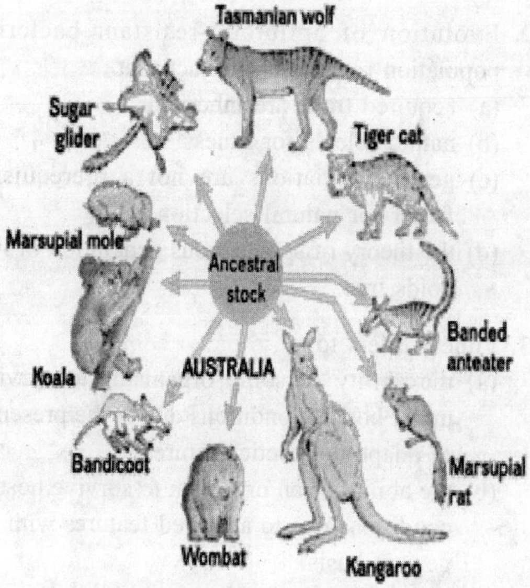

(a) Convergent evolution in Australian marsupials
(b) Gene flow
(c) Adaptive radiation in Australian marsupials
(d) Both (a) and (c) are correct

63. A population of a species invades a new area. Which of the following condition will lead to adaptive radiation?
(a) Area with many habitats occupied by a large number of species.
(b) Area with large number of habitats having very low food supply.
(c) Area with a single type of vacant habitat
(d) Area with many types of vacant habitats.

64. When more than one (i) appeared to have occurred in an isolated geographical area (representing different habitats), one can call this (ii).

Select the correct option regarding the given statement.
(a) (i) Adaptive radiation; (ii) Divergent evolution
(b) (i) Adaptive radiation; (ii) Convergent evolution
(c) (i) Convergent evolution; (ii) Adaptive radiation
(d) (i) Divergent evolution; (ii) Adaptive radiation

65. Australian marsupials correspond to the placental mammals present in North America because
(a) gene flow occurred between Australian marsupials and North American placental mammals.
(b) adaptive radiation in Australian marsupials paralleled the evolution of placental mammals in North America.
(c) some North American placental mammal served as founder population for Australian marsupials.
(d) some Australian marsupials served as founder population for North American placental mammals.

66. Match the placental mammals given in Column-I to their corresponding Australian marsupials given in Column- II.

	Column I		Column II
(a)	Mole	(1)	Marsupial mole
(b)	Anteater	(2)	Marsupial mouse
(c)	Mouse	(3)	Spotted cuscus
(d)	Lemur	(4)	Numbat

Select the correct option.

	a	b	c	d
(a)	1	4	2	3
(b)	4	1	2	3
(c)	4	2	1	3
(d)	4	1	3	2

67. Match the placental mammals given in column-I to their corresponding Australian marsupials given in column- II.

	Column I		Column II

(a)	Flying squirrel	(1)	Tasmanian tiger cat
(b)	Bobcat	(2)	Tasmanian wolf
(c)	Wolf	(3)	Long-eared bandicoot
(d)	Rabbit	(4)	Flying phalanger

Select the correct option.

	a	b	c	d
(a)	1	4	2	3
(b)	4	1	2	3
(c)	4	2	1	3
(d)	4	1	3	2

Topic 6 Biological Evolution and Mechanism of Evolution

68. According to Darwin, the driving force of evolution is
(a) mutation (b) natural selection
(c) gene flow (d) migration

69. Evolution by natural selection requires
(a) presence of acquired variations among the individuals
(b) presence of genetic variations among the individuals
(c) isolated habitat
(d) constant environmental conditions

70. Consider the following statements:
(a) Evolution by natural selection started with the origin of cellular life forms with differences in metabolic capability.
(b) The theory of natural selection was given by Lamarck.

Select the correct option.
(a) Both (A) and (B) are true.
(b) (A) is true but (B) is false.
(c) Both (A) and (B) is false.
(d) (A) is false but (B) is true.

71. The rate of evolution of new species in fishes is slower than that of in bacterial populations because
(a) the evolution of new species is linked to their lifespan.
(b) evolution of new species is linked to the habitat.
(c) both fishes and bacteria exhibit the same rate of evolution.
(d) species do not evolve at all.

72. Evolution of antibiotic-resistant bacterial population represents the fact that
(a) acquired traits are inherited.
(b) nature selects for fitness.
(c) genetic variations are not a prerequisite factor for natural selection.
(d) the theory of spontaneous generation of life holds true.

73. Fitness refers to
(a) the ability of some organisms to survive under hostile conditions due to the presence of adaptive genetic features.
(b) the ability of an organism to survive hostile conditions due to acquired features with no genetic basis.
(c) the ability of organisms to migrate.
(d) both a and b are true.

74. _____ is the end result of the ability to adapt and get selected by nature.
(a) Mutation (b) Fitness
(c) Adaptation (d) Acclimatization

75. Which of the following options correctly represent the two key features of Darwin's theory of evolution?
(a) Mutation and natural selection.
(b) Artificial and natural selection.
(c) Branching descent and natural selection.
(d) Branching descent and mutation.

76. Saltation is referred as
(a) Single step small mutation
(b) Multiple step small mutation
(c) Single step large mutation
(d) None of these

77. Match Column-I with Column-II.

	Column I		Column II
(a)	Branching descent	(1)	Unity and diversity of life
(b)	Natural selection	(2)	Source of new genes
(c)	Mutation	(3)	Lamarck
(d)	Inheritance of acquired features	(4)	Tendency of organisms to become adapted to the environment

Select the correct option.

	a	b	c	d
(a)	4	1	3	2
(b)	1	4	2	3
(c)	4	1	2	3
(d)	4	2	1	3

78. The first scientist to propose that evolution is a natural process rather than a divine intervention was

(a) Darwin (b) Lamarck
(c) Pasteur (d) Hugo de Vries

79. Select the correct option regarding Lamarck's principles of evolution

(a) Use and disuse; inheritance of acquired characters.
(b) Natural selection; fitness.
(c) Mutation; inheritance of acquired characters.
(d) Inheritance of acquired characters; gene flow.

80. Evolution of long and muscular neck of present day giraffe occurred as giraffes stretched their necks higher to reach the higher canopy of tall trees. The given statement represents the view of ………… about evolution.

(a) Darwin (b) Lamarck
(c) Louis Pasteur (d) Cuvier

81. Match Column-I with Column-II.

	Column I		Column II
(a)	Innate drive to become complex	(1)	Use and disuse
(b)	An Essay on the Principle of Population	(2)	Thomas Malthus
(c)	Evolution of longnecked giraffe population	(3)	Natural selection
(d)	Industrial melanism	(4)	Lamarck

Select the correct option.

	a	b	c	d
(a)	4	1	3	2
(b)	4	1	2	3
(c)	1	4	2	3
(d)	4	2	1	3

82. Natural selection is based on certain observations whish are factual. Such observations are

(a) Natural resources are limited.
(b) Populations are stable in size except for seasonal fluctuation.
(c) Members of a population vary in characteristics (infact no two individuals are alike) even though they look superficially similar, most of variations are inherited.
(d) All of these

83. Match the aspects of Darwin's theory of natural selection in Column-I with the correct description in Column-II.

	Column I		Column II
(a)	Variation	(1)	Competition for limited available resources
(b)	Overproduction	(2)	Increases the survival
(c)	Struggle for existence	(3)	More offspring
(d)	Differential reproductive success	(4)	Geometric increase in population size

Select the correct option.

	a	b	c	d
(a)	2	4	1	3
(b)	4	1	3	2
(c)	4	1	2	3
(d)	4	2	1	3

84. Hugo Devries explained mutations as a source of genetic variations while working on

(a) evening primrose (b) garden pea
(c) finches (d) fruit fly

85. Variations caused by mutation, as proposed by Hugo de Vries, are
(a) random and directional
(b) random and directionless
(c) small and directional
(d) small and directionless

86. According to Hugo Devries and Darwin, the driving factor for evolution is _____ and _____ respectively.
(a) minor variations, mutation
(b) natural selection, mutation
(c) mutation, minor variations
(d) mutation, acquired variations

87. Mutations are random and directionless while Darwinian variations are
(a) small and directional
(b) large and directional
(c) small and directionless
(d) large and directionless

88. Consider the following statements
(a) For Darwin, evolution was gradual and driven by small genetic variations.
(b) For de Vries, evolution was a single step process driven by large mutation.

Select the correct option.
(a) Both (a) and (b) are true.
(b) (a) is true but (b) is false.
(c) Both (a) and (b) are false.
(d) (a) is false but (b) is true.

89. The mutation is directionless as
(a) it can change any DNA sequence and is always beneficial for the individual.
(b) it can change any DNA sequence and may be beneficial or harmful for the individual.
(c) mutation imparts small variations.
(d) mutation imparts large variations.

90. Which of the following statements is incorrect?
(a) No variant is completely wiped out in the case of dark-winged and whitewinged moths in England.
(b) The essence of Darwinian theory about evolution is natural selection.
(c) Microbes that divide fast have the ability to multiply and become millions of individuals within hours.
(d) When we say that fitness of B is better than that of A under the new conditions, it means that A is better adapted than B under new conditions.

91. According to Hugo de Vries, large differences among the individuals evolve by _____ without any intermediate forms.
(a) gradual and small changes
(b) natural selection
(c) saltation
(d) gene flow

92. Assertion: Evolution refers to modification in the lines of descent.

Reason: Natural selection is the equal survival success of individuals of a population.
(a) Both Assertion and Reason are true and Reason is correct explanation of Assertion.
(b) Both Assertion and Reason are true, but Reason is not the correct explanation of Assertion.
(c) Assertion is true, but Reason is false.
(d) Assertion is false, but Reason is true.

93. Assertion: Darwin's theory of evolution was based on the inheritance of adaptive genetic variations.

Reason: He could not recognize the mechanism of inheritance.
(a) Both Assertion and Reason are true and Reason is correct explanation of Assertion.
(b) Both Assertion and Reason are true, but Reason is not the correct explanation of Assertion.
(c) Assertion is true, but Reason is false.

(d) Assertion is false, but Reason is true.

Topic 4 Hardy-Weinberg Principle

94. Select the incorrect statement among the following.

(a) $p^2 + 2pq + q^2 = 1$. This is binomial expansion of $(p + q)^2$.
(b) When frequency measured, differs from expected values, the difference (direction) indicates the extent of evolutionary change.
(c) Hardy-Weinberg principle says that phenotype frequencies in a population are stable and is constant from generation to generation.
(d) The gene pool (total genes and their alleles in a population) remains a constant. This is called genetic equilibrium. Sum total of all the allelic frequencies is 1.

95. How many of the following factors are known to affect Hardy-Weinberg equilibrium?

Gene migration, gene flow, genetic drift, mutation, genetic recombination, natural selection

(a) 4 (b) 5
(c) 6 (d) 3

96. Total genes and their alleles in a population make

(a) gene pool (b) genotype
(c) phenotype (d) mutation

97. Consider the following statements:

(a) Allele frequencies are supposed to remain constant over generations.
(b) Hardy Weinberg equation is the algebraic representation of the allele frequencies of a population.

Select the correct option.

(a) Both (a) and (b) are true.
(b) (a) is true but (b) is false.
(c) Both (a) and (b) are false.
(d) (a) is false but (b) is true.

98. Genetic equilibrium represents that

(a) the population is not evolving for the locus under study.
(b) the allele frequencies and genotype frequencies remain constant over generations.
(c) the allele frequencies remain constant over generations but genotype frequencies change.
(d) both (a) and (b) are correct.

99. The sum total of all allele frequencies for a particular locus is always ______

(a) 0
(b) 1
(c) more than 1
(d) cannot be determined

100. The frequency of an allele in a population ranges from

(a) 0 to 1
B) more than 1
(c) less than 0
(d) can be any number

101. Select the incorrect match:

(a) Frequency of a dominant allele in the population – p.
(b) Frequency of a recessive allele in the population – q.
(c) Frequency of heterozygous dominant genotype – 2pq.
(d) Frequency of homozygous dominant genotype – q^2.

102. In the Hardy-Weinberg equation, the frequency of heterozygous individual is represented by

(a) p^2 (b) 2pq
(c) pq (d) q^2

103. The binomial expansion of $(p + q)^2$ is

(a) $p^2 + 2pq + q^2 = 1$
(b) $p^2 + q^2 = 1$
(c) $p^2 + 2pq = 1$
(d) $p + q = 1$

104. Differences between expected and observed allele frequencies in a population represent that
(a) the population is at hardy-Weinberg equilibrium.
(b) the population is evolving.
(c) the population is deviating from hardy-Weinberg equilibrium.
(d) both (B) and (C) are correct.

105. A gene locus has two alleles A, a. If the frequency of dominant allele A is 0.4, then what will be the frequency of homozygous dominant, heterozygous and homozygous recessive individuals in the population?
(a) 0.36(AA); 0.48(Aa); 0.16(aa)
(b) 0.16(AA); 0.24(Aa); 0.36(aa)
(c) 0.16(AA); 0.48(Aa); 0.36(aa)
(d) 0.16(AA); 0.36(Aa); 0.48(aa)

106. In a population at Hardy-Weinberg equilibrium, the frequency of homozygous dominant genotype is 0.36. What are the frequencies of dominant and recessive alleles for the locus?
(a) $p = 0.4$; $q = 0.6$
(b) $p = 0.6$; $q = 0.4$
(c) $p = 0.36$; $q = 0.4$
(d) $p = 0.4$; $q = 0.36$

107. A population will not exist in Hardy- Weinberg equilibrium if
(a) there is no migration
(b) the population is large
(c) individuals mate selectively
(d) there are no mutations

108. Consider the following statements:
(a) Gene flow, genetic drift, mutation, natural selection and genetic recombination deviate the population from the Hardy-Weinberg equilibrium.
(b) Small population size is required to maintain constant allele frequencies over generations.

Select the correct option.
(a) Both (a) and (b) are true.
(b) (A) is true but (B) is false.
(c) Both (A) and (B) are false.
(d) (A) is false but (B) is true.

109. Select the correctly matched option.
(a) Gene flow: Changes in allele frequencies by a chance event.
(b) Mutation: Migration of individuals of the population.
(c) Random mating: Equal chances of each individual in a population to mate with any other individual.
(d) Natural selection: Selection of mate on the basis of morphological features.

110. Addition or removal of alleles from the gene pool of a population due to migration of individuals is known as
(a) genetic drift (b) gene flow
(c) natural selection (d) artificial selection

111. Genetic drift operates in
(a) a small isolated population
(b) large isolated population
(c) non-reproductive population
(d) slow reproductive population

112. Match the terms in Column-I with the correct description in Column-II.

	Column I		Column II
(a)	Genetic drift	(1)	Genetic drift due to small colonizing population
(b)	Founder effect	(2)	Change in allele frequency by a chance event
(c)	Bottleneck effect	(3)	A major factor to introduce variations in sexually reproducing populations
(d)	Genetic combination	(4)	Adverse environmental factors

Select the correct option.

	a	b	c	d
(a)	1	4	2	3
(b)	2	1	4	3
(c)	3	2	4	1
(d)	4	1	3	2

113. A small number of finches from the mainland were blown by a cyclone to a new island. The resultant new population of finches had distinct gene pool from the source population. It represents the

(a) gene flow (b) mutation
(c) founder effect (d) selective mating

114. Name the types of natural selection as depicted in images I, II and III. Select the correct option.

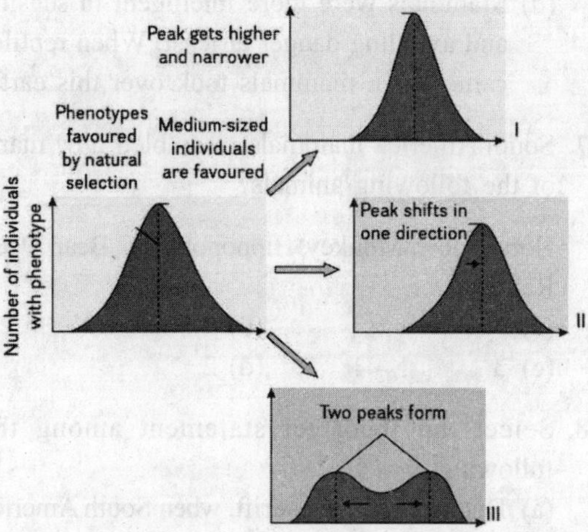

(a) I-Stabilizing; II-Disruptive; IIIDirectional
(b) I-Stabilizing; II-Directional; IIIDisruptive
(c) I-Directional; II-Stabilizing; IIIDisruptive
(d) I-Disruptive; II-Stabilizing; IIIDirectional

115. In a species, the weight of the newborn ranges from 2 to 5 kg. 97% of newborns with an average weight between 3 to 3.3 kg survive whereas 99% of the infants born with weight from 2 to 2.5 kg or 4.5 to 5 kg die. Which type of selection process is taking place?

(a) Directional selection
(b) Stabilizing selection
(c) Disruptive solution
(d) Cyclical Solution

116. Name the type of natural selection that tends to reduce the genetic variations in the population without affecting the mean.

(a) Disruptive selection
(b) Directional selection
(c) Stabilizing selection
(d) Both a and b are correct

117. The black-bellied finches with large beak sizes are able to feed on the most abundant food types in the habitat while the finches intermediate and small beak size die at a young age due to starvation.

This represents:

(a) Disruptive selection
(b) Directional selection
(c) Stabilizing selection
(d) Sexual selection

118. Natural selection in which more individuals acquire mean character value is known as ________ while the one wherein more individuals acquire peripheral character value at both ends of the distribution curve is ________

(a) stabilizing selection; disruptive selection
(b) disruptive selection; stabilizing selection
(c) disruptive selection; directional selection
(d) stabilizing selection; directional selection

119. Which of the given statement is incorrect?

(a) The peak of the distribution curve gets higher and narrower by stabilizing selection.
(b) Directional selection shifts the peak of the distribution curve in one direction.
(c) Disruptive selection does not affect the peak of the distribution curve.
(d) Stabilizing selection does not affect the mean of the phenotype.

120. Artificial selection to obtain cows yielding higher milk output represents

(a) directional selection as it pushes the mean of the character in one direction.
(b) disruptive selection as it splits the populati into two, one yielding higher output ar other lower output.
(c) stabilizing selection followed by selection as it stabilizes the

produce higher-yielding cows.

(d) stabilizing selection as it stabilizes this character in the population.

121. The northern elephant seal of North America and nearby islands was nearly hunted to extinction. The conservation efforts restored the population size. However, the restored population is vulnerable to extinction due to

(a) Stabilizing selection
(b) Bottleneck effect
(c) Founders effect
(d) Natural selection

Topic 5 A Brief Account of Evolution

122. Sea weeds and few plants existed probably around

(a) 0.32 bya (b) 400 mya
(c) 500 mya (d) 3.2 bya

123. Jawless fish probably evolved around -

(a) 400 mya (b) 350 mya
(c) 450 mya (d) 600 mya

124. Select the correct statement among the following.

(a) By the time of 500 mya, invertebrates were formed but were not active.
(b) The first organisms that invaded land were plants. They were widespread on land when animals invaded land.
(c) The amphibians evolved into reptiles. They lay thin-shelled eggs which do not dry up
sun unlike those of amphibians
amphibians in the next 2000 million
so, reptiles of different shapes and
ated on Earth.

ent but they all fell to form

Giant ferns
th B and C
ent among the following.

(a) Some of these land reptiles went back into water to evolve into amphibians like reptiles probably 200 mya (e.g. Ichthyosaurs).
(b) About 65 mya, the dinosaurs suddenly disappeared from the earth.
(c) The first mammals were like shrews. Their fossils are small sized. Mammals were viviparous and protected their unborn young inside the mother's body.
(d) Mammals were more intelligent in sensing and avoiding danger at least. When reptiles came down mammals took over this earth.

127. South America mammals resembled how many of the following animals?

Horse, Cow, Monkey, Hippopotamus, Bear, Deer, Rabbit

(a) 5 (b) 4
(c) 3 (d) 2

128. Select the incorrect statement among the following.

(a) Due to continental drift, when South America joined North America, these animals were overridden by North American fauna.
(b) Due to continental drift pouched mammals of Australia survived because of lack of competition from any other mammal.
(c) Evolution of horse, elephant, virus, etc., are special stories of evolution.
(d) The most successful story is the evolution of man with language skills and self-consciousness.

129. Select the correct statement about cells with a membranous envelope.

(a) The mechanism of how non-cellular aggregates of giant macromolecules could evolve into cells with membranous envelope is not known.
(b) Some of these cells had the ability to release O_2.
(c) The reaction could have been similar to the light reaction in photosynthesis where

water is split with the help of solar energy captured and channelised by appropriate light harvesting pigments.

(d) All of these

130. Fish with stout and strong fins could move on land and go back to water. This was about -

(a) 360 mya (b) 350 mya
(c) 0.4 bya (d) 3.5 bya

131. Select the correct statement about Coelacanth.

(a) In 1938, a fish caught in South America happened to be a Coelacanth which was thought to be extinct
(b) Coelacanth called lobefins evolved into the first reptile that lived on both land and water.
(c) There are no specimens of Coelacanth left with us.
(d) Coelacanth were ancestors of modern day frogs and turtles.

132. Select the correct statement about Tyrannosaurus rex?

(a) Tyrannosaurus rex had huge fearsome dagger like teeth.
(b) Tyrannosaurus rex was about 20 feet in height.
(c) Tyrannosaurus rex was biggest of all dinosaurs.
(d) All of these

133. The probable reason for disappearance of dinosaur was/were

(a) Climatic changes killed them
(b) Most of them evolved into birds.
(c) Both (a) and (b)
(d) Dinosaurs did not disappear.

134. How many of the following animals live wholly in water?

Whales, Shark, Dolphins, Turtles, Seals, Sea cows

(a) 5 (b) 4
(c) 3 (d) 2

Topic 6 Origin and Evolution of Man

135. Consider the following statements:

(a) Dryopithecus and Ramapithecus lived on Earth about 15 mya.
(b) They were primates with a hairy appearance and walked like gorillas.

Choose the correct option.

(a) Both (a) and (b) are true.
(b) (a) is true but (b) is false.
(c) Both (a) and (b) are false.
(d) (a) is false but (b) is true.

136. Ramapithecus was more _____ while Dryopithecus was more _____

(a) great ape-like; man-like
(b) man-like; ape-like
(c) ape-like; man-like
(d) both a and b are true

137. Which of the following is the immediate ancestor of genus Homo?

(a) Dryopithecus (b) Ramapithecus
(c) Australopithecines (d) Sahelanthropus

138. Match Column-I with Column-II.

	Column I		Column II
(a)	Australopithecus africanus	(1)	The oldest member of genus Homo
(b)	Homo habilis	(2)	Java man
(c)	Homo erectus	(3)	Cave painting
(d)	Neanderthal man	(4)	Lucy

Select the correct option.

	a	b	c	d
(a)	1	4	2	3
(b)	4	1	2	3
(c)	2	4	1	3
(d)	4	1	3	2

139. Consider the following statement.

(a) Homo habilis is also known as 'handy man' and could cook food using fire.

(b) Homo erectus was about 5 feet tall upright hominid and exhibited improved hunting skills. Select the correct option.
(a) Both (a) and (b) are true.
(b) (a) is true but (b) is false.
(c) Both (a) and (b) are false.
(d) (a) is false but (b) is true.

140. Which of the following hominid has a brain capacity of 1400cc and used hide to protect their body.
(a) Homo habilis
(b) Homo erectus
(c) Homo sapiens
(d) The Neanderthal man

141. The chronological order of human evolution from early to the recent is
(a) Australopithecus > Ramapithecus > Homo habilis > Homo erectus
(b) Ramapithecus > Australopithecus > Homo habilis > Homo erectus
(c) Ramapithecus > Homo habilis > Australopithecus > Homo erectus
(d) Australopithecus > Homo habilis > Ramapithecus > Homo erectus

142. Match the hominids with their correct brain size.

	Column I		Column II
(a)	Homo habilis	(1)	900 cc
(b)	Homo neanderthalensis	(2)	1350 cc
(c)	Homo erectus (3) 650–800 cc		
(d)	Homo sapiens	(4)	1400 cc

Select the correct option.

	a	b	c	d
(a)	4	1	3	2
(b)	3	4	1	2
(c)	4	1	2	3
(d)	2	4	1	3

143. Which of the following had the smallest brain capacity?
(a) Homo neanderthalensis
(b) Homo habilis
(c) Homo erectus
(d) Homo sapiens

144. Which of the following pair is incorrectly matched?
(a) Australopithecines – Fruit eater
(b) Pre-historic cave art –18000 years ago
(c) Agriculture and human settlements – 10000 years back
(d) Origin of modern Homo sapiens – 1,00,000–40,000 years back

145. Assertion: Ramapithecus was more manlike while Dryopithecus was more apelike.

Reason: Ramapithecus had a thick layer of enamel.
(a) Both Assertion and Reason are true and Reason is correct explanation of Assertion.
(b) Both Assertion and Reason are true, but Reason is not the correct explanation of Assertion.
(c) Assertion is true, but Reason is false.
(d) Assertion is false, but Reason is true.

146. Assertion: Evolution of modern humans exhibited a gradual increase in brain size.

Reason: Homo habilis had the smallest brain capacity.
(a) Both Assertion and Reason are true and Reason is correct explanation of Assertion.
(b) Both Assertion and Reason are true, but Reason is not the correct explanation of Assertion.
(c) Assertion is true, but Reason is false.
(d) Assertion is false, but Reason is true.

ANSWER KEY

1. (d)	2. (d)	3. (b)	4. (c)	5. (d)	6. (a)	7. (a)	8. (b)	9. (d)	10. (d)
11. (d)	12. (b)	13. (d)	14. (b)	15. (d)	16. (b)	17. (a)	18. (b)	19. (c)	20. (c)
21. (d)	22. (a)	23. (a)	24. (d)	25. (c)	26. (c)	27. (d)	28. (c)	29. (a)	30. (b)
31. (c)	32. (a)	33. (c)	34. (c)	35. (a)	36. (b)	37. (c)	38. (b)	39. (c)	40. (d)
41. (c)	42. (d)	43. (b)	44. (a)	45. (a)	46. (b)	47. (d)	48. (a)	49. (a)	50. (a)
51. (a)	52. (c)	53. (b)	54. (a)	55. (a)	56. (a)	57. (a)	58. (c)	59. (d)	60. (d)
61. (d)	62. (c)	63. (d)	64. (b)	65. (b)	66. (a)	67. (b)	68. (b)	69. (b)	70. (b)
71. (a)	72. (b)	73. (a)	74. (b)	75. (c)	76. (c)	77. (b)	78. (b)	79. (a)	80. (b)
81. (d)	82. (b)	83. (a)	84. (a)	85. (b)	86. (a)	87. (a)	88. (a)	89. (b)	90. (d)
91. (c)	92. (c)	93. (a)	94. (c)	95. (c)	96. (a)	97. (a)	98. (d)	99. (b)	100. (a)
101. (d)	102. (b)	103. (a)	104. (d)	105. (c)	106. (b)	107. (c)	108. (b)	109. (c)	110. (b)
111. (a)	112. (b)	113. (c)	114. (b)	115. (b)	116. (c)	117. (b)	118. (a)	119. (c)	120. (a)
121. (b)	122. (a)	123. (b)	124. (b)	125. (b)	126. (d)	127. (b)	128. (c)	129. (d)	130. (b)
131. (c)	132. (d)	133. (c)	134. (b)	135. (a)	136. (b)	137. (c)	138. (b)	139. (a)	140. (d)
141. (b)	142. (b)	143. (b)	144. (d)	145. (a)	146. (b)				

15 Chapter

Human Health and Diseases

MULTIPLE CHOICE QUESTIONS

1. Among the following, identify the infectious diseases.

(I) Cancer (II) Influenza
(III) Allergy (IV) Smallpox

(a) (I) and (II) (b) (II) and (III)
(c) (III) and (IV) (d) (II) and (IV)

2. Assertion: Diseases are characterized by the appearance of various signs and symptoms.

Reason: Disease always adversely affects only one organ or system.

(a) Both assertion and reason are true and reason is the correct explanation of assertion.
(b) Both assertion and reason are true but reason is not the correct explanation of assertion
(c) Assertion is true but reason is false.
(d) Both assertion and reason are false.

3. Assertion: AIDS is an infectious disease.

Reason: AIDS can be easily transmitted from one person to another.

(a) Both assertion and reason are true and reason is the correct explanation of assertion.
(b) Both assertion and reason are true but reason is not the correct explanation of assertion.
(c) Assertion is true but reason is false.
(d) Both assertion and reason are false.

4. The 'good humor' hypothesis of health was disproved by the

(a) discovery of blood circulation
(b) discovery of compound microscope
(c) demonstration of normal body temperature in persons with blackbile
(d) both (a) and (c)

5. The term health can be defined as

(a) the state of body and mind in a balanced condition
(b) the reflection of a smiling face
(c) a state of complete physical, mental and social well-being
(d) the symbol of economic prosperity

6. After diagnosis by a psychiatrist, an apparently healthy person was said to be unhealthy because the patient was

(a) inefficient at his work
(b) not prosperous economically
(c) not interested in sports
(d) showing behavioural and social Maladjustment

7. Choose the incorrect statement about health.

(a) Health can simply be defined as absence of disease.
(b) Healthy people are more efficient at work.
(c) Health increases productivity.
(d) Health reduces infant and maternal mortality.

8. Choose the correct statements about diseases.

(I) Disease adversely affects the functioning of one or more organs.
(II) A disease is characterized by the appearance of various signs and symptoms.
(III) AIDS is a fatal non-infectious disease.
(IV) Cancer is an infectious disease.

(a) (I) and (II) (b) (II) and (III)
(c) (III) and (IV) (d) (I) and (IV)

9. Diseases which are easily transmitted from one person to another are called

(a) non-infectious diseases

(b) genetic diseases
(c) infectious diseases
(d) none of these

10. Match Column-I with Column-II and choose the correct option from the codes given below. Column-I Column-II

	Column I		Column II
(a)	Health	(1)	AIDS
(b)	Genetic disorders	(2)	Physical, mental and social well-being
(c)	Infectious disease	(3)	Cancer
(d)	Noninfectious disease	(4)	Inherited from parents from birth

Select the correct option.

	A	B	C	D
(a)	2	4	3	1
(b)	2	4	1	3
(c)	3	2	4	1
(d)	1	3	2	4

Topic 1 Common Diseases in Humans

11. The malignant malaria is caused by
(a) Plasmodium vivax
(b) Plasmodium falciparum
(c) Plasmodium malaria
(d) None of these

12. Plasmodium enters the human body as
(a) Gametocyte (b) Haemozoin
(c) Sporozoite (d) None of these

13. Haemozoin is a toxin released from
(a) Streptococcus infected cells
(b) Plasmodium infected cells
(c) Homophilus infected cells
(d) None of these

14. The infectious stage of plasmodium, i.e., sporozoites are formed in
(a) RBCs of mosquito
(b) liver of the infected man
(c) gut of mosquito
(d) salivary glands of mosquito

15. How many hosts are required by the malarial parasite to complete its life cycle?
(a) One
(b) Two
(c) Three
(d) One or two according to environmental conditions

16. The person suffering from sickle cell anaemia is
(a) less prone to typhoid
(b) less prone to malaria
(c) more prone to typhoid
(d) more prone to malaria

17. Entamoeba histolytica is a protozoan parasite which infects
(a) stomach (b) small intestine
(c) large intestine (d) liver

18. The active form of Entamoeba histolytica feeds upon
(a) mucosa and submucosa of colon only
(b) food in intestine
(c) blood only
(d) erythrocytes, mucosa and submucosa of colon

19. Refer to the given figure showing stages in the life cycle of use plasmodium. In the figure, which type of reproduction is occurring at stages A and B respectively. Also, identify C and D in the figure.

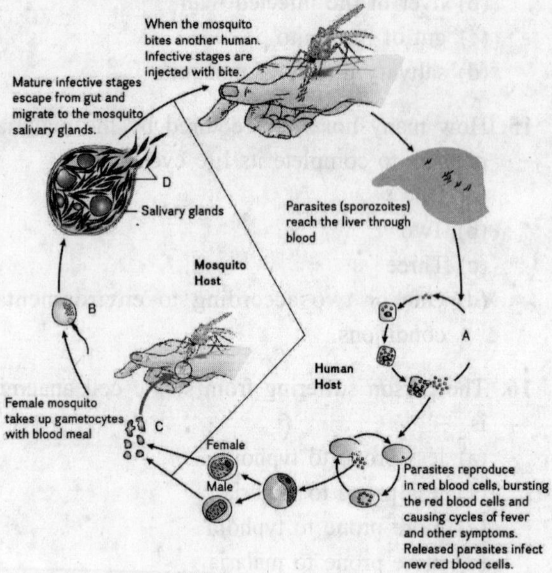

	A	B	C	D
A)	Asexual	Sexual	Gametocytes	Sporozoites
B)	Sexual	Asexual	Gametocytes	Sporozoites
C)	Asexual	Sexual	Sporozoites	Gametocytes
D)	Asexual	Sexual	Gametocytes	Ookinete

20. Match Column-I (Diseases) with Column-II (causative agents) and choose the correct option from the codes given below.

	Column I		Column II
(A)	Typhoid	(1)	Entamoeba histolytica
(B)	Malaria	(2)	Salmonella typhi
(C)	Amoebiasis	(3)	Rhino viruses
(D)	Common cold	(4)	Plasmodium vivax

Codes

	A	B	C	D
(a)	2	4	1	3
(b)	4	3	2	1
(c)	1	2	4	3
(d)	3	1	2	4

21. Choose the correct statements about amoebiasis.

(I) It is caused by the infection of Entamoeba histolytica.

(II) Its symptoms include loose motion, abdominal pain and cramps, stools with excess mucous and blood clots.

(III) Houseflies act as mechanical carriers for the parasite.

(IV) The main sources of its infection are drinking water and food contaminated by the fecal matter.

(a) (I) and (II) (b) (II) and (III)
(c) (I), (III) and (IV) (d) All of these

22. Diseases can be caused by the infection of
(a) bacteria (b) viruses
(c) helminths (d) all of these

23. The disease-causing organisms in plants and animals are called
(a) vectors (b) pathogens
(c) insects (d) worms

24. The pathogens can affect our body by
(a) interfering with normal vital activities
(b) resulting in morphological damage
(c) resulting in functional damage
(d) all of these

25. Most of the parasites are considered as
(a) vectors (b) pathogens
(c) worms (d) none of these

26. Production of digestive juices in the stomach in excessive amount results in
(a) ulcers (b) cirrhosis
(c) jaundice (d) diarrhoea

27. The inflammation of pancreas is
(a) pancreatitis (b) jaundice
(c) ulcer (d) none of these

28. Match Column-I with Column-II and choose the correct option from the codes given below.

	Column I		Column II
(A)	Lactose intolerance	(1)	Excessive passage of loose & watery feces
(B)	Celiac disease	(2)	Disability to synthesize lactase enzyme
(C)	Diarrhoea	(3)	Inflammation of pancreas

(D)	Pancreatitis	(4)	Inflammation of small intestine

Codes

	A	B	C	D
(a)	4	2	3	1
(b)	2	4	1	3
(c)	1	3	4	2
(d)	3	1	2	4

29. The pathogens that enter the gut can
(a) survive in the stomach at low pH
(b) resist the various digestive enzymes
(c) survive only at high temperature
(d) both (a) and (b)

30. In human beings, typhoid fever is caused by the infection of
(a) Salmonella typhi
(b) E. coli
(c) Plasmodium vivax
(d) Entamoeba histolytica

31. Salmonella typhi infects
(a) large intestine (b) stomach
(c) small intestine (d) liver

32. The pathogen of typhoid fever enters in the body through
(a) contaminated food
(b) contaminated water
(c) inhaling of air droplets released by diseased person
(d) both (a) and (b)

33. The symptoms of typhoid include
(a) high fever (b) stomach pain
(c) loss of appetite (d) all of these

34. The typhoid can be diagnosed by
(a) ELISA (b) PCR
(c) Widal test (d) ESR

35. Identify the correct pair representing the causative agent of typhoid fever and the confirmatory test for typhoid.
(a) Plasmodium vivax/UTI test
(b) Streptococcus pneumoniae/Widal test
(c) Salmonella typhi/Anthrone test
(d) Salmonella typhi/Widal test

36. The causative agents of pneumonia in humans is/are
(a) Streptococcus pneumoniae
(b) Haemophilus influenzae
(c) Bacillus anthracis
(d) Both (a) and (b)

37. In pneumonia, which part of respiratory system is affected?
(a) Alveoli
(b) Nose
(c) Respiratory passage
(d) All of these

38. The group of symptoms that is indicative of pneumonia
(a) constipation, abdominal pain, cramps, blood clots.
(b) difficulty in respiration, fever, chills, cough, headache.
C) nasal congestion and discharge, cough, constipation, headache.
(d) high fever, weakness, stomach pain, loss of appetite, constipation.

39. Among the following which one is not a bacterial disease?
(a) Dysentery (b) Plague
(c) Diphtheria (d) Common cold

40. Which of the following sets of diseases is caused by bacteria? **[NEET–2019]**
(a) Tetanus and mumps
(b) Herpes and influenza
(c) Cholera and tetanus
(d) Typhoid and small pox

41. Choose the incorrect statement from the following.
(a) In severe cases of typhoid, intestinal perforation and death may occur.
(b) Typhoid fever could be confirmed by Widal test.

(c) Streptococcus pneumoniae infects respiratory passage.
(d) Dysentery and plague are bacterial diseases.

42. Rhino viruses cause
(a) pneumonia (b) plague
(c) common cold (d) typhoid

43. Rhino viruses infect
(a) nose (b) respiratory passage
(c) lungs (d) both (a) and (b)

44. Which one is not a symptom of common cold in humans?
(a) Nasal congestion
(b) Sore throat
(c) Headache
(d) Grey to bluish colour of lips

45. Match Column-I (diseases) with Column-II (causative agent) and choose the correct option from the codes given below.

	Column I		Column II
(A)	Typhoid	(1)	*Streptococcus pneumoniae*
(B)	Pneumonia	(2)	*Rhino viruses*
(C)	Common cold	(3)	*Salmonella typhi*
		(4)	*Haemophilus influenzae*

Codes

	A	B	C
(a)	3	1,4	2
(b)	3	1,2	4
(c)	2	3	1,4
(d)	1	4	2,3

46. Ascaris, which causes ascariasis is an/a
(a) intestinal parasite (b) stomach parasite
(c) liver parasite (d) none of these

47. The symptoms of ascariasis include
(a) internal bleeding (b) anemia
(c) muscular pain (d) all of these

48. In which disease does mosquito transmitted pathogen cause chronic inflammation of lymphatic vessels?
(a) Elephantiasis (b) Ascariasis
(c) Ringworm disease (d) Amoebiasis

49. Refer to the given figure.

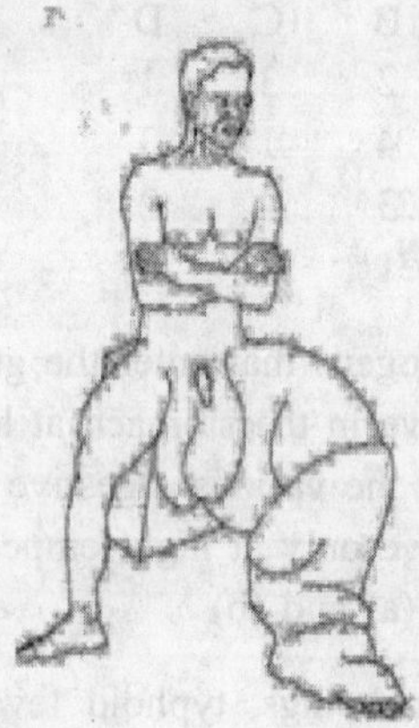

Which disease is shown in the figure?
(a) Amoebiasis (b) Filariasis
(c) Elephantiasis (d) Both (b) and (c)

50. Among the following which does not cause ringworm disease in humans?
(a) Microsporum
(b) Macrosporum
(c) Epidermophyton
(d) Trichophyton

51. Appearance of dry, scaly lesions on various parts of the body is symptom of which infectious disease?
(a) Filaria (b) Ascarisis
(c) Ringworm (d) None of these

52. Match the pathogens given in Column-I to the body organs to which they affect in Column-II. Choose the correct answer from the codes given below.

	Column I		Column II
(A)	Ascaris	(1)	Lymphatic vessels of lower limbs
(B)	Wuchereria	(2)	Intestine
(C)	Trichopyton	(3)	Skin, scalp and nails

(D)	Streptococcus pneumoniae	(4)	Lungs

Codes

	A	B	C	D
(a)	2	1	3	4
(b)	1	2	4	3
(c)	3	2	1	4
(d)	4	3	2	1

53. Ringworms are generally acquired from
(a) soil (b) infected towels
(c) infected combs (d) all of these

54. Match Column-I with Column-II and choose the correct option from the codes given below.

	Column I		Column II
(A)	Entamoeba histolytica	(1)	Sporozoa
(B)	Plasmodium vivax	(2)	Rhizopoda
(C)	Ascaris lumbricoides	(3)	Deuteromycetes
(D)	Trichophyton rubrum	(4)	Nematoda

Codes

	A	B	C	D
(a)	2	1	4	3
(b)	1	2	3	4
(c)	4	3	2	1
(d)	3	4	1	2

55. Trichophyton feeds on
(a) goblins of blood (b) keratin of skin
(c) cellulose of leaves (d) none of these

56. Choose the incorrect statement from the following.
(a) Wuchereria bancrofti causes chronic inflammation of the lymphatic vessels.
(b) The pathogens of filaria are transmitted to a healthy person through houseflies.
(c) Trichophyton is responsible for ringworm.
(d) Common cold is a viral disease.

57. Read the following statements carefully and choose the option that correctly identifies the true statements.
(I) Many infectious diseases can be prevented and controlled by maintaining personal and public hygiene
(II) Proper disposal of waste and excreta is particularly essential for the airborne diseases.
(III) Malaria can be prevented by eliminating its vector and their breeding places.
(IV) Chikungunya is a vector-borne disease.
(a) (I), (II) and (III) (b) (II), (III) and (IV)
(c) (IV), (II) and (I) (d) (I), (III) and (IV)

58. The most important measure to control malaria and filariasis is to control or eliminate the
(a) vectors
(b) breeding places of vectors
(c) causal organism
(d) both (a) and (b)

59. Match Column-I with Column-II and choose the correct option from the codes given below

	Column I		Column II
(A)	Food-borne disease	(1)	Pneumonia
(B)	Air-borne disease	(2)	Amoebic dysentery
(C)	Vector-borne disease	(3)	Malaria
		(4)	Typhoid

Codes

	A	B	C
(a)	2,4	1	3
(b)	1,4	2	3
(c)	3	2,4	1
(d)	2	3	4,1

60. Which fish is introduced in ponds that feed on mosquito larva?
(a) Rohu (b) Katla
(c) Gambusia (d) None of these

61. The vector that transmits the disease chikungunya is
(a) Housefly (b) Cockroach
(c) Aedes mosquito (d) Female Anopheles

62. Dengue is transmitted through
(a) female Anopheles (b) housefly

(c) Gambusia (d) Aedes mosquito

63. A deadly disease that has been eradicated from India is
(a) typhoid (b) smallpox
(c) dengue (d) cancer

64. Among the following diseases, for which one is vaccine available?
(a) Polio (b) Pneumonia
(c) Tetanus (d) All of these

65. Match Column-I with Column-II and choose the correct option from the codes given below.

	Column I		Column II
(a)	Amoebic dysentery	(1)	Female Anopheles
(b)	Dengue	(2)	Housefly
(c)	Malaria	(3)	Aedes mosquito

Codes

	A	B	C
(a)	3	2	1
(b)	2	3	1
(c)	1	2	3
(d)	2	1	3

66. Assertion: Most of the parasites are pathogens.

Reason: Disease causing organisms are called pathogens hosts as parasite.
(a) Both assertion and reason are true and reason is the correct explanation of assertion.
(b) Both assertion and reason are true but reason is not the correct explanation of assertion.
(c) Assertion is true but reason is false.
(d) Both assertion and reason are false.

67. Assertion: Pneumonia is caused by the infection of Streptococcus pneumoniae.

Reason: Streptococcus pneumoniae bacteria infect respiratory passage.
(a) Both assertion and reason are true and reason is the correct explanation of assertion.
(b) Both assertion and reason are true but reason is not the correct explanation of assertion.
(c) Assertion is true but reason is false.
(d) Both assertion and reason are false.

68. Assertion: The malarial parasite requires two hosts to complete its lifecycle.

Reason: These two hosts are human and mosquito.
(a) Both assertion and reason are true and reason is the correct explanation of assertion.
(b) Both assertion and reason are true but reason is not the correct explanation of assertion.
(c) Assertion is true but reason is false.
(d) Both assertion and reason are false.

69. Assertion: Wuchereria bancrofti affects the blood vessels of the lower limbs.

Reason: This pathogen is transmitted to a healthy person through the bite of male mosquito vectors.
(a) Both assertion and reason are true and reason is the correct explanation of assertion.
(b) Both assertion and reason are true but reason is not the correct explanation of assertion.
(c) Assertion is true but reason is false.
(d) Both assertion and reason are false.

Topic 2 Immunity

70. The ability of the host to fight against disease causing organism is known as
(a) pathogenicity (b) immunity
(c) immunisation (d) none of these

71. Innate immunity
(a) is non-specific type of defence.
(b) is present at the time of birth.
(c) consists of four types of barriers.
(d) all of these

72. Which type of immunity is present from the birth?
(a) Acquired (b) Innate
(c) Specific (d) None of these

73. Identify the physical barrier of immunity from the following.
(a) Skin on our body
(b) Acid in the stomach
(c) PMNL-neutrophils
(d) Both (a) and (b)

74. Saliva in the mouth is an example of
(a) physical barrier of immunity.
(b) physiological barrier of immunity.
(c) cellular barrier of immunity.
(d) cytokine barrier of immunity.

75. Humans have acquired immune system that produces antibodies to neutralize pathogens. Still innate immune system is present at the time of birth because it
(a) provides passive immunity
(b) is very specific and uses different macrophages
(c) produces memory cells for mounting fast secondary response
(d) has natural killer cells which can phagocytose and destroy microbes

76. A substance produced by a virus infected cell that can protect other cells from further infection is
(a) colostrum (b) serotonin
(c) interferon (d) histamine

77. Interferons, produced by virus-infected cells are
(a) enzymes (b) proteins
(c) lipids (d) none of these

78. Match Column-I with Column-II and choose the correct answer from the codes given below.

	Column I		Column II
(a)	Physical barrier	(1)	Acid in the stomach
(b)	Physiological barrier	(2)	Monocytes
(c)	Cellular barrier	(3)	Interferon
(d)	Cytokine barrier	(4)	Mucus coating of the epithelium lining of urogenital tract

Codes

	A	B	C	D
(a)	4	1	2	3
(b)	1	3	4	2
(c)	2	4	3	1
(d)	3	2	1	4

79. The pathogen specific immunity is
(a) innate immunity (b) acquired immunity
(c) physical strength (d) none of these

80. Subsequent encounter with the same pathogen elicits a/an
(a) secondary response
(b) highly intensified response
(c) anamnestic response
(d) all of these

81. Antibodies are produced by
(a) T-lymphocytes (b) B-lymphocytes
(c) monocytes (d) both (a) and (b)

82. An antibody is represented as
(a) H3L3 (b) H1L1
(c) H2L2 (d) None of these

83. Refer to the given figure showing structure of an antibody. In the figure some parts are labelled as P, Q, R and S. Identify the part which binds with antigen.

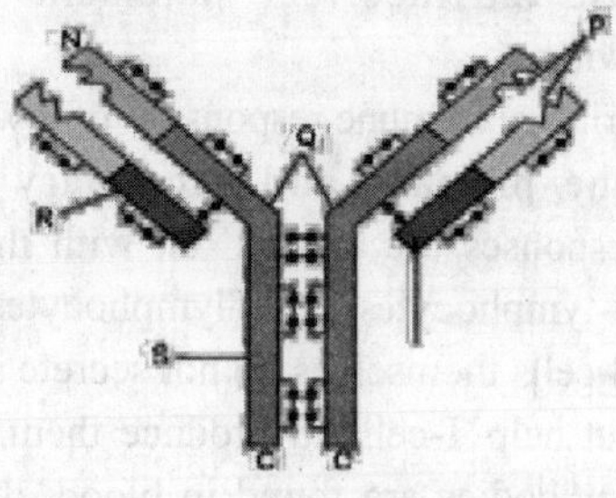

(a) Q (b) P
(c) R (d) S

84. Which of the following immune responses is responsible for rejection of kidney graft?
(a) Auto-immune response
(b) Humoral immune response
(c) Inflammatory immune response
(d) Cell-mediated immune response

85. Transplantation of tissues/organs fails often due to non-acceptance by the patient's body. Which type of immune response is responsible for such rejections?
(a) Cell-mediated immune response
(b) Humoral immune response
(c) Physiological immune response
(d) Auto immune response

86. Match Column-I with Column-II and choose the correct option from the codes given below.

	Column I		Column II
(a)	Innate immunity	(1)	Antibodies
(b)	Acquired immunity	(2)	Non-specific immune response
(c)	Humoral immune response	(3)	T-lymphocytes
(d)	Cellmediated immunity	(4)	Pathogen specific immune response

Codes

	A	B	C	D
(a)	4	2	3	1
(b)	2	4	1	3
(c)	2	4	3	1
(d)	1	3	4	2

87. Choose the incorrect statement from the following.
(a) Primary immune response is of low intensity.
(b) The primary and secondary immune responses are carried out with the help of B-lymphocytes and Tlymphocytes.
(c) B-cells themselves do not secrete antibodies but help T-cells to produce them.
(d) Antibodies are found in blood, therefore it is called humoral immune response.

88. When a host is exposed to antigens, antibodies are produced in the host body. This type of immunity is called
(a) active immunity (b) passive immunity
(c) innate immunity (d) none of these

89. When readymade antibodies are directly given to protect the body against foreign agents, it is called
(a) cell-mediated immunity
(b) passive immunity
(c) active immunity
(d) innate immunity

90. Colostrum, the yellowish fluid, secreted by mother during the initial days of lactation is very essential to impart immunity to the newborn infants because it contains
(a) natural killer cells (b) monocytes
(c) macrophages (d) immunoglobulin A

91. Which type of antibodies are found in colostrum?
(a) O = IgG type (b) O = IgD type
(c) O = IgA type (d) O = IgE type

92. Consider the following statements and choose the correct statements.
(I) Active immunity is slow and takes time to give its full effective response.
(II) In passive immunity, ready-made antibodies are directly given
(III) Colostrum contains IgE antibodies.
(IV) The foetus also receives some antibodies from its mother.
(a) (I), (II) and (III) (b) (I), (II) and (IV)
(c) (I), (III) and (IV) (d) All of these

93. The principle of immunisation is based on which property of immune system?
(a) Discriminate between self and nonself
(b) Memory
(c) Production of antibodies
(d) All of these

94. Match each disease with its correct type of vaccine.

	Column I		Column II
(a)	Tuberculosis	(1)	Harmless virus
(b)	Whooping cough	(2)	Inactivated toxin
(c)	Diphtheria	(3)	Killed bacteria
(d)	Polio	(4)	Harmless bacteria

Codes

	a	b	c	d
(a)	3	2	4	1
(b)	4	3	2	1
(c)	1	2	4	3
(d)	2	1	3	4

95. In case of snakebites, the injection which is given to the patients, contains
(a) antigens
(b) antigen – antibody complexes
(c) antibodies
(d) enzymes

96. In vaccination, what is introduced in the body?
(a) Antigenic proteins of pathogen
(b) Inactivated pathogen
(c) Weakened pathogen
(d) All of these

97. Vaccine of hepatitis B is produced from
(a) Yeast (b) Rhizobium
(c) Agrobacterium (d) Azadirachta

98. Choose the incorrect statement about vaccination.
(a) In passive immunization, preformed antibodies are injected in the body.
(b) Vaccines can be produced using recombinant DNA technology.
(c) Vaccines generate memory B-cells and T-cells.
(d) Vaccines given in case of snakebite contains preformed antigens.

99. The exaggerated response of the immune system to certain antigens in the environment is called
(a) immunisation (b) allergy
(c) vaccination (d) none of these

100. The substances which produce allergy are called
(a) antigens (b) pathogens
(c) allergens (d) antibodies

101. The antibodies produced against allergens are of
(a) O = IgA type (b) O = IgE type
(c) O = IgM type (d) O = IgG type

102. Match Column-I with Column-II and choose the correct option from the codes given below.

	Column I		Column II
(a)	Colostrum	(1)	IgE
(b)	Allergy	(2)	IgA
(c)	Graft rejection	(3)	Passive immunisation
(d)	Preformed antibodies	(4)	Cell-medicated immunity

Codes

	A	B	C	D
(a)	3	2	4	1
(b)	4	3	2	1
(c)	1	2	4	3
(d)	2	1	4	3

103. Symptoms of allergic reactions include
(a) sneezing (b) watery eyes
(c) running nose (d) all of these

104. The chemical which is released during allergic reaction?
(a) Histamine (b) Serotonin
(c) Steroid (d) Both (a) and (b)

105. Which of the following is not an autoimmune disease?
(a) Psoriasis (b) Rheumatoid arthritis
(c) Alzheimer's disease(d) Vitiligo

106. Which of the following diseases is an autoimmune disorder?
(a) Gout (b) Myasthenia gravis
(c) Arthritis (d) Osteoporosis

107. Rheumatoid arthritis is a/an
(a) infectious disease
(b) genetic disorder
(c) autoimmune disease
(d) non-infectious disease

108. Choose the correct reason(s) for rheumatoid arthritis.
(I) Body attacks self-cells
(II) The ability of immune system to differentiate between self and nonself increases.
(III) The production of antibodies increases.

(IV) Immune system fails to discriminate between self and non-self cells.

(a) (I) and (IV) (b) (II) and (III)
(c) (II) and (IV) (d) (I) and (III)

109. Our immune system is unique because it can
(a) recognise foreign antigens
(b) respond to antigens
(c) remember antigens
(d) all of these

110. The immune system plays an important role in
(a) allergic reactions
(b) autoimmune diseases
(c) organ transplantation
(d) all of these

111. Match Column-I with Column-II and choose the correct option from the codes given below.

	Column I		Column II
(a)	Allergy	(1)	Inability to discriminate self cells from nonself- cells
(b)	Autoimmunity	(2)	Introduction of killed/ Weakened pathogen
(c)	Active immunisation	(3)	Immune response against allergens
(d)	Passive immunisation	(4)	Introduction of antibodies

Codes

	A	B	C	D
(a)	3	1	2	4
(b)	4	3	2	1
(c)	3	1	4	2
(d)	2	4	3	1

112. Identify the primary lymphoid organ(s).
(a) Bone marrow (b) Thymus
(c) Spleen (d) Both (a) and (b)

113. In primary lymphoid organs,
(a) lymphocytes become mature
(b) lymphocyte interact with antigens
(c) lymphocytes become effecter cells
(d) both (a) and (b)

114. Among the following which one is not a lymphoid tissue?
(a) Tonsils (b) Spleen
(c) Thymus (d) ancreas

115. The figure given below is a diagrammatic representation of lymph nodes. In the figure some parts are labeled as A, B, C and D. Identify the part that serves to trap the antigens.

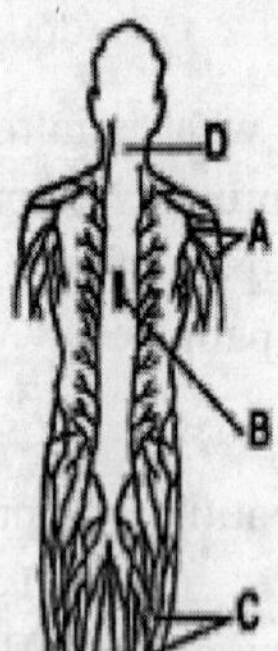

(a) B (b) A
(c) C (d) D

116. The lymphoid organ where all blood cells including lymphocytes are produced
(a) Spleen (b) Thymus
(c) Bone narrow (d) Payer's patches

117. The lymphoid organ that keeps reducing in size with age is
(a) tonsils (b) bone marrow
(c) thymus (d) spleen

118. Which one is known as the reservoir of erythrocytes?
(a) Spleen (b) Thymus
(c) Lymph nodes (d) None of these

119. The organ(s) that provide(s) microenvironments for the development and maturation of T-lymphocytes is/are
(a) Bone marrow (b) Thymus
(c) Spleen (d) Both (a) and (b)

120. The lymphoid tissue located within the lining of digestive tract is called

(a) lymphatic vessels (b) lymph nodes
(c) MALT (d) none of these

121. MALT constitutes about ______ percent of the lymphoid tissue in human body.
(a) 20% (b) 70%
(c) 10% (d) 50%

122. Match Column-I with Column-II and choose the correct option from the codes given below.

	Column I		Column II
(a)	Bone marrow	(1)	Maturation of T-Lymphocytes
(b)	Thymus	(2)	Production of blood cells
(c)	Spleen	(3)	Serve to trap antigens
(d)	Lymph Nodes	(4)	Reservoir of erythrocytes

Codes

	A	B	C	D
(a)	1	2	3	4
(b)	2	1	4	3
(c)	4	3	2	1
(d)	3	4	1	2

123. Assertion: The immune response in which antibodies are formed is called humoral immune response.

Reason: Antibodies are found in blood.
(a) Both assertion and reason are true and reason is the correct explanation of assertion.
(b) Both assertion and reason are true but reason is not the correct explanation of assertion.
(c) Assertion is true but reason is false.
(d) Both assertion and reason are false.

124. Assertion: In passive immunization, preformed antibodies are given to the patients.

Reason: Polio vaccine also contains preformed antibodies.
(a) Both assertion and reason are true and reason is the correct explanation of assertion.
(b) Both assertion and reason are true but reason is not the correct explanation of assertion.
(c) Assertion is true but reason is false.
(d) Both assertion and reason are false.

125. Assertion: Bone marrow is a secondary lymphoid organ.

Reason: In secondary lymphoid organs maturation of lymphocytes occur.
(a) Both assertion and reason are true and reason is the correct explanation of assertion.
(b) Both assertion and reason are true but reason is not the correct explanation of assertion.
(c) Assertion is true but reason is false.
(d) Both assertion and reason are false.

Topic 3 AIDS

126. The word AIDS stands for 'Acquired Immuno Deficiency Syndrome', here syndrome means
(a) group of diseases
(b) group of symptoms
(c) group of antigens
(d) none of these

127.HIV is a member of a group of viruses called
(a) rota virus (b) rhino virus
(c) retro virus (d) none of these

128. Which of the following is correct regarding AIDS causative agent HIV?
(a) HIV is an unenveloped retrovirus.
(b) HIV does not escape but attacks the acquired immune response.
(c) HIV is an enveloped virus containing one molecule of single-stranded RNA and one molecule of reverse transcriptase.
(d) HIV is an enveloped virus that contains two identical molecules of single-stranded RNA and two molecules of reverse transcriptase.

129. HIV is not transmitted by
(a) transfusion of contaminated blood
(b) sharing of infected needles
(c) sexual contact with infected persons

(d) shaking hands with infected person

130. HIV/AIDS spreads through
(a) droplets resulting from cough
(b) body fluids
(c) mere touch
(d) All of these

131. HIV that causes AIDS, first starts destroying
(a) leucocytes (b) helper T-cells
(c) thrombocytes (d) B-lymphocytes

132. AIDS is diagnosed through which technique?
(a) ELISA (b) PCR
(c) PAGE (d) Electrophoresis

133. Refer to the given figure showing replication of retrovirus. How is it different from the replication of other viruses?

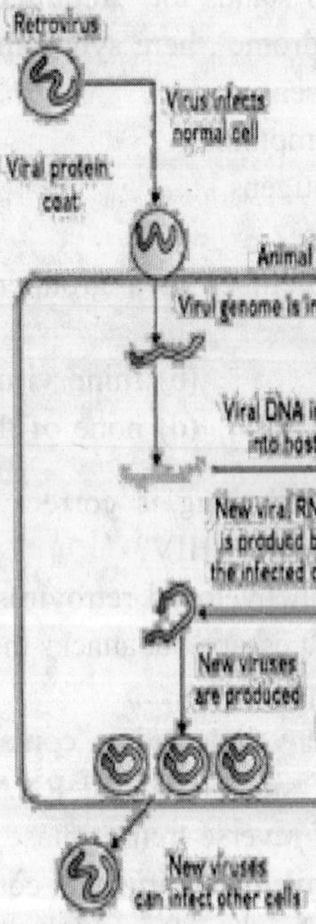

(a) Viral DNA is produced from viral RNA by reverse transcriptase.
(b) Viral RNA produces dsRNA by RNA polymerase.
(c) Viral DNA is produced from viral RNA by DNA polymerase.
(d) Viral DNA is produced from host DNA by DNA polymerase.

134. Match Column-I with Column-II and choose the correct option from the codes given below

	Column I		Column II
(a)	AIDS	(1)	Retrovirus
(b)	HIV	(2)	Enzyme
(c)	Reverse transcriptase	(3)	Diagnostic technique
(d)	ELISA	(4)	Syndrome

Codes

	A	B	C	D
(a)	4	1	2	3
(b)	1	2	4	3
(c)	3	4	1	2
(d)	2	3	4	1

135. Choose the incorrect statement about AIDS.
(a) AIDS is caused by HIV.
(b) It can be diagnosed using ELISA technique.
(c) HIV destroys B-lymphocytes.
(d) HIV infected people need help and sympathy instead of being shunned by the society.

136. Assertion: AIDS is caused by Human Immuno Deficiency Virus (HIV).

Reason: It is a member of group 'retroviruses'.
(a) Both assertion and reason are true and reason is the correct explanation of assertion.
(b) Both assertion and reason are true but reason is not the correct explanation of assertion.
(c) Assertion is true but reason is false.
(d) Both assertion and reason are false.

137. Assertion: Transmission of HIV infection generally occurs by sexual contact with infected person.

Reason: HIV is not transmitted by the transfusion of contaminated blood and blood products.
(a) Both assertion and reason are true and reason is the correct explanation of assertion.
(b) Both assertion and reason are true but reason is not the correct explanation of assertion.

(c) Assertion is true but reason is false.
(d) Both assertion and reason are false.

138. Assertion: AIDS leads to a progressive decrease in number of helper Tlymphocytes in the infected person.

Reason: HIV virus replicates and produces progeny virus in helper Tlymphocytes which are released in blood.
(a) Both assertion and reason are true and reason is the correct explanation of assertion
(b) Both assertion and reason are true but reason is not the correct explanation of assertion.
(c) Assertion is true but reason is false.
(d) Both assertion and reason are false.

Topic 4 Cancer

139. Cancer cells do not show this property.
(a) Metastasis (b) Growth
(c) Contact inhibition (d) Both (a) and (C)

140. Cancer cells divide continuously and give rise to a mass of cells called
(a) fibroid (b) tumor
(c) oncogene (d) none of these

141. Tumours that remain confined to their original location and cause little damage are
(a) benign (b) malignant
(c) carcinogen (d) none of these

142. Mass of neoplastic cells is called
(a) benign tumour
(b) fibroid
(c) cyst
(d) malignant tumour

143. The cells of malignant tumour
(a) grow very rapidly
(b) invade and damage other normal tissues
(c) show metastasis
(d) all of these

144. Cells sloughed off from malignant tumour move to other parts of the body to form new tumours. This stage of disease is called
(a) teratogenesis (b) metastasis
(c) mitosis (d) metagenesis

145. Match Column-I with Column-II and choose the correct option from the codes given below.

	Column I		Column II
(a)	Contact inhibition	(1)	Consists of neoplastic cells
(b)	Benign tumour	(2)	Property of normal cells to inhibit uncontrolled growth of other cells
(c)	Malignant tumour	(3)	Property of cancerous cells to form new tumour at distant sites
(d)	Metastasis	(4)	Remains confined to original location

Codes

	A	B	C	D
(a)	1	3	4	2
(b)	4	2	3	1
(c)	2	4	1	3
(d)	3	1	2	4

146. Choose the incorrect statement about Malignant tumours.
(a) These tumours consist of neoplastic cells.
(b) They show the property of metastasis.
(c) The cells of this tumour have the property of contact inhibition.
(d) The cells of malignant tumour starve the normal cells by competing for vital nutrients.

147. Transformation of normal cells into cancerous neoplastic cells may be induced by
(a) physical agents (b) chemical agents
(c) biological agents (d) all of these

148. The cancer-causing agents are called
(a) carcinogens (b) teratogens
(c) mutagens (d) none of these

149. X-rays lead to neoplastic transformation by

causing damage to
(a) enzymes (b) hormones
(c) DNA (d) all of these

150. The genes that cause cancer are called
(a) expressor genes (b) oncogenes
(c) regulatory genes (d) structural genes

151. Match Column-I with Column-II and choose the correct option from the codes given below.

	Column I (Type of Carcinogen)		Column II (Example)
(a)	Physical agent	(1)	Oncogenic virus
(b)	Chemical agent	(2)	UV rays
(c)	Biological agent	(3)	Tobacco
		(4)	Gamma rays

Codes

	A	B	C
(a)	4,2	3	1
(b)	4,1	2	3
(c)	2	4,3	1
(d)	3	2	4,1

152. Cancer causing viruses are called
(a) retrovirus (b) rhinovirus
(c) oncogenic virus (d) none of these

153. Choose the correct statements about carcinogens.
(I) Carcinogens transform normal cells into cancerous cells.
(II) These carcinogens could be physical, chemical or biological.
(III) Ionizing radiations like UV-rays damage DNA leading to neoplastic transformation.
(IV) Several proto-oncogenes have been identified in neoplastic cells that get activated under certain conditions.
(a) (I) and (III) (b) (I) and (II)
(c) (III) and (IV) (d) all of these

154. Which technique can be used for the detection of cancer of internal organs?
(a) Radiography (b) CT
(c) MRI (d) All of these

155. Computed tomography uses 'A' to generate a three-dimensional image of the internals of an object. Here 'A' is
(a) X-rays (b) γ-rays
(c) α-rays (d) UV rays

156. Match Column-I with Column-II and choose the correct option from the codes given below.

	Column I		Column II
(a)	Biopsy	(1)	Threedimensional image using Xrays
(b)	Radiograp hy	(2)	Histopathologic al study
(c)	Computed Tomograp hy	(3)	Use of strong magnetic fields and nonionising radiations
(d)	MRI	(4)	Use of X-rays

Codes

	A	B	C	D
(a)	4	2	3	1
(b)	2	4	1	3
(c)	3	1	4	2
(d)	1	3	2	4

157. The common approaches for the treatment of cancer is/are
(a) surgery (b) radiation therapy
(c) immunotherapy (d) all of these

158. Which substance is given to cancer patients to activate their immune system?
(a) Carcinogens (b) Cytokinin
(c) α-interferon (d) None of these

159. Assertion: X-rays and γ-rays are called carcinogens.

Reason: Carcinogens transform normal cells into cancerous neoplastic cells.
(a) Both assertion and reason are true and reason is the correct explanation of assertion.
(b) Both assertion and reason are true but reason is not the correct explanation of assertion.
(c) Assertion is true but reason is false.
(d) Both assertion and reason are false.

160. Assertion: Computed tomography can be used for the early detection of cancer of internal organs.

Reason: Computed tomography uses UV-rays to generate a three-dimensional image of the internals of an object.

(a) Both assertion and reason are true and reason is the correct explanation of assertion.
(b) Both assertion and reason are true but reason is not the correct explanation of assertion.
(c) Assertion is true but reason is false.
(d) Both assertion and reason are false.

161. Assertion: The patients of cancer are given α-interferon.

Reason: α-interferon is a biological response modifier.

(a) Both assertion and reason are true and reason is the correct explanation of assertion.
(b) Both assertion and reason are true but reason is not the correct explanation of assertion.
(c) Assertion is true but reason is false.
(d) Both assertion and reason are false.

Topic 5 Drug and Alcohol Abuse

162. Opioid receptors are found in
(a) central nervous system
(b) reproductive system
(c) gastrointestinal tract
(d) both (a) and (c)

163. Heroin is commonly called
(a) smack (b) cocaine
(c) crack (d) none of these

164. Drug called 'Heroin' is synthesised by
(a) methylation of morphine
(b) acetylation of morphine
(c) glycosylation of morphine
(d) nitration of morphine

165. 'Smack' is obtained from the
(a) leaves of Cannabis sativa
(b) latex of Papaver somniferum
(c) fruits of Erythroxylum coca
(d) flowers of Datura

166. Refer to the given chemical structure. It is

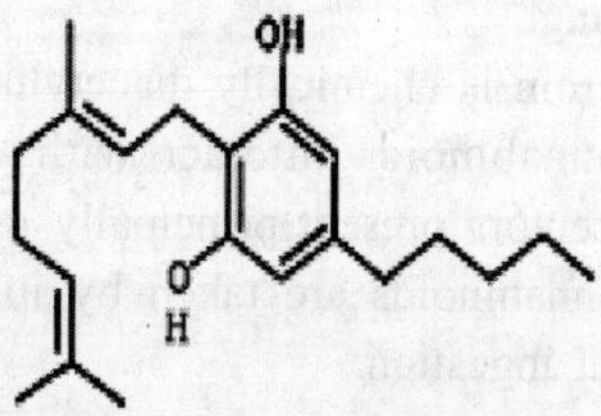

(a) morphine (b) cocaine
(c) cannabinoid (d) none of these

167. Refer to the given figure.

The drug obtained from this plant affects
(a) reproductive system
(b) respiratory system
(c) nervous system
(d) none of these

168. Natural cannabinoids are obtained from the
(a) inflorescence of Cannabis sativa
(b) latex of Cannabis sativa
(c) fruits of Cannabis sativa
(d) leaves of Cannabis sativa

169. Cannabinoids are generally taken by
(a) inhalation (b) oral ingestion
(c) snorting (d) both (a) and (b)

170. Cannabinoids affect
(a) cardiovascular system
(b) nervous system
(c) digestive system
(d) none of these

171. Choose the incorrect statement from the following.
(a) Heroin is chemically diacetylmorphine
(b) Cannabinoids interact with cannabinoid receptors present principally in the gut.
(c) Cannabinoids are taken by inhalation and oral ingestion.
(d) Heroin is a depressant.

172. Cocaine is obtained from
(a) Papaver somniferum
(b) Cannabis sativa
(c) Atropa belladona
(d) Erythroxylum coca

173. Coca alkaloid or cocaine is obtained from
(a) Datura
(b) Papaver somniferum
(c) Atropa belladonna
(d) Erythroxylum coca

174. Crack is usually
(a) ingested orally (b) injected
(c) inhaled (d) snorted

175. Refer to the given figure. This plant causes

(a) hallucinations (b) insomnia
(c) depression (d) all of these

176. Among the following which plant does not process hallucinogenic properties?
(a) Atropa belladonna (b) Datura
(c) Cannabis sativa (d) Erythroxylum coca

177. Among the following which one is abused by some sportspersons?
(a) Heroin (b) Barbiturates
(c) Cannabinoids (d) Amphetamines

178. Match Column-I with Column-II and choose the correct option from the codes gives below.

	Column I		Column II
(a)	Smack	(1)	Hallucination
(b)	Cocaine	(2)	Depressant
(c)	Datura	(3)	Pain killer
(d)	Morphine	(4)	Stimulant

Codes

	A	B	C	D
(a)	2	4	1	3
(b)	3	2	4	1
(c)	1	3	2	4
(d)	4	1	3	2

179. Morphine is a very effective
(a) sedative (b) pain killer
(c) stimulant (d) both (a) and (b)

180. Which chemical substance of tobacco stimulates adrenal grand to release adrenaline and noradrenaline?
(a) Tannic acid (b) Nicotine
(c) Curamin (d) Catechin

181. Whose concentration is increased in blood by smoking?
(a) Carbon dioxide (CO2)
(b) Oxygen (O2)
(c) Carbon monoxide (CO)
(d) Water (H2O)

182. Choose the correct statements.
(I) Tobacco is smoked, chewed or used as a snuff
(II) Tobacco contains nicotine, an alkaloid.
(III) Smoking decreases heart rate.
(IV) Tobacco chewing is associated with increased risk of cancer of the oral cavity.
(a) (I) and (II) (b) (III) and (IV)

(c) (I), (II) and (IV) (d) All of these

183. Among the following which motivates youngsters towards drug and alcohol abuse?
(a) Need for adventure (b) Need for excitement
(c) Experimentation (d) All of these

184. Withdrawal syndrome is characterised by
(a) anxiety (b) shakiness
(c) nausea (d) all of these

185. When drugs are taken intravenously, there are increased chances of having
(a) AIDS (b) Hepatitis B
(c) Polio (d) Both (a) and (b)

186. Match Column-I with Column-II and choose the correct answer from the codes given below.

	Column I		Column II
(a)	Adolescence	(1)	Oral cancer
(b)	Addiction	(2)	Abrupt discontinuation of regular close of drug
(c)	Smoking	(3)	Bridge linking childhood and adulthood
(d)	Withdrawal syndrome	(4)	Psychological euphoria associated with drugs

Codes

	A	B	C	D
(a)	4	2	3	1
(b)	3	4	1	2
(c)	1	3	4	2
(d)	2	1	3	4

187. The side effects of the use of anabolic steroids in females include
(a) masculinisation
(b) increased aggressiveness
(c) depression
(d) all of these

188. The measure(s) useful for the prevention and control of alcohol and drug abuse among adolescents is/are
(a) avoid undue peer pressure
(b) education and counseling
(c) looking for danger sign
(d) all of these

189. Assertion: Cocaine is obtained from coca plant.

Reason: It has a potent stimulating action on central nervous system.
(a) Both assertion and reason are true and reason is the correct explanation of assertion.
(b) Both assertion and reason are true but reason is not the correct explanation of assertion.
(c) Assertion is true but reason is false.
(d) Both assertion and reason are false.

190. Assertion: Withdrawal syndrome is characterised by anxiety, shakiness, nausea and sweating.

Reason: Withdrawal syndrome is not relieved even when use of drugs is resumed again.
(a) Both assertion and reason are true and reason is the correct explanation of assertion.
(b) Both assertion and reason are true but reason is not the correct explanation of assertion.
(c) Assertion is true but reason is false.
(d) Both assertion and reason are false.

191. Assertion: Those who take drugs intravenously are much more likely to acquire AIDS and Hepatitis B.

Reason: The viruses of AIDS and Hepatitis B are transferred from one person to another by sharing of infected needles and syringes.
(a) Both assertion and reason are true and reason is the correct explanation of assertion.
(b) Both assertion and reason are true but reason is not the correct explanation of assertion.
(c) Assertion is true but reason is false.
(d) Both assertion and reason are false.

ANSWER KEY

1. (d) 2. (c) 3. (a) 4. (d) 5. (c) 6. (d) 7. (a) 8. (a) 9. (c) 10. (b)
11. (b) 12. (c) 13. (b) 14. (c) 15. (b) 16. (b) 17. (c) 18. (d) 19. (a) 20. (a)
21. (c) 22. (d) 23. (b) 24. (d) 25. (b) 26. (a) 27. (a) 28. (b) 29. (d) 30. (a)
31. (c) 32. (d) 33. (d) 34. (c) 35. (d) 36. (d) 37. (a) 38. (b) 39. (d) 40. (c)
41. (c) 42. (c) 43. (d) 44. (d) 45. (a) 46. (a) 47. (d) 48. (a) 49. (d) 50. (b)
51. (c) 52. (a) 53. (d) 54. (a) 55. (b) 56. (b) 57. (d) 58. (d) 59. (a) 60. (c)
61. (c) 62. (d) 63. (b) 64. (d) 65. (b) 66. (a) 67. (c) 68. (b) 69. (d) 70. (b)
71. (d) 72. (b) 73. (a) 74. (b) 75. (d) 76. (c) 77. (b) 78. (b) 79. (b) 80. (d)
81. (b) 82. (c) 83. (b) 84. (d) 85. (a) 86. (b) 87. (c) 88. (a) 89. (b) 90. (d)
91. (c) 92. (b) 93. (b) 94. (b) 95. (c) 96. (d) 97. (a) 98. (d) 99. (b) 100. (c)
101. (b) 102. (d) 103. (d) 104. (d) 105. (c) 106. (b) 107. (c) 108. (a) 109. (d) 110. (d)
111. (a) 112. (d) 113. (a) 114. (d) 115. (b) 116. (c) 117. (c) 118. (a) 119. (d) 120. (c)
121. (d) 122. (b) 123. (a) 124. (c) 125. (d) 126. (b) 127. (c) 128. (d) 129. (d) 130. (b)
131. (b) 132. (a) 133. (a) 134. (a) 135. (c) 136. (b) 137. (c) 138. (a) 139. (c) 140. (b)
141. (a) 142. (d) 143. (d) 144. (b) 145. (c) 146. (c) 147. (d) 148. (a) 149. (c) 150. (b)
151. (a) 152. (c) 153. (b) 154. (d) 155. (a) 156. (b) 157. (d) 158. (c) 159. (a) 160. (c)
161. (a) 162. (d) 163. (a) 164. (b) 165. (b) 166. (c) 167. (c) 168. (a) 169. (d) 170. (a)
171. (b) 172. (d) 173. (d) 174. (d) 175. (a) 176. (c) 177. (c) 178. (a) 179. (d) 180. (b)
181. (c) 182. (c) 183. (d) 184. (d) 185. (d) 186. (b) 187. (d) 188. (d) 189. (b) 189. (c)
191. (a)

16 Chapter

Strategies for Enhancement in Food Production

MULTIPLE CHOICE QUESTIONS

Topic 1 Animal Husbandry

1. Select the incorrect statement.
(a) Inbreeding is essential to evolve pure line in any animal.
(b) Inbreeding selects harmful recessive genes that reduce fertility and productivity.
(c) Inbreeding helps in accumulation of superior genes and elimination of undesirable genes.
(d) Inbreeding increases homozygosity.

2. Homozygous pure lines in cattle can be obtained by
(a) Mating of related individuals of the same breed.
(b) Mating of unrelated individuals of the same breed.
(c) Mating of individuals of different breed.
(d) Mating of individuals of different species.

3. The results of inbreeding are not always desirable because
(a) the selected superior bull produces progeny with improved traits
(b) crossing the related animals of the same breed produces pure lines
(c) the progeny exhibits increased the frequency of both undesirable and desirable genes
(d) it increases milk production in cows

4. Match the Column-I with Column-II.

	Column I		Column II
(a)	Outbreeding	(1)	Mating of animals of the same breed but no common ancestors for 4–6 generations.
(b)	Outcrossing	(2)	Mating of animals of different species
(c)	Crossbreeding	(3)	Includes outcrossing and cross-breeding.
(d)	Interspecific hybridization	(4)	Mating of superior males and females of different breeds.

Select the correct option

	(I)	(II)	(III)	(IV)
(a)	1	4	2	3
(b)	4	1	2	3
(c)	4	2	1	3
(d)	3	1	4	2

5. Most of the mating done by animal breeders are outcrossing because
(a) it reduces the expression of harmful genes by masking them in heterozygous genotype
(b) it helps in bringing the desirable traits into the progeny
(c) it increases homozygosity
(d) it produces pure lines Select the correct option.
(a) a, b, c, d are true (b) a and b are true
(c) a and d are true (d) c and d are true

6. The offspring of crossbreeding is
(a) pure line

(b) hybrid
(c) homozygous genotype
(d) inbred lines

7. Hisardale is the breed of sheep developed by crossing
(a) Cochin ram and Ghagus ewe
(b) Bikaneri ewes and Marino rams
(c) Bikaneri rams and Marino ewes
(d) Cochin ewe and Marino ram

8. Mule is the hybrid produced by crossing
(a) male donkey and a female horse
(b) female donkey and a male horse
(c) male hinny and a female horse
(d) stallion and mare

9. Interspecific hybridization is the mating of
(a) superior males and females of different breeds
(b) more closely related individuals within the same breed for 4–6 generations
(c) animals within the same breed without having common ancestors
(d) two different related species

10. Interspecific hybridization between stallion and female donkey produces the hybrids called
(a) mule (b) hinny
(c) jack (d) jennet

11. The process of placing the sperms in the female reproductive tract by artificial means is known as
(a) artificial insemination
(b) interspecific hybridization
(c) asexual reproduction
(d) parthenogenesis

12. The process of artificial insemination is advantageous over normal mating. Select the incorrect statement about the process.
(a) It permits the fertilization of a large number of female animals from the semen collected in one ejaculation of a superior bull.
(b) Collected semen is cooled slowly and stored at −195.5 degrees Celsius for a longer period.
(c) It permits the easier use of exotic breed bulls as superior males.
(d) The collected semen should not be frozen as it kills sperms.

13. The technique of controlled breeding experiments that includes superovulation in cows to make them produce 6 8 eggs per ovarian cycle is known as
(a) artificial insemination
(b) hormonal induction
(c) multiple ovulation embryo transfer technology
(d) embryo transfer technology

14. MOET (Multiple Ovulation Embryo Transfer technology) includes the use of ______ to stimulate superovulation in cows.
(a) LH
(b) FSH and prostaglandins
(c) GnRH
(d) gonadotropins

15. Consider the following event:
(I) Superovulation in cows
(II) Fertilization of eggs
(III) Mating with elite bull
(IV) Transfer of eggs to surrogate mothers
(V) The second round of superovulation in genetic mother

Arrange the events of MOET in the correct order and select the correct option.
(a) I, II, IV, III, V (b) I, III, II, IV, V
(c) I, III, IV, II, V (d) II, III, I, IV, V

16. Practical applications of biological and biotechnological principles include
(a) animal husbandry and plant breeding to increase the production of food products
(b) plant breeding to increase dairy production
(c) tissue culture techniques
(d) Both (a) and (c) are correct

17. Selective breeding of livestock is known as

(a) animal husbandry (b) plant breeding
(c) poultry farming (d) fisheries

18. Consider the following statements:
(a) The practices of animal husbandry include raising and breeding the livestock, fisheries and poultry farming.
(b) More than 70% of the world's livestock population is in India and China.

Select the correct option.
(a) Both (a) and (b) are true.
(b) (a) is true but (b) is false.
(c) Both (a) and (b) are false.
(d) (a) is false but (b) is true.

19. Which of the following sets of organisms does not represent livestock?
(a) Cows, pigs, horses, and fishes
(b) Sheep, pigs, camels, and fishes
(c) Cows, pigs, camel, and goats
(d) Poultry, fishes, and elephants

20. Fisheries include rearing, catching and selling of
(a) fishes, mollusks, and crustaceans
(b) fishes only
(c) fishes and shell-fish only
(d) fishes and crustaceans only

21. Raising and breeding of animals for milk and milk products are known as
(a) fisheries (b) dairying
(c) poultry farming (d) plant breeding

22. Which of the following set of animals is not expected to be found in dairy farms?
(a) Cattle, buffalo, goat, and sheep
(b) Camel, buffalo, goat, and sheep
(c) Cattle, buffalo, goat, and cattle
(d) Poultry, cattle, goat, buffalo

23. Which of the following set of products is not obtained from a dairy farm?
(a) Milk, butter, cheese, yogurt
(b) Milk, condensed milk, cheese, yogurt
(c) Butter, egg, cheese, yogurt
(d) Ice cream, yogurt, milk, cheese

24. Milk production at dairy farms is dependent on the quality of breeds. Which of the following is not a criterion for the selection of dairying breeds?
(a) High yielding potential under the local climatic conditions
(b) Disease resistance
(c) Color of purebred for registering purpose and dairy type
(d) Expression of foreign genes

25. Which of the following measures are taken to realize the yield potential of cattle breeds at dairy farms?
(a) Proper housing, adequate water supply
(b) Cleanliness and hygiene of both cattle and handler
(c) Feeding cattle in a scientific manner
(d) All of these

26. Consider the following statements:
(a) Animal husbandry refers to the domesticated birds used for food and/or eggs.
(b) Chicken, ducks, turkey, geese are some examples of poultry.

Select the correct option.
(a) Both (a) and (b) are true.
(b) (a) is true but (b) is false.
(c) Both (a) and (b) are false.
(d) (a) is false but (b) is true.

27. Match the Column-I with Column-II.

	Column I		Column II
(I)	Raised for egg production	(1)	Turkey, geese and ducks
(II)	Raised for meat	(2)	Ayrshire, Guernsey, Holstein-Friesian and Jersey
(III)	Cattle breed with high milk production	(3)	Important factors for success with poultry
(IV)	Proper feeding, good management and sanitation	(4)	Hens

Select the correct option

	(i)	(ii)	(iii)	(iv)
(a)	1	4	2	3
(b)	4	1	2	3
(c)	4	2	1	3
(d)	3	1	4	2

28. The causative agent of Avian Influenza is
(a) H5N1 virus (b) HIV
(c) E. coli (d) Clostridium

29. Which of the following measures is/are required to prevent the spread of H5N1 virus from birds to a human?
(a) Consumption of poultry and eggs above the temperature of 100°C
(b) Influenza vaccination
(c) Maintain personal hygiene
(d) All of thesc

30. Leghorn, Minorca and Andalusia are the small-sized breeds of chickens. The most population breed for egg production is
(a) Leghorn (b) Minorca
(c) Andalusia (d) None

31. Consider the following statements:
(a) A breed is a group of related animals that are true to the genetic traits characteristic of the breed.
(b) Animals of the same breed do not share a common ancestor. Select the correct option.
(a) Both (a) and (b) are true.
(b) (a) is true but (b) is false.
(c) Both (a) and (b) are false.
(d) (a) is false but (b) is true.

32. Use of selective mating for the production of breeds of domesticated animals with desired traits is known as
(a) plant breeding (b) animal breeding
(c) poultry (d) farming

33. Consider the following statements about animal breeding. Which of the given statements is incorrect?
(a) Animal breeding aims to improve the growth rate and production of useful products from the animals.
(b) It includes the production of improved breeds of domesticated animals to obtain milk and meat of superior quality.
(c) It does not aim to improve disease resistance in animals.
(d) Methods of animal breeding are based on selective breeding.

34. Match the terms in Column-I with a suitable description in Column-II.

	Column I		Column II
(I)	Inbreeding	(1)	Overcomes inbreeding depression
(II)	Outbreeding	(2)	Increased homozygosity
(III)	Inbreeding depression	(3)	Crossing the different breeds.
(IV)	Outcrossing	(4)	Reduced productivity due to inbreeding

Select the correct option

	(i)	(ii)	(iii)	(iv)
(a)	4	1	2	3
(b)	2	3	4	1
(c)	1	2	4	3
(d)	2	3	4	1

35. The straight-breeding technique of crossing the related animals to increase the genetic purity and homozygosity of progeny is
(a) outbreeding (b) inbreeding
(c) outcrossing (d) crossbreeding

36. The scientific name of the Indian honey bee is
(a) Apis indica (b) Apis Indica
(c) Apis Indiana (d) Apis Indica

37. Maintenance of hives of honeybees for honey production is called
(a) bee-keeping (b) apiculture
(c) bee-breeding (d) both (a) and (b)

38. Which of the following does not represent the importance of apiculture?
(a) Obtain nutritious hone

(b) Provides bee wax
(c) Honey bees are pollinating agents
(d) All are the importance of apiculture

39. Which of the following sets of industries use the products obtained from apiculture?
(a) Indigenous system of medicines, cosmetics and polishes
(b) Allopathic medicines, cosmetics
(c) Medicines, leather production
(d) The food industry, dairy products

40. Honey bees are pollinator of
(a) corn, sunflower, apple and oats
(b) barley, corn, apple and sunflower
(c) sunflower, Brassica, apple and pear
(d) wheat, rye, apple and pear

41. Which of the following species are most commonly domesticated to obtain honey?
(a) Apis dorsata and Apis indica
(b) Apis dorsata and Apis mellifera
(c) Apis florea and Apis indica
(d) Apis indica and Apis mellifera

42. Match Column-I with Column-II

	Column I		Column II
(a)	Edible freshwater fishes	(1)	Prawn, lobster, oyster
(b)	Edible marine fishes	(2)	Catla, Rohu and common carp
(c)	Seafood	(3)	Improved production of useful products from aquaculture
(d)	Blue revolution	(4)	Hilsa, Sardines, Mackerel and Pomfrets

Select the correct option.

	A	B	C	D
(a)	2	4	1	3
(b)	1	3	4	2
(c)	4	2	1	3
(d)	3	1	4	2

43. Among the following edible fishes, which one is a marine fish having a rich source of omega-3 fatty acids?
(a) Mrigala (b) Mackerel
(c) Mystus (d) Mangur

44. Fisheries have an important place in Indian economy because
(a) it provides food to the population
(b) it serves as the only source of livelihood in many coastal regions
(c) it obtains fish oil, pearls, fish protein, etc.
(d) all the given options are correct.

45. Select the incorrect match from the following
(a) Pisciculture: fish farming
(b) Aquaculture: raising aquatic animals to obtain useful products
(c) Fishes: a rich source of vitamin D, riboflavin, omega-3 fatty acid and minerals
(d) Honey: a rich source of sugars, fats, and fibers

46. Father of Blue revolution in India is
(a) Dr. Arun Krishnan (b) Nirpakh Tutej
(c) Vishal Shekhar (d) Durgesh Patel

47. Assertion: Bird flu is a viral disease and is caused by the H5N1 virus.

Reason: Bird flu is transmitted from affected birds to humans through direct contact or consumption of their eggs.
(a) Both assertion and reason are true but reason is the correct explanation of assertion.
(b) Both assertion and reason are true but reason is not the correct explanation of assertion.
(c) Assertion is true but reason is false.
(d) Both assertion and reason are false.

48. Assertion: Inbreeding is required to obtain pure line in any animal.

Reason: Mendel obtained pure line of pea plants by cross-pollination.
(a) Both assertion and reason are true but reason is the correct explanation of assertion.
(b) Both assertion and reason are true but reason

is not the correct explanation of assertion.
(c) Assertion is true but reason is false.
(d) Both assertion and reason are false.

49. Assertion: Controlled breeding experiments are done using interspecific hybridization.

Reason: Outcrossing increases homozygosity in the progeny.
(a) Both assertion and reason are true but reason is the correct explanation of assertion.
(b) Both assertion and reason are true but reason is not the correct explanation of assertion.
(c) Assertion is true but reason is false.
(d) Both assertion and reason are false.

50. Assertion: A group of bees is called swarm.

Reason: Honey bees are pollinators of many crop plants.
(a) Both assertion and reason are true but reason is the correct explanation of assertion.
(b) Both assertion and reason are true but reason is not the correct explanation of assertion.
(c) Assertion is true but reason is false.
(d) Both assertion and reason are false.

51. A tremendous increase in crop and food production as an outcome of the application of plant breeding and production technology is known as
(a) Green revolution (b) White revolution
(c) Blue revolution (d) Grey revolution

52. Father of the green revolution in India is
(a) Verghese Kurien (b) Vikram Sarabhai
(c) MS Swaminathan (d) Homi J Bhabha

53. Consider the following statements about plant breeding.
(a) It is the deliberate manipulation of plant genome to create or impart the desired traits in the plants.
(b) It aims to obtain plant types with better productivity and disease resistance.

Select the correct option.
(a) Both (a) and (b) are true.
(b) (a) is true but (b) is false.
(c) Both (a) and (b) are false.
(d) (a) is false but (b) is true.

54. A true breeding plant is
(a) near homozygous and produces offspring of its own kind
(b) always homozygous recessive in its genetic constitution
(c) one that is able to breed on its own
(d) produced due to cross-pollination among unrelated plants

55. Classical breeding approach uses the proven tools of
(a) hybridization of pure lines and artificial selection of desired genotypes.
(b) hybridization of pure lines and genome manipulation of selected progeny.
(c) incorporation of desired genes and artificial selection of progeny.
(d) genome manipulation only.

56. Which of the following is not an objective of plant breeding?
(a) To improve crop productivity and quality.
(b) To impart stress and pathogen resistance in crop plants.
(c) To increase tolerance of crop plants for insect pests.
(d) All are the objectives of plant breeding.

57. Which of the following set of factor cause environmental stress in plants?
(a) Pathogens, drought and flood
(b) Salinity, extreme temperatures and drought
(c) Parasites, extreme temperatures and drought
(d) Parasites, pathogens and flood

58. Consider the following steps in plant breeding:
(I) Testing, release and commercialization of new cultivars
(II) Collection of variability
(III) Selection and testing of superior recombinants
(IV) Cross hybridization among the selected parents

(V) Evaluation and selection of parents Arrange the steps in correct order and selection the correct option.
(a) I, V, IV, II, III (b) II, V, III, IV, I
(c) II, V, IV, III, I (d) II, IV, V, III, I

59. In the plant breeding programs, the entire collection (plants/seeds) having all the diverse alleles for all genes of a given crop is called
(a) Germplasm collection
(b) Selection of superior recombinants
(c) Cross-hybridization among the selected parents
(d) Evaluation and selection of parents

60. Sum total of all the hereditary material belonging to single species is known as
(a) genotype (b) germplasm
(c) hybrid (d) cultivar

61. Consider the following statements about germplasm collection:
(a) The gene of interest should be present in the base population to initiate a breeding program.
(b) Genetic variability is a prerequisite to develop a new cultivar by breeding programs.

Select the correct option.
(a) Both (a) and (b) are true.
(b) (a) is true but (b) is false.
(c) Both (a) and (b)are false.
(d) (a) is false but (b) is true.

62. The selected superior recombinants in plant breeding program are selfpollinated for several generations so as to
(a) increase the homozygosity to prevent segregation of the desired trait in the progeny.
(b) increase the heterozygosity to prevent segregation of the desired trait in the progeny.
(c) increase the homozygosity to allow segregation of the desired trait in the progeny.
(d) increase the heterozygosity to allow segregation of the desired trait in the progeny.

63. The new cultivars produced by plant breeding programs are evaluated for
(a) yield
(b) morphological and quality traits
(c) resistance to diseases and stress
(d) all the given choices are correct

64. Around _____ of the Indian population is employed in agricultural activities which in turn accounts for _____ of the country's GDP.
(a) 62%, 33% (b) 33%, 62%
(c) 32%, 63% (d) 30%, 62%

65. Which of the following factors were responsible for limited agricultural production after the independence of India?
(a) Limited land for agriculture and scarce resources
(b) Seasonal rainfall in deserts
(c) Lower temperature conditions in Northern plains
(d) A small fraction of the population involved in agricultural activities

66. The key strategies targeted by Dr. Norman E. Borlaug that resulted in the Green Revolution in the world were
(a) development of sugarcane cultivars with insect pest resistance
(b) development of high yielding wheat cultivars with desired agronomic traits to realize the maximum productivity
(c) development of maize cultivars with disease resistance
(d) all the given options are correct

67. Which of the following set of the traits correctly represent the features of semidwarf varieties developed by Dr. Norman E. Borlaug?
(a) Better crop production and lodging resistance
(b) Adapted to local climatic conditions and

lodging resistance

(c) High yielding, adapted to local climatic conditions, lodging resistance

(d) Lodging resistance and better crop yield

68. Nobel laureate Norman E. Borlaug was the director of Wheat Program at_____ and developed semidwarf varieties of wheat.

(a) Center for Plant Breeding and Genetics

(b) Indian Society of Genetics and Plant Breeding

(c) Centro Internationale de Mejoramiento de Maiz y Trigo

(d) International Centre for Plant Breeding Education and Research

69. Match Column-I with Column-II.

	Column I		Column II
(a)	Kalyan Sona, Sonalika	(1)	Pearl millet
(b)	Jamnagar Giant and Improved Ghana	(2)	Wheat
(c)	Pusa Lal and Pusa Sunhari	(3)	Tomato
(d)	Pusa Ruby	(4)	Sweet potato

Select the correct option.

	A	B	C	D
(a)	1	4	2	3
(b)	2	1	4	3
(c)	4	2	1	3
(d)	3	1	4	2

70. Which of the following crop cultivars is incorrect matched?

(a) Sonora 64 and Lerma Rojo: Wheat

(b) TN 1, IR8, IR 28: Rice

(c) P 1542, Rachna: Linseed

(d) C251, K12: Barley

71. Which of the following rice cultivar is incorrectly matched with its land of origin?

(a) IR 8: International Rice Research Institute (IRRI), Philippines

(b) Taichung Native-1: Taiwan

(c) Jaya: India

(d) Ratna: Mexico

72. The _____ and _____ were the rust resistant high yielding varieties of wheat introduced in India from Mexico.

(a) Kalyan Sona and Sonalika

(b) TN-I and Sonalika

(c) IR-8 and Kalyan Sona

(d) IR-8 and TN-1

73.

(a) The semi dwarf wheat and rice varieties that made India selfsufficient in food grains were lodging resistant, responsive to the application of fertilizer and high yielding.

(b) The rice varieties were resistant to all three rusts and other prevalent diseases.

Select the correct option.

(a) Both (a) and (b) are true.

(b) (a) is true but (b) is false.

(c) Both (a) and (b) are false.

(d) (a) is false but (b) is true.

74. The photoperiod insensitive wheat and rice varieties are beneficial because

(a) they are disease resistant.

(b) It allows the late planting of the crop.

(c) these varieties could be grown in nontraditional regions.

(d) both (b) and (c)

75. The scientific name of Indian canes is

(a) Saccharum officinarum

(b) Saccharum spontaneum

(c) Saccharum munja

(d) Saccharum barberi

76. Match the Column-I with Column-II.

	Column I		Column II
(a)	Saccharum barberi	(1)	South Indian cane with thicker stems and higher sugar content
(b)	Saccharum officinarum	(2)	High yield, thick stems, high sugar and adapted to grow in North India

(c)	Noblized canes	(3)	Resistant to water stress
(d)	Hybrid millets	(4)	North Indian cane with poor sugar content and yield

Select the correct option

	A	B	C	D
(a)	4	1	2	3
(b)	2	1	4	3
(c)	2	4	1	3
(d)	2	3	4	2

77. The objective/s of development of disease resistance in crop plants is/are
(a) to reduce or prevent the invasion, growth, and development of pathogen
(b) to reduce dependence on the use of fungicides and bacteriocides
(c) to realize the maximum crop production
(d) all the given choices are correct

78. Which of the following sets of plant diseases include all fungal diseases?
(a) Turnip mosaic, black rot of crucifers and brown rust of wheat
(b) Black rot of crucifers, brown rust of wheat and red rot of sugarcane
(c) Brown rust of wheat, red rot of sugarcane and late blight of potato
(d) Tobacco mosaic, black rot of crucifers and brown rust of wheat

79. Which of the following statement correctly differentiates conventional breeding techniques for the disease resistance in plants from the mutational breeding?
(a) Mutational breeding screens germplasm for the source of disease resistance genes.
(b) Conventional breeding includes the introduction of disease resistance genes in plants by induced mutations.
(c) Mutational breeding induces mutations in plants to introduce disease resistance in them.
(d) Mutational breeding cannot be applied to crop plants.

80. Breeding for disease resistance in crop plants is carried out by conventional techniques or by mutational breeding. Which of the following crop was not bred by conventional techniques for disease resistance?
(a) Resistance to white rust in Brassica
(b) Resistance to bacterial blight in cowpea
(c) Resistance to hill bunt in wheat
(d) Resistance to powdery mildew in mung bean

81. Which of the following is a wheat variety bred by conventional breeding techniques to develop resistance to leaf and stripe rust in them?
(a) Himgiri (b) Pusa swarnim
(c) Pusa Shubhra (d) Pusa Snowball K-1

82. Pusa komal variety of cowpea is resistant to ____ while Pusa sadabahar variety of chilly is resistant to ____
(a) Bacterial blight and Leaf curl
(b) White rust and Tobacco mosaic virus
(c) Black rot and Chilly mosaic virus
(d) Bacterial blight and hill bunt

83. Plant breeding for disease resistance by conventional techniques has limited success due to
(a) Time consuming screening of germplasm
(b) Limited availability of disease resistance genes the collected germplasm
(c) The tedious process of evaluation of developed cultivars
(d) Limited knowledge about the pathology of plant disease

84. When a source of disease resistance gene is not available or not known, ____ is followed to produce disease resistant mutants plants.
(a) Conventional breeding
(b) Mutagenesis
(c) Plant breeding
(d) Germplasm screening

85. Consider the following statements:

(a) Disease resistant somaclonal variants may serve as a source of disease resistance for plant breeding.
(b) Recombinant DNA technology develops the disease resistant transgenic crop plants by transferring the disease resistance gene in crops from any other source.

Select the correct option.
(a) Both (a) and (b) are true.
(b) (a) is true but (b) is false.
(c) Both (a) and (b)are false.
(d) (a) is false but (b) is true.

86. Which of the following is a yellow mosaic virus resistant variety of Abelmoschus esculentus?
(a) TN-1 (b) Prabhani kranti
(c) Himgiri (d) Pusa komal

87. Which of the following set of examples represent insect resistance due to morphological features?
(a) Resistance to jassids in cotton and cereal leaf beetle in wheat
(b) Stem borer resistance in maize
(c) Rust resistance in wheat
(d) Rot resistance in cauliflower

88. Cereals are the staple source of nutrition in human diet. Which of the following is a man made cereal?
(a) Triticum (b) Triticale
(c) Sorghum (d) Bajra

89. Select the incorrect statement about insect pest resistance in crop plants.
(a) Solid stems of wheat are not preferred by stem sawflies.
(b) The presence of smooth leaves and no nectar makes the cotton varieties resistant to bollworms.
(c) High aspartic acid and low nitrogen in maize impart stem borer resistance.
(d) Maize varieties with high sugar content are resistant to maize stem borers.

90. Pusa gaurav is the ____ resistant variety of ____ plants bred by conventional hybridization techniques.
(a) Wheat, stem borer
(b) Jassids, cotton
(c) Aphids, rapeseed mustard
(d) Jassids, beans

91. Pusa sem 2 and Pusa Sem 3 varieties of the flat bean are resistant to
(a) bollworms and jassids
(b) stem sawfly and aphids
(c) leaf beetle and fruit borer
(d) jassids, aphids and fruit borer

92. Which of the following are the shoot and fruit borer resistant varieties of Okra?
(a) Pusa Gaurav (b) Pusa Sem 3
(c) Pusa Sem 2 (d) Pusa A-4

93. Which of the following set correctly represents the three major food crops that feed most of the world population?
(a) Maize, wheat, and rice
(b) Maize, jowar, and bajra
(c) Corn, soybean, and wheat
(d) Corn, soybean, and rice

94. Which of the given statements is incorrect about the nutritional quality of food crops?
(a) Cereals are generally low in protein content.
(b) Legumes tend to be low in tryptophan amino acid.
(c) Corn, wheat, and rice are low in lysine amino acids.
(d) Rice is a rich source of proteins and vitamin A.

95. Select the option that correctly represents some of the essential micronutrients required by the human body.
(a) Iron, nitrogen, oxygen and phosphorus
(b) Iron, vitamin A, iodine and zinc
(c) Iron, vitamin A, carbon and potassium
(d) Manganese, copper, nitrogen and carbon

96. Parents often complain about fatigue and weakness in their children despite the proper

intake of food. Which of the following could be a reason behind the same?
(a) Hidden hunger
(b) Over consumption of proteins
(c) Obesity
(d) Over consumption of carbohydrates

97. Which of the following micronutrients is correctly matched with its respective deficiency disorder?
(a) Vitamin A – anemia
(b Iron – night blindness
(c) Iodine – goiter
(d) Zinc – beriberi

98. Application of breeding or biotechnological processes to improve the nutrient levels of crop plants is known as
(a) biogeochemistry
(b) biofortification
(c) biomagnification
(d) plant breeding

99. Which of the following components determine the nutritional quality of food crops?
(a) Protein content and balance of amino acids
(b) Oil content and fatty acid composition
(c) Vitamin and mineral content
(d) All the given choices are correct

100. To improve the protein content of cultivated wheat, the high protein content gene from _____ was transferred into _____. This improved wheat variety exhibited higher protein content with no reduction in its yield.
(a) Atlas 56, Lancota (b) Atlas 66, Lancota
(c) Lancota, Atlas 66 (d) Lancota, Atlas 56

101. Which of the following food/vegetable crop is incorrectly matched with the nutrients for which they were bred?
(a) Maize: Lysine and tryptophan
(b) Carrots, spinach, pumpkin: Vitamin A
(c) Bitter gourd, bathua, mustard, tomato: Vitamin C
(d) Spinach and bathua: Lysine and phenylalanine

102. Assertion: Limited land availability for agriculture was the major reason for food production in India before the green revolution.

Reason: High yielding and disease resistant varieties of cereal crops made India self-sufficient in food production.
(a) Both assertion and reason are true but reason is the correct explanation of assertion.
(b) Both assertion and reason are true but reason is not the correct explanation of assertion.
(c) Assertion is true but reason is false.
(d) Both assertion and reason are false.

103. Assertion: The presence of genetic variability is a prerequisite for plant breeding techniques.

Reason: Conventional breeding use existing genes for desired traits as parents for hybridization.
(a) Both assertion and reason are true but reason is the correct explanation of assertion.
(b) Both assertion and reason are true but reason is not the correct explanation of assertion.
(c) Assertion is true but reason is false.
(d) Both assertion and reason are false.

104. Assertion: Saccharum barberi was the south Indian sugarcane with higher sugar content and yield.

Reason: Saccharum officinarum was grown in north India and had thicker stems but poor sugar content.
(a) Both assertion and reason are true but reason is the correct explanation of assertion.
(b) Both assertion and reason are true but reason is not the correct explanation of assertion.
(c) Assertion is true but reason is false.
(d) Both assertion and reason are false.

105. Assertion: The random changes in the genome of living beings are called mutations.

Reason: Mutations introduce new genes/alleles and add variations.
(a) Both assertion and reason are true but reason is the correct explanation of assertion.
(b) Both assertion and reason are true but reason is not the correct explanation of assertion.

(c) Assertion is true but reason is false.
(d) Both assertion and reason are false.

106. Assertion: Mutation breeding uses artificial mutations to obtain the plants with desired genetic traits.

Reason: Yellow mosaic virus resistance variety of mung bean was developed by mutation breeding.
(a) Both assertion and reason are true but reason is the correct explanation of assertion.
(b) Both assertion and reason are true but reason is not the correct explanation of assertion.
(c) Assertion is true but reason is false.
(d) Both assertion and reason are false.

107. Assertion: Cereals are a poor source of carbohydrates.

Reason: Legumes are rich in tryptophan amino acid.
(a) Both assertion and reason are true but reason is the correct explanation of assertion.
(b) Both assertion and reason are true but reason is not the correct explanation of assertion.
(c) Assertion is true but reason is false.
(d) Both assertion and reason are false.

Topic 2 Single Cell Protein

108. Which of the following microorganisms serve in the production of single-cell protein?
(a) Bacteria (b) Yeast
(c) Algae (d) All of these

109. A bulk of dead and dry cell microbes that possess high levels of proteins and is grown on varieties of carbon sources is known as
(a) hyphae (b) single cell protein
(c) colony (d) microbial mount

110. Consider the following statements:
(a) Single cell proteins are rich sources of essential amino acids such as lysine and tryptophan which are scarce in plant and animal proteins.
(b) Around 60%–82% of dry cell weight of single cell protein is protein.

Select the correct option.
(a) Both (a) and (b) are true.
(b) (a) is true but (b) is false.
(c) Both (a) and (b) are false.
(d) (a) is false but (b) is true.

111. Consider the following statements.
(a) A shift from grain to meat diet creates more demand for grains.
(b) Only 10% energy of one trophic level is available for the next trophic level.

Select the correct option.
(a) Both (a) and (b) are true.
(b) (a) is true but (b) is false.
(c) Both (a) and (b) are false.
(d) (a) is false but (b) is true.

112. Single cell proteins as an alternative to human food sources is an environmentfriendly approach because
(a) microbes are a good source of protein
(b) microbes have higher reproduction rates
(c) microbes are grown on the organic waste matter
(d) conventional breeding programs cannot meet the demand for food

113. Single cell proteins are being produced using waste organic matter. Which of the following set correctly represents the organic waste materials used for the purpose?
(a) Wastewater from potato processing plants, straw, molasses
(b) Animal manure, sewage, antibiotics
(c) Sewage, industrial waste, waste from nuclear reactors
(d) Hydrocarbon, straw, volcanic eruptions

114. Consider the following statements.
(a) More than 50% of the human population is suffering from hunger and malnutrition.

(b) Single cell protein is also known as microbial protein or bio protein.

Select the correct option.
(a) Both (a) and (b) are true.
(b) (a) is true but (b) is false.
(c) Both (a) and (b) are false.
(d) (a) is false but (b) is true.

115. Match Column-I with Column-II.

	Column I		Column II
(a)	Cucumber and orange peel	(1)	Bacterial
(b)	Methanomonas	(2)	Algae
(c)	Spirulina	(3)	Fungus
(d)	Aspergillus	(4)	Production of single cell proteins

Select the correct option.

	A	B	C	D
(a)	1	4	2	3
(b)	4	1	2	3
(c)	4	2	1	3
(d)	3	1	4	2

116. Which of the following represents the production of single cell proteins?
(a) Production of Saccharomyces cerevisiae from molasses in World War I.
(b) Production of Torula yeast on sulfite liquor waste from paper production plants during World War II.
(c) Growing cell biomass of Saccharomyces cerevisiae on fruit peels.
(d) All the given choices are correct.

117. 250 g of Methylophilus methylotrophus can obtain as much as _____ of proteins as compared to 250 kg cow that produces only 200 g of proteins.
(a) 250 tonnes (b) 25 tonnes
(c) 2.5 tonnes (d) 12 tonnes

118. Select the incorrect match from the given examples of single cell protein microbes.
(a) Bacteria: Methanomonas, Pseudomonas, Bacillus
(b) Yeast: Candida utilis, S. cereviseae, Pichia pastoris
(c) Fungi: Fusarium, Aspergillus, Penicillium
(d) Algae: Spirullina, Chiarella, Hydrogenomonas

119. Assertion: Microbes have higher growth rates and produce more biomass in less time.

Reason: Some microbial species are rich sources of essential amino acids.
(a) Both assertion and reason are true but reason is the correct explanation of assertion.
(b) Both assertion and reason are true but reason is not the correct explanation of assertion.
(c) Assertion is true but reason is false.
(d) Both assertion and reason are false.

120. Assertion: Algae are autotrophs and produce organic matter by the process of photosynthesis.

Reason: Spirulina and Scenedesmus are the most commonly used bacterial sources of single cell proteins.
(a) Both assertion and reason are true but reason is the correct explanation of assertion.
(b) Both assertion and reason are true but reason is not the correct explanation of assertion.
(c) Assertion is true but reason is false.
(d) Both assertion and reason are false.

Topic 3 Tissue Culture

121. Growth of tissues or cells in an artificial medium under sterile conditions is known as
(a) Callus
(b) Tissue culture
(c) Somatic hybridization
(d) Somatic hybrid

122. The excised plant tissue or organ is grown in a test tube under aseptic conditions to generate whole plants is known as
(a) meristem (b) explant

(c) hybrids (d) stem cells

123. Which of the following plant parts serve as source of explant for tissue culture?
(a) Petal, leaves and flower buds
(b) Ovaries and anther
(c) Seeds and nodal segment
(d) All of these

124. Consider the following statement:
(a) A totipotent cell contains a complete set of genetic information to direct the development of an entire organism.
(b) A pluripotent cell is a stem cell that can produce many but not all the cell types in an organism.

Select the correct option.
(a) Both (a) and (b) are true.
(b) (a) is true but (b) is false.
(c) Both (a) and (b) are false.
(d) (a) is false but (b) is true.

125. Concept of totipotency was given by
(a) Morgan (b) Haberlandt
(c) MS Swaminathan (d) Norman Borlaug

126. Given below are the various steps of plant tissue culture. Arrange them in correct order and select the correct option.
(I) Preparation of instrument and nutrient culture medium
(II) Preparation of explant
(III) Sterilization of culture medium
(IV) Acclimatization of plantlets and transfer to pots
(V) Inoculation of explant and incubation for growth
(a) I, III, II, V, IV (b) II, I, III, V, IV
(c) I, II, III, V, IV (d) I, III, II, IV, V

127. Sterilization of tissue culture apparatus is done by
(a) autoclave only
(b) autoclave and washing with chromic acid and detergent
(c) autoclave and washing with detergent
(d) surface treatment with chromic acid

128. During the 1950s, _____ and _____ performed various experiments that led to the development of synthetic growth medium to stimulate growth and division in explants.
(a) Miller and Morgan
(b) Miller and Skoog
(c) Morgan and Mendel
(d) Hugo de Vries and Morgan

129. The basic requirements for tissue culture techniques are
(a) Aseptic conditions
(b) Synthetic growth medium
(c) Explant
(d) All of these

130. Consider the following statements about tissue culture.
(a) A tissue culture medium provides minerals and growth regulators to the growing cells.
(b) It serves as a source of organic compounds but does not provide physical support.

Select the correct option.
(a) Both (a) and (b) are true.
(b) (a) is true but (b) is false.
(c) Both (a) and (b) are false.
(d) (a) is false but (b) is true.

131. Which of the following growth regulators is incorrectly matched with its effect on the growing explant in a synthetic medium?
(a) Naphthalene acetic acid (NAA) and indole-3-butyric acid (IAA): Induce rooting
(b) 2, 4-diclorophenoxyacetic acid (2, 4-D): Induce rooting
(c) Kinetin: Induces shoot formation
(d) Higher auxin to cytokinin ratio: Promotes shoot formation

132. A callus is
(a) undifferentiated mass of cells formed on an explant.
(b) aggregation of totipotent cells that can be manipulated to develop into any plant part.

Select the correct option.
(a) Both (a) and (b) are true.
(b) (a) is true but (b) is false.
(c) Both (a) and (b) are false.
(d) (a) is false but (b) is true.

133. Based on their ability to give rise to new cell types, how would you classify zygote and spermatogonia in humans?
(a) Totipotent and pluripotent respectively
(b) Totipotent and unipotent respectively
(c) Unipotent and pluripotent respectively
(d) Pluripotent and pluripotent respectively

134. A synthetic growth medium should provide all the nutrients required for the development of a new plant. Select the nutrient category that is correctly matched with its representative.
(a) Carbon source: vitamins
(b) Inorganic nutrients: Sucrose
(c) Growth regulators: Minerals
(d) Salts: Sulfates

135. Match Column-I with Column-II.

	Column I		Column II
(a)	Micro-propagation	(1)	Apical and axillary
(b)	Somaclones	(2)	Protoplast fusion
(c)	Somatic hybrids	(3)	In vitro clonal propagation of plants
(d)	Meristem	(4)	Genetically identical plants produced by tissue culture

Select the correct option.

	A	B	C	D
(a)	4	1	2	3
(b)	2	3	4	2
(c)	3	4	2	1
(d)	3	1	4	2

136. Rapid clonal propagation of explant to obtain genetically identical plants is known as
(a) somatic hybridization
(b) micropropagation
(c) protoplasts
(d) meristem culture

137. Micropropagation is advantageous over sexual reproduction in orchids as
(a) It is a rapid process and reduces the dependency on seeds for reproduction.
(b) It maintains the desirable genetic traits present in the parent plant.

Select the correct option.
(a) Both (a) and (b) are true.
(b) (a) is true but (b) is false.
(c) Both (a) and (b) are false.
(d) (a) is false but (b) is true.

138. Production of virus-free plants from a virus-infected plant is done by meristem culture because
(a) meristem culture is a technique of rapid clonal propagation.
(b) some of the progeny from the meristem culture may be virus-free.
(c) it produces a large number of plants from a small explant.
(d) meristems are virus-free plant tissues.

139. Protoplast is
(a) a plant cell without a cell wall
(b) a plant cell without a cell membrane
(c) a plant cell undergoing division
(d) a plant cell without a nucleus

140. Which of the following options represents the correct sequence of steps in somatic hybridization?
(a) Isolation of plant cells → Fusion of protoplasts from different plant varieties → Production of somatic hybrids → Digestion of cell wall.
(b) Isolation of plant cells → Digestion of cell wall → Fusion of protoplasts from different plant varieties → Production of somatic hybrids.
(c) Isolation of plant cells → Fusion of

protoplasts from different plant varieties → Digestion of cell wall → Production of somatic hybrids.

(d) Isolation of plant cells → Fusion of protoplasts from different plant varieties → Production of somatic hybrids.

141. A technique of micropropagation is
(a) somatic embryogenesis
(b) protoplast fusion
(c) embryo rescue
(d) somatic hybridization

142. Which of the following enhances or induces the fusion of protoplasts?
(a) IAA and kinetin
(b) IAA and gibberellins
(c) Sodium chloride and potassium chloride
(d) Polyethylene glycol and sodium nitrate

143. Select the mismatch from the given options.
(a) Tissue culture: Jaya and Ratna
(b) Somatic hybridization: Pomato
(c) Micropropagation: Tomato, banana, apple
(d) Meristem culture: Banana, sugarcane, potato

144. Assertion: Totipotency is the ability of explants to give rise while plant.

Reason: The cells of explants contain a complete set of genetic information.

(a) Both assertion and reason are true but reason is the correct explanation of assertion.
(b) Both assertion and reason are true but reason is not the correct explanation of assertion.
(c) Assertion is true but reason is false.
(d) Both assertion and reason are false.

145. Assertion: Meristems are the localized regions of active cell division in a plant body.

Reason: Somaclones are genetically identical plants.

(a) Both assertion and reason are true but reason is the correct explanation of assertion.
(b) Both assertion and reason are true but reason is not the correct explanation of assertion.
(c) Assertion is true but reason is false.
(d) Both assertion and reason are false.

146. Assertion: Pomato is an intergeneric somatic hybrid.

Reason: Cybrids are the somatic hybrids with the nuclear genome from both the parent plants.

(a) Both assertion and reason are true but reason is the correct explanation of assertion.
(b) Both assertion and reason are true but reason is not the correct explanation of assertion.
(c) Assertion is true but reason is false.
(d) Both assertion and reason are false.

ANSWER KEY

1. (b)	2. (a)	3. (c)	4. (d)	5. (b)	6. (b)	7. (b)	8. (a)	9. (d)	10. (b)
11. (a)	12. (d)	13. (c)	14. (b)	15. (b)	16. (d)	17. (a)	18. (a)	19. (d)	20. (a)
21. (b)	22. (d)	23. (c)	24. (d)	25. (d)	26. (d)	27. (b)	28. (a)	29. (d)	30. (a)
31. (b)	32. (b)	33. (c)	34. (b)	35. (b)	36. (a)	37. (d)	38. (d)	39. (a)	40. (c)
41. (d)	42. (a)	43. (b)	44. (d)	45. (d)	46. (a)	47. (b)	48. (c)	49. (d)	50. (b)
51. (a)	52. (c)	53. (a)	54. (a)	55. (a)	56. (d)	57. (b)	58. (c)	59. (a)	60. (b)
61. (a)	62. (a)	63. (d)	64. (a)	65. (a)	66. (b)	67. (c)	68. (c)	69. (b)	70. (c)
71. (d)	72. (a)	73. (b)	74. (d)	75. (d)	76. (a)	77. (d)	78. (c)	79. (c)	80. (d)
81. (a)	82. (a)	83. (b)	84. (b)	85. (a)	86. (b)	87. (a)	88. (b)	89. (d)	90. (c)
91. (d)	92. (d)	93. (a)	94. (d)	95. (b)	96. (a)	97. (c)	98. (b)	99. (d)	100. (a)
101. (d)	102. (b)	103. (a)	104. (d)	105. (a)	106. (b)	107. (d)	108. (d)	109. (b)	110. (a)
111. (a)	112. (c)	113. (a)	114. (d)	115. (b)	116. (d)	117. (b)	118. (d)	119. (b)	120. (c)
121. (b)	122. (b)	123. (d)	124. (a)	125. (b)	126. (a)	127. (b)	128. (b)	129. (d)	130. (b)
131. (d)	132. (a)	133. (b)	134. (d)	135. (c)	136. (b)	137. (a)	138. (d)	139. (a)	140. (b)
141. (d)	142. (d)	143. (a)	144. (a)	145. (b)	146. (d)				

17 Chapter

Biotechnology: Principles & Processes

MULTIPLE CHOICE QUESTIONS

1. EFB refers to
   (a) English Federation of Biology
   (b) European federation of Biology
   (c) English Federation of Biotechnology
   (d) European federation of Biotechnology Biosphere

Topic 1 Principles of Biotechnology

2. Plasmid is-
   (a) autonomously replicating, extra chromosomal
   (b) non- autonomously replicating extra chromosomal
   (c) autonomously replicating chromosomal
   (d) non-autonomously replicating extrachromosomal

3. Plasmid is
   (a) Linear RNA (b) Circular RNA
   (c) Linear DNA (d) Circular DNA

4. First recombinant DNA involved native plasmid of
   (a) Escherichia coli
   (b) Salmonella typhimurium
   (c) Streptococcus pneumonia
   (d) Clostridium butylicom

5. First recombinant DNA was made by
   (a) Herbert Cohen & Stanley Boyer, 1972
   (b) Stanley Cohen & Herbert Boyer, 1992
   (c) Stanley Cohen & Herbert Boyer, 1972
   (d) Herbert Cohen & Stanley Boyer, 1992

6. The recombinant DNA was made
   (a) before discovery of DNA cutting restriction enzymes
   (b) after discovery of DNA cutting restriction enzymes
   (c) after discovery of DNA cutting Ligases
   (d) before discovery of DNA cutting Ligases

7. The plasmid DNA linked with cut piece of DNA acts as
   (a) host
   (b) vector
   (c) medium to transfer the DNA piece
   (d) more than one option

8. Linking of antibiotic resistance gene with plasmid is done using enzyme
   (a) Ligase (b) Lyase
   (c) Hydrolase (d) Nuclease

9. The plasmid joined with required DNA of interest is transferred into........ by Boyer.
   (a) Escherichia coli
   (b) Salmonella typhimurium
   (c) Streptococcus pneumonia
   (d) Clostridium butylicom

10. Two core techniques that enabled birth of modern biotechnology are
    (a) Physical & biological engineering
    (b) Bioprocess & genetic engineering
    (c) Molecular & cellular genetics
    (d) None of these

11. Biotechnology uses techniques to alter chemistry of
    (a) Protein & Lipid (b) Protein & RNA
    (c) Lipid & DNA (d) RNA & DNA

12. In chemical engineering processes, it is important to maintain
(a) maintain microbe-free environment
(b) microbe-full environment
(c) sterile environment
(d) more than one option

13. Unique combinations of genetic setup is naturally provided by
(a) Sexual reproduction
(b) Asexual reproduction
(c) Biotechnology
(d) More than one option

14. All genetic changes occurring naturally are
(a) harmful to organism & its population
(b) beneficial for organism & its population
(c) not harmful for organism & its population
(d) Both (a) & (c)

15. Genetic information is preserved by
(a) sexual reproduction (b) asexual reproduction
(c) Both of these (d) none of these

16. When a piece of DNA is transferred to an alien organism as it is
(a) it will multiply itself
(b) it will not be able to multiply itself
(c) it will be present in progeny cells of organism.
(d) Both (a) & (c)

17. Chromosome replication is initiated at
(a) gateway of replication a specific RNA sequence
(b) origin of replication a specific RNA sequence
(c) path of replication a specific RNA sequence
(d) None of these

18. For alien DNA to replicate it needs to be a part of
(a) chromosome without origin of replication site
(b) mitochondrial DNA with origin of replication site
(c) chromosome with origin of replication site
(d) cytoplasmic DNA with origin of replication site

Topic 2 Tools of Recombinant DNA Technology

19. The key tools for recombinant DNA technology are
(a) Restrication enzyme, polymerase, hydrolase, vectors
(b) Recognition enzyme, polymerase, ligase, vector
(c) Restriction endonuclease, polymerase, ligase, vector
(d) Restriction enzyme, polymerase, dehydrogenase vector

Topic 3 Restriction Enzyme

20. In 1963, two restriction endonucleases were isolated in E. Coli that restricted growth of bacteriophage by
(a) cutting DNA
(b) adding methyl group to DNA
(c) removing methyl group to DNA
(d) more than one option

21. The first restriction endonuclease was
(a) Hind-III (b) Hind-II
(c) Hind-I (d) Hind-IV

22. EcoRI comes from
(a) genus Eichhonia (b) species coli
(c) genus Echinus (d) species crispus

23. Recognition sequence is
(a) Specific sugar sequence in DNA which is recognized by restriction endonuclease
(b) Specific protein sequence which is recognized by restriction endonuclease
(c) Specific lipase sequence which is recognized by restriction endonuclease
(d) Specific base sequence in DNA which is recognized by restriction enconulcease

24. The convention for naming restriction endonucleases is
(a) First two letters come from genus & third from species of prokaryotic cell from which they were isolated.
(b) First two letters come from species & third from genus of prokaryotic cell from which they were isolated.
(c) First letter come from genus & second two from species of prokaryotic cell from which they were isolated.
(d) First letter come from species & second two from genus of prokaryotic cell from which they were isolated

25. Roman number indicate
(a) order in which enzyme were isolated
(b) strain of bacteria
(c) lab number in which enzyme was isolated
(d) none of these

26. Restriction enzymes belong to
(a) Exonucleases (b) Endonucleases
(c) Both (d) None

27. Exonuclease cuts DNA from
(a) specific position within DNA
(b) ends of DNA
(c) Both (a) & (b)
(d) None of these

28. Restriction enzyme recognize
(a) Paleondromic sequence of nucleoside in DNA
(b) Palindromic sequence of nucleoside in DNA
(c) Paleondromic sequence of nucleotide in DNA
(d) Palindromic sequence of nucleotide in DNA

29. ECoRI cuts DNA at

(a) 5' G|AATTC 3'
3' CTTAA|G 5'

(b) 3' GAAT|TC 5'
5' CTTA|AG 3'

(c) 3' G|AATTC 5'
5' CTTTAA|G 3'

(d) All of these

30. Which of the following is a palindrome?
(a) 5' – GAATAC – 3'
3' – CTTATG – 5'
(b) 5' – GATATAC – 3'
3' – CTATATG – 5'
(c) 5' – GAATTC – 3'
3' – CTTAAG – 5'
(d) All of these

31. Restriction enzyme cuts DNA
(a) between same two bases on opposite strands, in centre of DNA sequence recognized
(b) between same two bases on opposite strands, a little away from centre of DNA sequence recognized
(c) between different two bases on opposite strands, in centre of DNA sequence recognized
(d) between different two bases on opposite strands, living away from centre of DNA sequence recognized.

32. Same restriction enzyme produce
(a) same kind sticky ends joined using endonucleases
(b) different kinds of sticky ends joined using ligase
(c) same kind of sticky ends joined using ligase
(d) different kind of sticky ends joined using endonucleases

33.

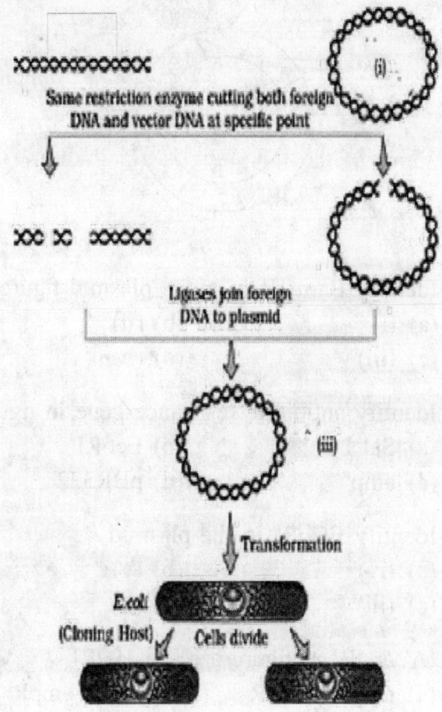

Identify correct labelling

	(i)	(ii)	(iii)
(a)	vector plasmid	Recombina nt DNA	Foreign DNA
(b)	Foreign DNA	vector plasmid	Recombinant DNA
(c)	Recombina nt DNA	vector plasmid	Foreign DNA
(d)	vector plasmid	Foreign DNA	Recombinant DNA

34. The process of 'Transformation' is taking place when
(a) bacteria replicates and makes copies of rDNA with it
(b) bacteria picks up rDNA II
(c) foreign gene is added to cloning host prokaryote cell
(d) more than one option

Topic 4	Separation & Isolation of Dna Fragments

35. Technique used for separation of DNA fragments are
(a) Gel electrophoresis (b) DNA fingerprinting
(c) PCR (d) DNA cloning

36. DNA fragments are
(a) negatively charged (b) positively charged
(c) neutral (d) none of these

37. In gel electrophoresis, DNA are forced to move towards
(a) anode under magnetic field
(b) cathode under magnetic field
(c) anode under electric field
(d) cathode under electric field

38. Matrix used in electrophoresis is
(a) ethidium bromide
(b) agarose gel
(c) natural polymer extracted from sea weeds
(d) more than one option

39. Ethidium bromide is used to stain because
(a) DNA fragments are visible without staining
(b) DNA fragments are not visible under staining
(c) DNA fragments are not visible without staining
(d) DNA fragments are visible under staining

40. Stained DNA is exposed to
(a) visible light (b) UV light
(c) IR light (d) Radio wave

41. Colour of DNA visible under UV light after Ethidium bromide staining is
(a) blue (b) black
(c) orange (d) green

42. The extraction of separated bands of DNA from agarose gel are
(a) Dilution (b) Elition
(c) Elution (d) Delution

43.

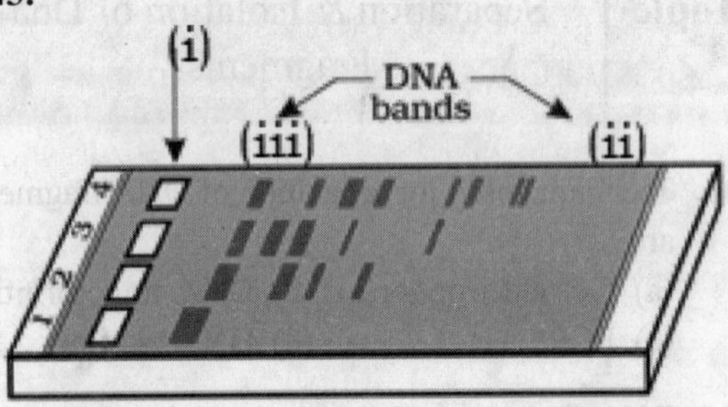

Identify labels correctly

	(i)	(ii)	(iii)
(a)	Largest DNA band	Smallest DNA band	Wells
(b)	Wells	Largest DNA bands	Smallest DNA bands
(c)	Smallest DNA bands	Largest DNA bands	Wells
(d)	Smallest DNA bands	Wells	Largest DNA bands

Topic 5 Cloning Vectors

44. Plasmids in bacterial cells replicate
(a) depending on chromosomal DNA
(b) independent of chromosomal DNA
(c) depending on extra-nuclear DNA
(d) more than one option

45. Bacteriophages
(a) replicate independent of other organisms
(b) replicate inside bacterial cell, controlled by chromosomal DNA of bacteria.
(c) replicate inside bacterial cell autonomously
(d) more than one option

46. Bacteriophages serve as ____ in biotechnology.
(a) host (b) vector
(c) molecular marker (d) enzyme

Figure for question. (47 to 53)

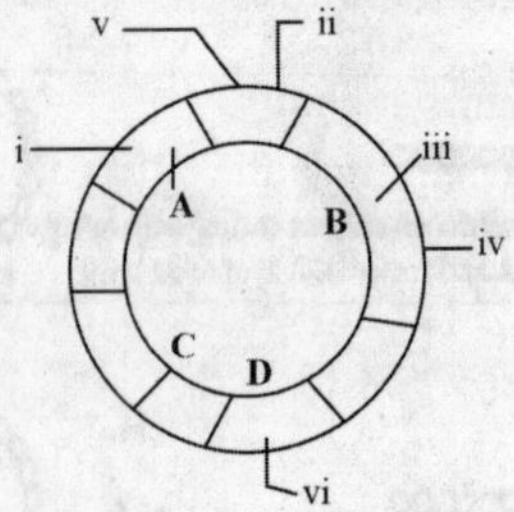

47. Identify Bam HI in given plasmid figure
(a) (i) (b) (ii)
(c) (iii) (d) (iv)

48. Identify antibiotic resistance gene in figure
(a) Sal I (b) EcoRI
(c) amp^R (d) pBR322

49. Identify ECoRI in the plasmid
(a) (iv) (b) (v)
(c) (iii) (d) (ii)

50. 'A' & 'B' in figure are [Pg-199E]
(a) ampR & tetR (b) ori & ampR
(c) tetR & ampR (d) rop & tetR

51. 'rop' codes for i & is shown in figure by ii
(a) proteins involved in replication; D
(b) proteins involved in transcription, C
(c) proteins involved in transcription, D
(d) proteins involved in replication, C

52. 'Ori' means ____ & is shown in figure by
(a) origin of translocation; C
(b) origin of replication ; D
(c) origin of translation ; D
(d) origin of replication; C

53. Identify pvu II in given figure of plasmid
(a) i (b) ii
(c) vi (d) iv

54. Which of the following is correct?
(a) Any piece of DNA linked to ori gene will be replicated
(b) Number of replication copies is under control of recognition site
(c) Vector should not be chosen based on number of copies supported by it

(d) More than one option

55. Transformants include
(a) cells which have picked vector with foreign DNA ligated to it.
(b) cells which have picked up vector without foreign DNA ligated to it
(c) cells which have not picked up vector
(d) Both (a) & (b)

56. Recombinants are
(a) cells which have picked vector with foreign DNA ligated to it.
(b) cells which have picked up vector without foreign DNA ligated to it
(c) cells which have not picked up vector
(d) Both (a) & (b)

57. Which is true about recombinant & transformant?
(a) All transformants are recombinants
(b) All recombinants are transformants
(c) no relation between these two
(d) Both are same thing

58. Normal E.coli cell-
(a) Carries resistance against antibiotics ampicillin, tetracycline and kanamycin
(b) Does not carry resistance against antibiotics ampicillin, tetracycline and kanamycin
(c) Carries resistance against ampicillin but not tetracycline and kanamycin
(d) Carries resistance against tetracycline but not ampicillin and kanamycin

59. In order to link alien DNA, vector needs to have ____ recognition sites for commonly used restriction enzymes.
(a) very few
(b) preferably single
(c) many
(d) more than one option

60. Assertion- Vector should have many recognition sites for commonly used restriction enzymes.

Reason- Lot of recognition sites generate several fragments, which make gene cloning easy.

(a) Assertion and Reason are both correct and Reason is correct explanation for Assertion
(b) Assertion and Reason are both correct but Reason is not correct explanation for Assertion
(c) Assertion and Reason both are incorrect
(d) Assertion is correct but Reason is incorrect

61. If a foreign gene is ligated at Bam HI site of vector PBR322, then the resistance for ____.
(a) tetracycline is lost
(b) ampicillin is lost
(c) tetracycline is not lost
(d) more than one option

62. The recombinants mentioned previous question non-recombinants by-
(a) Plating the transformants on tetracycline
(b) Planting the transformants on ampicillin
(c) Both of these are necessary
(d) None of these

63. Recombinants mentioned in 'If a foreign gene is ligated at Bam HI site of vector PBR322' will-
(a) Grow in ampicillin and tetracycline both
(b) Grow in ampicillin but not tetracycline
(c) Grow in tetracycline but not ampicillin
(d) Grow neither in tetracycline nor in ampicillin

64. Non-recombinants transformants will
(a) Grow in ampicillin and tetracycline both
(b) Grow in ampicillin but not tetracycline
(c) Grow in tetracycline but not ampicillin
(d) Grow neither in tetracycline nor in ampicillin

65. Non-transformants E.coli will-
(a) Grow in ampicillin and tetracycline both
(b) Grow in ampicillin but not tetracycline
(c) Grow in tetracycline but not ampicillin
(d) Grow neither in tetracycline nor in ampicillin

66. When rDNA is inserted in coding sequence of β-galactosidase,
(a) The enzyme gets synthesized
(b) Blue coloured colonies are produced
(c) Colourless colonies are produced
(d) Orange colonies are produced

67. Ti-plasmid stands for ____ and are present in ____.
(a) Tumor inhibiting, Agrobacterium speciense
(b) Tumor inducing, Agrobacterium speciense
(c) Tumor inhibiting, Agrobacterium tumifaciens
(d) Tumor inducing, Agrobacterium tumifaciens

68. The Ti-plasmid being used as cloning vector-
(a) causes crown gall disease
(b) is not pathogenic
(c) is pathogenic
(d) More than one option

Topic 6 Competent Host (For transformation with recombinant DNA)

69. DNA is-
(a) hydrophilic and can pass through cell membrane
(b) hydrophobic and can pass through cell membrane
(c) hydrophilic and cannot pass through cell membrane
(d) hydrophobic and cannot pass through cell membrane

70. Bacterial host cells are made competent to take up rDNA by-
(a) Treating with Na^+
(b) Treating with Al^{3+}
(c) Treating with Ca^{2+}
(d) More than one options

71. Choose the correct sequence to be followed to enable bacteria to take up rDNA.
(i) Treating with divalent cation.
(ii) Heat shock (42ºC).
(iii) Incubating on ice.
(a) i-ii-iii-ii (b) i-iii-ii-iii
(c) ii-iii-i-ii (d) iii-ii-i-iii

72. Other methods for introducing foreign DNA into host cells are-
(a) Micro-injection for animal cells
(b) Gene gun for plant cells
(c) Disarmed pathogens
(d) All of these

73. In micro-injection technique, rDNA is injected into-
(a) Cytoplasm (b) Nucleus
(c) Cell membrane (d) Lysosomes

74. In biolistics, cells are bombarded with high velocity-
(a) Micro-particles of iron
(b) Macro-particles of tungsten
(c) Micro-particles of gold
(d) More than one option

Topic 7 Processes of Recombinant DNA Technology

75. Identity correct sequence of process of rDNA technology:
(i) transferring rDNA into host
(ii) isolation of DNA fragment desired
(iii) isolation of DNA
(iv) culturing host cells in medium at large scale
(v) fragmentation of DNA by restriction enzyme
(vi) ligation of DNA fragment into a vector
(vii) extraction of desired product
(a) (iii) – (ii) – (v) – (vi) – (i) – (iv) – (vii)
(b) (iii) – (v) – (i) – (vi) – (ii) – (iv) – (vii)
(c) (iii) – (v) – (ii) – (vi) – (i) – (iv) – (vii)
(d) (iii) – (v) – (vi) – (i) – (ii) – (iv) – (vii)

Topic 8 Isolation of the Genetic Material (DNA)

76. Nucleic acid is genetic material of:
(a) some organisms
(b) no organism
(c) all organisms without exception
(d) most organisms with some exception

77. How many of given enzymes involved ir

extraction of genetic material from cell of organisms are:

(i) cellulase (ii) chitinase
(iii) lysozyme (iv) Ribonuclease
(v) protease (vi) deoxyribonuclease
(a) 3 (b) 2
(c) 5 (d) 6

78. Match the following:

A	B
(i) cellulase	I. plant
(ii) chitinase	II. Bacteria
(iii) lysozyme	III. Fungi

	(i)	(ii)	(iii)
(a)	I	III	II
(b)	II	III	I
(c)	III	I	II
(d)	I	II	III

79. Purified DNA is precipitated out by addition of:
(a) warm acetic acid (b) chilled acetic acid
(c) warm ethanol (d) chilled ethanol

80.

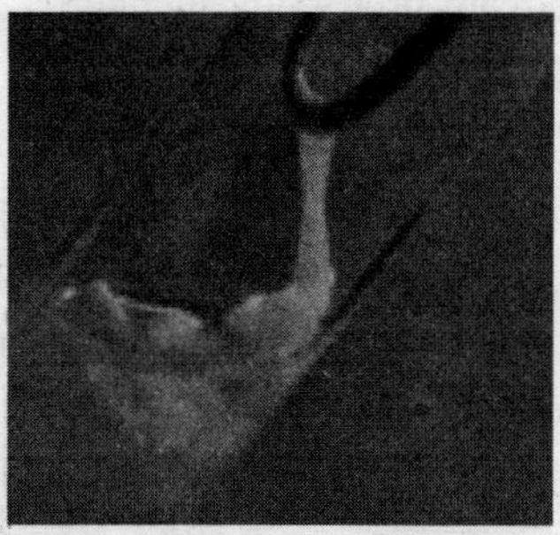

The figure shows DNA separated out, removed by:
(a) spooning (b) spooling
(c) spilling (d) speeling

81. The precipitated DNA is seen as:
(a) collection of fine threads in suspension
(b) collection of fine threads in solution
(c) coagulated mass in suspension
(d) coagulated mass in solution

Topic 9 Cutting of DNA at Specific Location

82. To check the progression of restriction enzyme digestion, _______ is used.
(a) PCR
(b) gel electrophoresis
(c) DNA fingerprinting
(d) Selectable marker gene

83. Preparation of rDNA involves the enzymes:
(a) specific restriction enzyme
(b) gene of interest
(c) vector DNA
(d) all of these

Topic 10 Amplification of Gene of Interest using PCR

84. PCR stands for:
(a) Polynuclease chain reaction
(b) Polylipase chain reaction
(c) Polyamide chain reaction
(d) None of these

85. PCR is an:
(a) in vitro process (b) in vivo process
(c) both (d) none

86. How many sets of primers are used in PCR?
(a) 1 (b) 2
(c) 3 (d) 4

87. Enzyme involved in PCR is:
(a) DNA endonuclease (b) RNA polymerase
(c) DNA polymerase (d) DNase

88. The enzyme involved in PCR with thermostability is isolated from:
(a) Thermus aquaticus fungi
(b) Escherechia coli bacteria
(c) Agrobacterium tumefaciense bacteria
(d) None of these

89.

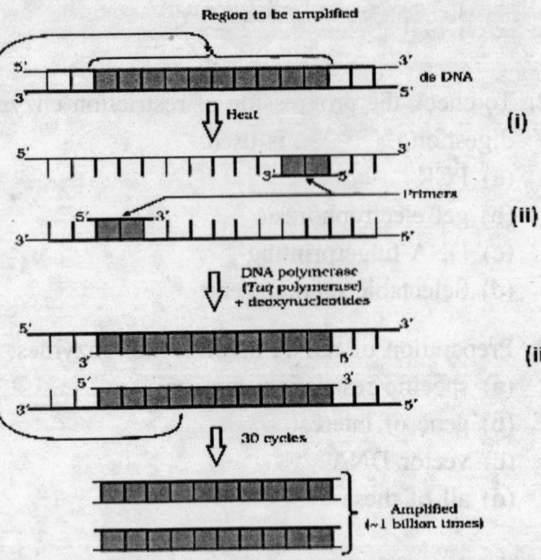

Identify correct labeling of sequence:

	(i)	(ii)	(iii)
(a)	Annealing	Denaturation	Extension
(b)	Denaturation	Extension	Annealing
(c)	Denaturation	Annealing	Extension
(d)	Extension	Annealing	Denaturation

Topic 11 Insertion of Recombinant DNA into the Host Cell / Organism

90. A-Ampicillin resistance gene is called selectable marker in case E.coli is made to take up rDNA bearing ampicillin resistance gene.

B-Such E.coli coil grow on amplicillin containing agar plates.

Choose right option with regards to above statements.

(a) Both are correct (b) Only A is correct
(c) Only B is correct (d) None is correct

Topic 12 Obtaining Foreign Gene Product

91. If a protein encoding gene is expressed in a heterologous host, it is called:

(a) secondary protein (b) recombinant protein
(c) transmitted protein (d) tertiary protein

92. In continuous culture system:

(a) used medium is drained at the end
(b) used medium is drained twice in the whole process
(c) used medium is continuously drained out
(d) none of these

93. Bioreactors are:

(a) large vessels
(b) used for large quantity production
(c) used for biological conversion of raw materials into products
(d) all of these

Topic 13 Downstream Processing

94. Downstream processing includes :

(a) separation (b) purification
(c) both the above (d) none of these

95. A- Suitable preservatives are added

B- These formulations need clinical trials.

C- Quality control testing is uniform for all the products.

How many of the above statements is incorrect?

(a) 0 (b) 1
(c) 2 (d) 3

96. Optimal conditions for growth include. How many of the following-

pH, Salt, Temperature, Vitamin, Oxygen

(a) 5 (b) 6
(c) 7 (d) 4

97.

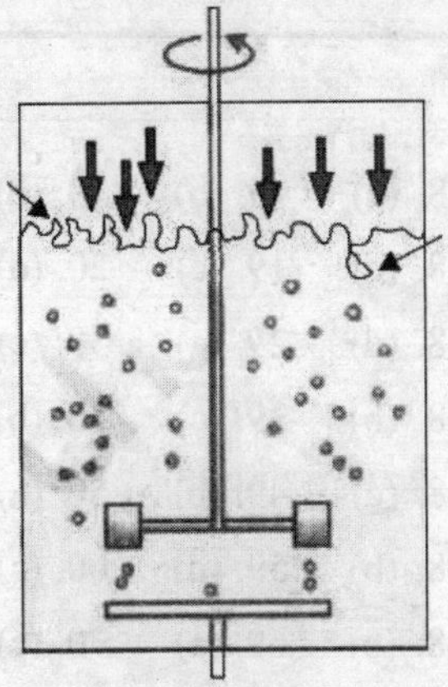

(i)

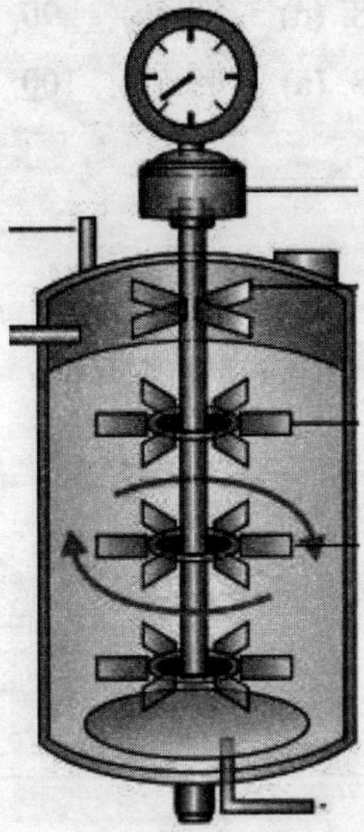

(ii)

Identify types of stirred-tank bioreactor-

	(i)	(ii)
(a)	Simple stirred-tank bioreactor	complex stirredtank bioreactor
(b)	Complex stirred-tank bioreactor	simple stirredtank bioreactor
(c)	Simple	Sparged
(d)	Sparged	Simple

98.

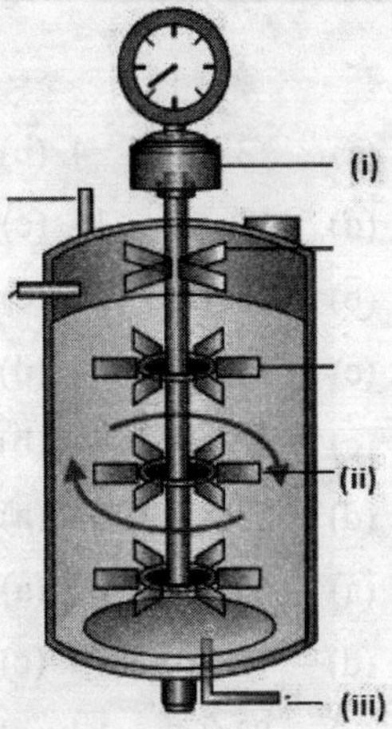

Identify the correct labels-

	(i)	(ii)	(iii)
(a)	Motor	Culture broth	Sterile air
(b)	Culture broth	Motor	Sterile air
(c)	Motor	Sterile air	Culture broth
(d)	Sterile air	Culture broth	Motor

99. Samling ports are mainly required to-
(a) Keep adding samples into Bioreactors
(b) Withdraw small volume of culture
(c) Add Acid/Base for pH control
(d) All of these

100. Sterile air bubbles are sprayed in the biovector in a type of bioreactor. That is because-
(a) air bubbles makes it easier to agitate the system
(b) air bubbles increase surface area for oxygen transfer
(c) air bubbles enable microbes to grow
(d) none of these

ANSWER KEY

1. (d)	2. (a)	3. (d)	4. (b)	5. (c)	6. (b)	7. (b)	8. (a)	9. (b)	10. (b)
11. (d)	12. (d)	13. (a)	14. (c)	15. (b)	16. (b)	17. (d)	18. (c)	19. (c)	20. (d)
21. (b)	22. (b)	23. (d)	24. (c)	25. (a)	26. (b)	27. (b)	28. (d)	29. (a)	30. (c)
31. (b)	32. (c)	33. (d)	34. (d)	35. (a)	36. (a)	37. (c)	38. (b)	39. (c)	40. (b)
41. (c)	42. (c)	43. (b)	44. (b)	45. (c)	46. (b)	47. (c)	48. (c)	49. (b)	50. (a)
51. (a)	52. (d)	53. (c)	54. (a)	55. (d)	56. (a)	57. (b)	58. (b)	59. (d)	60. (c)
61. (a)	62. (a)	63. (b)	64. (a)	65. (d)	66. (b)	67. (d)	68. (b)	69. (a)	70. (c)
71. (b)	72. (d)	73. (b)	74. (c)	75. (c)	76. (c)	77. (c)	78. (a)	79. (d)	80. (b)
81. (a)	82. (b)	83. (d)	84. (d)	85. (a)	86. (b)	87. (c)	88. (d)	89. (c)	90. (a)
91. (b)	92. (c)	93. (d)	94. (c)	95. (b)	96. (a)	97. (c)	98. (a)	99. (b)	100. (b)

18 Chapter

Biotechnology and its Application

MULTIPLE CHOICE QUESTIONS

Topic 1 Basis of Classification

1. Application like bioremediation, processed food, therapeutics and diagnostics are related to

(a) Biochemistry (b) Microbiology

(c) Biotechnology (d) Medical Science

2. ____ is/are the critical research area(s) of biotechnology.

(a) Creating optimals conditions for catalyst function

(b) Providing best catalyst

(c) Developing down streaming processing technique

(d) All of the above

3. Biotechnology mainly deals with

(a) Industrial scale production of biopharmaceutical

(b) Biological use of genetically modified microbes, fungi, plants and animals

(c) Both (a) and (b)

(d) None of these

4. Which of the following is not included in the application of biotechnology-

(a) Waste treatment

(b) Conventional hybridisation

(c) Energy production

(d) Genetically modified crops

Topic 2 Biotechnology Applications in Agriculture

5. A Novel strategy was adopted to present Meloidiogyne incognita infection in tobacco plants that was based on the process of

(a) DNA interference (b) RNA interference

(c) RNA initiation (d) DNA initiation

6. Resistance against a Nematode was introduce by implying RNA in ____ plants.

(a) Tomato (b) Bt corn

(c) Bt cotton (d) Tobacco

7. RNAi stand for

(a) RNA inteteron (b) RNA interference

(c) RNA inactivation (d) RNA initiation

8. RNAi take place in all ____ organisms as method of ____.

(a) prokaryotes, insect resistant

(b) eukaryotes, insect resistant

(c) eukaryotes, cellular defence

(d) prokaryotes, cellular defence

9. ____ is used for silencing of an unwanted gene

(a) RNA (b) DNA polymerase

(c) Restriction enzyme (d) All of these

10. Silencing of mRNA molecule in order to control the production of a harmful protein has been used in the protection of plants from

(a) Beetles (b) Armywork

(c) Budworm (d) Nematodes

11. Transposons are also known as
(a) Silenced gene
(b) Plesotropic genes
(c) Mobile genetic elements
(d) Both (a) and (c)

12 Tobacco plant resistant to a nemotode have been developed by the introduction of DNA and it is produced in the lost cells as
(a) A particular hormone
(b) Toxic protein
(c) Both sense and antisense RNA
(d) An antifeedant

13. Which of the following is not for increasing food production?
(a) Agrochemical based agriculture
(b) Organic agriculture
(c) Genetic engineered crop-based agriculture
(d) None of these

14. Organic agriculture is a technique of raising crops for
(a) increased food production
(b) reduction in required labour
(c) increasing the use of agrochemicals
(d) Both (a) & (c)

15. Use of genetically modified crops in crop field may
(a) reduce the harmful effects of fertilizers
(b) maximize yield
(c) be environment friendly
(d) All of the above

16. Plants bacteria, fungi and animals whose genes have been altered by manipulation are called
(a) Pest resistant organism
(b) Hybrid organisms
(c) Genetically modified organism
(d) Insect resistant organism

17. Golden rice is genetically modified crop plant with incorporate gene meant for biosynthesis of
(a) Vitamin E (b) Vitamin K
(c) Omega-3 (d) Vitamin A

18. ____ produced by Bacillus thuringiensis
(a) t- toxin (b) Bt toxin
(c) An acid (d) All of these

19. The bacterium Bacillus thuringiensis produce ____ plants which reduces the amount of ____ used.
(a) disease resistant, insecticide
(b) insect resistant, fertilizers
(c) disease resistant, industrial enzyme
(d) insect resistant, insecticide

20. Which of the following crops are modified using Bacillus thuringiensis?
(a) Corn and cotton
(b) Tomato and rice
(c) Potato and soyabean
(d) All of the above

21. Which of the following is being grown in India by farmers as Bt crop?
(a) Maize (b) Brinzal
(c) Cotton (d) Soyabean

22. By inserting a piece of DNA from ____ insect resistant transgenic cotton has been produced.
(a) a wild relative of cotton
(b) bacterium
(c) an insect
(d) virus

23. Some strains of Bacillus thuringiensis produce proteins that will insect like
(a) Lepidopterans (b) Coleopterans
(c) Dipterans (d) All of these

24. Coleopterans examples are/is-
(a) Flies (b) Mosquitoes
(c) Beetles (d) All of the above

25. Bacillus thuringiensis forms protein crystals which contain a-
(a) Simple protein
(b) Non-toxic insecticidal protein
(c) Toxic insecticidal protein
(d) Simple lipids

26. Why does Bt toxin protein crystal not kill the Bacillus? Because-
(a) Bacteria encloses toxins in special sac
(b) Bacteria are resistant to toxin
(c) Toxin occurs as inactive protoxins in bacteria
(d) All of the above

27. Bt toxin kills insect by-
(a) Inhibiting protein synthesis
(b) Generating excessive heat
(c) Creating pores leading to cell swelling and lysis in the mid gut epithelial cells
(d) None of these

28. The choices of genes of Bacillus thuringiensis, incorporated in to crop depends upon
(a) Crop (b) Targeted pest
(c) Both (a) and (b) (d) Toxin

29. The crops having cry genes need
(a) Small amount of fungicide
(b) Large amount of pesticide
(c) Small amount of insecticide
(d) None of the above

30. The Bt toxin protein
(a) Obstruct a biosynthetic pathway
(b) Causes death of the insect
(c) Stops egg laying of adult
(d) Generating excessive heat

31. Cotton bollworm controlled by-
(a) Cry I Ac, Cry II Ab
(b) Cry I Ac, Cry II Ac, Cry I Ab
(c) Cry II Ac, Cry I Ab
(d) Cry I Ab

32. Bt corn has been made resistant to corn borer by the introduction of gene
(a) Cry I Ac (b) Cry II Ab
(c) Cry I Ab (d) Cry II Ac

33. Cry II Ab and Cry I Ab produces toxins that control
(a) Cotton bollworms and corn borer resp.
(b) Cotton bollworm and budworms of tobacco resp.
(c) Corn borer and cotton bollworms resp.
(d) Nematodes and tobacco budworms resp.

34. Which of the following nematodes infects the root of the tobacco plants which reduces the production of tobacco?
(a) Melodiogyne incognitia
(b) Ascaris
(c) Wuckereria
(d) Interobious

Topic 3 Biotechnology Applications in Medicine

35. The first human hormone produced by recombinant technology is
(a) Oestrogen (b) Progesterone
(c) Thyroxine (d) Insulin

36. The demerits of using bovine insulin (from cow) and porcine insulin (from pig) in diabetic patients is-
(a) It leads to hypercalcemic
(b) It may cause allergic reaction
(c) It is expensive
(d) All of the above

37. The two polypeptides of human insulin are linked together by
(a) Phosphodiester bonds
(b) Disulphide bridge
(c) Hydrogen bonds
(d) None of the above

38. ____ is removed during the maturation of proinsulin to insulin.
(a) A-chain (b) B-chain
(c) C-chain (d) Both (a) and (b)

39. The main challenge for production of insulin using rDNA techniques was
(a) Splitting A and B- peptide chains
(b) Addition of C- peptide to proinsulin
(c) Getting insulin assembled to mature form
(d) Removal of C- peptide from active insulin

40. Which of the following companies prepared human insulin in 1983?
(a) Monsanto (b) Eli Lily
(c) Genetech (d) GEAC

Topic 4 Gene Therapy

41. Treatment of genetic disorder by manipulating gene is called-
(a) Gene therapy
(b) rDNA technology
(c) Bone marrow transplantation
(d) Enzyme replacement therapy

42. For the first time, therapy was tried on a 4 year old girl in 1990 to treat ____.
(a) Cytosine Deaminase (CDA)
(b) Adenosine Deaminase (ADA)
(c) Tyrosine oxidase
(d) Glutamate tryhydrogenase

43. Which kind of therapy was given in 1990 to 4 year old girl with enzyme deficiency?
(a) Gene therapy (b) Chemotherapy
(c) Immunotherapy (d) Radiation therapy

44. Adenosine deaminase (ADA) deficiency can be treated by ____ and ____ but it is not fully curative. Here A and B can be
(a) A- gene therapy, B- radiation therapy
(b) A- bone marrow transplantation, Benzyme replacement therapy
(c) A- organ transplantation, Bhormone replacement
(d) A- radiation therapy, B- enzyme replacement therapy

45. The advantage of beginning gene therapy prior to birth is-
(a) The body would not reject it as it has not yet recognised 'self'.
(b) This would give the body plenty of time.
(c) The cell being extremely young are more receptive to gene therapy.
(d) None of these

Topic 5 Molecular Diagnosis

46. Why using conventional method for diagnosis is not very relevant?
(a) Early detection is not possible
(b) Not reliable
(c) Results are incorrect
(d) All of these

47. Which of the following molecular diagnostic technique is used to detect the presence of a pathogen in its early stage of infection-
(a) Angiography
(b) Radiography
(c) Enzyme replacement technique
(d) Polymerase chain reaction

48. Why PCR is used?
(a) to detect HIV in suspected AIDS patients
(b) to detect Mutation in the genes of suspected cancer patients
(c) Diagnose many genetic disorders
(d) All of the above

49. A single stranded Nucleic acid tagged with a radioactive molecule is called
(a) Plasmid (b) Probe
(c) Vector (d) Selectable market

50. In which of the following method, a probe is allowed hybridise to its complementary DNA in the clone of cells?
(a) Enzyme linked Immono sorbent Assay (ELISA)
(b) PCR
(c) Autoradiography
(d) Gene therapy

51. Technique used to detect mutation in genes is known as-
(a) Gel electrophoresis (b) PCR
(c) Gene therapy (d) Autoradiography

52. Which of the following technique is based on the principle of antigen – antibody interaction?

(a) PCR
(b) ELISA
(c) Recombinant DNA technology
(d) Gene therapy

Topic 6 Transgenic Animals

53. Animals whose DNA is manipulated to possess and express an extra (foreign) gene are known as
(a) Transgenic animals (b) Hybrid animals
(c) Transferrin animals (d) All of the above

54. Transgenic animals are those which have foreign?
(a) DNA in all of their cells
(b) Proteins in all of their cells
(c) RNA in all their cells
(d) RNA in some of their cells

55. 95% of all the existing transgenic animals are
(a) Pigs (b) Cows
(c) Mice (d) All of these

56. Transgenic animals can be used to
(a) Study normal physiology
(b) Study the biological effects
(c) Study the vaccine safety
(d) All of the above

57. Transgenic animals made to serve as models for human diseases. The disease are-
(a) Alzheimer's disease (b) Cancer
(c) Cystic fibrosis (d) All of these

58. Which of the following transgenic human protein products development are used to treat emphysema?
(a) α-1 antitrypsin (b) α-1 trypsin
(c) α-1 albumin (d) α-1 globulin

59. When was the first transgenic cow, Rosie produced?
(a) 1979 (b) 1997
(c) 1996 (d) 1999

60. ____ was introduced in the first trans genetic cow-
(a) α-1 antirypsin
(b) Human β-Lactalbumin
(c) β-1 antitrypsin
(d) None of these

61. The first transgenic cow, Rosie produced
(a) Human calcium enriched milk (2.4 g/l)
(b) Human protein enriched milk (2.4 g/l)
(c) Human calcium enriched milk (2.6 g/l)
(d) Human protein enriched milk (2.8 g/l)

62. ____ are used in testing safety of polio vaccine before they are used on human.
(a) Transgenic pig (b) Transgenic monkey
(c) Transgenic rabbits (d) Transgenic mice

63. ____ animals are made that carry genes which makes them more sensitive to toxic substances than non-transgenic animals.
(a) Transgenic (b) Mutaled
(c) Transverred (d) Transformed

Topic 7 Ethical Issue

64. Exploitation of bio resources of a nation by multinational companies without authorisation from the concerned country is referred to-
(a) Bioethics (b) Bioweapon
(c) Bio piracy (d) Bio-exploitation

65. Bio piracy is related with the-
(a) Stealing of bio resources
(b) Traditional knowledge and utilization
(c) Biomolecules and regarding bio resources exploitation
(d) Both (a) and (c)

66. ____ was taken by Indian parliament to meet and fulfill the requirements of patent terms and other emergency provisions in this regard?
(a) Indian patents bill (b) Bioethics act
(c) Bio piracy act (d) All of these

67. Basmati is unique for its aroma and flavour, whose __A__ documented verities cultivated in __B__.
(a) A-37, B-India (b) A-27, B-India
(c) A-27, B-USA (d) A-30, B-USA

68. Which committee takes decision regarding the validity of GM research and the safety of introducing GM-organisms for public services?
(a) Indian Council of Medical Research (ICMR)
(b) Genetic Engineering Approval committee (GEAC)
(c) Indian Institute of Science Education and Research (IISER)
(d) Genetic Engineering Appraisal Committee (GEAC)

69. A ____ granted to a person who has either invented a new and useful product, made improvement existing product or invented a new process of making a product is called-
(a) bioethics
(b) patent
(c) bio piracy
(d) genetic recombination

70. Bio patent means
(a) Right to use an invention
(b) Right to use application are processes
(c) Both (a) and (b)
(d) None of these

71. ____ have been present in India from long time yet foreign country got patent through the US patent and Trademark office.
(a) Brown rice (b) Basmati rice
(c) Co-667 (d) All of these

72. Bioethics is-
(a) Process of discovery and commercialisation of new products.
(b) Use of bio resources with proper authorisation.
(c) Standards used to regulate human activities in relation to the biological world.
(d) All of these

ANSWER KEY

1. (c)	2. (d)	3. (c)	4. (b)	5. (b)	6. (d)	7. (b)	8. (c)	9. (a)	10. (d)
11. (c)	12. (c)	13. (d)	14. (a)	15. (d)	16. (c)	17. (d)	18. (b)	19. (d)	20. (d)
21. (c)	22. (b)	23. (d)	24. (c)	25. (c)	26. (c)	27. (c)	28. (c)	29. (d)	30. (b)
31. (a)	32. (c)	33. (a)	34. (a)	35. (d)	36. (b)	37. (b)	38. (c)	39. (c)	40. (b)
41. (a)	42. (b)	43. (a)	44. (b)	45. (c)	46. (a)	47. (d)	48. (d)	49. (b)	50. (c)
51. (b)	52. (b)	53. (a)	54. (a)	55. (c)	56. (d)	57. (d)	58. (a)	59. (b)	60. (d)
61. (b)	62. (d)	63. (a)	64. (c)	65. (d)	66. (a)	67. (b)	68. (b)	69. (b)	70. (c)
71. (b)	72. (a)								